100 years +

100 Jahre Wilkhahn – 100 Jahre nachhaltiges Gestalten
100 years Wilkhahn – 100 years of sustainable design

avedition edited by Wilkhahn

Dass ein Unternehmen 100 Jahre alt wird, ist in unserer schnelllebigen Zeit schon etwas Besonderes. Wenn es einem Unternehmen des Mittelstandes jedoch darüber hinaus auch noch gelingt, sich über Jahrzehnte hinweg mit einer eigenständigen Haltung an der Weltspitze zu behaupten und in einem globalen Markt immer wieder neue Maßstäbe zu setzen, dann verdient dies allen Respekt und Anerkennung. Dann lohnt es sich auch, genauer hinzusehen und zu hinterfragen, was die Besonderheit und letztlich den Erfolg eines solchen Unternehmens ausmacht.

An erster Stelle stehen bei Wilkhahn die Menschen. Sie sind Garant dafür, dass es dem Unternehmen immer wieder gelingt, uns mit ebenso innovativen wie nützlichen, schönen und werthaltigen Einrichtungslösungen zu überraschen. Sie sind aber auch Garant für die Kontinuität und Konsequenz dessen, was heute als Corporate Identity bezeichnet wird: Auf der Grundlage der Überzeugungen des Deutschen Werkbunds, des Bauhauses und der legendären Ulmer Hochschule für Gestaltung hat Wilkhahn über Jahrzehnte hinweg seine eigene Haltung entwickelt, der das Unternehmen allen Versuchungen zum Trotz treu geblieben ist. Das bezieht sich auf die Produktgestaltung, auf die Kommunikation und auf die Architektur ebenso wie auf die Gestaltung der zwischenmenschlichen Beziehungen innerhalb des Unternehmens, zu Kunden und Marktpartnern.

Lange bevor die „Corporate Social Responsibility" in den Chefetagen deutscher Unternehmer zum Thema wurde, hat sich der Pionierunternehmer Fritz Hahne über das Erwirtschaften von Gewinnen hinaus in der Pflicht für das Gemeinwohl gesehen. Das schloss frühzeitig auch den verantwortlichen Umgang mit Ressourcen und Umweltgütern ein. Zukunftsweisend in Bezug auf Erfolgspotenziale mittelständischer Unternehmen mit Standort in Deutschland. Zukunftsweisend aber auch für den Standort Deutschland mit seiner gewachsenen Kultur der Sozialpartnerschaft.

Ich kenne und schätze die Menschen bei Wilkhahn seit vielen Jahren. Der Ehrgeiz, die Dinge anders und vor allem besser zu machen, ist auch in der dritten Generation des familiengeführten Unternehmens tief verankert. Offenheit, Neugier und Begeisterungsfähigkeit, aber auch kritisches Hinterfragen, sorgsames Abwägen zeichnen die Vergangenheit aus und sind fruchtbare Basis für die Zukunft.

Vieles, was von diesem Unternehmen und seinen Mitarbeiterinnen und Mitarbeitern bewegt und angeschoben wurde, ist zu neuen, international gültigen Trends geworden. Das darf Sie bei Wilkhahn zum 100sten Geburtstag zu Recht mit Stolz erfüllen und das kann uns auch Vorbild in anderen Bereichen sein.

Ich wünsche dem Unternehmen für die Zukunft alles erdenklich Gute und Ihnen viel Freude und Anregungen beim Blättern durch das Jubiläumsbuch!

Frank-Walter Steinmeier

Dr. Frank-Walter Steinmeier

Bundesminister des Auswärtigen
Bundesrepublik Deutschland

Grußwort
Word of welcome

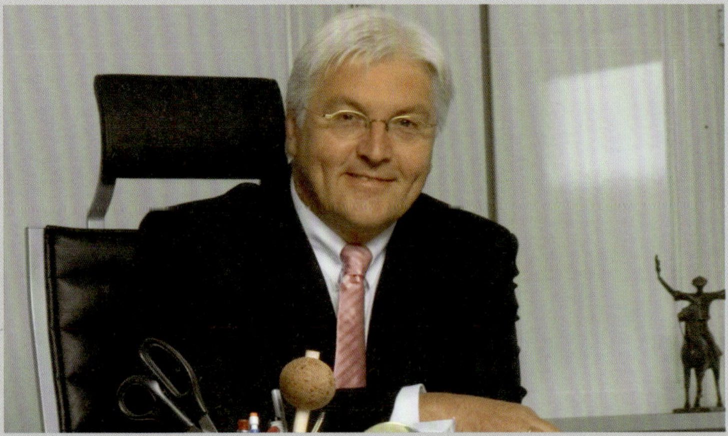

The fact that a company is 100 years old is quite a special achievement in these fast-moving times. Moreover, a medium-sized company that has succeeded in maintaining a leading international position over decades with its consistent approach and in continuing to set new standards in a global market certainly deserves a great deal of respect and recognition. It is surely worth taking a closer look to find out what accounts for the uniqueness and also the success of such a company.

At Wilkhahn, people have top priority. They guarantee that the company repeatedly succeeds in surprising us with furnishing solutions that are not only innovative, but also useful, aesthetic and sustainable in terms of value. They also serve as a guarantee for continuity and consistency of corporate identity: based on the convictions of the Deutscher Werkbund, Bauhaus and the legendary Hochschule für Gestaltung in Ulm, over the decades Wilkhahn has developed its own approach to which the company has remained loyal and thus resisted any temptation to change its course. This applies to product design, communication and architecture, as well as to the design of relationships among people within the company and with customers and market partners.

Long before "corporate social responsibility" had become a topic of interest on management levels of German companies, pioneer entrepreneur Fritz Hahne had recognized his duty to strive towards the common good over and above making a profit. At an early stage, this way of thinking also included responsible handling of resources and environmental goods. It marked a groundbreaking approach in terms of the success potential of medium-sized companies located in Germany. It was also groundbreaking for Germany as an industrial and commercial base with its growing culture of social partnership.

I have known and highly valued the people at Wilkhahn for many years now. The ambition to do things differently and, in particularly, better is also deeply rooted in the third generation of this family-managed business. The past is characterized by openness, curiosity and capacity for enthusiasm, but also by critical questioning and careful evaluation, all of which are values that form a fruitful basis for the future.

Much of what has been set in motion and initiated by this company and its employees has resulted in new, internationally recognized trends. Such an achievement may quite rightly make you feel proud at Wilkhahn on this 100th anniversary and serve as a model of excellence in other fields.

I wish the company all the very best for the future, and hope you will enjoy looking through this centenary book and above all feel inspired.

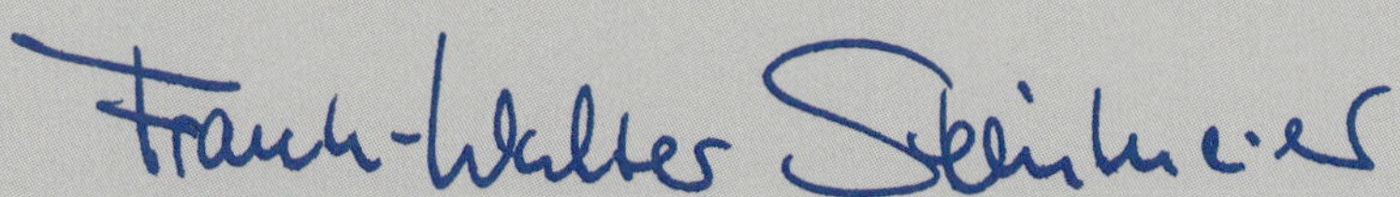

Dr. Frank-Walter Steinmeier

Minister for Foreign Affairs
Federal Republic of Germany

Inhalt
Contents

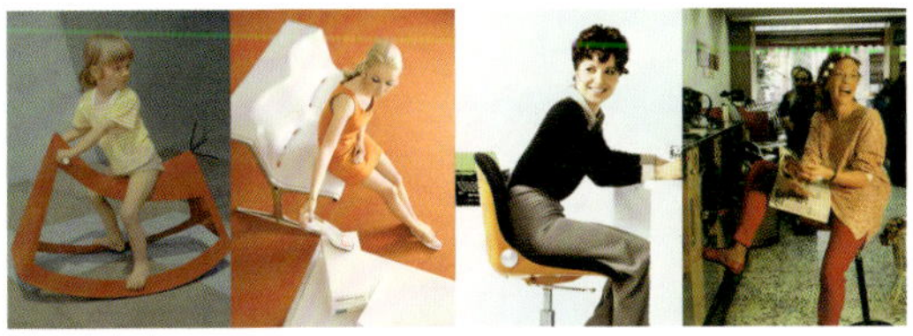

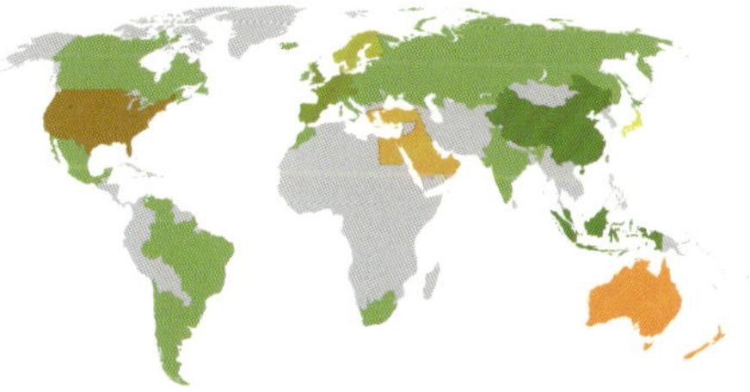

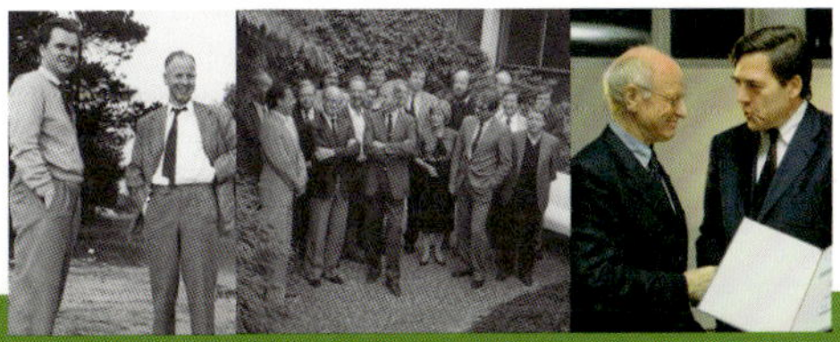

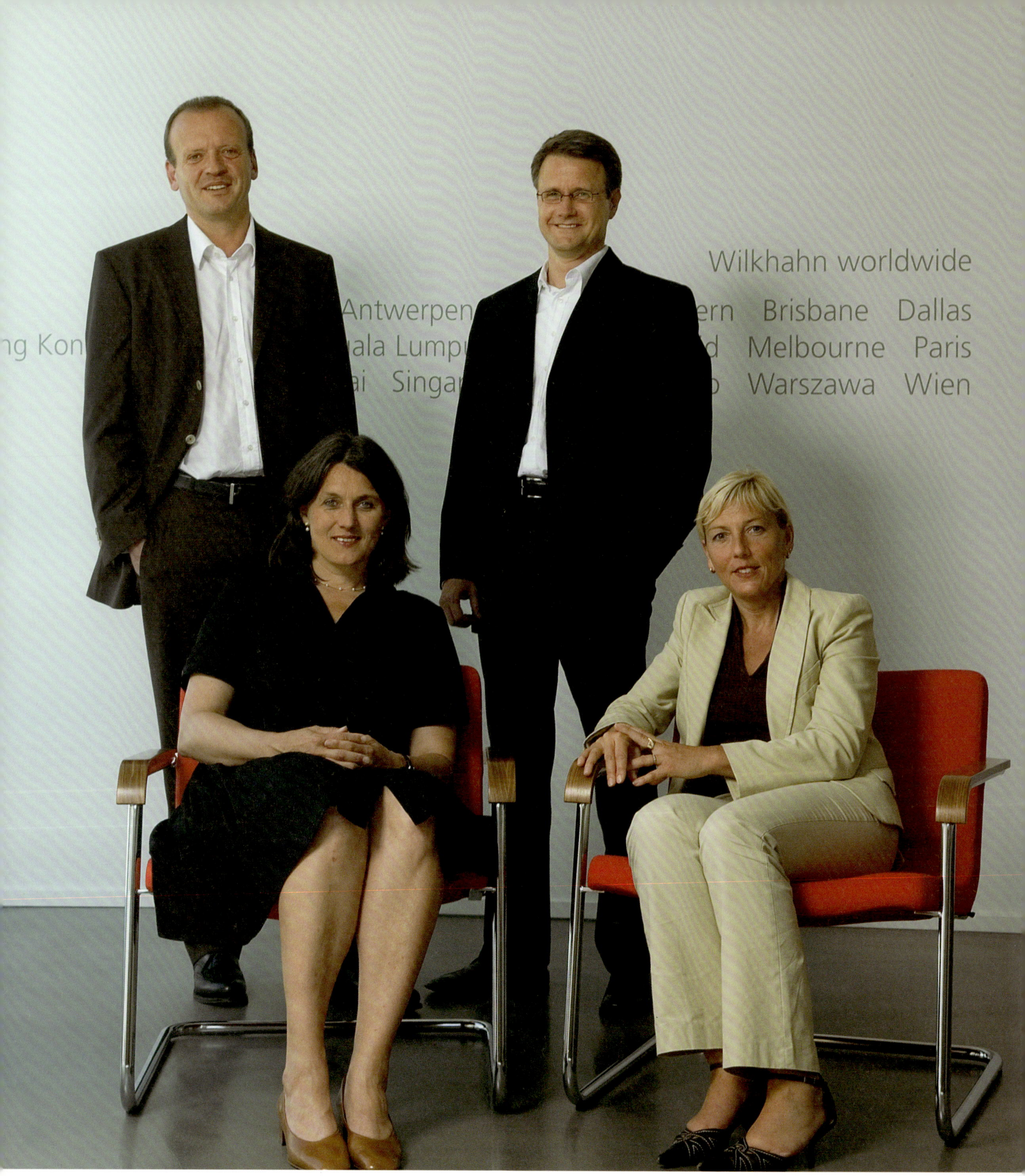
Wilkhahn worldwide
Antwerpen
Brisbane Dallas
Melbourne Paris
Warszawa Wien

Gestaltung als Ausdruck der Haltung
Design as an expression of corporate mission

Auf 100 Jahre Unternehmensgeschichte zurückzublicken, ist eine besondere Freude – zumal, wenn es sich dabei um eine Erfolgsgeschichte handelt. Wilkhahn hatte in der Vergangenheit schon häufig Anlässe, zu feiern: runde Jubiläen, Geburtstage, wichtige internationale Designpreise, Unternehmensauszeichnungen wie den Deutschen Marketingpreis, den Deutschen Umweltpreis oder die Auszeichnung von Fritz Hahne mit dem Bundespreis für Förderer des Designs. Daneben gab es immer wieder feierliche Eröffnungen etwa der bemerkenswerten Fertigungsgebäude, die von Frei Otto und Thomas Herzog entworfen wurden, oder des weltweiten EXPO-2000-Projekts.

Alle diese und viele weitere Höhepunkte der Unternehmensgeschichte haben vor allem zwei Dinge gemeinsam: Sie waren und sind geprägt vom unbedingten Willen, eine umfassend hohe Gestaltungsqualität zu erreichen – sei es bei den Produkten, bei der Umweltgestaltung, bei der Kommunikation oder bei den Unternehmensprozessen. Die beiden Gründersöhne Adolf Wilkening und Fritz Hahne haben bei Wilkhahn nach dem 2. Weltkrieg die Grundlagen für dieses umfassende Verständnis von Unternehmensgestaltung gelegt. Der eine als begnadeter Techniker und Ingenieur, der andere als „Menschenfänger" und begeisterter Visionär mit dem Gespür für das Wichtige und Richtige. Damit sind wir bei der zweiten Gemeinsamkeit, die sich als „Roter Faden" durch Geschichte und Gegenwart von Wilkhahn zieht: bei den Menschen. Denn erst sie geben dem Unternehmen die Seele: die Mitarbeiterinnen und Mitarbeiter, die sich für Wilkhahn einsetzen, die Impulsgeber und Freunde, die uns immer wieder inspirieren und anspornen, die Partner, Architekten und Innenarchitekten, die sich engagiert für eine hohe und werthaltige Gestaltungsqualität einsetzen, und vor allem unsere Kunden in aller Welt, die den Mut zur besseren Lösung honorieren!

Wir widmen dieses Buch deshalb all jenen Menschen in der Welt, ohne die es das Unternehmen Wilkhahn so nicht gäbe. Dass sie unsere Ideale teilen und dass sie dafür ihre Kraft, ihre Kreativität und ihre Kompetenz einsetzen, erfüllt uns mit Stolz. Und mit der Gewissheit, auch zukünftig unseren Beitrag zu einer besseren Gestaltung der Welt leisten zu können.
100 years + kann nur eine kleine Auswahl der Wegbegleiter und ihrer Leistungen zeigen. Wir wünschen uns sehr, dass sich auch alle diejenigen in dem Buch wiederfinden mögen, die nicht mit Bild und Wort zur Sprache kommen konnten, dass es Freude macht, durch die Wilkhahn-Welt zu blättern, und dass weiterhin so viele Menschen mit Leidenschaft, Mut und Herz an der zukünftigen Geschichte des Unternehmens mitschreiben werden!

It is a special pleasure to be able to look back on 100 years of corporate history – particularly if it is the story of success. In the past, Wilkhahn had many occasions for celebration: 10-year anniversaries, birthdays, key international design awards, corporate awards such as the German Marketing Prize, the German Ecology Prize or the German Award for Design Leadership presented to Fritz Hahne. In addition, there have been many festive inaugurations of outstanding production buildings such as those designed by Frei Otto and Thomas Herzog, or of the worldwide EXPO 2000 project.

All these highlights in corporate history and many others have two things in common: they have been and still are guided by the unswerving will to achieve a coherently high quality of design – for products, environmental design, communications or corporate processes. Adolf Wilkening and Fritz Hahne, the sons of the two founders of the company, laid the foundations for this coherent approach to corporate design after the Second World War. The former was an exceptionally gifted technician and engineer and the latter was a charismatic and enthusiastic visionary with a feeling for what was important and right. And that leads us to the second common denominator that forms a central theme throughout Wilkhahn's past and present: to the human factor. It is after all only people who form the real life and soul of the company, the spirit of Wilkhahn: the employees who are committed to Wilkhahn, impetus-givers and friends who have continued to inspire us and spur us on, architects and interior designers as active supporters of high and sustainable design quality, and last but not least, our customers throughout the world who appreciate the courage to find a better solution.

We would therefore like to dedicate this book to all those throughout the world without whom the company would not exist. We are proud that they share our ideals and that they have invested their strength, creativity and competence to this end. And with the certainty that we will be able to make our contribution to designing the world in a better way in the future too. 100 years + can only present a limited selection of our formative friends and their achievements. We hope that all those who have not had the opportunity of being mentioned in the book in the form of images or words will be able to identify with it. That you will enjoy browsing through the world of Wilkhahn, and that so many people will help to write the future history of the company with passion, courage and heart and soul.

Dr. Jochen Hahne, Vorsitzender der Geschäftsführung Chairman
Holger Jahnke, Geschäftsführer Int. Marketing und Vertrieb Managing Director Int. Marketing and Sales
Rosine Hölsch, Leiterin Technik und Produktion Head of Engineering and Production
Dr. Dorothea Giesecke-Kuppe, Leiterin Finanzen und Personal Head of Finance and Human Resources

Als die beiden verschwägerten Tischlermeister Christian Wilkening und Friedrich Hahne 1907 im „Stuhldorf" Eimbeckhausen bei Bad Münder eine gemeinsame Möbeltischlerei gründeten, konnte niemand ahnen, dass daraus später ein Unternehmen mit Weltgeltung entstehen würde: Im waldreichen Deister-Süntel-Tal, 30 km südwestlich von Hannover, gab es an die 100 Betriebe, die konventionelle Holzstühle und -sessel für den kleinbürgerlichen Massengeschmack produzierten. Das Tal hatte sich zu einem Zentrum der Stuhlproduktion in Deutschland entwickelt. Gründersohn Adolf Wilkening erwies sich nach seinem Eintritt in das Unternehmen (1920) als begnadeter Techniker und Produktionsleiter, etwa mit der Umstellung der bis dahin dampfgetriebenen Holzbearbeitungsmaschinen auf einen elektrischen Produktionsbetrieb. Wilkhahn florierte und gab vor dem 2. Weltkrieg etwa 40 Beschäftigten Arbeit und Brot, bevor die Produktion kriegsbedingt stillgelegt wurde.

Als nach dem Krieg sein wesentlich jüngerer Vetter Fritz Hahne 1946 in das elterliche Unternehmen eintrat, verband sich die technoloische Innovation mit einer grundlegenden Neuausrichtung des Unternehmens: In der Zusammenarbeit mit führenden Architekten und Gestaltern der Nachkriegszeit, die durchweg vom Deutschen Werkbund und vom Bauhaus geprägt waren, wurde Wilkhahn zum Pionierunternehmen für moderne Möbelgestaltung. Der ganzheitliche Gestaltungsansatz der Ulmer Hochschule für Gestaltung (HfG Ulm), 1953 von Inge Aicher-Scholl, Otl Aicher, Max Bill und anderen gegründet, legte die Grundlagen für die weitere Unternehmensentwicklung: Produktgestaltung, grafisches Erscheinungsbild und Unternehmensorganisation wurden als Einheit begriffen, die in allen Bereichen sichtbar werden sollte. „Ziel ist es, langlebige Produkte zu entwickeln, den Gebrauchswert zu erhöhen und die Verschwendung zu reduzieren" – der Anspruch aus dem Gründungsmanifest der HfG Ulm formulierte bereits vier Jahrzehnte vor dem Umweltgipfel in Rio de Janeiro (1992) die Prinzipien einer nachhaltigen Produktgestaltung!

Fritz Hahne, dessen humanistisches Menschenbild von existentiellen, persönlichen Erfahrungen aus der Kriegszeit geprägt war, trieb neben der innovativen Entwicklung langlebiger Produkte auch die Entwicklung der Unternehmenswerte voran: „Fairness in der Zusammenarbeit" wurde seit Mitte der 50er Jahre, „ökologische Verantwortung" schon seit den 80er Jahren zum festen Bestandteil der Unternehmenskultur. Wie zukunftsweisend dieses Unternehmensverständnis war, zeigte das Ausstellungsprojekt bei Wilkhahn zur Weltausstellung EXPO 2000 zum Thema „Zukunft der Arbeit": mit einer besonders sozialverantwortlichen Unternehmensorganisation, mit stabilen Beziehungen zu Markpartnern und Kunden, mit den beispielhaften Fabrikationsgebäuden, die Frei Otto und Thomas Herzog für Wilkhahn gestaltet haben, und vor allem mit über Jahrzehnte hinweg innovativen Produkten, die Langlebigkeit in Form, Funktion und Qualität sowie ökologische Anforderungen auf einen Nenner bringen. Mit einem der ersten Nachhaltigkeitsberichte leistete Wilkhahn einen wichtigen Beitrag für das globale Konzept einer umfassenden Corporate Social Responsibility, die ökonomischen Erfolg mit sozialer Orientierung und ökologischer Verantwortung verbindet.

Seit dem Jahr 2000 steht mit Dr. Jochen Hahne ein Vertreter der dritten Unternehmergeneration mit einem leistungsfähigen Team an der Spitze des Unternehmens. Auf dem zukunftsträchtigen Fundament der Vergangenheit werden die Stärken des Unternehmens weiterentwickelt und durch neue Kompetenzen ergänzt: mit der Schärfung des Markenbilds, mit der Konzentration auf die Entwicklung ebenso innovativer wie werthaltiger Konferenz- und Büroeinrichtungen und mit dem dynamischen Ausbau der internationalen Märkte. Die vielfach prämierte Corporate Website, zahlreiche Produkt- und Prozessinnovationen und eine zukunftsfähige Ausgestaltung der sozialen und ökologischen Verantwortung markieren den weiteren Weg in die Zukunft. Fortsetzung folgt …

Produkte und Themen: Meilensteine der Entwicklung
Products and themes: milestones of development

When the two master carpenters and brothers-in-law Christian Wilkening and Friedrich Hahne founded a cabinet making business in the "chair village" of Eimbeckhausen near Bad Münder in 1907, nobody could possibly have guessed that a company with international standing would grow out of this: in the woody Deister-Süntel Valley, 30 km southwest of Hanover, there were some 100 businesses that were making conventional wooden chairs and armchairs to suit mass taste. The valley had developed into a centre of chair production in Germany. Adolf Wilkening, son of one of the founders, proved to be a gifted technician and production manager after joining the business (1920), for example with the changeover of the woodcutting machines that had previously been driven by steam to electricity-powered production. Wilkhahn thrived and provided work for some 40 people before production was suspended due to the outbreak of the war.

When his younger cousin, Fritz Hahne, joined the family business after the war in 1946, technological innovation combined forces with the completely new approach of the company: in collaboration with leading architects and designers of the post-war period, who were to a large extent influenced by the Werkbund and Bauhaus, Wilkhahn became a pioneer of modern furniture design. This holistic design approach that emanated from the Ulmer Hochschule für Gestaltung (HfG Ulm, Academy of Design) which was founded by Inge Aicher-Scholl, Otl Aicher, Max Bill et al in 1953, laid the foundations for further corporate development: product design, graphic corporate identity and corporate organization were regarded as a single entity that was to be communicated by all areas of the company. "Our aim is to develop sustainable products, to increase their utility value and to reduce waste" – a mission statement from the Founding Manifesto of the HfG Ulm had already formulated the principles of sustainable product design four decades before the Conference on Environment and Development in Rio de Janeiro (1992).

Fritz Hahne, whose humanistic idea of man was shaped by existential, personal experience from the war period, was not only driven by the innovative development of sustainable products, but also by the development of corporate values: "Fairness in cooperation" had been an integral part of corporate culture since the mid-1950s and focus had been on "ecological responsibility" since the 1980s. The degree to which this corporate approach was groundbreaking was demonstrated by the exhibition concept at Wilkhahn on the occasion of the World Exposition EXPO 2000 on the theme of the "future of work": with a corporate organization with special focus on social responsibility, with stabile relations with market partners and customers, with exemplary production halls designed for Wilkhahn by Frei Otto and Thomas Herzog, and particularly with the development of innovative products over the decades that achieved a synthesis of sustainability in terms of form, function and quality, as well as ecological requirements. With one of the first sustainability reports ever, Wilkhahn made a key contribution to the global concept of all-embracing corporate social responsibility that combines economic success with social orientation and ecological responsibility.

Representing the third generation of the family of entrepreneurs, Dr Jochen Hahne has been managing the company together with an efficient team since 2000. Based on the firm foundations that have made Wilkhahn fit for the future, the strengths of the company are being further developed and new fields of competence added: by sharpening the brand image, concentration on development, as well as by means of innovative and value-preserving conference and office furniture, and the dynamic expansion of international markets. The Wilkhahn corporate website, winner of many awards, numerous product and process innovations and the application of the principles of social and ecological responsibility point the way to the future. To be continued ...

Solide Sitzmöbel für den Massengeschmack: Die Ursprünge von Wilkening & Hahne, abgekürzt Wi-Ha, glichen denen zahlloser anderer Stuhlhersteller im Deister-Süntel-Tal, nahe Hannover. Polstertechnik, Material und Form waren entweder sehr einfach und preisgünstig oder aber sie nahmen die Geschmacksvorbilder längst vergangener Stilepochen auf, die von Biedermeier bis Chippendale reichten.
Als der junge Kriegsheimkehrer Fritz Hahne 1947 in die elterliche Firma eintrat, erforderte der allgegenwärtige Mangel ein hohes Maß an Improvisationstalent. Stühle wurden an die Küste geliefert und gegen Heringe eingetauscht, die wiederum als Zahlungsmittel für Schrauben, Leim, Backsteine und Zement dienten. Besonders lukrativ war in dieser Zeit der „Tauschwirtschaft" die Fertigung von Radiogehäusen für das Unternehmen Sennheiser, die nach einer festgelegten Quote mit Rundfunkgeräten abgegolten wurden und Tauschobjekte von unschätzbarem Wert bildeten. So rettete sich das Unternehmen über die ersten Jahre der Nachkriegszeit und konnte die Stuhlproduktion aufrechterhalten, bis Anfang der 50er Jahre eine vollkommen neue Ära eingeläutet wurde.

Solid chairs for mass consumption: the origins of Wilkening & Hahne, abbreviated as Wi-Ha, resembled those of limitless other chair manufacturers in the Deister-Süntel Valley, near Hanover. Upholstery technique, material and form were either extremely simple and low-priced, or they adopted the preferences in taste of former eras, ranging for example from Biedermeier to Chippendale. When young Fritz Hahne returned home from the war and joined the family business, the general shortage of everything called for a high degree of improvisational talent. Chairs were delivered to coastal areas and exchanged for herring, which in turn served as a means of payment for screws, glue, bricks and cement. In times of a "barter economy", it was highly lucrative to manufacture radio casing for Sennheiser, which was paid for by means of a fixed quota of radios, themselves becoming exchangeable objects of inestimable value. The company thus managed to survive in the initial post-war years and could keep chair production running until the early 1950s heralded a completely new age.

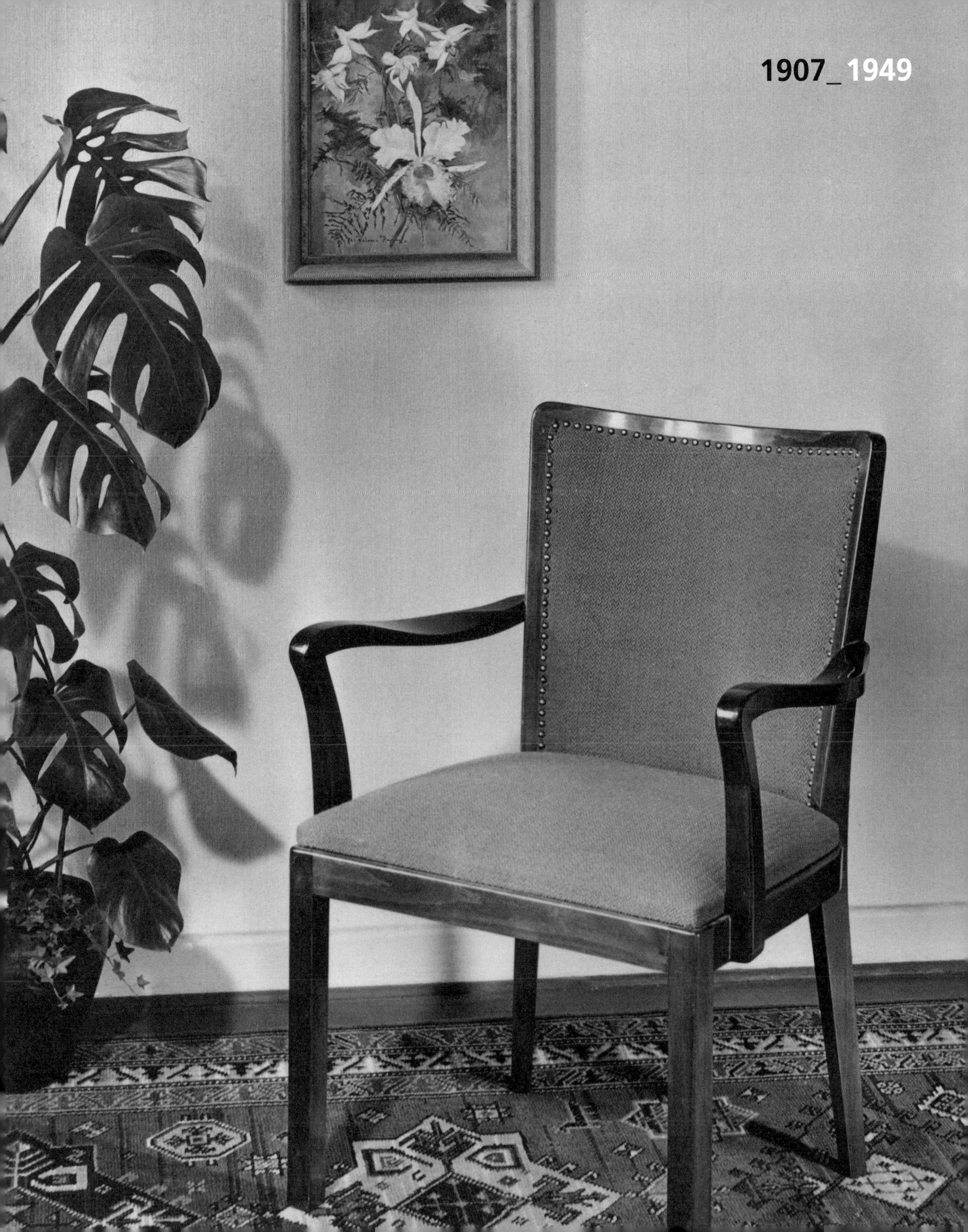
1907_ 1949

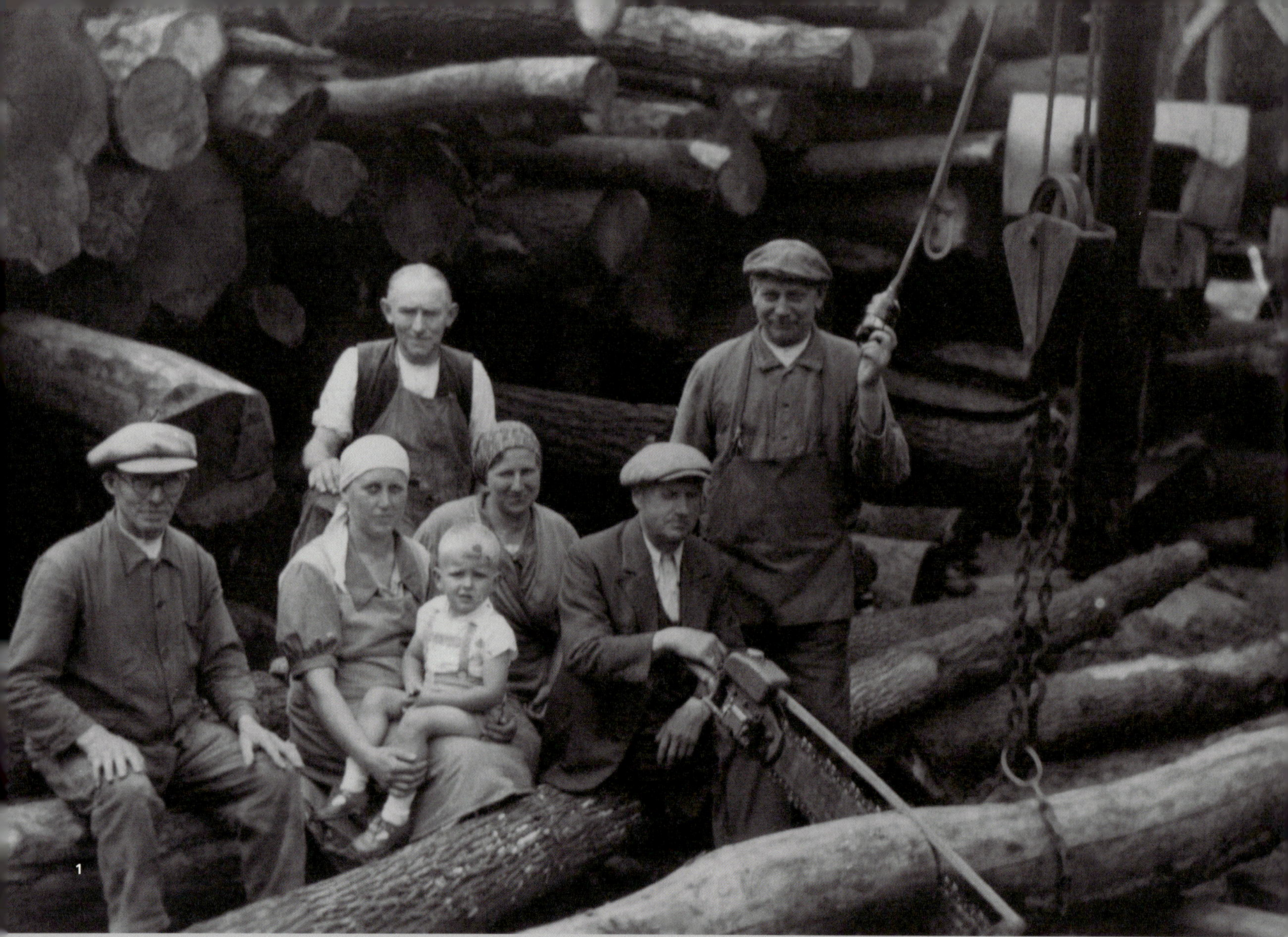

1

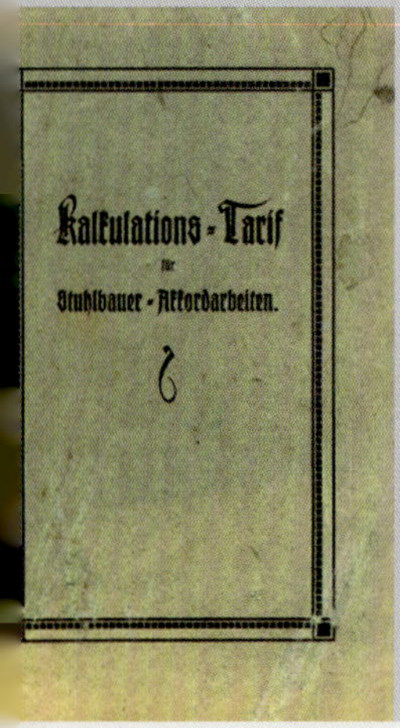

3

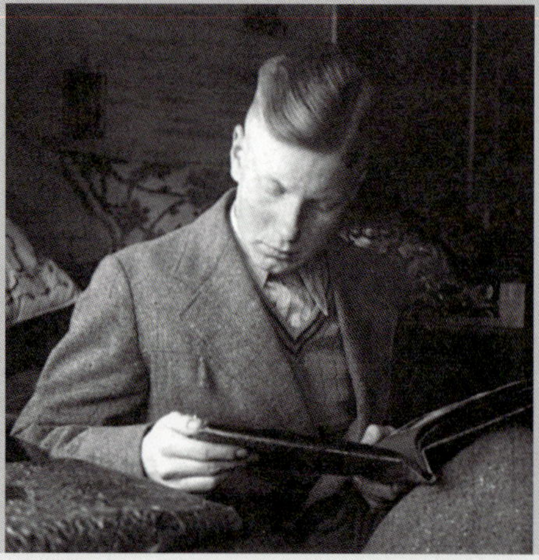

4

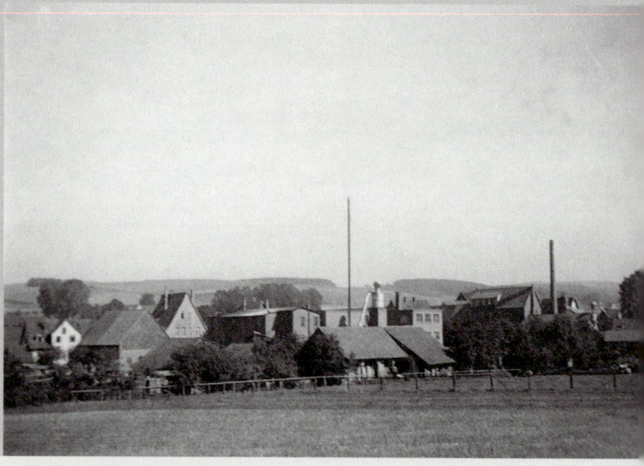

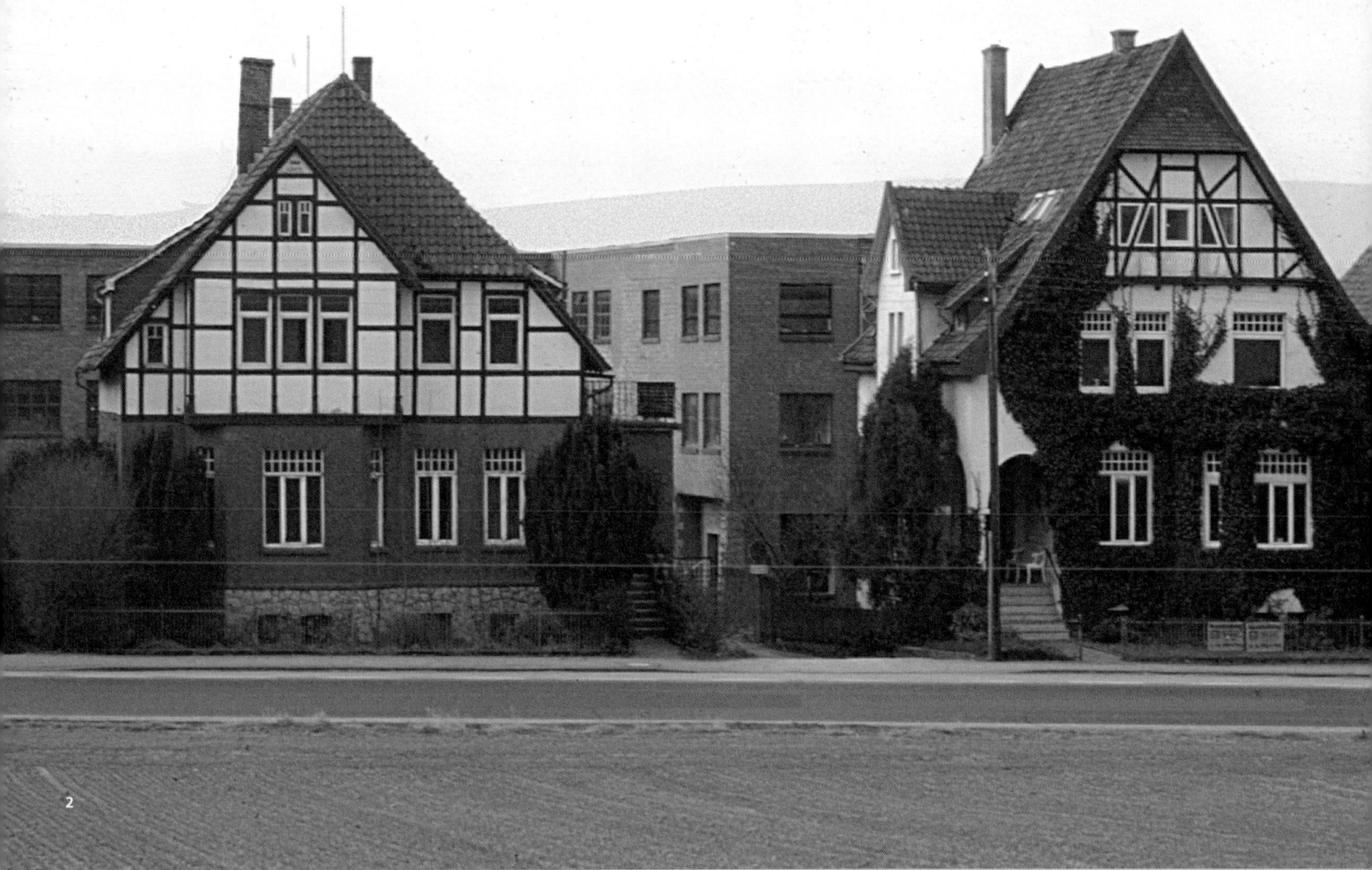

2

5

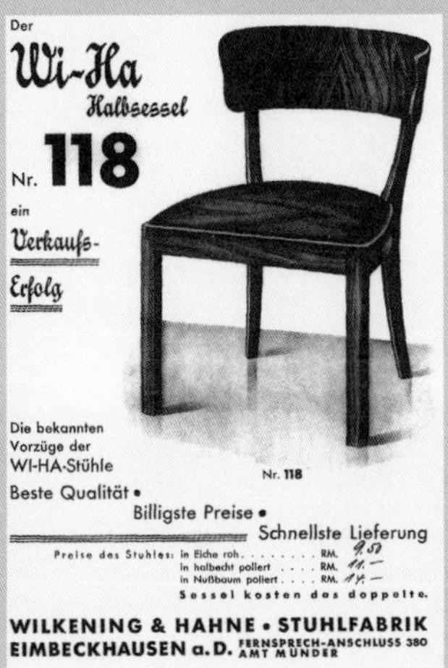

6

1_Fertigungstiefe bis zum Baumstamm. Die Bearbeitung erfolgte anfangs durch Maschinen, die von einer Lokomobile mit Transmissionsriemen angetrieben wurden. 2_Die Wohnhäuser der beiden Gründerfamilien Wilkening und Hahne im landestypischen Fachwerkstil gebaut und direkt vor der eigentlichen Fabrik gelegen. 3_Preisliste aus den 20er Jahren: „Kalkulations-Tarif". 4_Der junge Fritz Hahne beim Studieren der Buchhaltung. 5_Die Fabrikansicht am Dorfrand des „Stuhldorfs" Eimbeckhausen. 6_Anzeigen-Motiv: „Wi-Ha-Stühle mit bester Qualität und billigsten Preisen".

1_Manufacturing penetration right down to the tree trunk. Such work was initially carried out using machines that were driven by a locomobile with a transmission belt. 2_The homes of the two founding families of Wilkening and Hahne, built in the local timber-framed style and located in the direct vicinity of the factory. 3_Price list from the 1920s: "price calculator". 4_Young Fritz Hahne studying the books. 5_View of the factory on the outskirts of the "chair village" of Eimbeckhausen. 6_Advertising image: "Wi-Ha chairs with best quality and lowest prices".

Aus einer Kooperation mit den Deutschen Werkstätten entstanden die ersten Kontakte zur „Moderne“ und die langjährige Zusammenarbeit mit der HfG Ulm. Namhafte Architekten und Gestalter wie Jupp Ernst, Roland Rainer, Hans Bellmann, Walter Papst, Herbert Hirche und Georg Leowald entwarfen für Wilkhahn Möbel, die heute zu begehrten Sammlerstücken zählen. Bald hielten neben dem traditionellen Werkstoff Holz neue Materialien im Möbelbau Einzug, wie Metall und glasfaserverstärkter Kunststoff, der selbsttragende, unzerstörbare und witterungsbeständige Freiformen ermöglichte. Zu einer Ikone der Wilkhahn-Gestaltung wurde die Schaukelplastik von Walter Papst: Die abstrakte, auf das Wesentliche reduzierte Formgebung, das innovative Material und der „neutrale“ Name reflektierten den gestalterischen Anspruch ebenso wie die emanzipatorisch aufgeklärte, soziale Haltung der Designer im Umfeld der HfG Ulm. Zur sozialen Haltung bei Wilkhahn gehörte auch die Einführung der betrieblichen Altersversorgung durch Fritz Hahne – zu einer Zeit, als der Wiederaufbau des zerstörten Deutschland geleistet wurde, ohne dass die Aussicht auf angemessene Renten bestand.

Cooperation with the Deutsche Werkstätten resulted in initial encounters with the “Modern Age” and longstanding collaboration with the Academy of Design in Ulm (HfG Ulm). Renowned architects and designers such as Jupp Ernst, Roland Rainer, Hans Bellmann, Walter Papst, Herbert Hirche and Georg Leowald designed furniture units for Wilkhahn that are still sought-after collectors’ items today. Besides the traditional material of wood, new materials found their way into furniture manufacturing: such as metal and fibre glass-reinforced plastic that allowed indestructible, cantilevered and weather-resistant freely shaped forms. Walter Papst’s “Rocking Sculpture” became an icon of Wilkhahn design: the abstract design, reduced to essentials, innovative material and the “neutral” name reflect the design approach as well as the emancipatory, enlightened social approach of the designers in the HfG Ulm circles. This social focus at Wilkhahn also included Fritz Hahne’s introducing a company pension scheme – at a time when the reconstruction of Germany in the aftermath of war had started and there were no prospects of adequate pensions being paid.

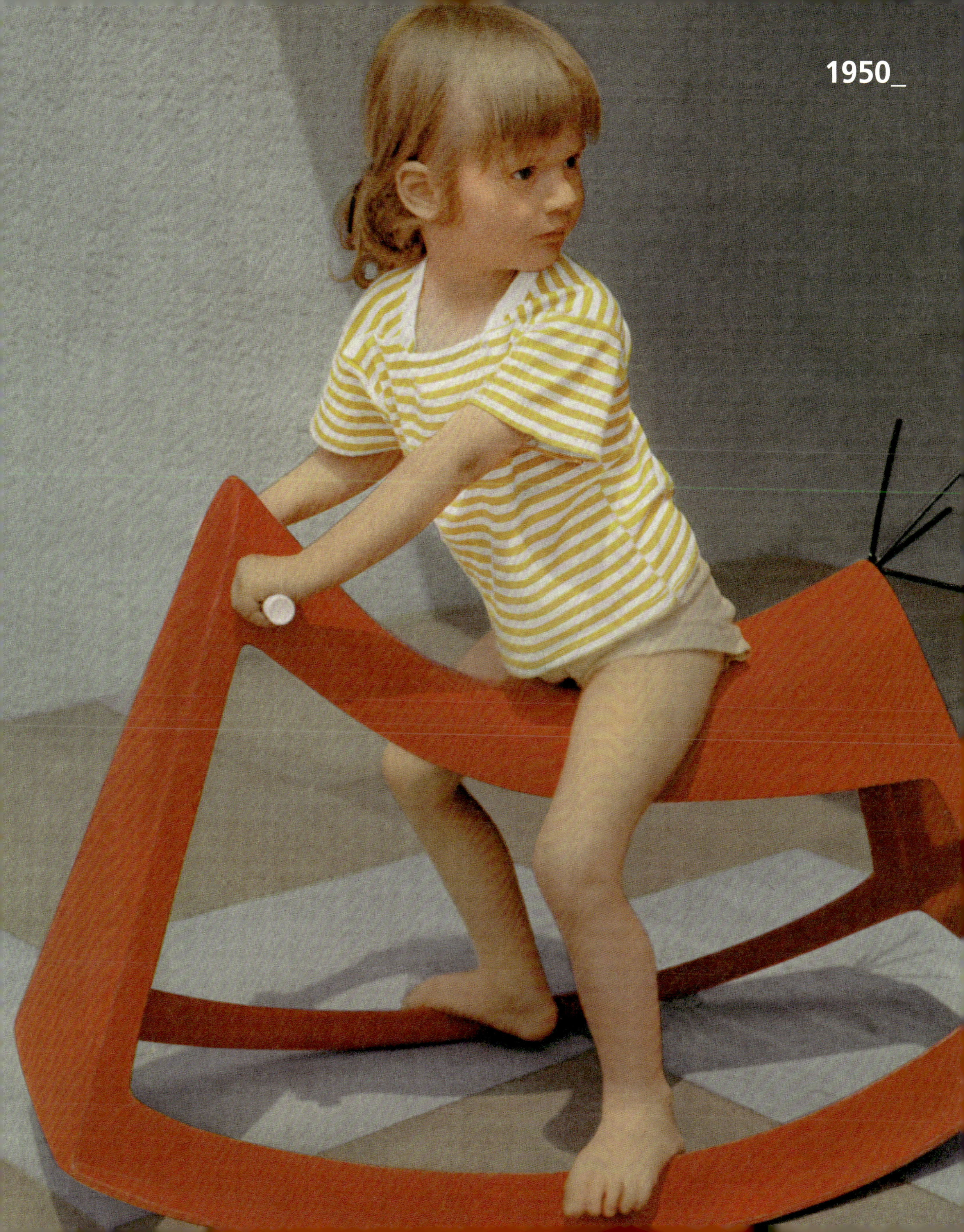
1950_

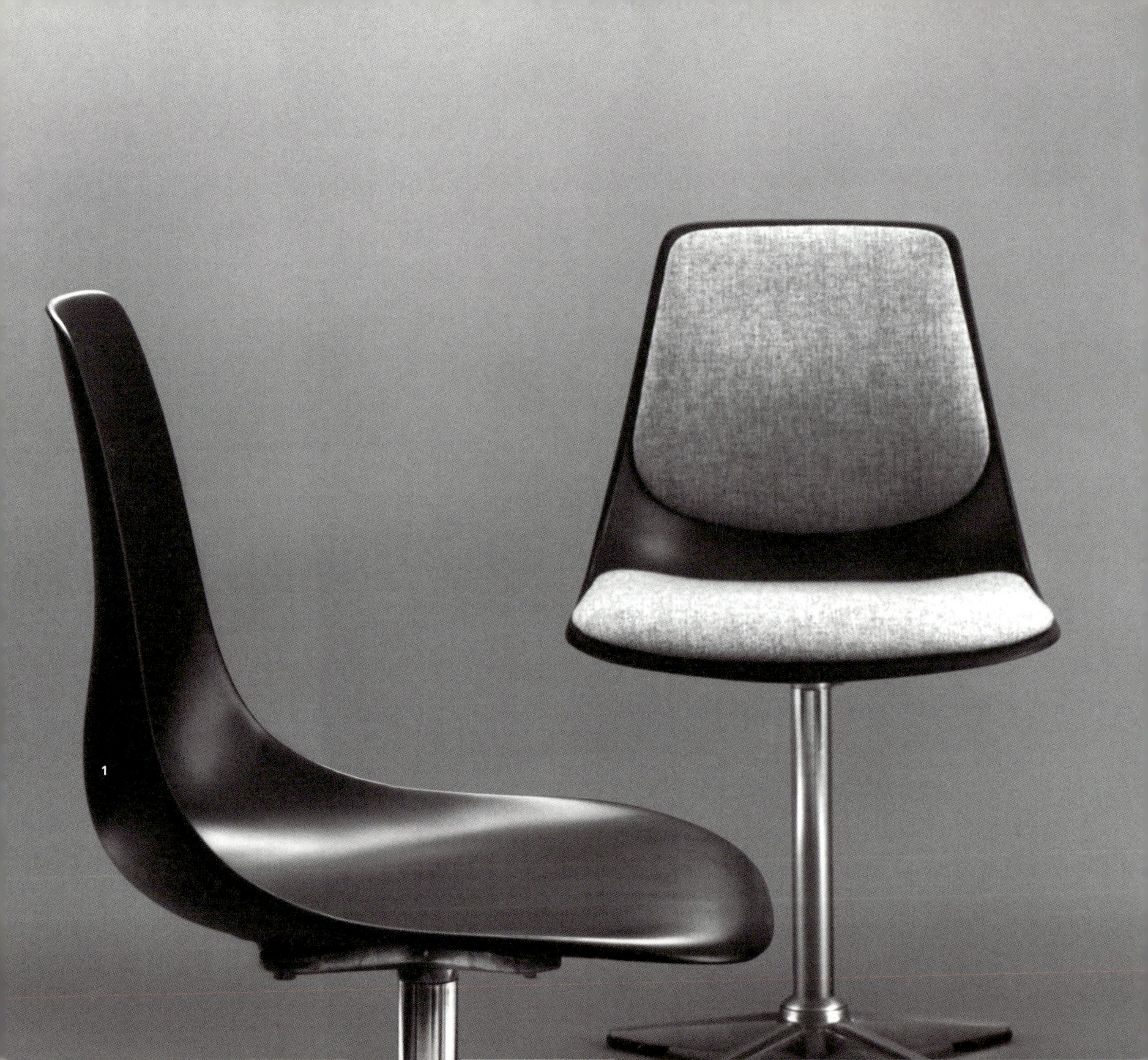

1

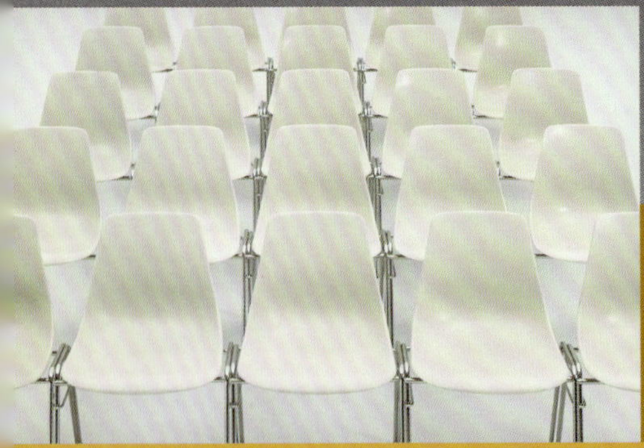

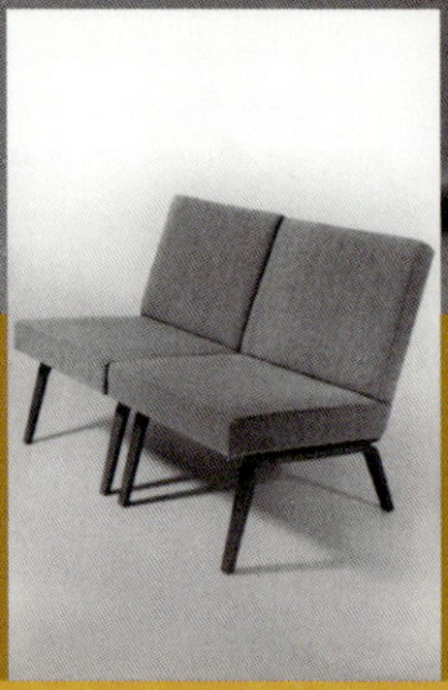

3

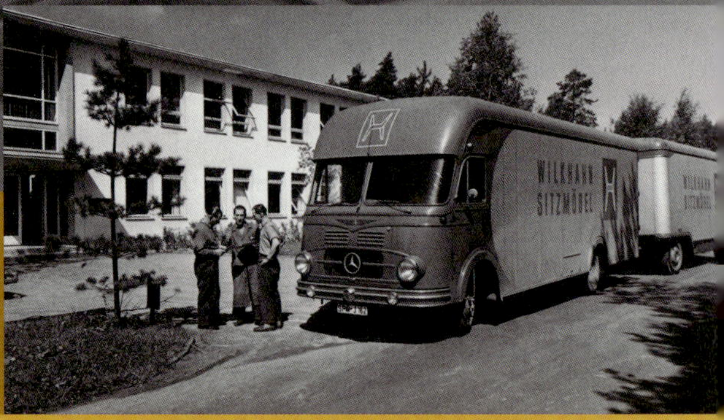

4

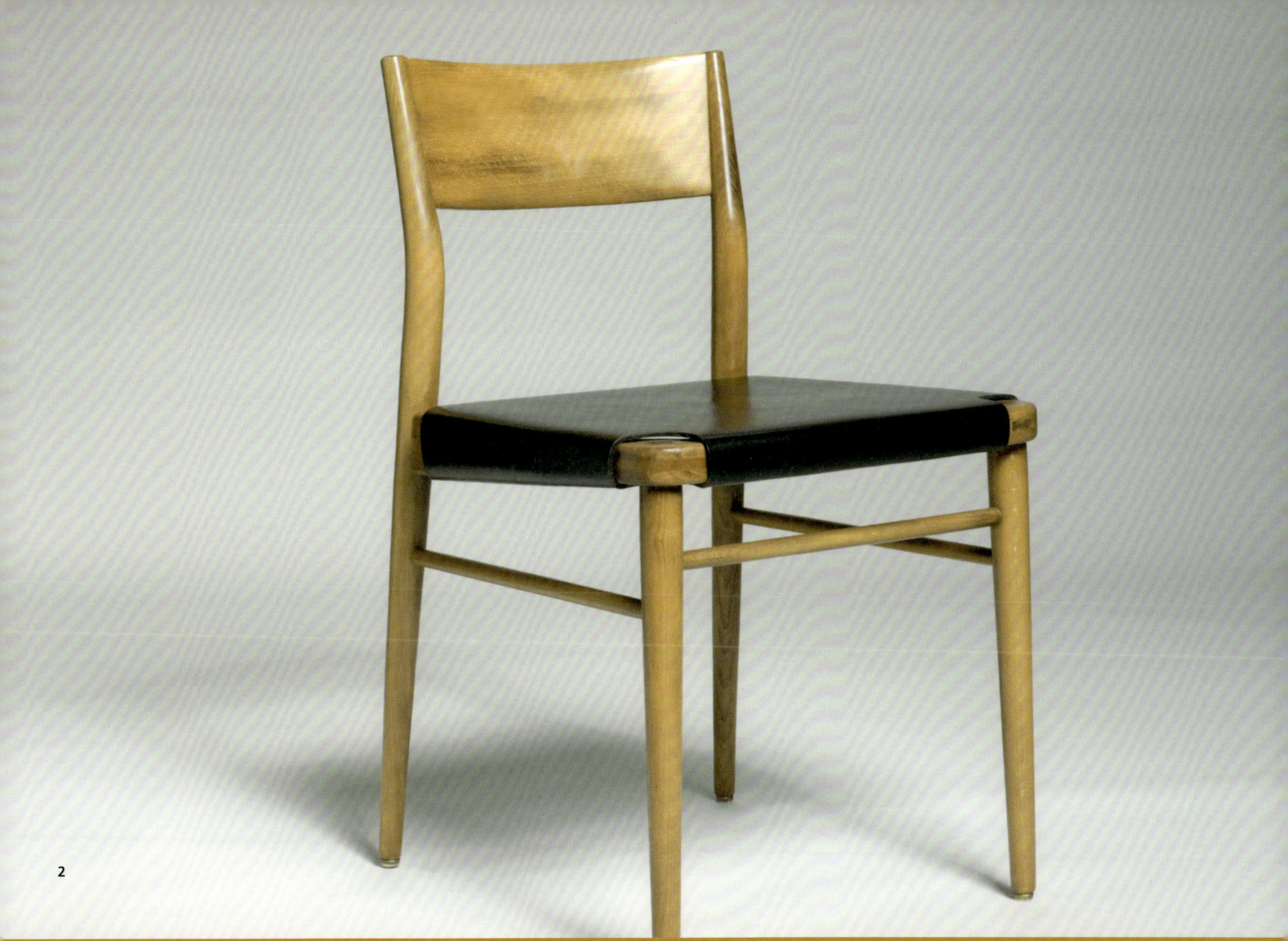

2

1_Die ersten GFK-Schalenstühle für Wilkhahn: Auf Basis der Grundmodelle aus den 50er Jahren entwickelte Georg Leowald eine breite Programmfamilie, die vierbeinige Stühle, Drehstühle mit und ohne Polster und Drehsessel umfasste. 2_Traditionelles Material, handwerkliche Qualität und moderne Formensprache: der vierbeinige Holzstuhl war wahlweise mit Flechtwerk- oder Ledersitz zu haben, Programm 351/3, 1955–1971, Design: Georg Leowald. 3_GFK-Schalenstuhl mit vierbeinigem Fußgestell: Modell 225/1, 1954–1966, später als Modell 220/5 auch in Reihe verkettbar. 4_Im Wohnbereich und in Sozialbauten zu Hause: geradliniger Polstersessel mit gebogenem Holzgestell, Modell 457, 1959–1961, Design: Herbert Hirche. 5_Wilkhahn-LKW vor dem zeitweiligen Zweigwerk in Espelkamp mit einem abstrahierten Sitzmöbel als Wilkhahn-Signet, das intern „Backenzahn" genannt wurde. 6_Katalog der Deutschen Werkstätten (DeWe), für die Wilkhahn Anfang der 50er Jahre einen Teil der Kollektion produzierte.

1_The first chairs with fibre glass-reinforced shells for Wilkhahn: Georg Leowald developed an extensive product family comprising four-legged chairs, upholstered and non-upholsered swivel chairs according to basic models from the 1950s. 2_Traditional material, craft quality and modern design: the four-legged wooden chair was optionally available with a seat in wicker work or leather, 351/3 range, 1955–1971; designed by Georg Leowald. 3_Chair with fibre glass-reinforced shell with four-legged frame: model 225/1, 1954–1966, subsequently as model 220/5 also with online connectors. 4_For the home and often found in council housing: linearly designed chairs with soft upholstery and a curved wooden frame, model 457, 1959–1961; designed by Herbert Hirche. 5_Wilkhahn lorry in front of branch facilities in Espelkamp with the abstraction of a chair as the Wilkhahn design mark, which was called the "back tooth" at the company. 6_Catalogue of Deutsche Werkstätten (DeWe) for whom Wilkhahn produced part of its collection in the early 1950s.

5

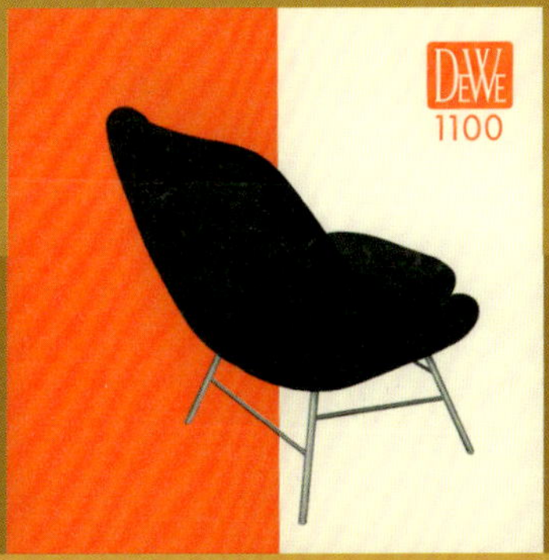

6

Die Zusammenarbeit mit den „Ulmern" erstreckte sich nun auch auf das Grafikdesign: Tomás Gonda entwickelte ein neues, geradliniges Signet, Bernd Franck gestaltete die Wilkhahn-Broschüren. Wilkhahn hatte als eines der ersten Unternehmen der Branche eine „Kommunikationsabteilung", die über 35 Jahre hinweg vom gelernten Journalisten Rudolf Schwarz geleitet wurde. Industriell zu produzierende Holzstühle oder aber GFK-Stühle, die für gewerbliche Projekte in Verwaltungs- oder Sozialbauten in großen Stückzahlen gefertigt wurden, sorgten endgültig für den Sprung vom handwerklichen Betrieb zum modernen Industrieunternehmen. Und: Der rasant steigende Platzbedarf am Stammsitz erforderte den Neubau weiterer Gebäude, mit denen die Architekten Herbert Hirche und Georg Leowald sichtbare Zeichen auch der Baukultur setzten. Auch die Entwicklung der Unternehmenskultur wurde mit innovativen Modellen vorangetrieben: „Co-Management" und „Kooperativer Führungsstil" waren die Schlagworte, die damals unerhört modern waren und den Unternehmer Fritz Hahne ebenso bekannt wie anderen Unternehmern verdächtig machte.

Collaboration with the Ulm circles also extended to graphic design: Tomàs Gonda developed a new, linear design mark; Bernd Franck designed Wilkhahn product brochures. Wilkhahn was one of the first companies in the industry to set up a "Communications Department" that was managed by qualified journalist Rudolf Schwarz for over 35 years. Wooden chairs that could be produced industrially, but also fibre glass-reinforced chairs manufactured in large numbers for commercial projects in administration or public, communal and social welfare facilities finally secured the transition from a craft business to a modern industrial company. In addition, the rapidly increasing shortage of space at head office required further buildings to be erected. Architects Herbert Hirche and Georg Leowald were commissioned, and their projects left a visible mark on industrial architecture too. The development of corporate culture was also pursued by using innovative models of management: "co-management" and "cooperative management" became the order of the day: these terms had an incredibly modern ring at the time and served to make entrepreneur Fritz Hahne well known while other entrepreneurs became suspicious of what was going on.

1960_

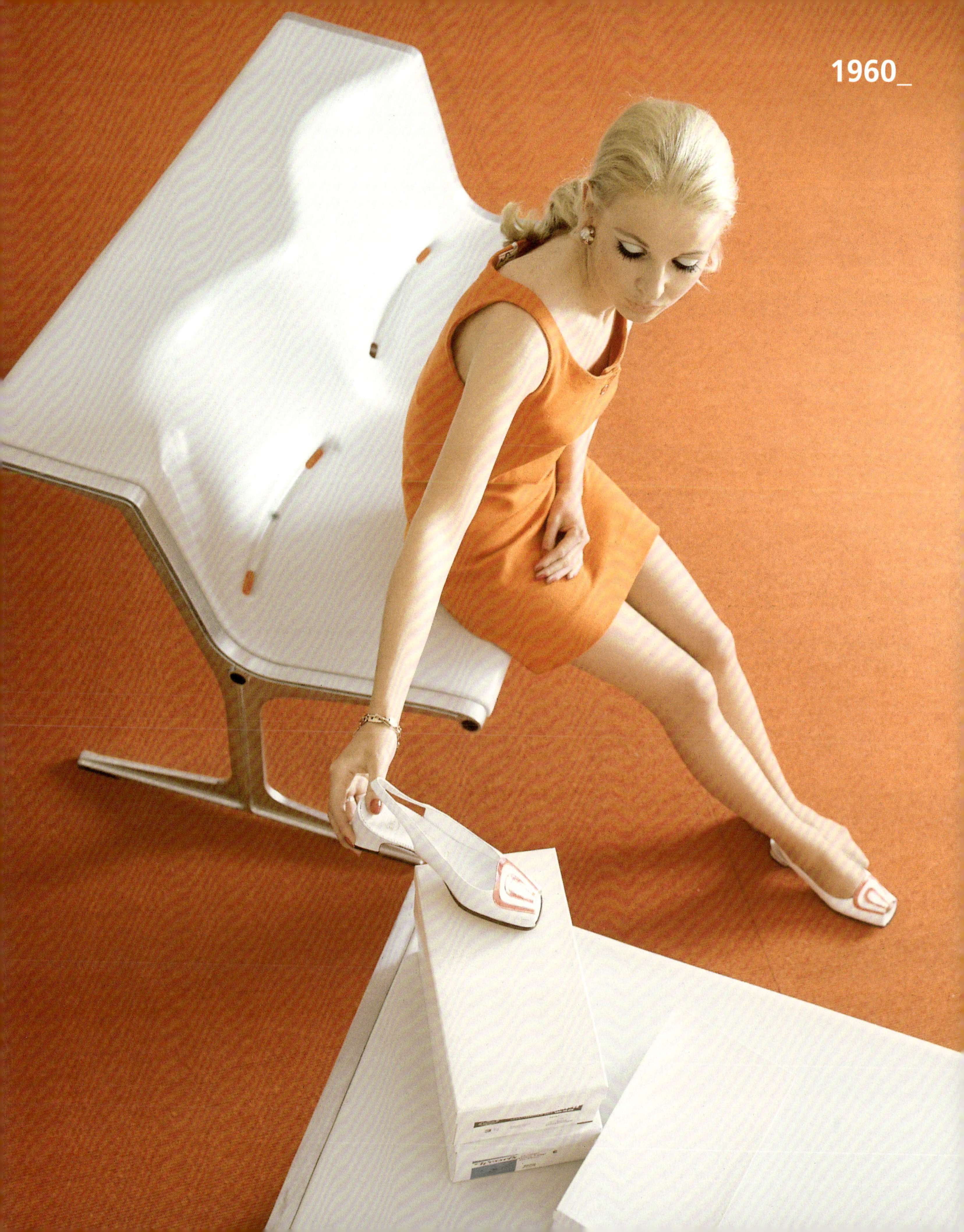

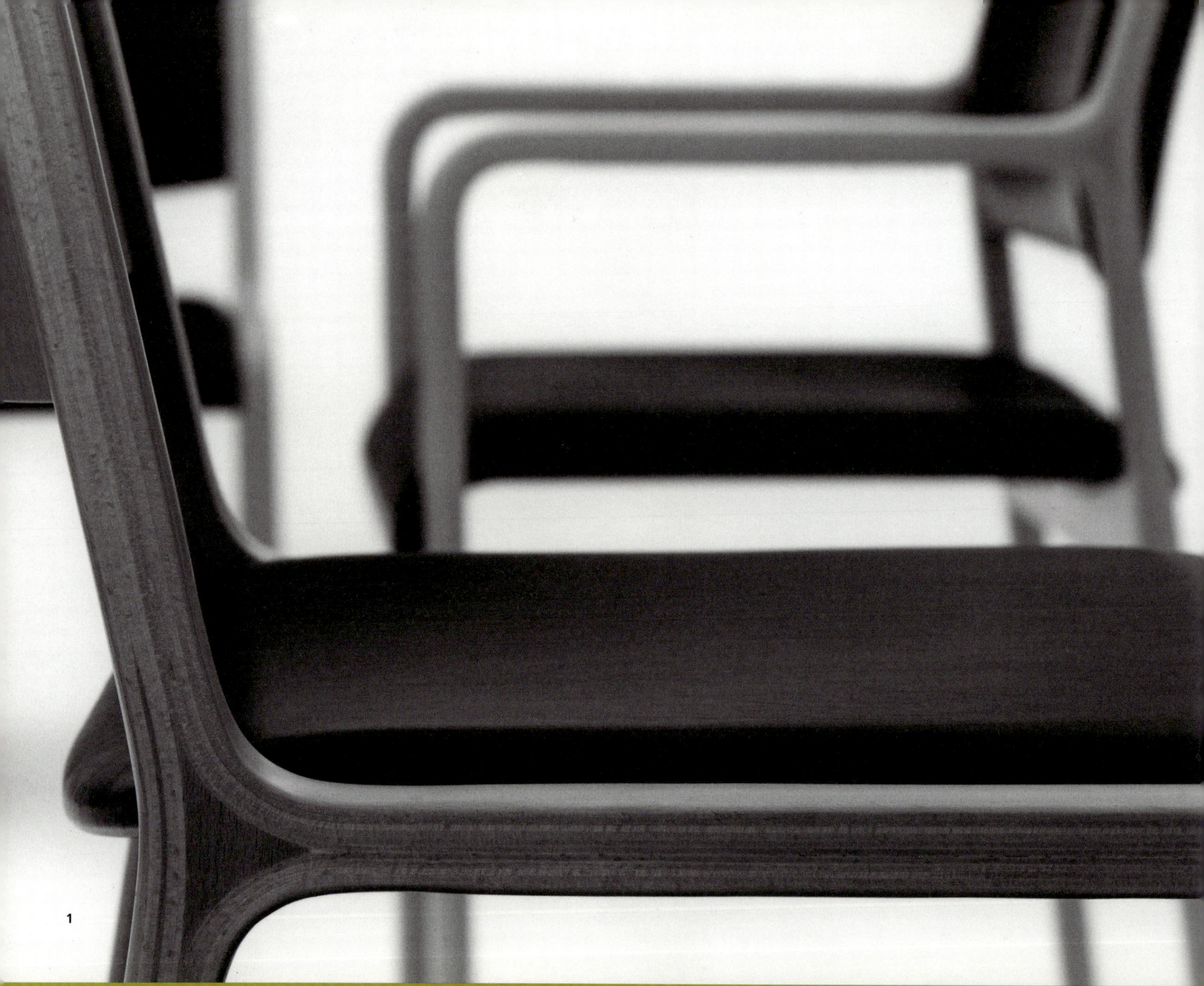

1

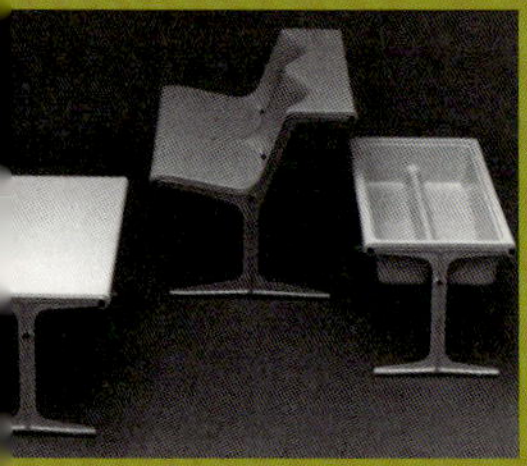

3

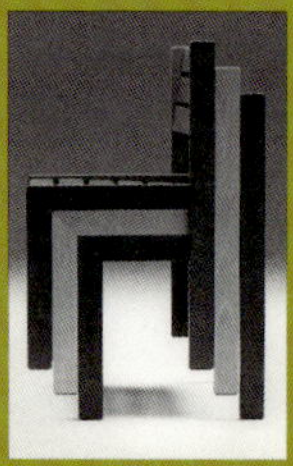

4

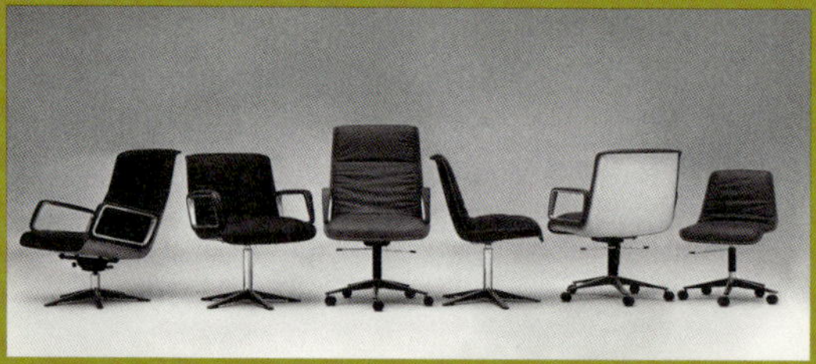

5

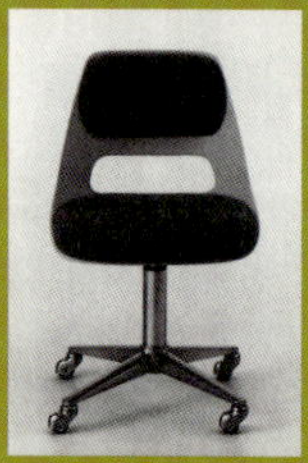

6

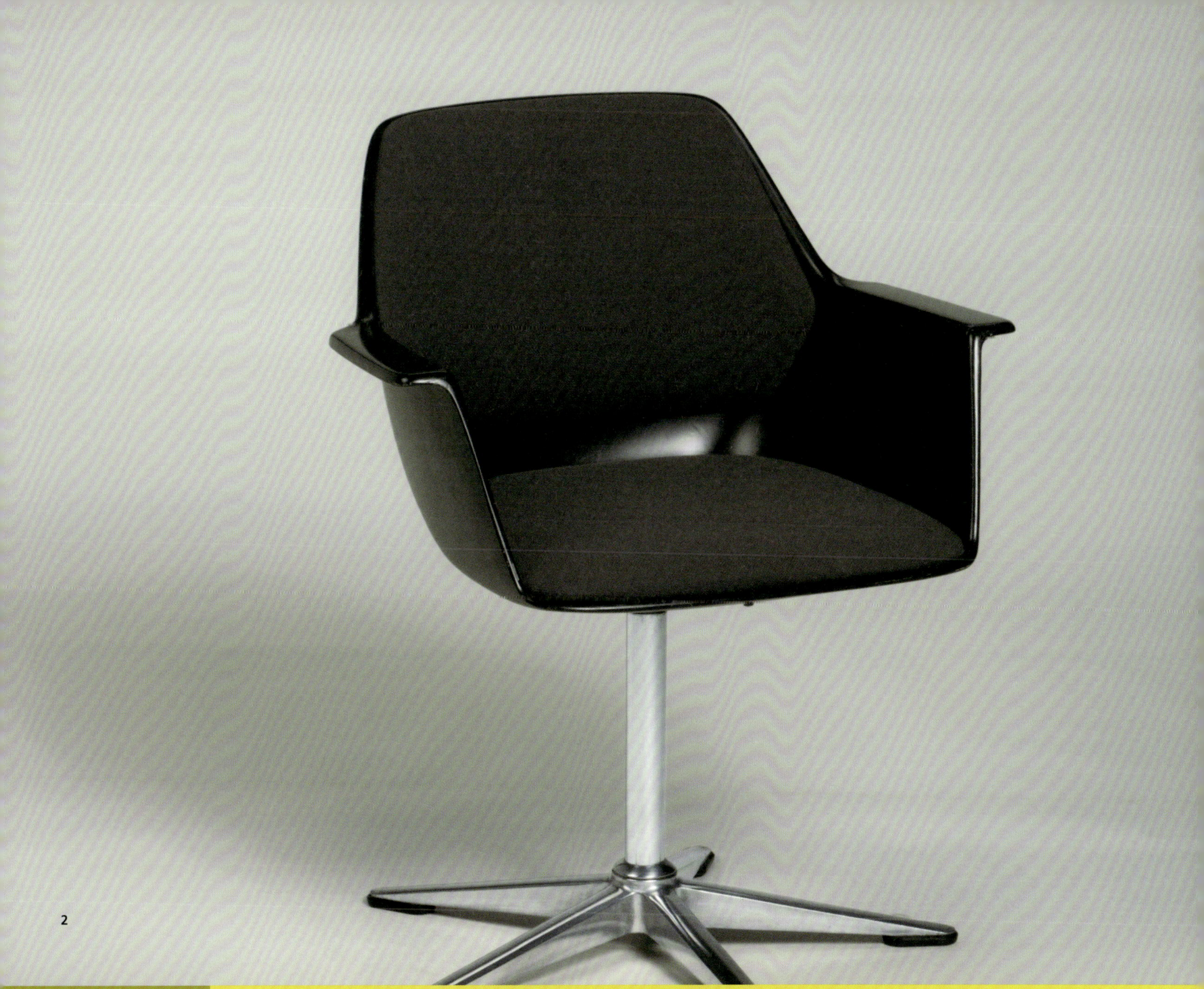

2

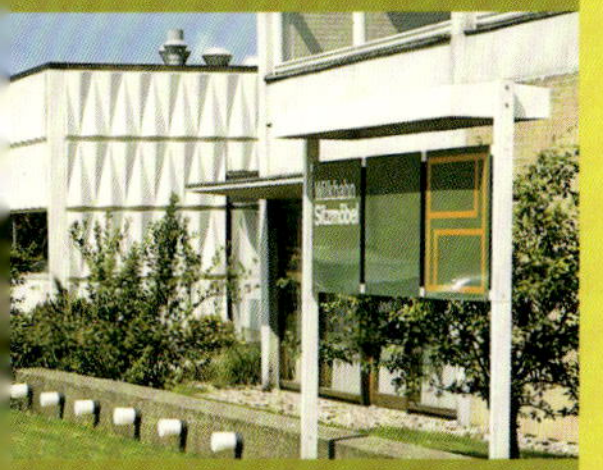

7

1_Von 1961–1998 in Produktion: Mit der charakteristischen „Tennisschläger-Verbindung" wurde das Schichtholzprogramm 400 als leichter, flexibler und unzerstörbarer Mehrzweckstuhl zum Bestseller, Design: Wilhelm Ritz, 1961–1998. 2_Ursprünglich glaubte Georg Leowald, alle Bedürfnisse mit nur zwei Schalentypen und verschiedenen Gestellstrukturen lösen zu können. Der Wunsch des Menschen nach Status und Differenzierung führte dann zu einer breiten Programmfamilie. 3_Bänke, Tische, Tröge aus GFK und Aluminium: Das Banksystem 1200 des niederländischen Designers Friso Kramer für Wartelandschaften war der Einstieg für die Einrichtungen von Transitzonen, produziert von 1968–1988 und noch immer in vielen U-Bahn-Stationen zu Hause. 4_Der Archetyp eines Holzstuhls: Programm 300, Design: Hans Peter Piehl. Der stapelbare „Latten-Stuhl" wurde von 1965–1990 zu Hunderttausenden produziert und verkauft. 5_Das Chef-Programm mit den bis dahin größten produzierbaren Sitzschalen: Sesselprogramm 2000 von Delta-Design, 1968–1989. Damals wie heute luxuriöse und begehrte Sitzmöbel in Repräsentationsbereichen. 6_Materialminimierung und bessere Hinterlüftung: Die letzte Modellvariante der Leowald-Stühle, Programm 230/5, 1967–1972 markierte den Einstieg in das Marktsegment Bürostühle. 7_Bürogebäude und Ausstellungsneubau: Herbert Hirche realisierte 1960 den Verwaltungsneubau in der klaren Formensprache der Moderne. Die später angebaute Ausstellung wurde mit Wilkhahn-Fassaden aus GFK ausgestattet – einem der vielen Experimentierfelder dieser Zeit.

1_In production from 1961–1998: with a characteristic "tennis racquet" frame, the laminated wood 400 range marked a light, flexible and indestructible multi-purpose chair that became a bestseller; designed by Wilhelm Ritz, 1961–1998. 2_Georg Leowald originally thought he could satisfy all needs with just two types of seat shell and various frame structures. However, people's desire for status and differentiation resulted in an extensive product family. 3_Bench seating, tables, chests in glass fibre-reinforced plastic and aluminium: the 1200 bench system for lounge applications by Dutch designer Friso Kramer marked the starting point for furnishing transit zones; it was produced from 1968–1988 and can still be found in many underground stations. 4_The archetype of a wooden chair: the 300 range; designed by Hans Peter Piehl. Hundreds of thousands of the stackable "slatted-frame chair" were produced and sold from 1965–1990. 5_The executive range with the largest seat shells that could be produced at the time: the 2000 range by Delta-Design, 1968–1989. Luxurious and coveted seating for prestigious areas as is still popular today. 6_Minimum material input and improved rear ventilation: the last model variation of the Leowald chairs, 230/5, 1967–1972, marked the move into the office chair market segment. 7_Office building and new showroom: administration building by Herbert Hirche from 1960 using the clear design language of the Modern Age. The showroom, which was added later on, used Wilkhahn front cladding in fibre glass-reinforced plastic – one of the many experimental fields at the time.

Die 70er Jahre standen im Zeichen der Professionalisierung, insbesondere des Designmanagements. Das Multitalent Klaus Franck, ebenfalls ein „Ulmer", ordnete die Geschäftsfelder für die Produktentwicklung, entwickelte das Grafik-Design weiter, gestaltete Messeauftritte, Ausstellungen und Werbung, betätigte sich als Autor und entwickelte mit seinem kongenialen Designpartner Werner Sauer zahlreiche erfolgreiche Sitzmöbel. Nick Roericht mit seinem Ulmer Designbüro ProduktEntwicklung Roericht (PER) war die zweite herausragende Persönlichkeit, die mit ihren Bürostuhlentwicklungen neue Impulse für Wilkhahn setzte. Die Basis für den Erfolg im Bereich Büroarbeitsstühle war das Bürostuhlprogramm 232 von Wilhelm Ritz. Der heute als Designikone heiß begehrte Klassiker mit der zweigeteilten Sitzschale und dem markanten Drehgelenk eroberte im Sturm die Verwaltungen und Kassenhallenarbeitsplätze, insbesondere bei Finanzinstituten und Versicherungen, die sich zu dieser Zeit ein modernes Erscheinungsbild geben wollten.
Und Wilkhahn machte von sich reden, indem es unter dem Leitbegriff „Soziale Gerechtigkeit" eine Gewinnbeteiligung der Mitarbeiterinnen und Mitarbeiter einführte.

The 1970s were characterized by professionalism, especially in terms of design management. Multi-talent Klaus Franck, also from the Ulm Academy, organized the business segments for product development, took graphic design even further, designed performance concepts for trade fairs, exhibitions and advertising, and worked as an author, and in collaboration with his congenial partner, Werner Sauer, he designed and developed successful seating models. Nick Roericht, with his design studio in Ulm, ProduktEntwicklung Roericht (PER), was the second outstanding personality who gave new impetus to Wilkhahn with his office chair designs. The 232 office chair range by Wilhelm Ritz became the basis of success in the office chair segment. This design classic with the two-section seat shell and distinctive swivel joint – a design icon that is still in great demand today – took administrations and cashier workplaces, especially in financial institutions and insurance offices, by storm at a time when such businesses were keen on adopting a modern corporate image. And Wilkhahn became the talk of the town when it introduced profit-sharing for employees in line with the guiding principle of "social justice".

1970_

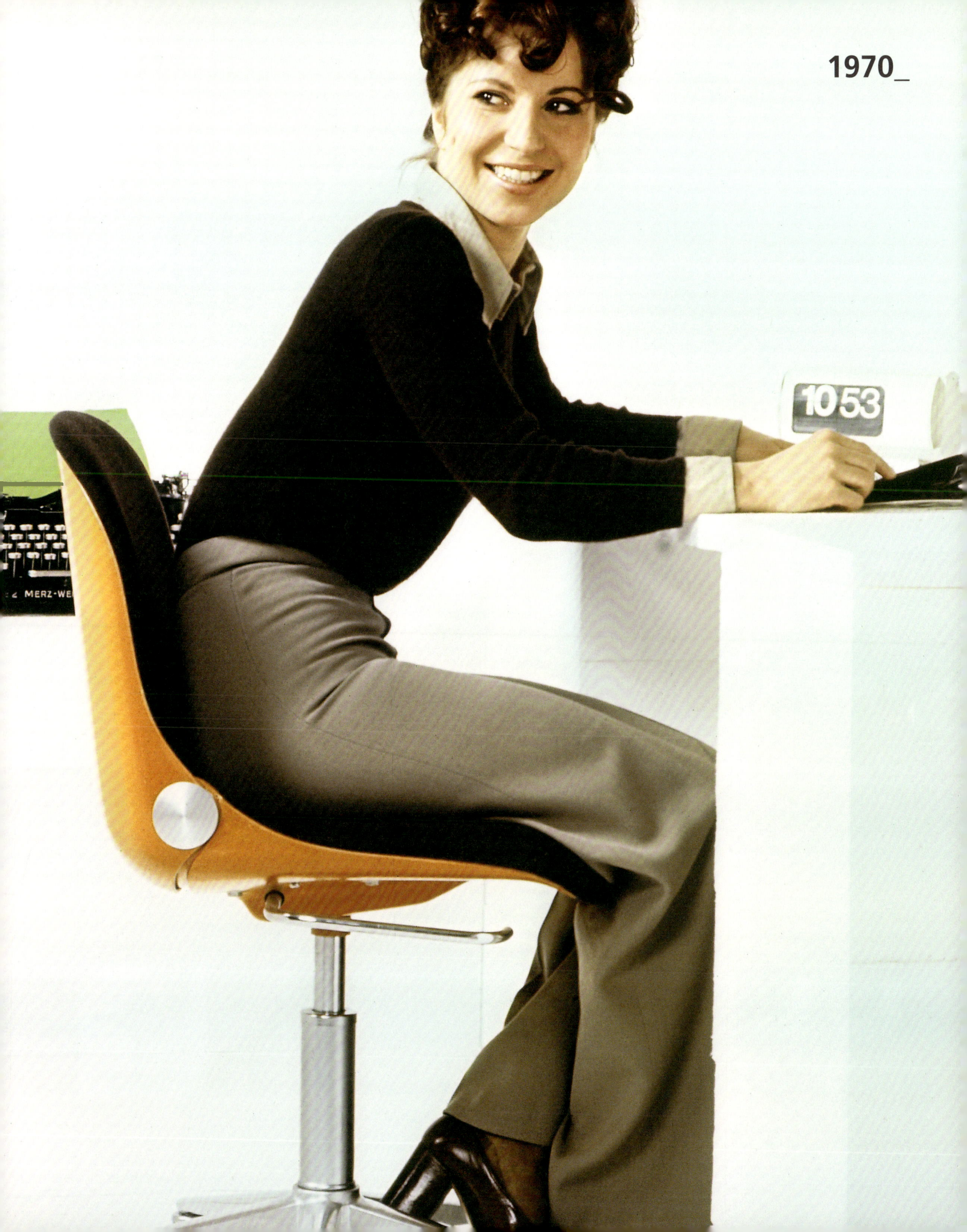

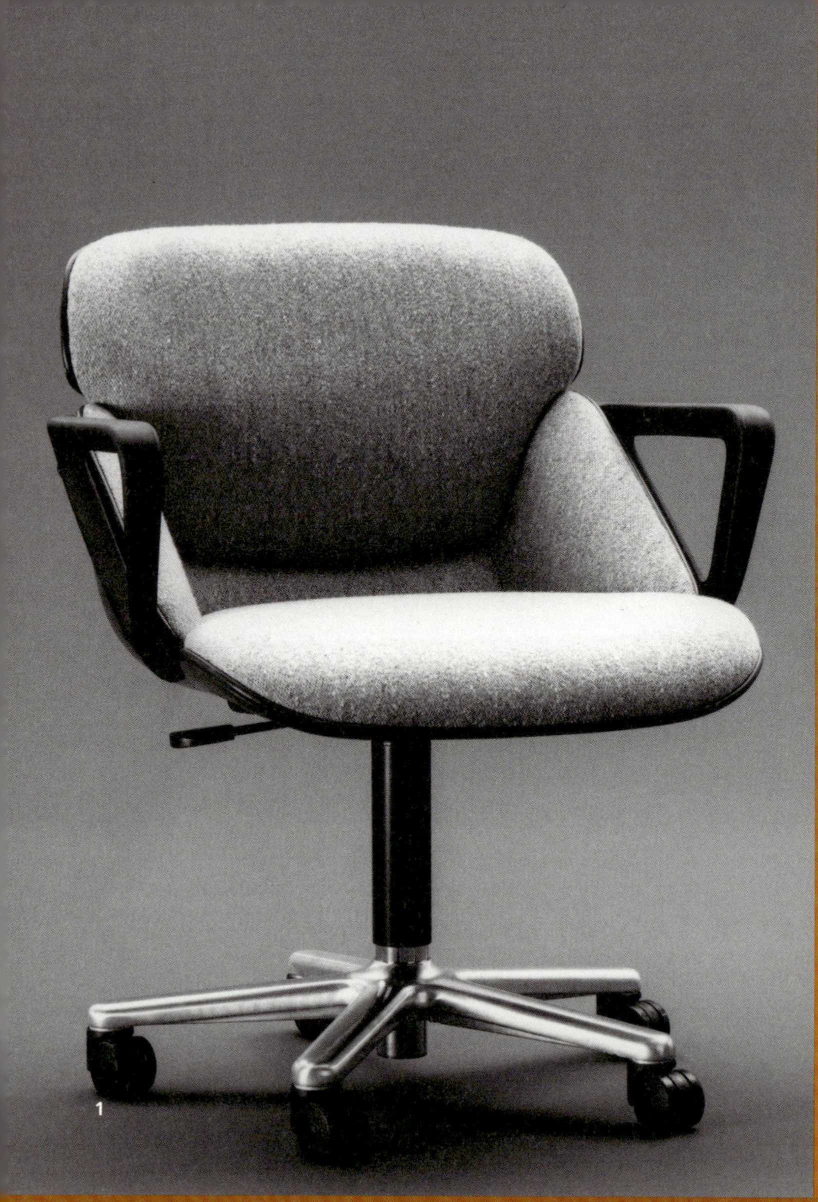

1

1_Bürostuhlprogramm 190, Design: ProduktEntwicklung Roericht. Die in der Horizontalen flexibel in drei Teilen ausgebildete Sitz- und Rückenschale aus glasfaserverstärktem Polyester war der erste Schritt zum „dynamischen" Sitzen, produziert von 1976–1983. 2_Das Bürostuhlprogramm 232, Design: Wilhelm Ritz, wurde intern „Sparkassen-Stuhl" genannt und zum Bürostuhlbestseller, der von 1971–1990 verkauft wurde. 3_Für Kundenhallen und Wartezonen konzipiert, als Einzelsessel und Mehrsitzer oder mit Tischelementen zu „Sitzlandschaften" verkettbar: Programm 840, ProduktEntwicklung Roericht, 1979–1998. 4_Der Zeit weit voraus war die Programmfamilie 238 von Delta-Design, die 1976–1985 produziert wurde. Schulterstütze, Seitenflügel und Sitzschale aus Polypropylen und eine ausgeklügelte Mechanik machten das Modell 238/7 zu einem echten „Hingucker", das arbeitsmedizinischen Erkenntnissen sehr nahe kam und deshalb auch „DIN-Stuhl" genannt wurde. 5_Design als Ergebnis genauer Forschung und Entwicklung: Funktions- und Designmodell des vielfach prämierten Programms 190. 6_Die hervorragende Ergonomie und das futuristische Design bescherten dem Modell 238/7 zwar nur verhältnismäßig geringe Stückzahlen, aber einen eindrucksvollen Auftritt im James-Bond-Film „Moonraker".

1_190 office chair range; designed by ProduktEntwicklung Roericht: the seat and back shell, comprising three horizontal sections in fibre glass fibre-reinforced plastic marked the first step towards "dynamic sitting". It was produced from 1976–1983. 2_The 232 office chair range, designed by Wilhelm Ritz, was called the "savings bank chair" at the factory; it became a bestseller in the office chair segment and was sold from 1971–1990. 3_Designed for customer service halls and lounge areas as an individual chair and a two or three seater or linkable by means of table segments to configure "sitting landscapes": the 840 range, ProduktEntwicklung Roericht, 1979–1998. 4_The 238 product family by Delta-Design, produced from 1976 to 1985, was ahead of its time. Shoulder supports, wing-type armrests, a seat shell in polypropylene and an ingenious mechanism ensured that the 238/7 model turned quite a few heads. It also complied to a large extent with occupational therapy knowledge and therefore became known as the "DIN chair" (German industrial standard). 5_Design as the result of research and development: functional and design model of the 190 range – winner of numerous awards. 6_Although excellent ergonomics and futuristic design resulted in relatively low prduction figures for the 238/7 model, it did have an impressive showing in the James Bond "Moonraker" film.

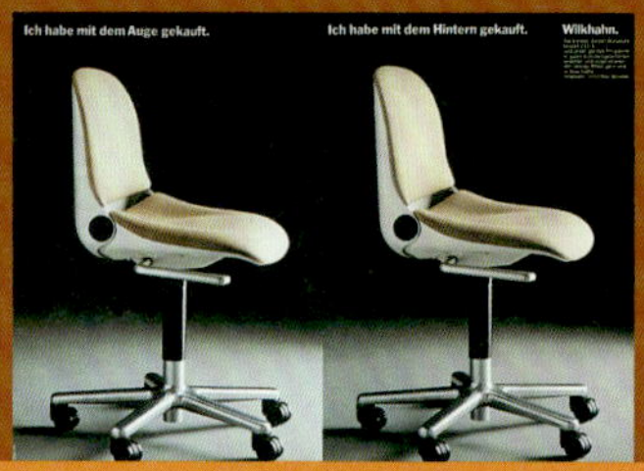

2

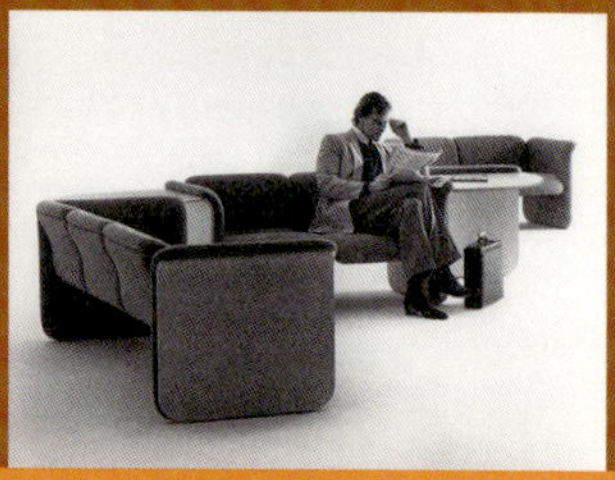

3

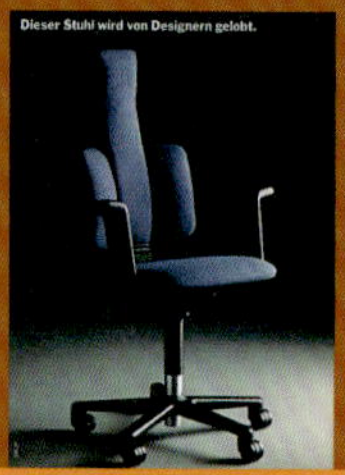

4

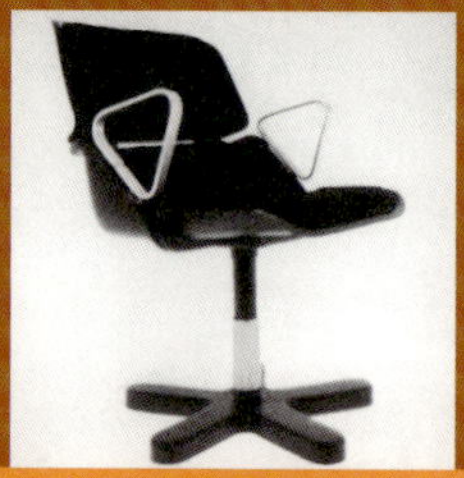

5

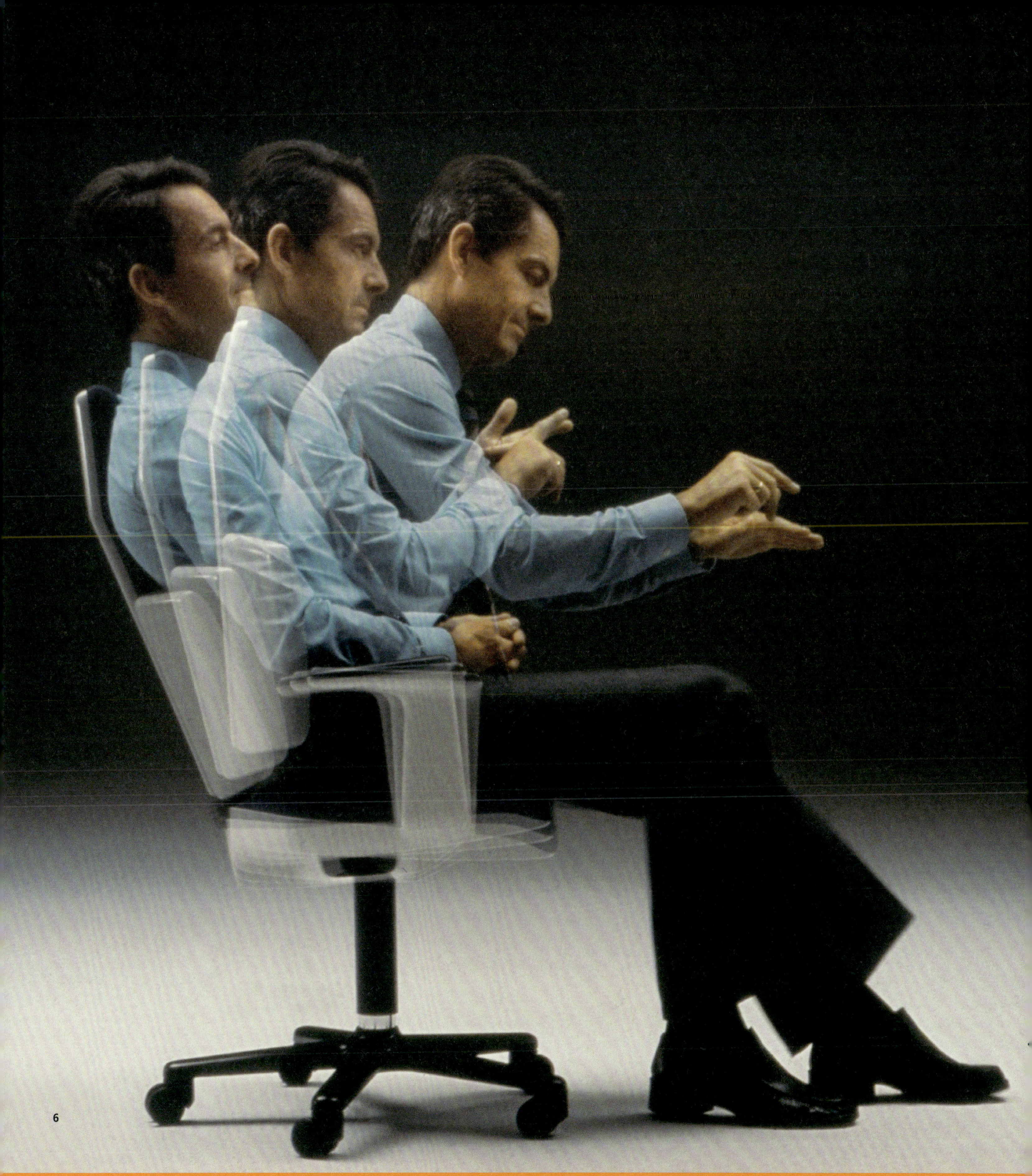

6

Der Start in das neue Jahrzehnt wurde von der Vorstellung der FS-Linie geprägt, die nach den beiden Wilkhahn-Designern Klaus Franck und Werner Sauer benannt wurde. Gegen den Zeitgeist, der Bürostühle zu regelrechten Sitzmaschinen werden ließ, setzten die beiden konsequent auf Reduktion: Wie eine flexible Hängematte ist die Sitzschale zwischen eine Gabel und die vordere Drehachse eingespannt. Doch was so selbstverständlich und in Sachen Selbstanpassung genial einfach erscheint, war das Ergebnis langjähriger Bewegungsstudien, Materialtests und neuer Fertigungsverfahren. Zunächst von aller Welt belächelt, wurde die FS-Linie unter der Überschrift „Sitzen ohne Führerschein" bis heute zum auch international meistverkauften Bürostuhl-Klassiker. Begleitet wurde die Erfolgsgeschichte durch einen Wechsel im Top-Management: Getreu seinen Überzeugungen zog sich Fritz Hahne in den Verwaltungsrat zurück und überließ das operative Feld dem erfahrenen Industriemanager Theodor Diener, der die Internationalisierung mit dem Aufbau neuer Märkte vehement vorantrieb. Die Wilkhahn Designabteilung wurde als eigenes Tochterunternehmen unter dem Namen wiege ausgegründet, um Technologietransfer zu fördern und externen Designern den Zugang zu Wilkhahn zu erleichtern.

Das gewachsene ökologische Bewusstsein schließlich führte zum Bau der in dieser Form einzigartigen Produktionspavillons des Architekten Frei Otto und am Ende des Jahrzehnts zu einem denkwürdigen Verwaltungsratsbeschluss bei Wilkhahn: „Im Zweifelsfall hat das ökologische Anliegen Vorrang vor schnellem Gewinn." Die Ära eines umfassenden ökologisch orientierten Wandels des Unternehmens war eingeläutet.

The onset of the new decade was marked by the launch of the FS-Line that was named after the two designers Klaus Franck and Werner Sauer. Contrary to the zeitgeist, which had office chairs transformed into sheer sitting machines, the two designers focused on reduction to essentials: the seat shell is suspended between a fork and the front swivel joint like a hammock. But what appears quite obvious and ingeniously simple in terms of adjustment was the result of many years of motion studies, material tests and new manufacturing procedures. Initially ridiculed by everyone, the FS-Line has remained the office chair classic that has been sold most frequently until today, even internationally, using the design signature of "sitting without a driving licence". This success story was accompanied by a change in top management: true to his convictions, Fritz Hahne retired to the Administrative Board and left the operative business to Theodor Diener, an experienced industrial manager, who was the driving force behind internationalization by building up new markets. The Wilkhahn Design Department became an independent subsidiary company under the name of wiege in order to promote the transfer of technology and to facilitate access to Wilkhahn on the part of external designers.

The ecological awareness that had developed led to the building of the unique production pavilions by architect Frei Otto, and at the end of the decade to a memorable resolution by the Administrative Board at Wilkhahn: "In case of doubt, the ecological aspect has priority over quick profit". The era of an all-embracing, ecologically oriented change within the company had dawned.

1980_

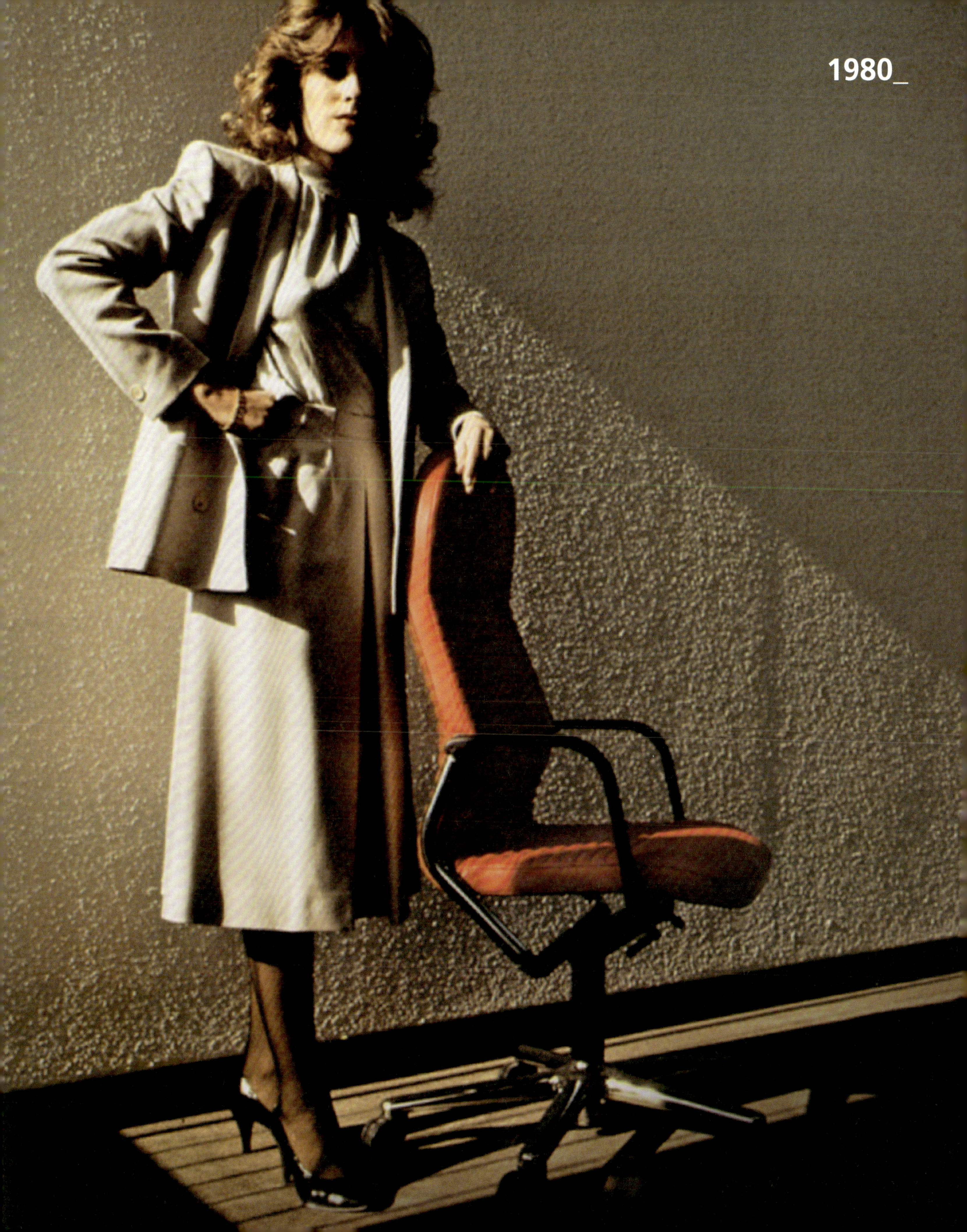

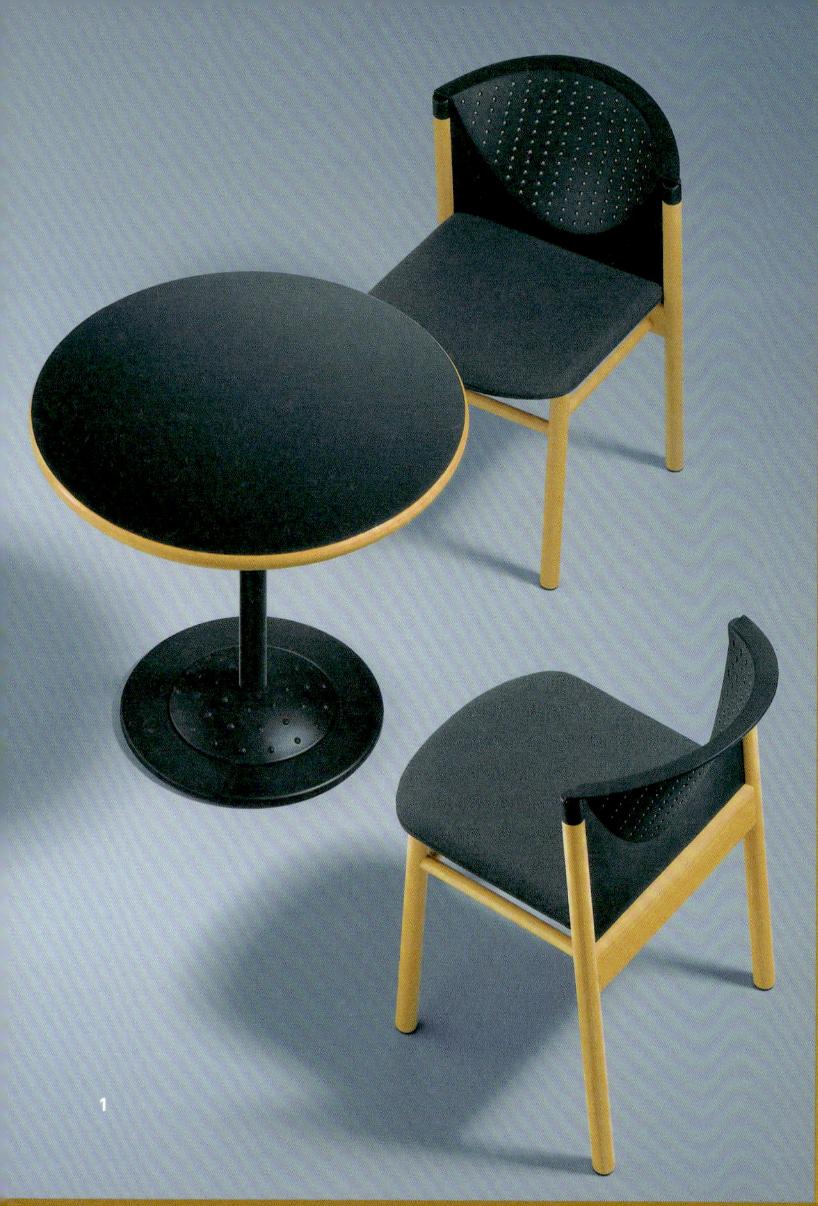

1

1_Innovative Kombination aus Holzgestell und ergonomisch ausgebildeter Rückenschale aus perforiertem Kunststoff: das Programm Kendo, Design: ProduktEntwicklung Roericht, für das Marktsegment „Pausen, Speisen, Aufenthalt", produziert von 1988–2006. 2_„Sitzen auf Netzen" – mit der O-Linie des Designers und letzten Direktors der HfG Ulm, Herbert Ohl, setzte Wilkhahn neue Akzente: Statt voluminöser Polstermöbel dominierte filigrane Leichtigkeit und Transparenz in organischen Formen, produziert von 1982–1995. 3_Das Programm Basis, Design: Klaus Franck, Werner Sauer, wiege, löste das Banksystem 1200 von Friso Kramer für Transitbereiche ab. Produziert von 1984–1998. 4_Ergebnis langjähriger Forschungs- und Entwicklungsarbeit: Bewegungsstudien und zahllose Modelle führten schließlich zum einzigartigen Konzept der FS-Linie. 5_Ökonomisch sinnvoll, ästhetisch langlebig, aus Holz als nachwachsendem Rohstoff gefertigt und als Beitrag zur Humanisierung der Arbeitswelt gestaltet: Die Produktionspavillons von Frei Otto stehen beispielhaft für Innovation, Wirtschaftlichkeit und Ästhetik im Einklang mit Mensch und Natur. 6_Bis heute in Sachen Selbstanpassung, Robustheit und formaler Reduktion unerreicht: Die FS-Linie gehört seit 1980 zu den Bestsellern.
Und: Durch einfaches Austauschen der Schale kann jeder Stuhl auch nach Jahrzehnten wieder auf den neuesten Stand gebracht werden – ein Musterbeispiel für nachhaltiges Design.

1_Innovative combination of wooden frame and ergonomically shaped back shell in perforated plastic: the Kendo range, designed by ProduktEntwicklung Roericht, for the market segment of "Relaxing, Dining, Waiting", produced from 1988–2006. 2_"Sitting on netting" – with the O-Line by Herbert Ohl, designer and last director of the HfG in Ulm, Wilkhahn set new trends: instead of voluminous chairs with soft upholstery, graceful lightness and transparency with organic forms were the dominating features. It was produced from 1982–1995. 3_The Basis range, designed by Klaus Franck, Werner Sauer and wiege, succeeded Friso Kramer's 1200 bench system for transit areas. Produced from 1984–1998. 4_The result of many years of research and development: motion studies and numerous models finally led to the unique concept of the FS-Line. 5_Efficient economically, enduring in terms of aesthetics, made from the renewable raw material of wood and designed as a contribution to the humanization of the working world: the production pavilions by Frei Otto are models of excellence for innovation, economy and aesthetics in harmony with people and nature. 6_Still unrivalled in terms of self-adjustment, sturdiness and reduction to design essentials: the FS-Line has been one of the bestsellers since 1980. What is more: by simply exchanging the shell, any chair can be updated even after decades – a model of excellence for sustainable design.

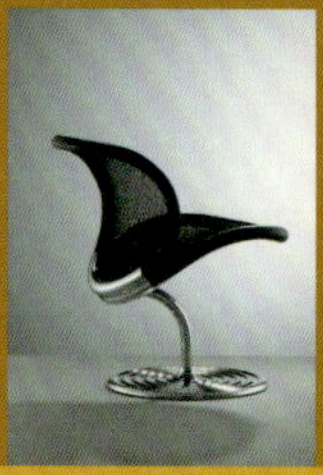

2

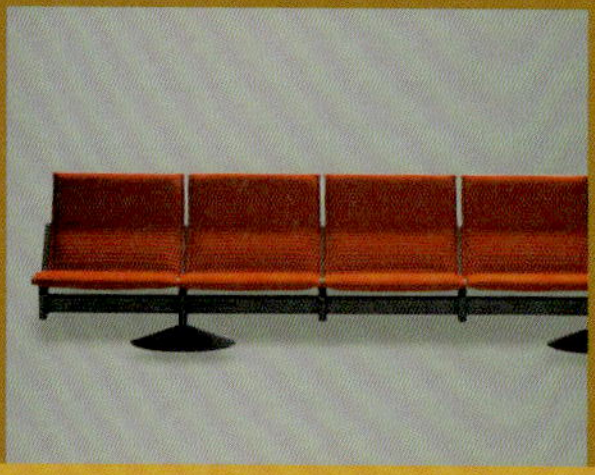

3

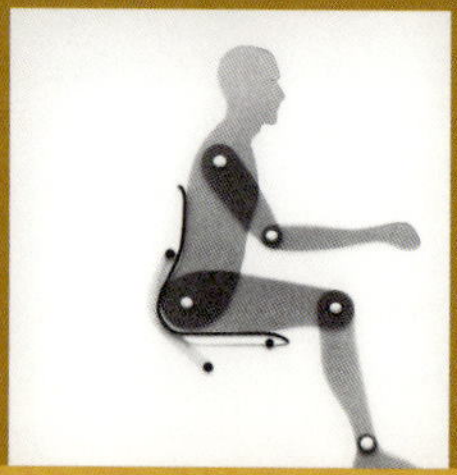

4

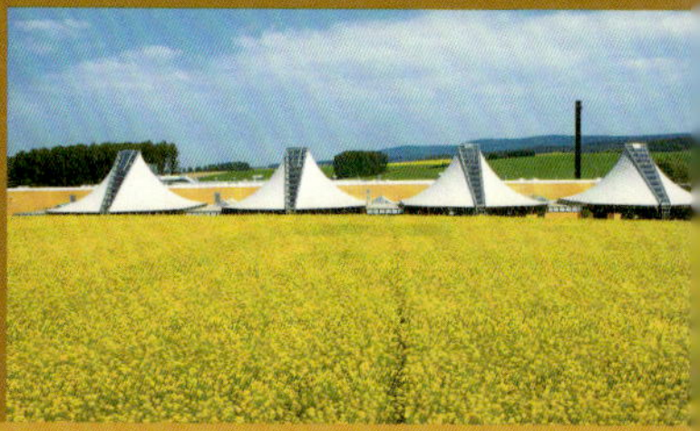

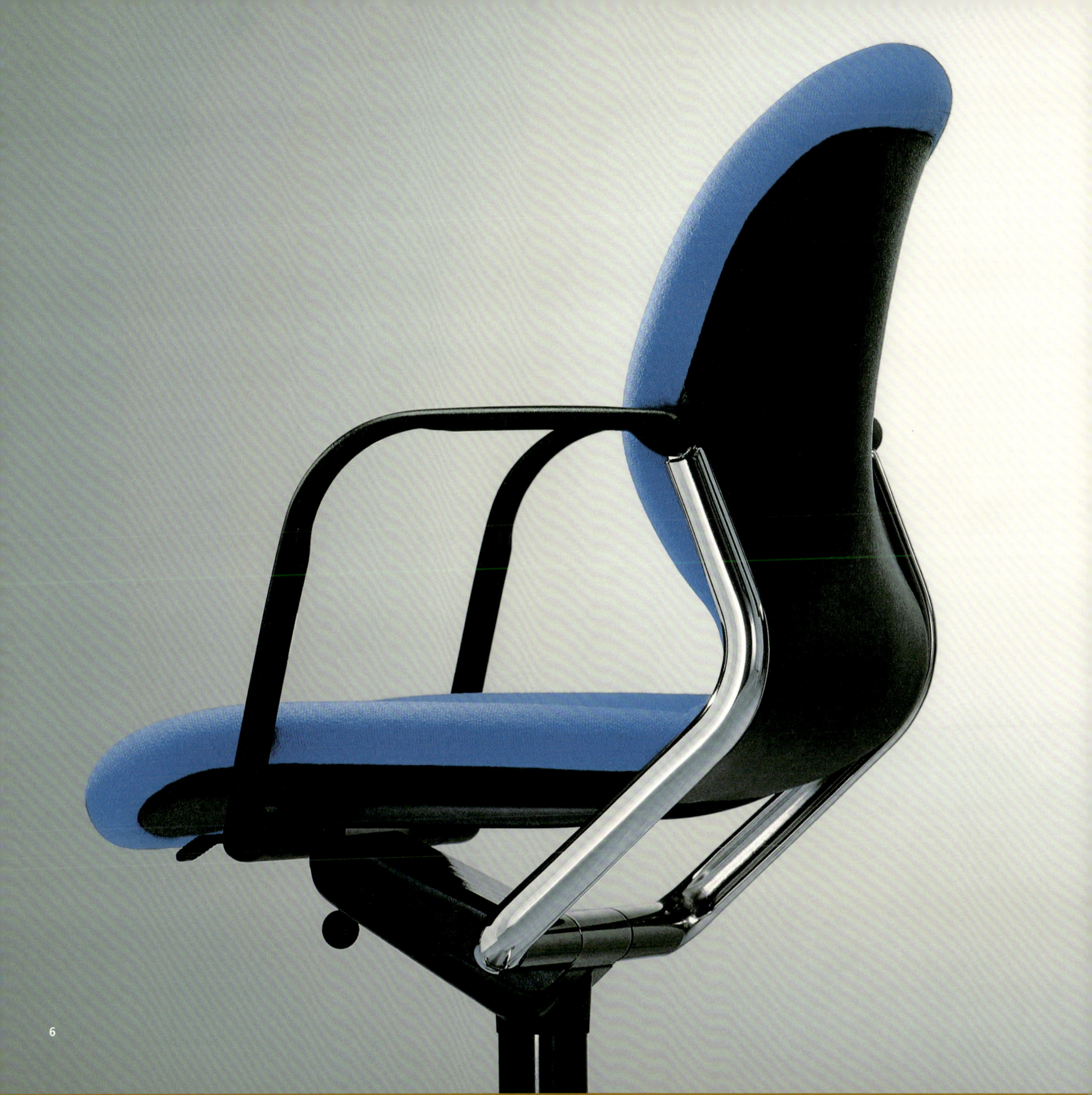

6

5

Unter der Überschrift „Wilkhahn Grün“ umfasste der ökologische Wandel nahezu alle Geschäftsbereiche. Was in den Designgrundlagen der Moderne implizit enthalten war, wurde jetzt systematisch analysiert und bewertet. Wilkhahn entwickelte als erstes Unternehmen der Branche die Grundlagen für eine Ökobilanzierung und ein umweltorientiertes Designkonzept, das den gesamten Produktlebenszyklus beschreibt. Neue Technologien, wie die Lackierung mit lösemittelarmen Wasserlacken, Energie- und Wassersparmaßnahmen, ein Mehrwegverpackungssystem sowie ein neues Abfallkonzept ergänzten die Ausrichtung auf Seiten der Produktionsprozesse. Das Bürostuhlprogramm Picto wurde zur Ikone eines modernen Öko-Designs. Zeitgleich machten die neuen Fertigungshallen, entworfen von Thomas Herzog, die Ausrichtung auch in der Architektur begreifbar. Der Auszeichnung mit dem Deutschen Marketingpreis 1992 folgte der Deutsche Umweltpreis1996, die höchstdotierte Umweltauszeichnung in Europa.

Unter der Leitung des damaligen Betriebsratsvorsitzenden Horst Knigge wurde eine moderne, beteiligungsorientierte Arbeitsorganisation entwickelt. Die Stehhilfe Stitz zeigte neue Wege für gesunde Haltungswechsel am Arbeitsplatz und das filigrane Programm Modus wurde mit seiner bespannten Rückenlehne zum Vorbild einer völlig neuartigen Bürostuhlgeneration. Schließlich wurden mit Confair, dem ersten dynamischen Konferenzprogramm weltweit, die Weichen für den Weg zum führenden Konferenzeinrichter gestellt. Mitte der 90er Jahre übernahm mit Manfred Schmitz ein ausgewiesener Topmanager und Branchenprofi den Vorsitz der Geschäftsführung, um den Generationswechsel vorzubereiten: Die Auslandsaktivitäten wurden professionalisiert, interne Prozesse gestrafft und der Schwerpunkt auf das Premiumsegment der Objekteinrichtung gelegt, um sich im zunehmenden globalen Wettbewerb eindeutig als Spitzenunternehmen zu positionieren.

Entitled "Wilkhahn Green", the ecological change process embraced virtually all business areas. What had been implied in the design guidelines of the Modern Age, was now analyzed and evaluated systematically. Wilkhahn was the first company in the industry to develop the principles of an eco-balance and an environmentally-oriented design concept that covers the entire product life cycle. New technologies, such as lacquering with low-solvent water-based lacquer, measures for saving energy and water, a system of multi-use packaging, as well as a new waste management concept supported the repositioning process on the production process side.
The Picto office chair range became the icon of modern eco-design. At the same time, the new production halls, designed by Thomas Herzog, made this new positioning visible in architecture too. The award of the German Marketing Prize 1992 was followed by the German Ecology Prize 1996, the highest award for environmental achievements in Europe.

A modern work organization, based on the principle of participation, was developed under the leadership of Horst Knigge, chairman of the Works Council at the time. The Stitz leaning aid marked a turning point for healthy changes in posture at the workplace, and the graceful Modus range with a back rest in highly resilient material became a model of excellence for a completely new kind of office chair generation. Last but not least, Confair, the first range for dynamic conferences worldwide, prepared the way for Wilkhahn's becoming one of the leading providers of conference furniture. Manfred Schmitz, a top manager with a good track record and a professional in his field, became Managing Director in mid-1990 with a view to preparing the change in the generation of leadership: activities abroad became more professional, internal processes were more tightly organized and the focus was placed on the premium segment of contract furniture in order to position oneself clearly as a top company in the wake of increasingly global competition.

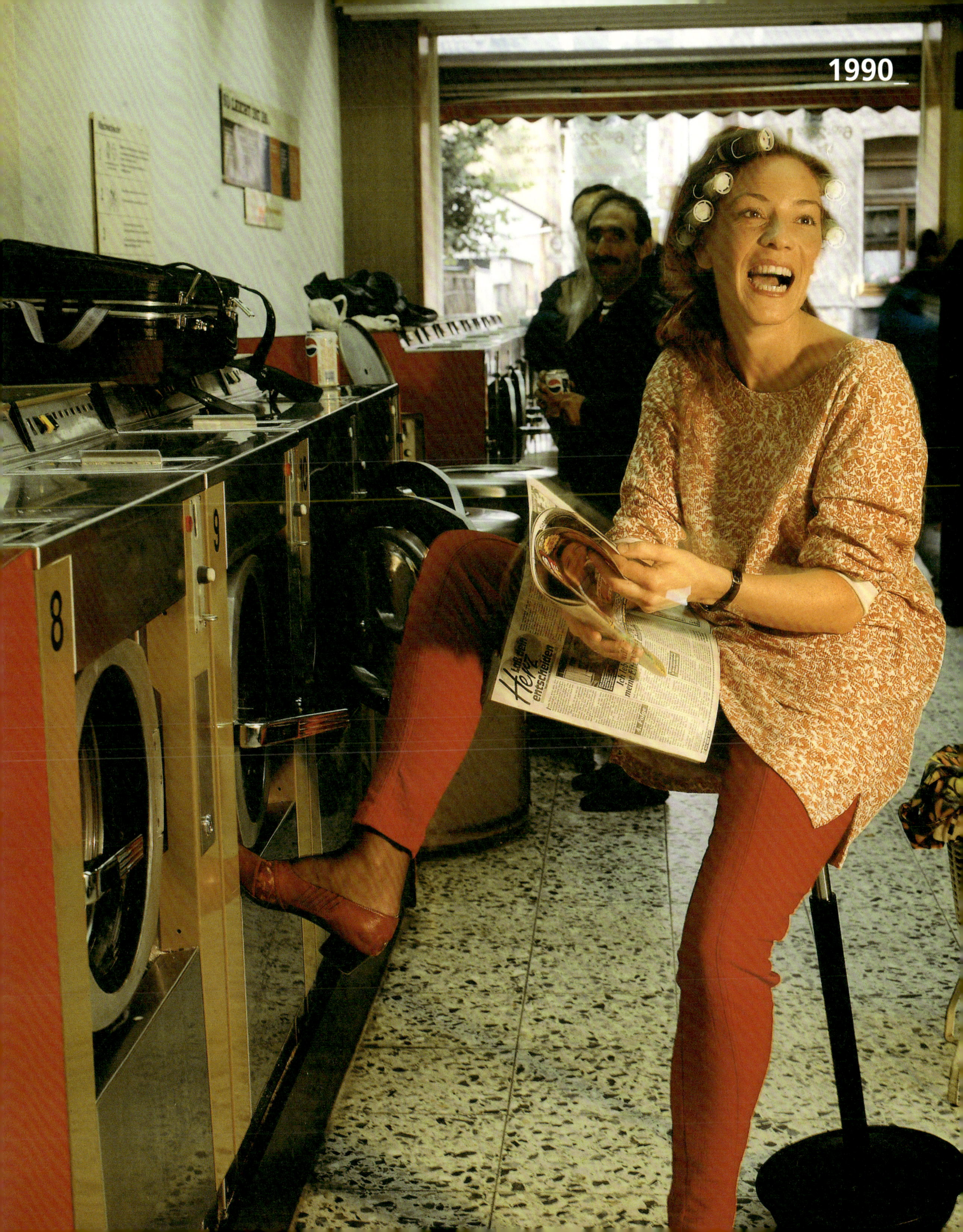
1990_

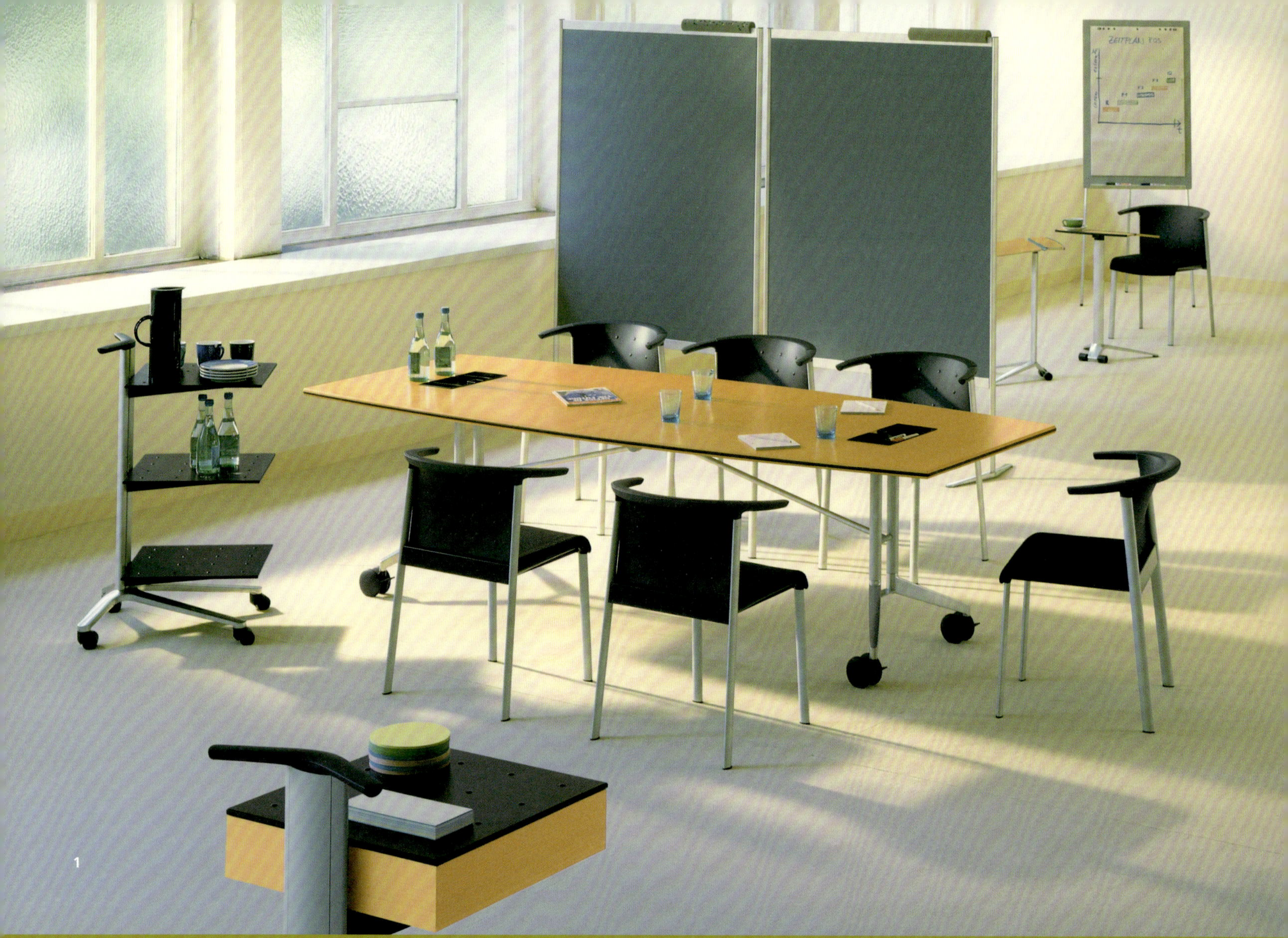

1

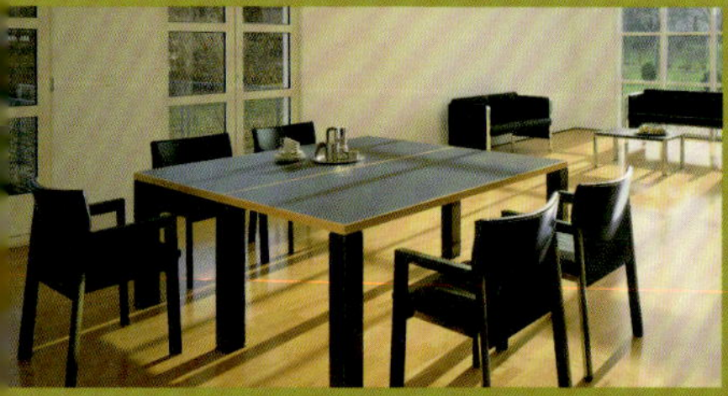

2

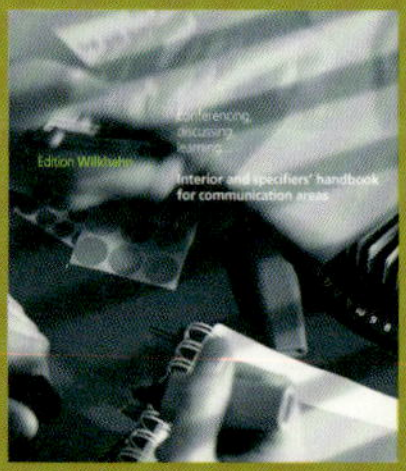

3

1_Weltweit ein neuer Maßstab für Konferenzräume: Programm Confair, Design: Fritz Frenkler, Justus Kolberg und wiege, und Confair-Falttisch, Design: Andreas Störiko, 1994. **2**_Wie für die Ewigkeit gemacht und dennoch variabel: Tischprogramm Palette, Design: Karlheinz Rubner, 1992. **3**_Einrichtungshandbuch für Kommunikationsräume, Guido Englich, Burkhard Remmers, 1997. **4**_Ikone für Öko-Design: Bürostuhlprogramm Picto, Design: ProduktEntwicklung Roericht, Burkhard Schmitz, Franz Biggel, 1992. **5**_Die neuen, ökologisch ausgelegten Fabrikhallen, Architekt: Thomas Herzog, 1992. **6**_Vorbild für das 21. Jahrhundert: Programm Modus, Design: Klaus Franck, Werner Sauer und wiege, 1994.

1_A new international standard for conference interiors: the Confair range, designed by Fritz Frenkler, Justus Kolberg and wiege, and the Confair folding table, designed by Andreas Störiko, 1994. 2_Enduring design, yet versatile: the Palette range, designed by Karlheinz Rubner, 1992. 3_Interior and specifiers' handbook for communication areas, Guido Englich, Burkhard Remmers, 1997. 4_Icon of eco-design: the Picto office chair, designed by ProduktEntwicklung Roericht, Burkhard Schmitz and Franz Biggel, 1992. 5_The new ecologically designed factory buildings by architect Thomas Herzog, 1992. 6_Model of excellence for the 21st century: the Modus range, designed by Klaus Franck, Werner Sauer and wiege, 1994.

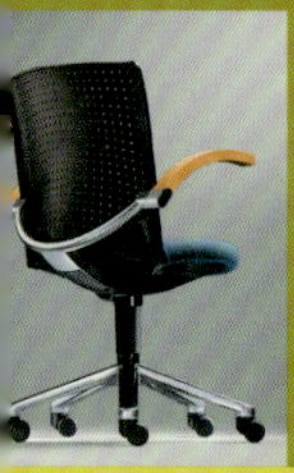

4

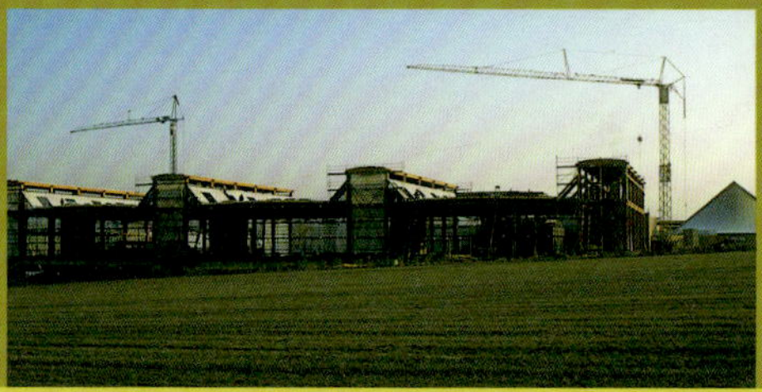

5

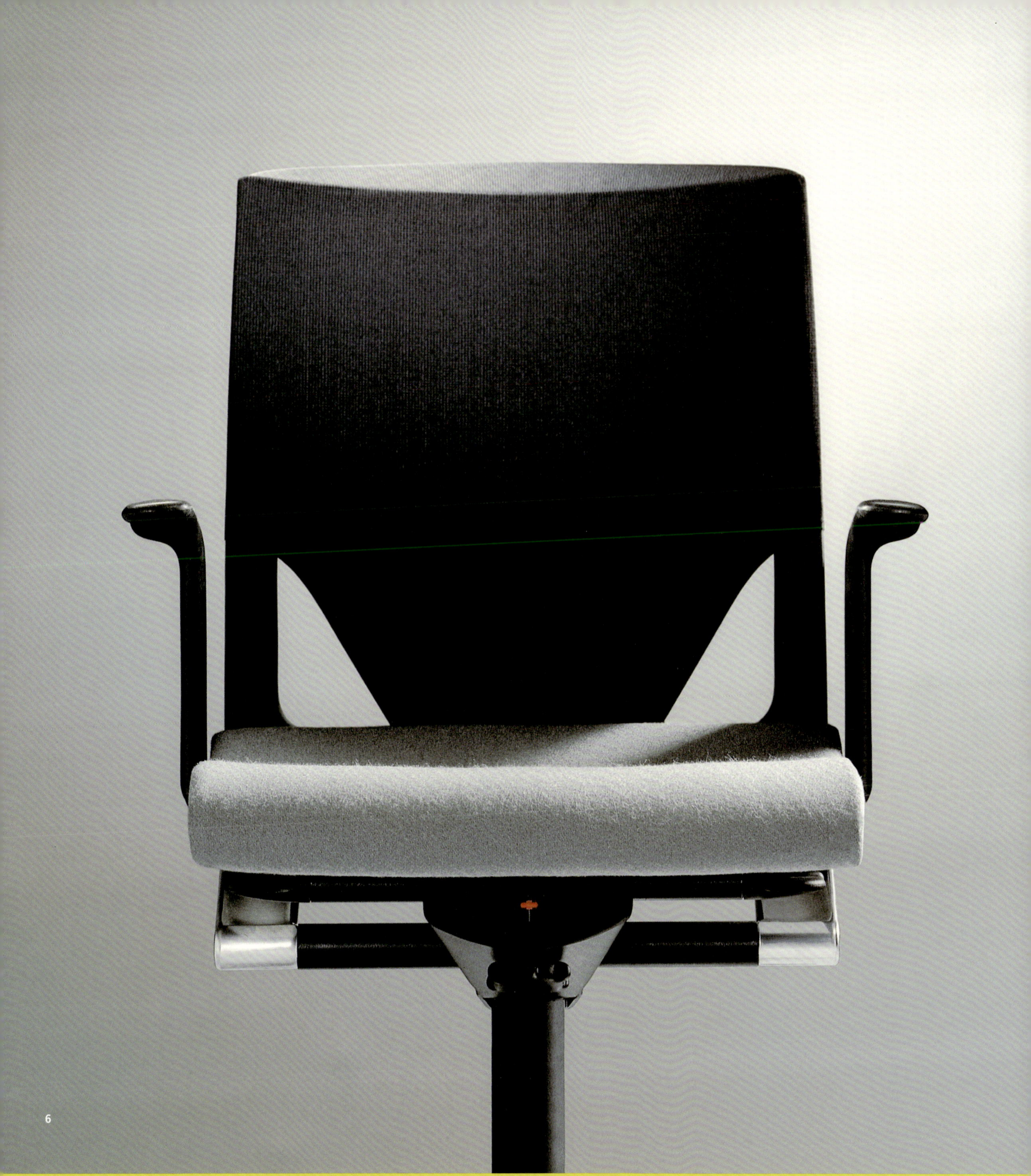

6

Das neue Jahrtausend wurde bei Wilkhahn mit dem Generationswechsel und einem Großereignis eingeläutet: Dr. Jochen Hahne, der sich seit 1992 seine Meriten erst im Export, dann als kaufmännischer Leiter und schließlich als kaufmännischer Geschäftsführer verdient hatte, wurde zum Vorsitzenden der Geschäftsführung. Und mit dem weltweiten EXPO-2000-Projekt „Zukunft der Arbeit" wurde der Wilkhahn-Beitrag zur Weltausstellung eröffnet. Die Aspekte Mensch, Natur, Technik und Markt mit ihren Wechselwirkungen und Abhängigkeiten waren für die rund 7.500 internationalen Besucher bei Wilkhahn konkret begreifbar. Höhepunkt der Ausstellung war die „Roomware®-Umgebung", ein Konferenzraum, bei dem großformatige, interaktive Displays in Wände und Tische integriert waren, um den realen Kommunikationsraum mit der digitalen Welt zu verschmelzen. Lange bevor Software-Riesen wie Microsoft interaktive Tische entdeckten, entstanden bei Wilkhahn Einrichtungselemente wie der InteracTable®, die InterWall® und das CommBoard®. Mit dem Tochterunternehmen foresee™ gründete Wilkhahn 2005 ein Beratungs-, Entwicklungs- und Vertriebsunternehmen, das auf die Schnittstelle zwischen Mensch, Prozess, Raum und innovativer Informationstechnologie spezialisiert ist und weltweit Projekte mit interaktiven Einrichtungen realisiert.

Der Designer Andreas Störiko steuerte für das Kerngeschäftsfeld Konferenz zwei sehr erfolgreiche Programme bei: den Timetable (2000), einen mobilen Tisch mit schwenkbarer Tischplatte, und das Programm Aline mit einem luftig leichten Mehrzweckstuhl (2004) und Tischen, Stehhockern, Dreh- und Besucherstühlen (2006) für informelle Kommunikationsbereiche. Vom Start weg mit höchsten internationalen Innovations- und Designpreisen ausgezeichnet, entwickelte sich das Programm innerhalb kurzer Zeit zu einem der Wilkhahn-Bestseller. Die wiege unter der Leitung von Michael Englisch schließlich schuf mit dem Freischwingerprogramm Sito (2000), den Bürostuhlfamilien Solis und Solis F (2002) und Neos (2004) ebenso markante wie Wilkhahn-typische Ergänzungen zum Dauerbrenner Modus und zum Evergreen FS-Linie. Seit 2002 verstärkt Geschäftsführer Holger Jahnke, inzwischen auch Wilkhahn-Gesellschafter, die Leitungsebene. Er verantwortet weltweit die Entwicklung der Märkte und die internationale Markenbildung. Und im Jubiläumsjahr 100 + sieht es ganz danach aus, dass es das bislang erfolgreichste Jahr der Unternehmensgeschichte werden wird …

At Wilkhahn, the new millennium was marked by a generation change and an important event: since 1992, Dr. Jochen Hahne, who had achieved a high standing, first in Export, then as Commercial Manager and later as Commercial Director, became Managing Director of Wilkhahn. And with the EXPO 2000 project, the "future of work", Wilkhahn's contribution to the World Exposition was opened. The aspects of humankind, nature, technology and the market with all the interconnections and interdependencies involved were presented as a hands-on experience to some 7,500 international visitors at Wilkhahn. The highlight of the exhibition comprised the Roomware® environment, a conference interior with the integration of large-format, interactive displays into walls and tables for the purpose of merging a real communication space with the digital world. Long before software giants such as Microsoft discovered interactive tables, Wilkhahn had designed furnishing elements such as the InteracTable®, the InterWall® and the CommBoard®. With its subsidiary company, foresee™, Wilkhahn founded a company for consultation, development and sales that is specialized in the interface between people, processes, space and innovative information technology, and carries out projects using interactive furniture throughout the world.

Designer Andreas Störiko contributed two highly successful ranges to the core segment of Conferencing: Timetable (2000), a mobile table with a pivotable table top and the Aline range, a light, graceful multi-purpose chair (2004) and tables, stools, swivel chairs and visitor chairs (2006) for informal communication spaces. Winner of the highest international awards for innovation and design right from the very start, the latter range soon became one the Wilkhahn best-sellers. And last but not least, with the Sito cantilever range (2000), the Solis and Solis F office chair families (2002) and Neos (2004), the wiege design studio, under the management of Michael Englisch, created some distinctive and typically Wilkhahn additions to long-selling Modus and the evergreen FS-Line. Since 2002, the management level has been strengthened by Holger Jahnke, now also a managing partner at Wilkhahn. He is responsible for the development of markets and international brand building worldwide. And for the 100 + anniversary year it looks as if it is going to be the most successful year in the history of the company …

2000_

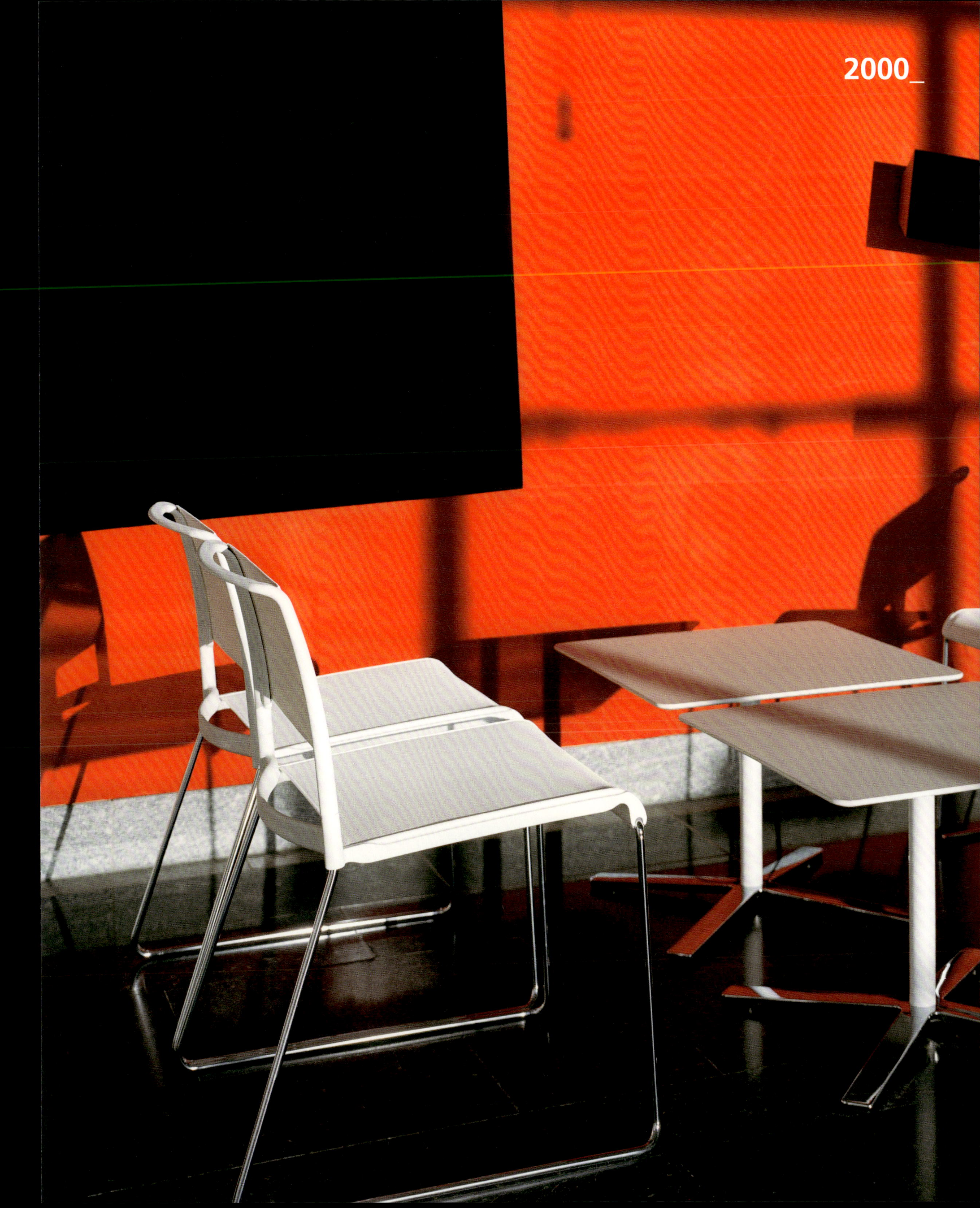

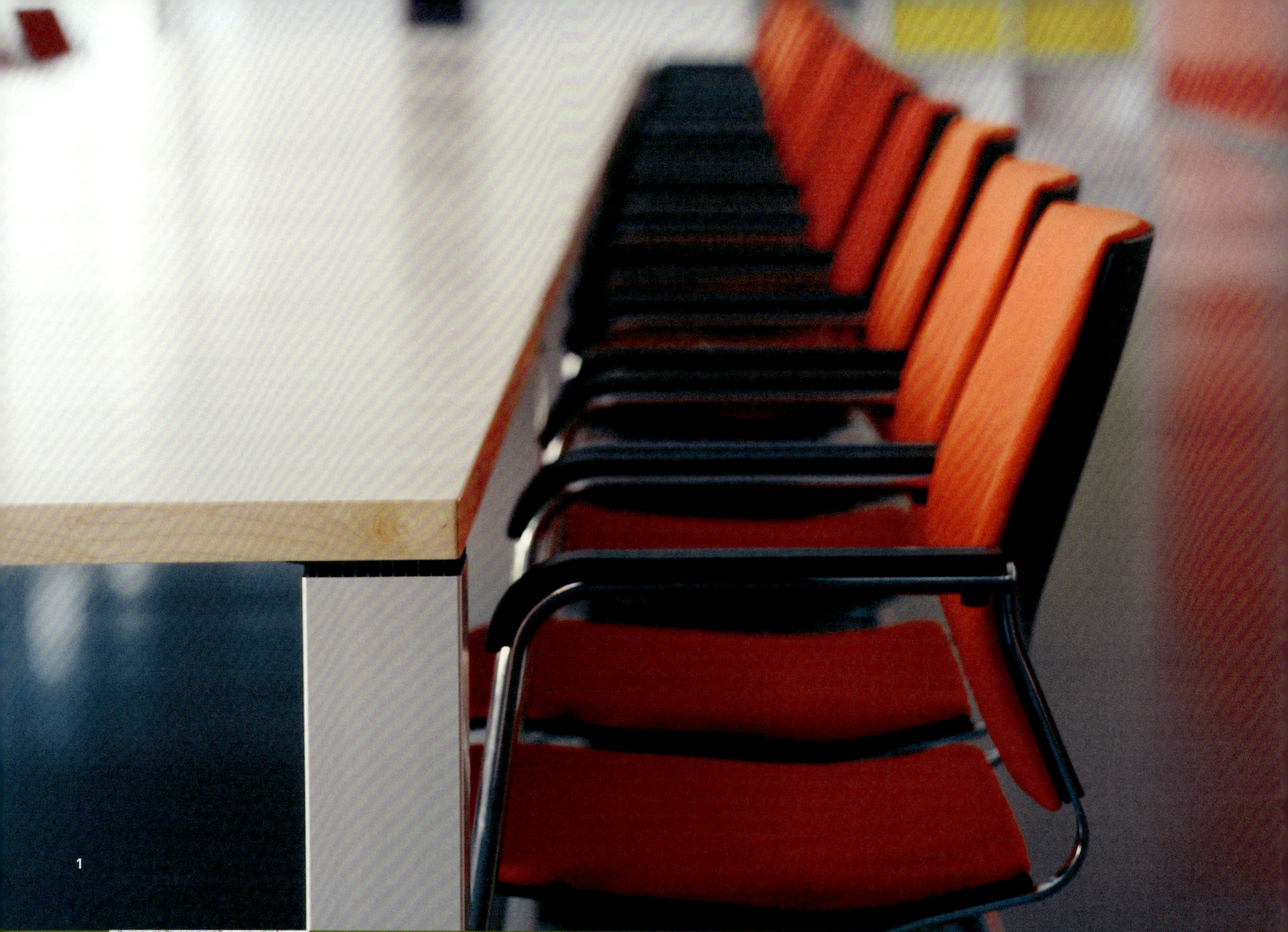

1

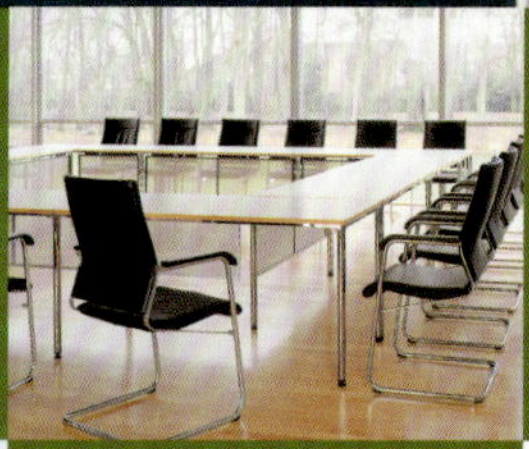

2

EXPO2000 HANNOVER

Registriertes Projekt der Weltausstellung

3

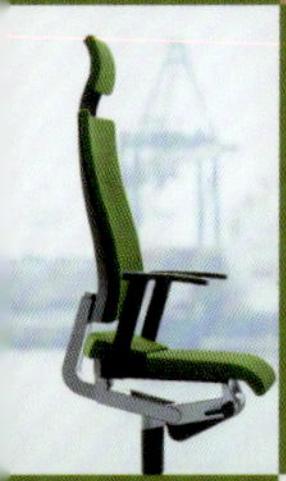

4

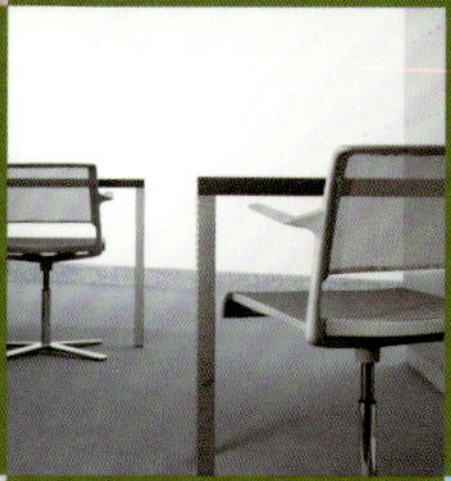

5

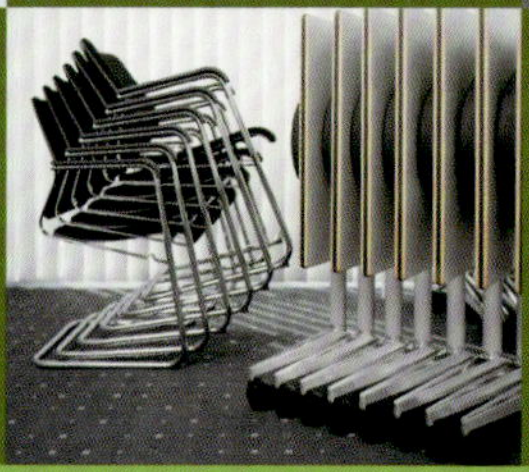

6

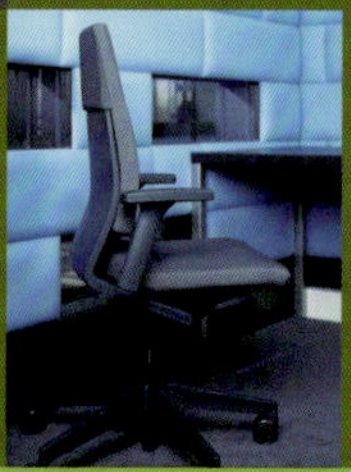

7

1_Die neue Interpretation des Themas Freischwinger: Programm Sito, Design: wiege, 2000. 2_Für variable Schulungs- und Konferenzräume: das Tischsystem Contas mit innovativer Leichtbauplatte, Design: wiege, Fritz Frenkler, Anette Ponholzer, 2000. 3_Symbol für den Beitrag zur Weltausstellung 2000: das Logo der EXPO 2000. 4_Markant und konsequent architektonisch gestaltet: Bürostuhlprogramm Solis, Design: wiege, 2002. 5_Luftig, leicht und verblüffend komfortabel: Besucherstuhl Aline, Design: Andreas Störiko, 2006, am Tischprogramm DinA, Design: Udo Schill, Timo Küchler, 2003. 6_Verblüffend funktional und hochwertig: Programm Timetable, Design: Andreas Störiko, 2000. 7_„Sitzen ohne Führerschein" – mit allen Einstelloptionen: Programm Neos, Design: wiege 2004. 8_Futuristischer Statist in der letzten Starwars-Episode: Programm Solis F, Design: wiege, 2002.

1_A new interpretation of the principle of the cantilever chair: the Sito range, designed by wiege, 2000. 2_For training and conference interiors with a variable character: the new Contas table system with an innovative table top comprising a lightweight sandwich construction, designed by wiege, Fritz Frenkler, Anette Ponholzer, 2000. 3_Symbol of Wilkhahn's contribution to the World Exposition: the EXPO 2000 logo. 4_Distinctively and consistently architecturally designed: the Solis office chair range (design: wiege, 2002). 5_Light, graceful and ingeniously comfortable: the Aline office chair range, designed by Andreas Störiko, 2006, at a table from the DinA range, designed by Udo Schill, Timo Küchler, 2003. 6_Ingeniously functional and high-quality: the Timetable range, designed by Andreas Störiko, 2000. 7_"Sitting without a driving licence" – with a host of adjustment features: the Neos range, designed by wiege 2004. 8_A futuristic "extra" in the most recent Starwars episode: the Solis F range, designed by wiege, 2002.

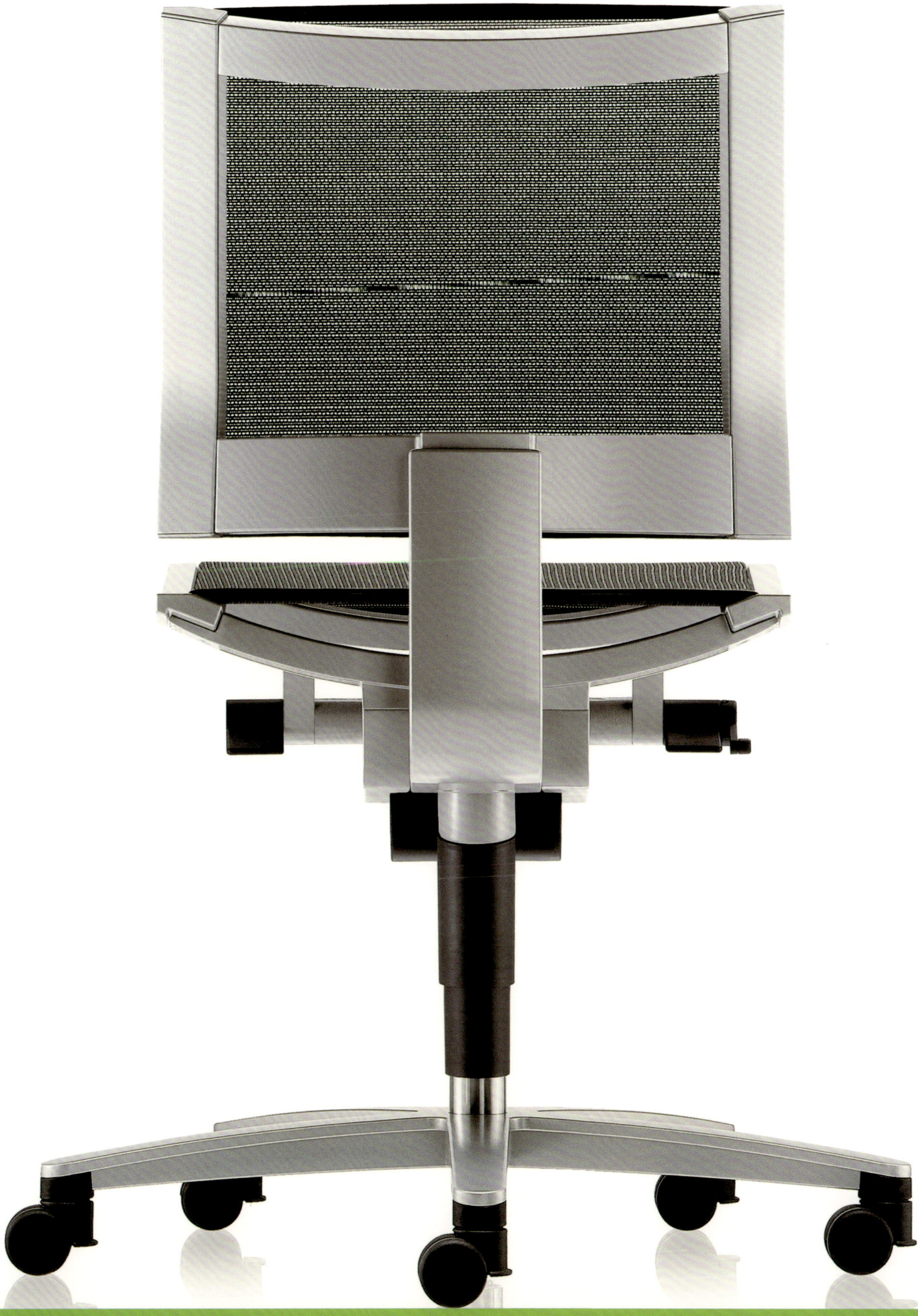

8

Die Internationalisierung von Wilkhahn reicht bis in die 50er Jahre des letzten Jahrhunderts zurück. Überzeugt von Qualität und Klasse des Wilkhahn-Angebots bereiste damals der polyglotte Exportleiter Dr. Robert Heinemann die Nachbarländer und gewann mit der Einrichtung der Rotterdamer Hafenverwaltung den ersten internationalen Großauftrag. Mit dem Aufbau von Geschäftsfreundschaften in den europäischen Nachbarländern stieg der Exportanteil bis 1975 bereits auf 26 %.

Ab Anfang der 80er Jahre wurde das Auslandsgeschäft zur erklärten Strategie: Geschäftsführer Theodor Diener und Exportleiter Karl Schneider erschlossen systematisch den Weltmarkt für Wilkhahn. War Wilkhahn in einem neuen Markt etabliert, wurde eine eigene Tochtergesellschaft oder eine Lizenzpartnerschaft gegründet, um Partner wie Kunden kompetent und ortsnah zu betreuen. Heute sind Wilkhahn-Programme in 91 Ländern der Welt verfügbar, von denen 59 direkt von Wilkhahn betreut werden. Zu den Kunden zählen leistungs-, qualitäts- und markenbewusste Unternehmen aus Industrie und Dienstleistung, aber auch öffentliche Verwaltungen und Einrichtungen, die mit Wilkhahn in der Vielfalt der Kulturen einen gemeinsamen Nenner der Gestaltungsqualität finden. Mittlerweile hat der internationale Geschäftsanteil 70 % erreicht. Gibt es einen schöneren Beweis dafür, dass die Programme ganz unterschiedliche Kultur- und Gestaltungskonzepte unterstützen, ohne sie zu dominieren?

Wilkhahn international: Partner, Kunden, Projekte
Wilkhahn international: partners, customers, projects

The internationalization of Wilkhahn goes back to the 1950s. Convinced of the quality and excellence of Wilkhahn's product portfolio, the polyglot head of export, Dr. Robert Heinemann, travelled back then to neighbouring countries and secured the first large international commission – the furnishing of the administration building at the port of Rotterdam. With the creation of business alliances in the neighbouring European countries, the export share of the business had already increased to 26 % by 1975.

As of the beginning of the 1980s, business abroad became a professed strategy: Manager, Theodor Diener and Head of Export, Karl Schneider systematically developed the world market for Wilkhahn. As soon as Wilkhahn was established in a new market, it founded its own subsidiary company or licence partnership, ensuring competent care of both partners and customers on location. Today, Wilkhahn ranges are available in 91 countries worldwide, fifty-nine of which are looked after by Wilkhahn direct. Wilkhahn's customers include performance, quality and brand-conscious companies in the industry and service sector, but also public administrations and facilities, which, together with Wilkhahn in the diversity of cultures, find a common denominator with regard to design quality. In the meantime, the share accounted for by international business has reached the 70 % mark. What better proof is there that the ranges support completely different cultural and design concepts without dominating them?

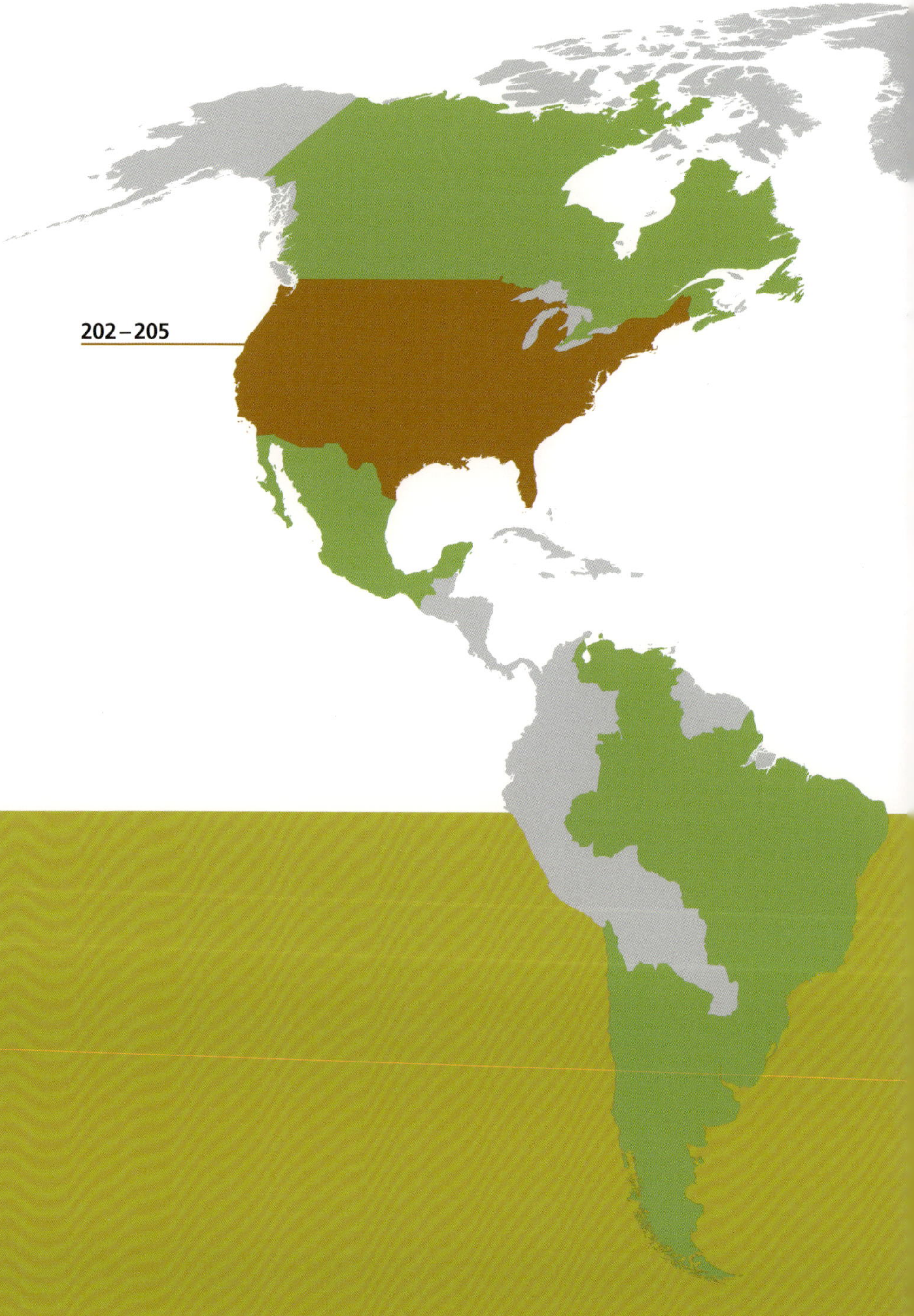
202 – 205

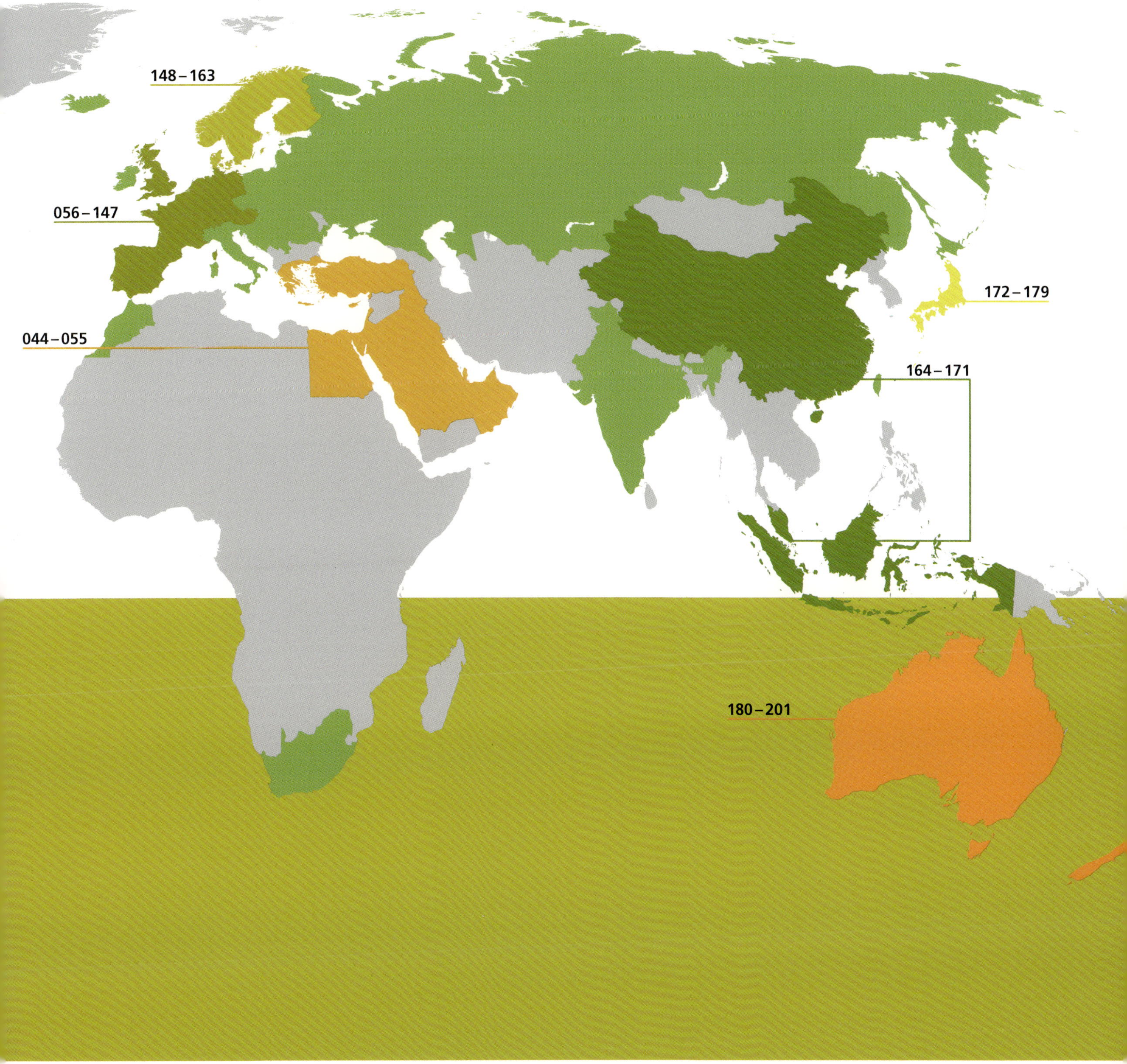
148 – 163
056 – 147
044 – 055
172 – 179
164 – 171
180 – 201

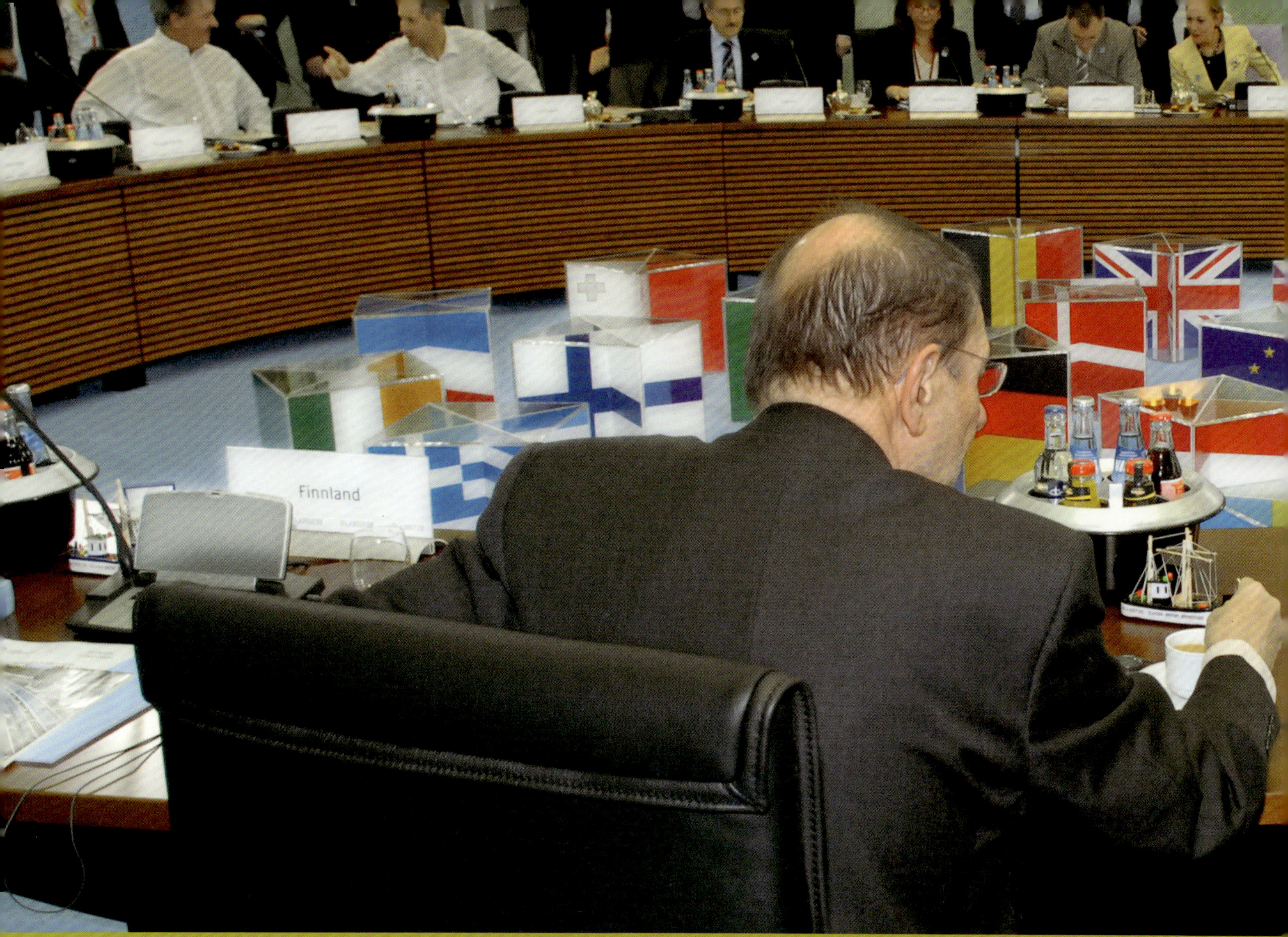
Finnland

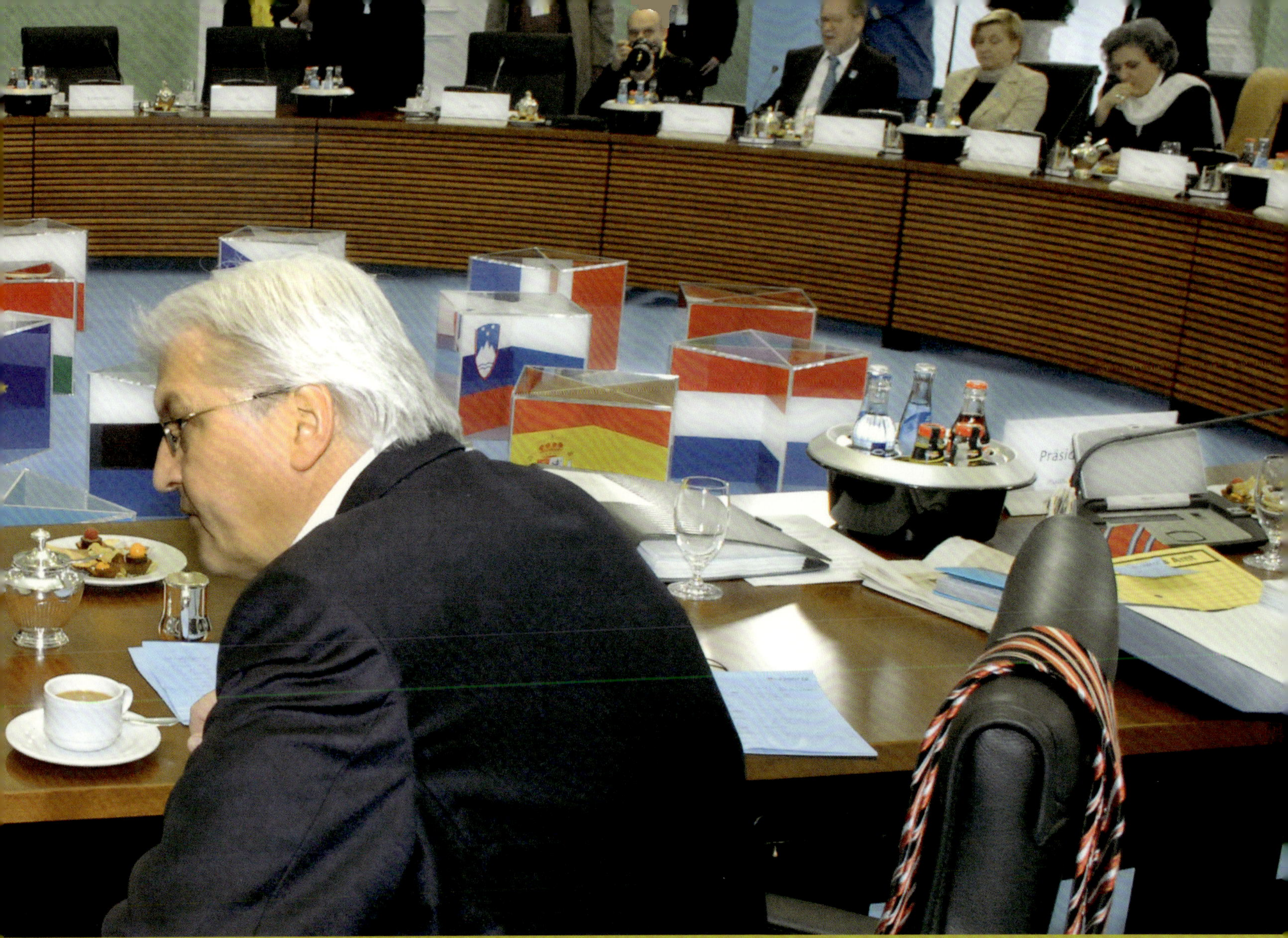

Auf internationalem Parkett: Die EU-Ratspräsidentschaft Deutschlands in 2007 tagte an einer variablen Palette-Tischanlage und auf Sesseln der Modellreihe FS 220, die dafür quer durch Deutschland tourten. Das gesamte Konferenzequipment kam innerhalb weniger Monate an 49 Tagungsorten zum Einsatz ...

In international circles: during Germany's presidency of the EU in 2007, meetings were held at a variable Palette table configuration, and chairs from the FS 220 range were part of the equipment which toured the whole of Germany. Within just a few months, the entire conference equipment was utilized at 49 conference venues.

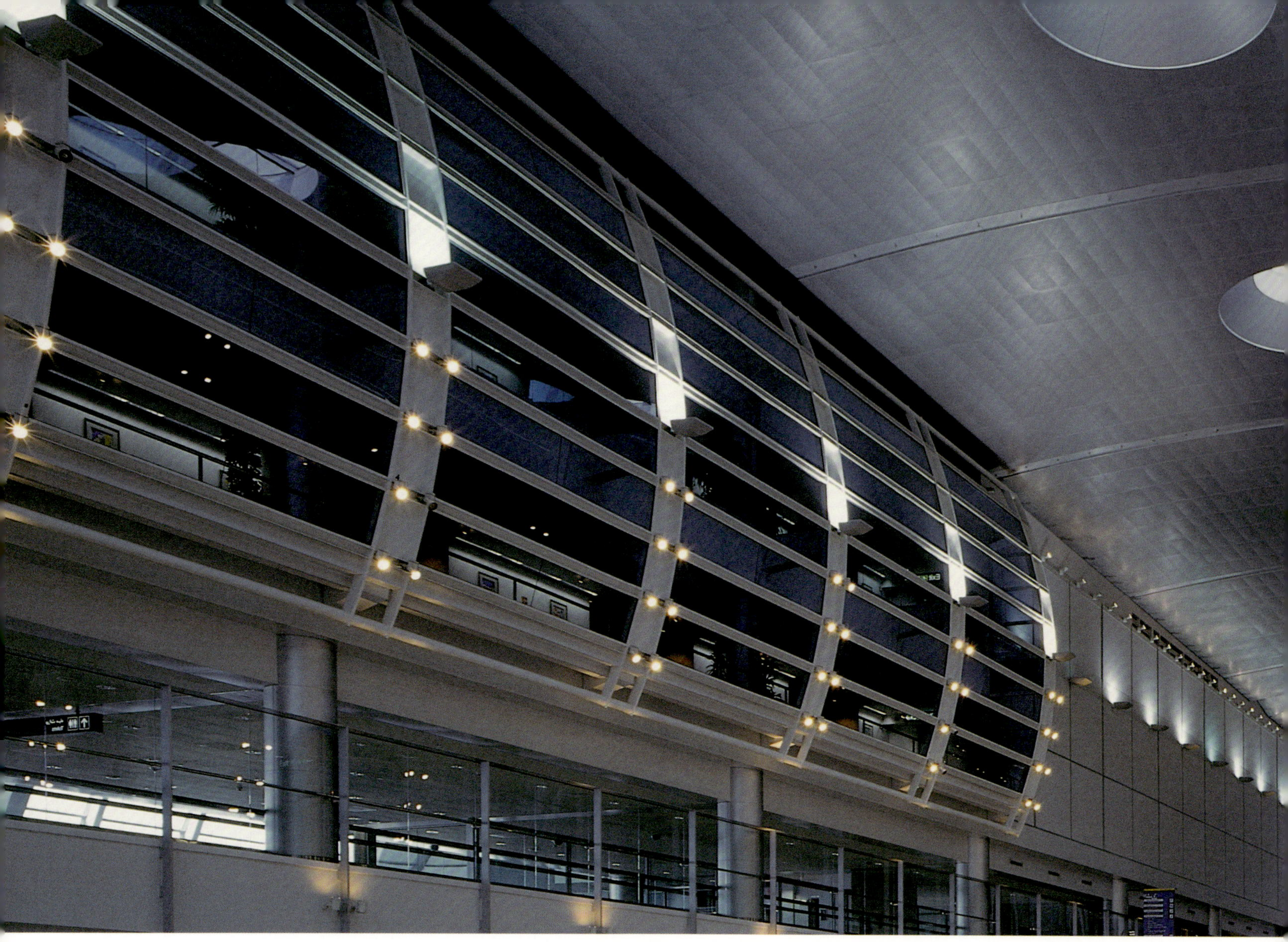

In Afrika ist Wilkhahn mit Export- und Lizenzpartnern wie Emergent Office Solutions (Pty) Ltd., Johannesburg, und Maroc Bureau, Casablanca seit vielen Jahren engagiert vertreten. Die Einrichtung der Terminals des International Airport in Dubai mit dem Banksystem Tubis und die boomende Entwicklung in den Vereinigten Arabischen Emiraten gaben den Impuls für die Gründung von Wilkhahn Middle East in Dubai, das neben der Arabischen Halbinsel auch den südlichen Mittelmeerraum bearbeitet.

Wilkhahn has been represented in Africa for many years now by such dedicated export and licence partners as Emergent Solutions (Pty) Ltd., Johannesburg, and Maroc Bureau, Casablanca. The furnishing of the terminals at Dubai International Airport with Tubis seating units, and the booming development in the United Arab Emirates gave the impetus to found Wilkhahn Middle East in Dubai, which looks after the Arabian Peninsula as well as the southern Mediterranean region.

Dubai International Airport, passenger terminals furnished with the Tubis range.

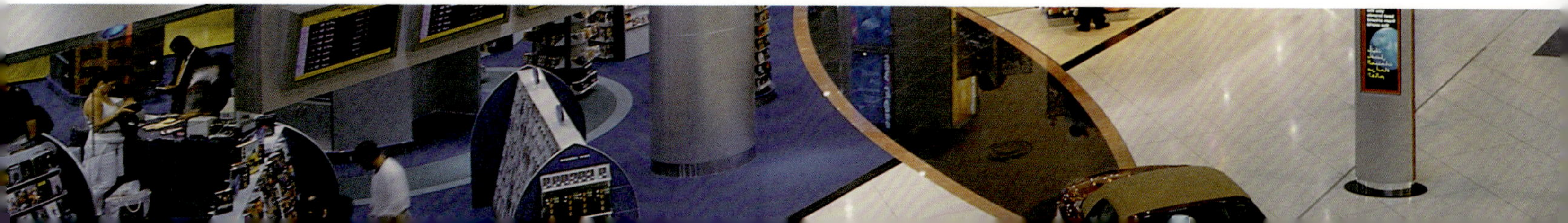

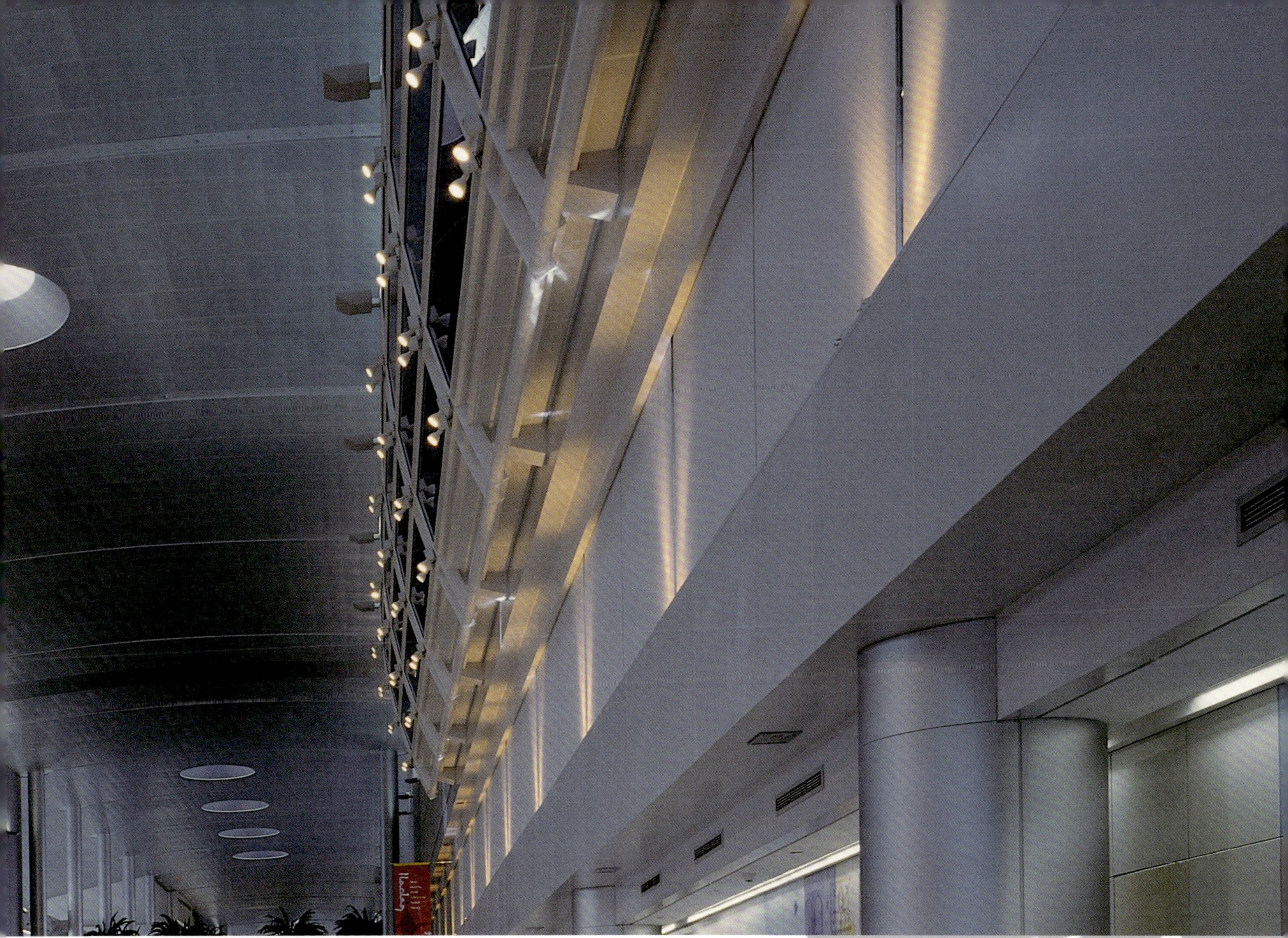

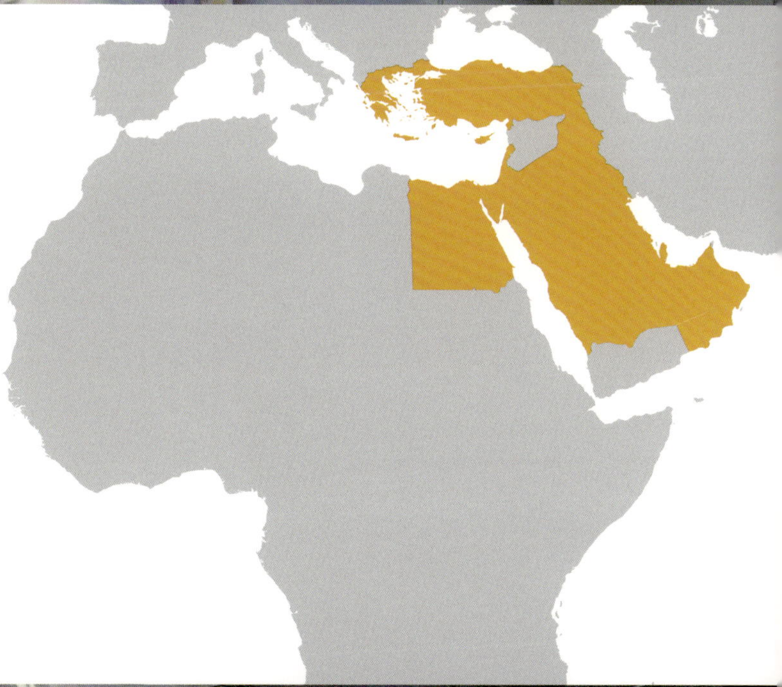

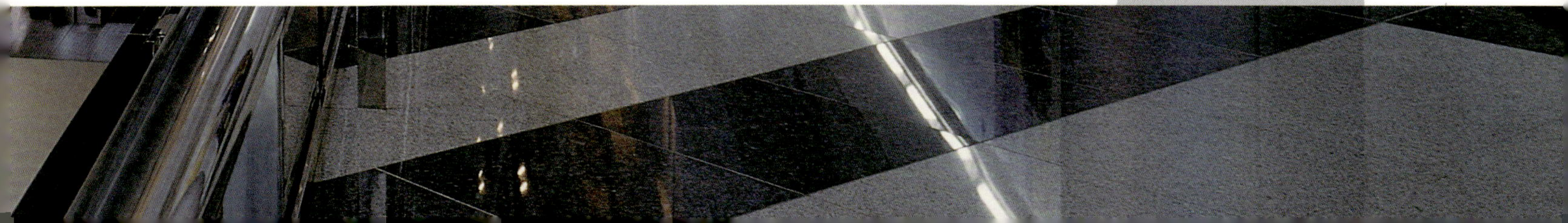

Hamdan Al Muhairbi

ويلكان أثاثات رائعة حيث التصميمات للمستقبل و الجودة تبقى للأبد. أحب هذا الكرسي لأنه صحي للظهر

"Wilkhahn furniture is fantastic because the design is so futuristic and the quality lasts forever. I love the chair because it is so healthy for the back."

Al Muhairbi, Real Estate, Dubai Lounge chair: Cubis range

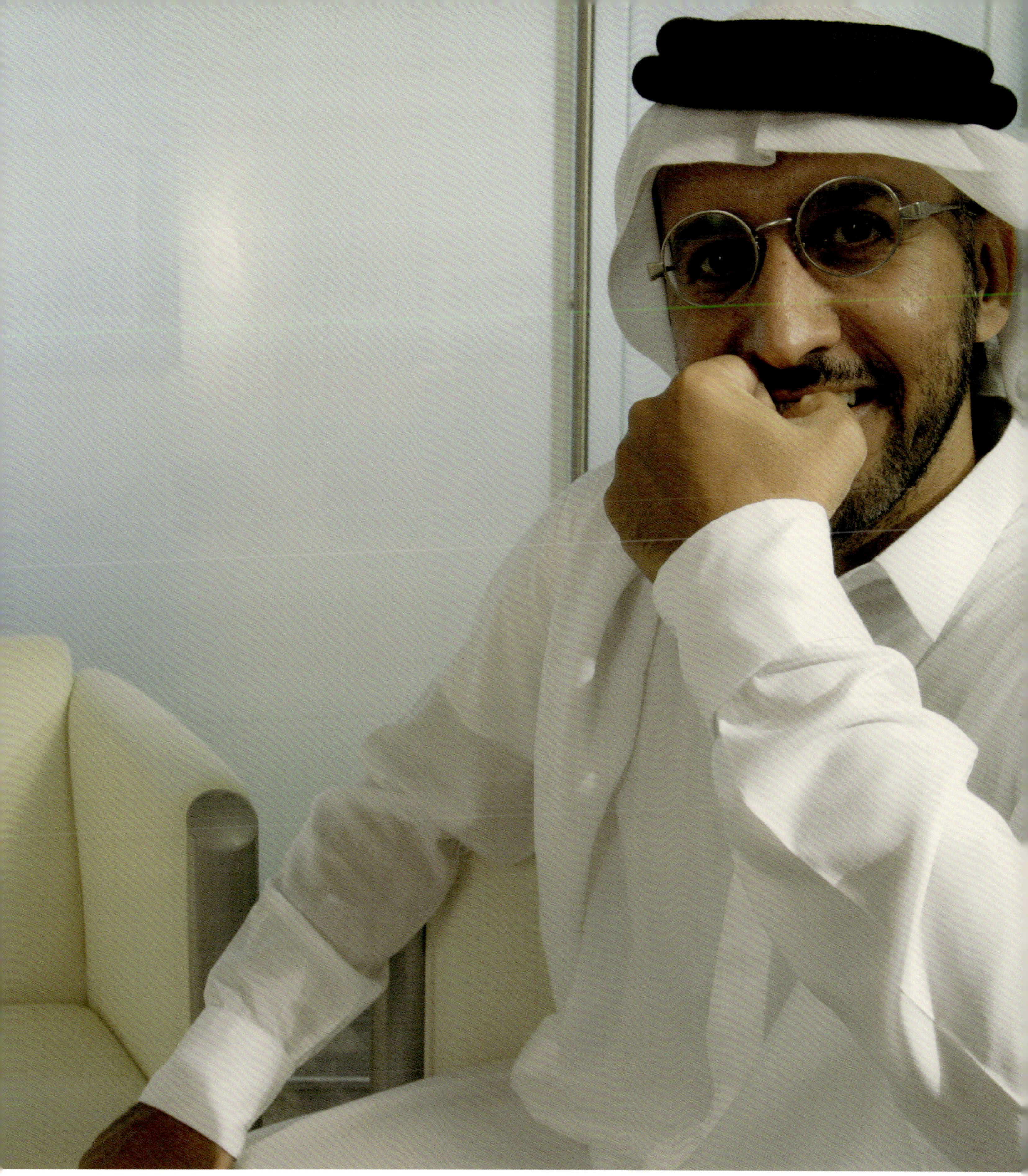

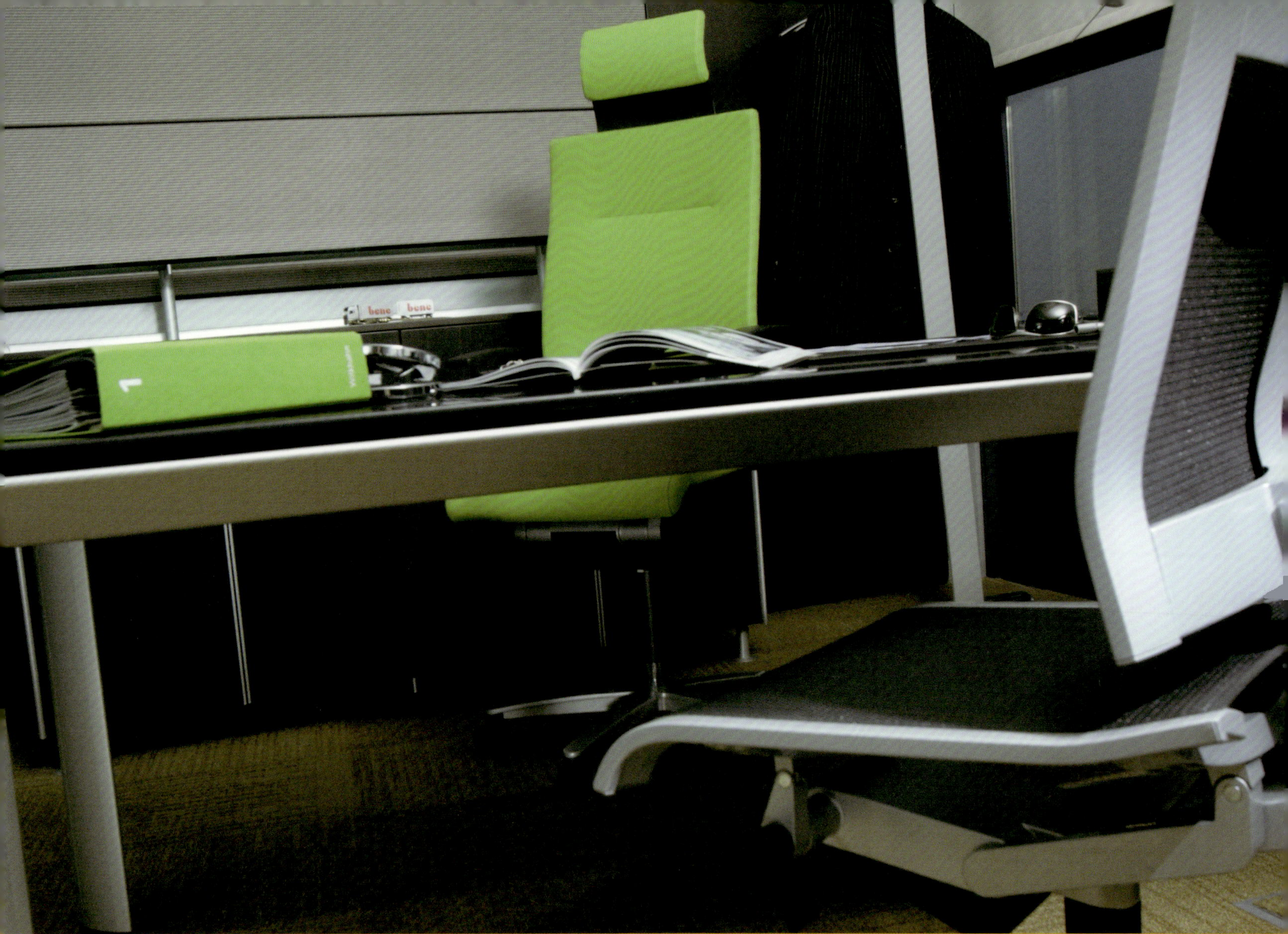

Ahmed Kandil
"Requirements in the Middle East have changed a lot in the past 15 years – towards high quality and simplicity of design. Ergonomics are in high demand as well. Like all multinational companies with regional offices in Dubai to serve clients and their needs, it was a decision to have Wilkhahn as a strategic partner since its products match ours as does the company philosophy and vision."

Bene AG Middle East, Dubai Swivel chairs: Solis and Solis F range

Swivel chair: Solis F range

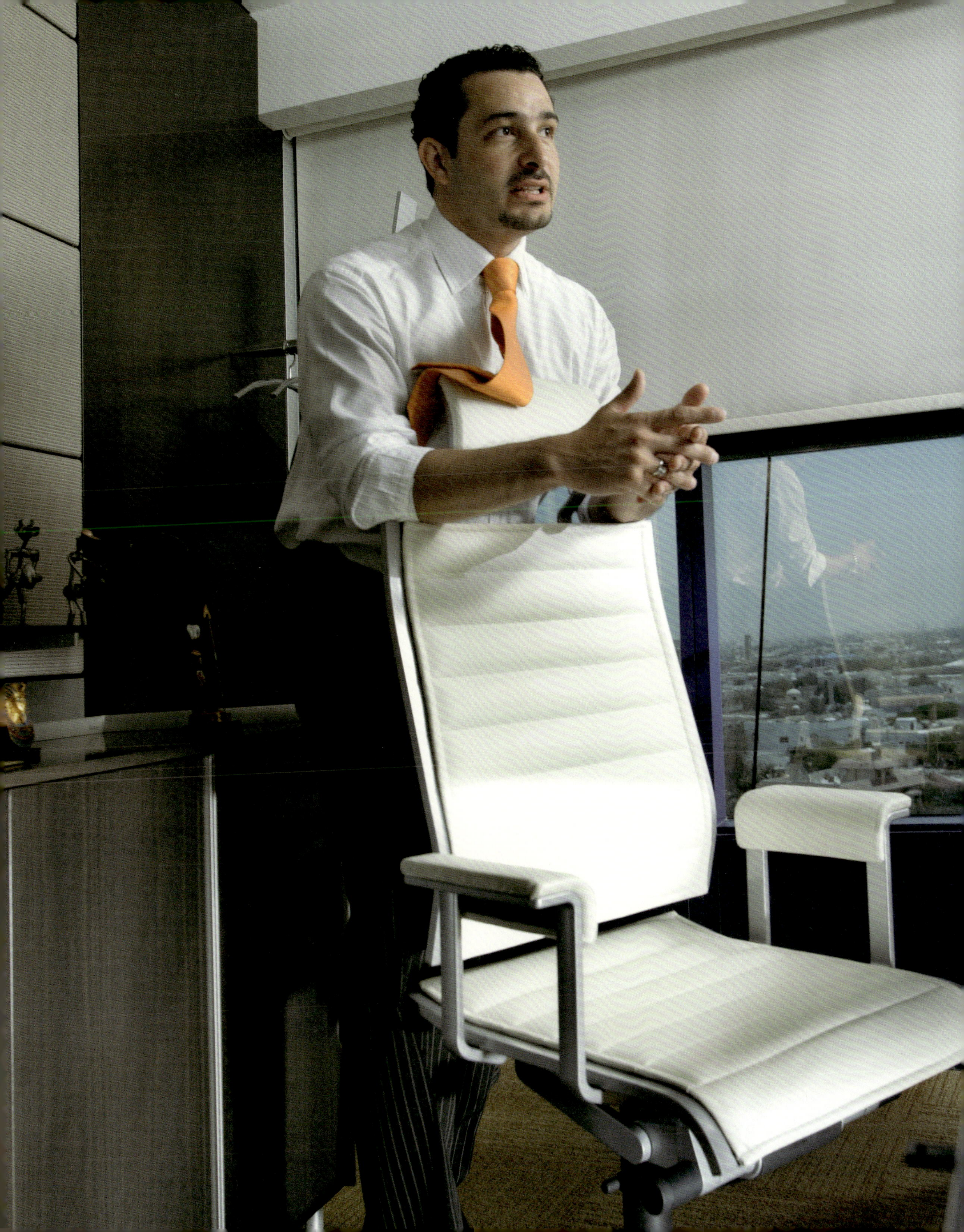

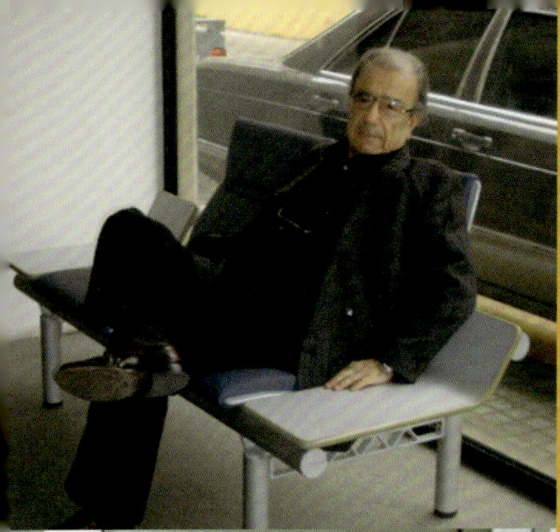

Camille EL-Khouri
"No wonder Wilkhahn is a leader in its field. Wilkhahn products are intelligently designed to respond to global demands for high quality, simplicity, versatility, functionality and ergonomics in addition to luxury. Our collaboration with Wilkhahn for almost three decades has given us an additional powerful marketing instrument and success in the region."

Interdesign, Beirut Settee: Tubis range

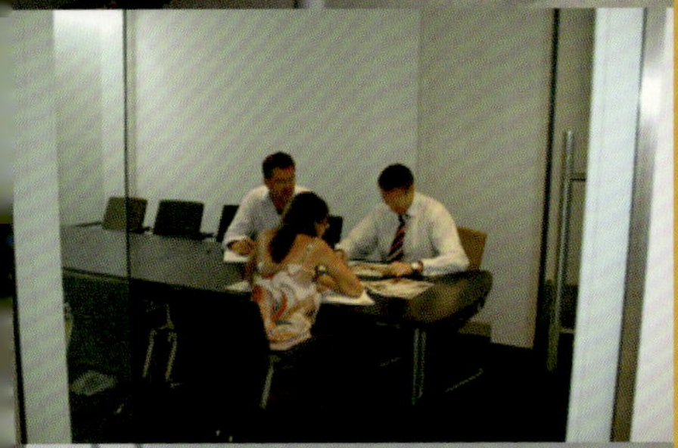

Richard Hamilton, Architect
"Ability may get you to the top, but it takes a lot to keep you there. Congratulations Wilkhahn."

Hamilton Design, Dubai

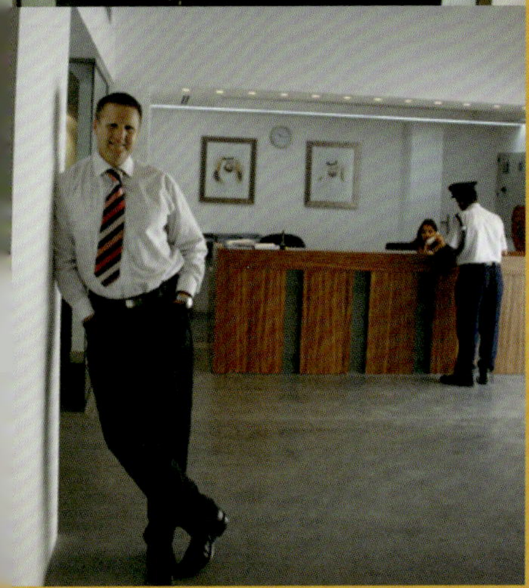

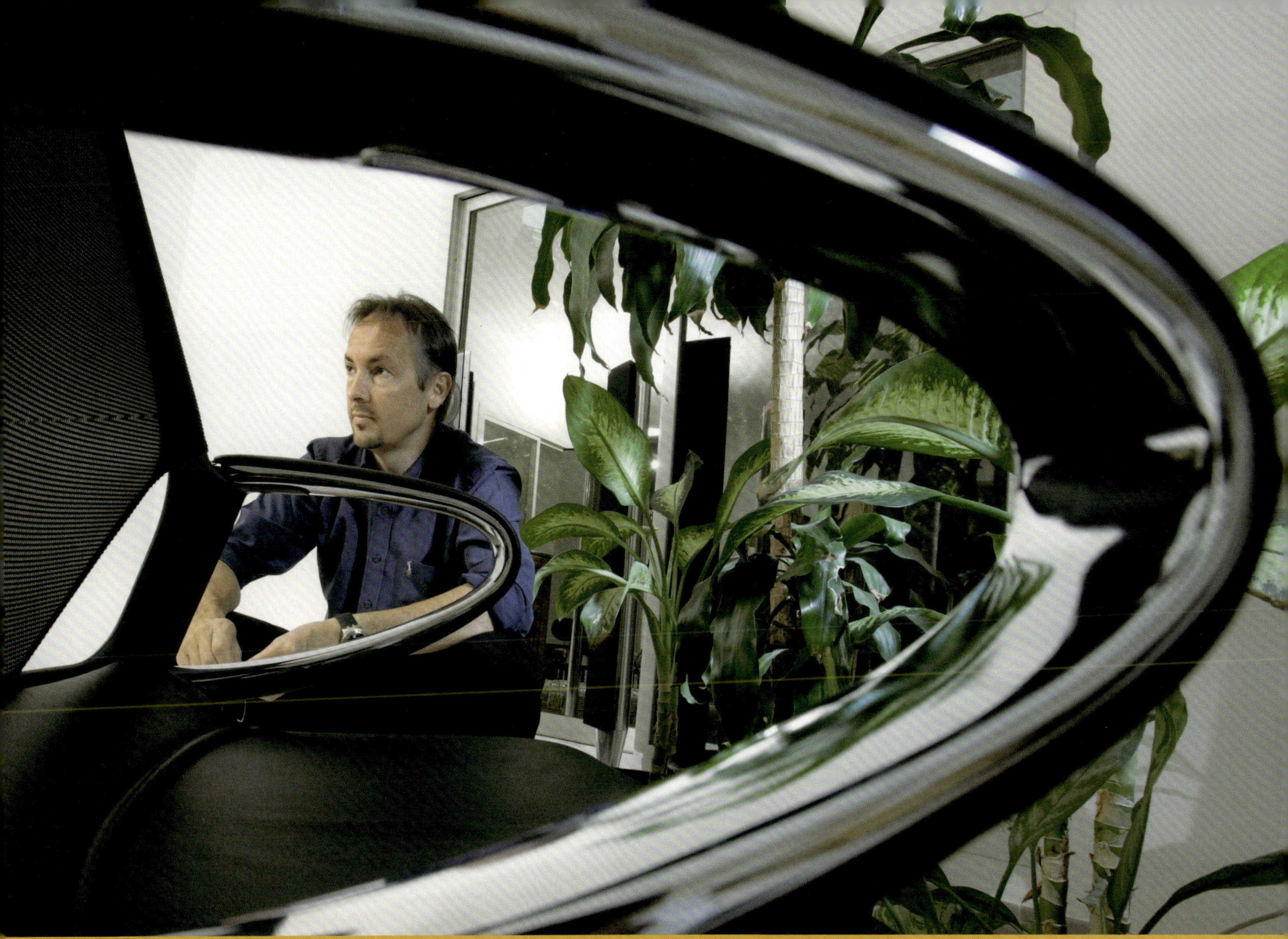

Henry Stevens
"Today´s work environment is frenetic and complex. Many of our clients instinctively respond to simplicity. Simplicity of design, simplicity of space and simplicity of operation. Many of our most successful past projects are carefully designed to subtly manipulate mood and expectation. They are calming stage sets upon which complex activities are acted out. The understated lines of Wilkhahn have often played a large part in creating our successes."

ISG Asia FZE, Dubai Swivel chair: Modus Medium range

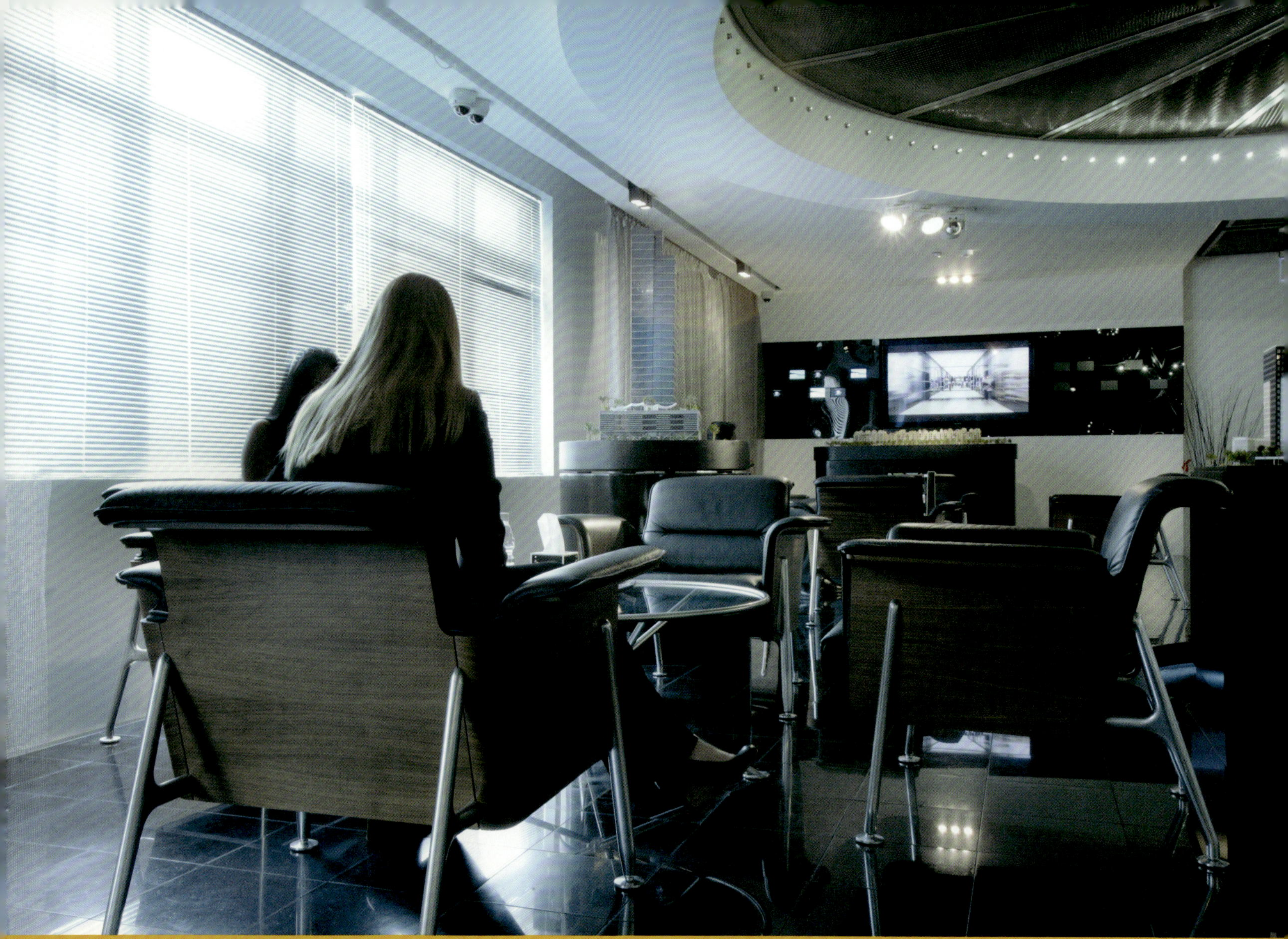

Hannes Werner, Architect
"All concepts need an emotional starting point – a strong idea. Once it's sketched on the paper, it is discussed and continuously improved; it will guide you through the whole project."

Omniyat Properties, Dubai Lounge chairs: Avera range, swivel chair: Solis F

Spielmann's Office Design, Dubai

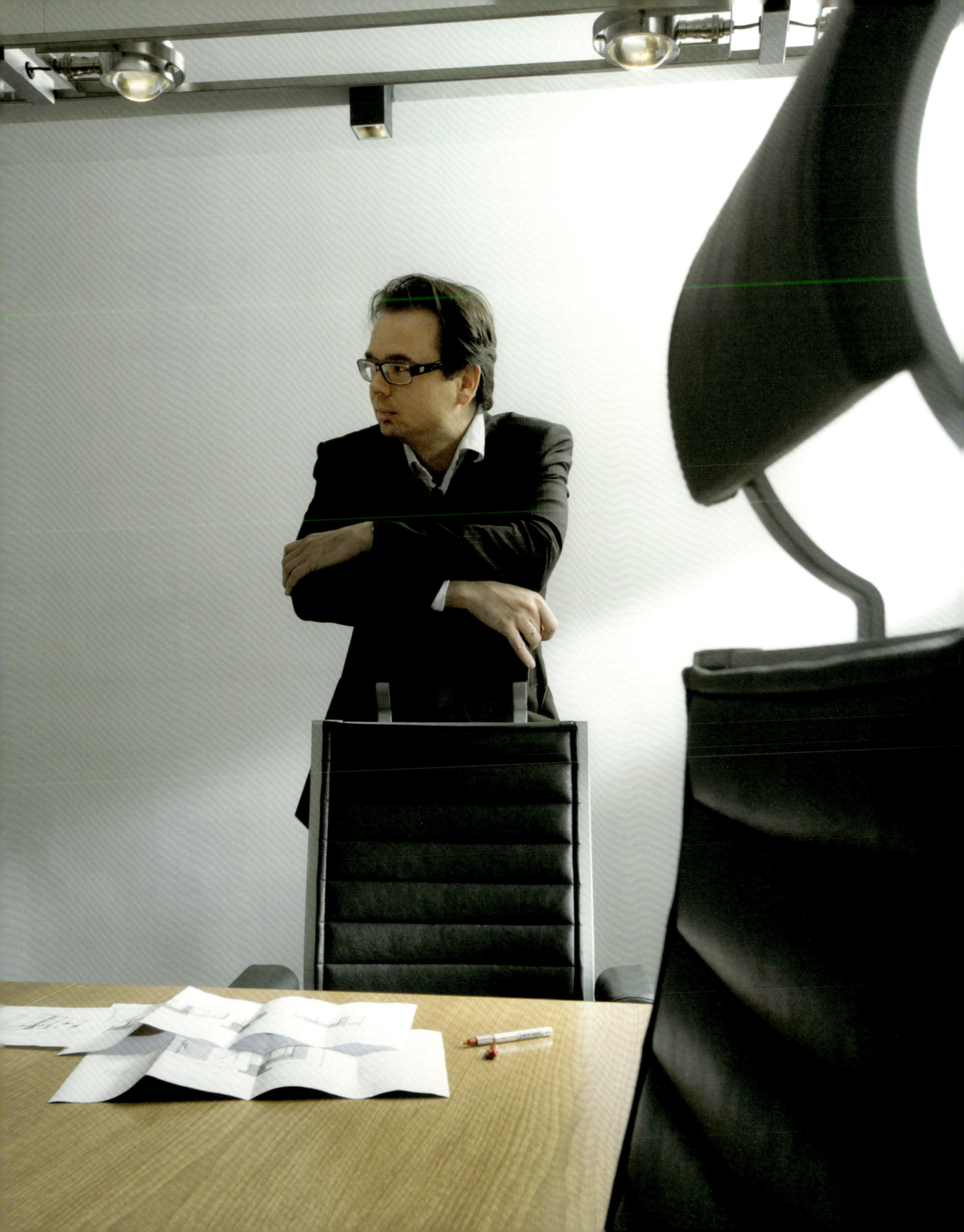

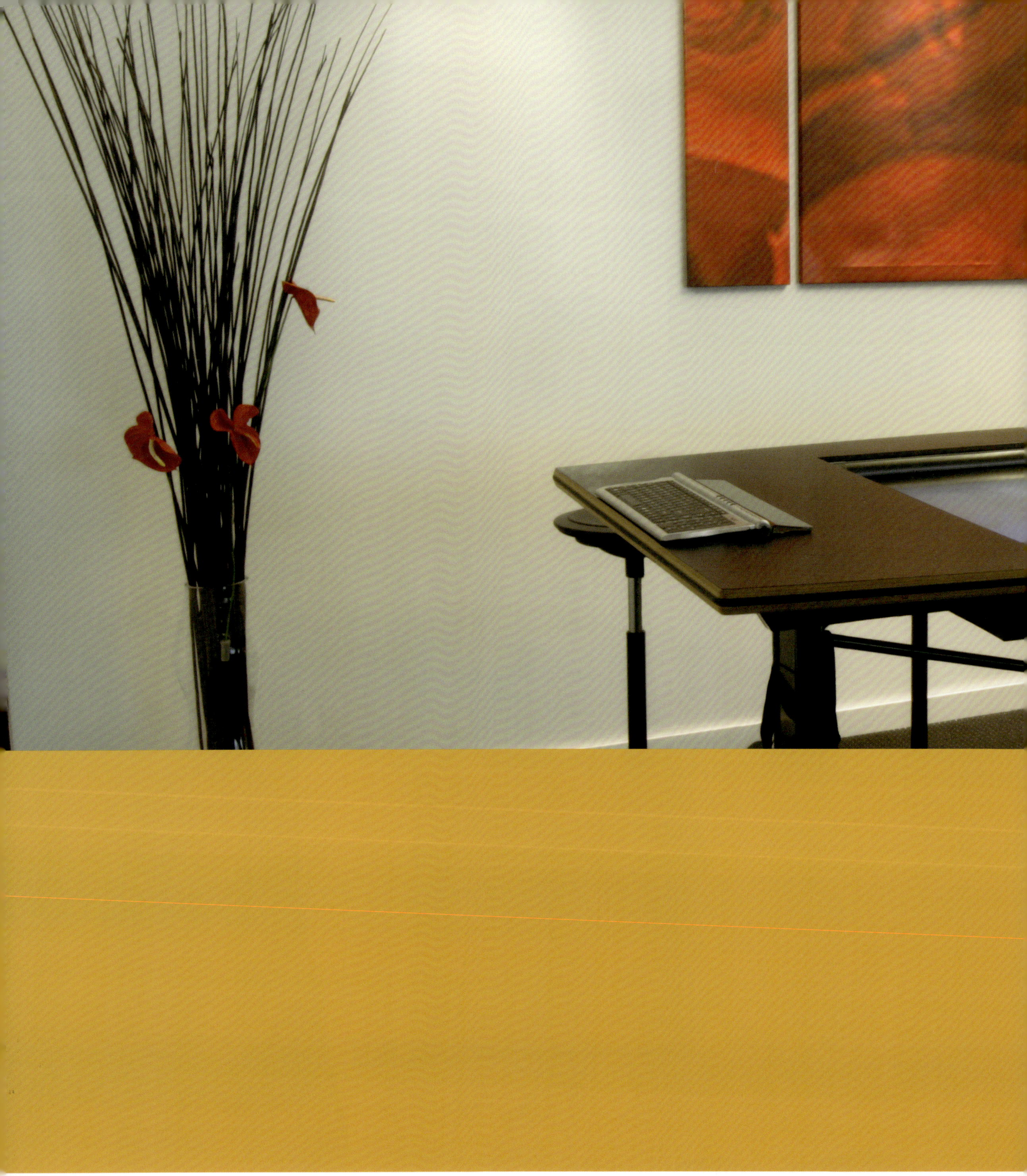

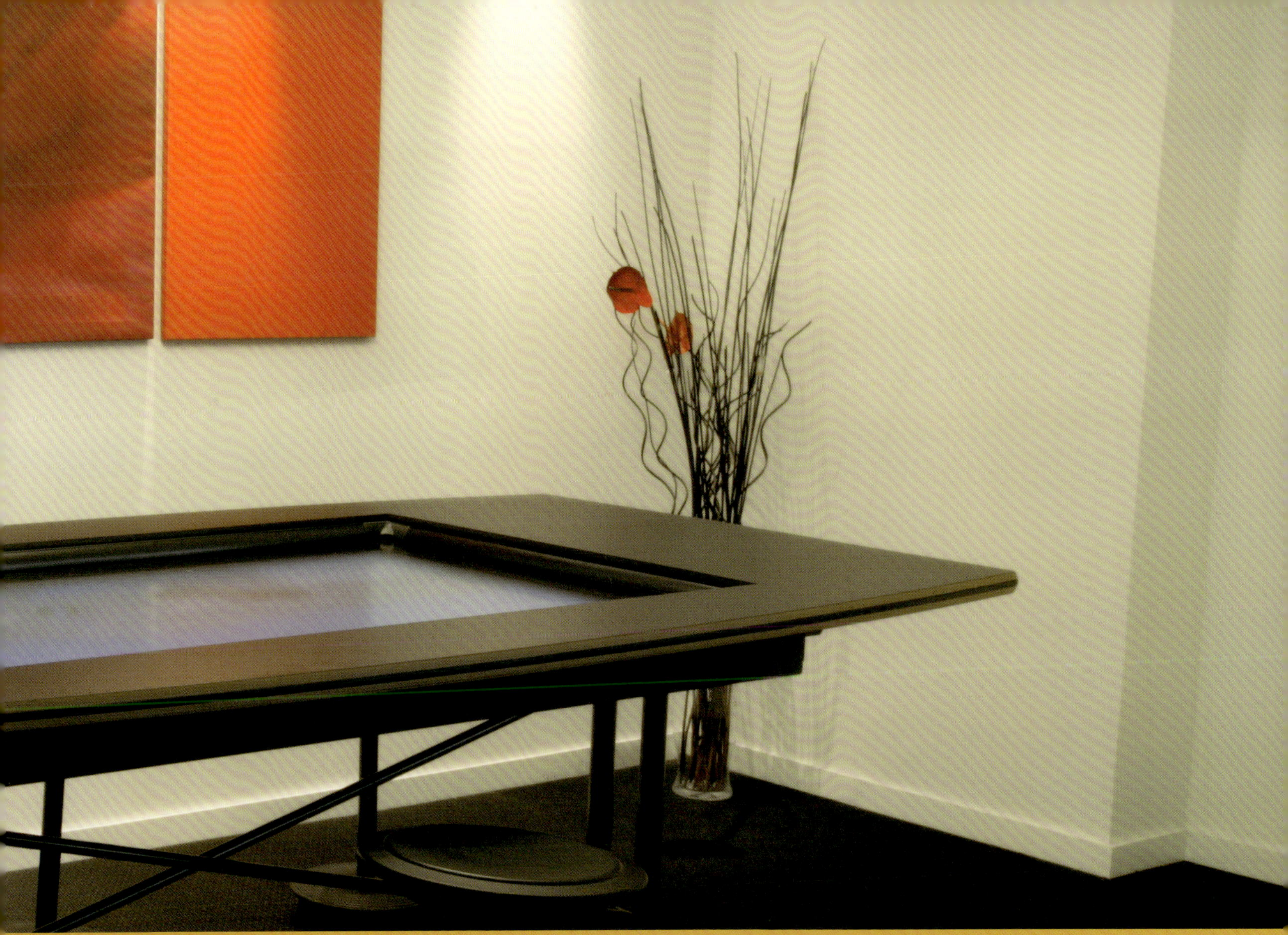

Omniyat Properties, Dubai Interactive high table: InteracTable®

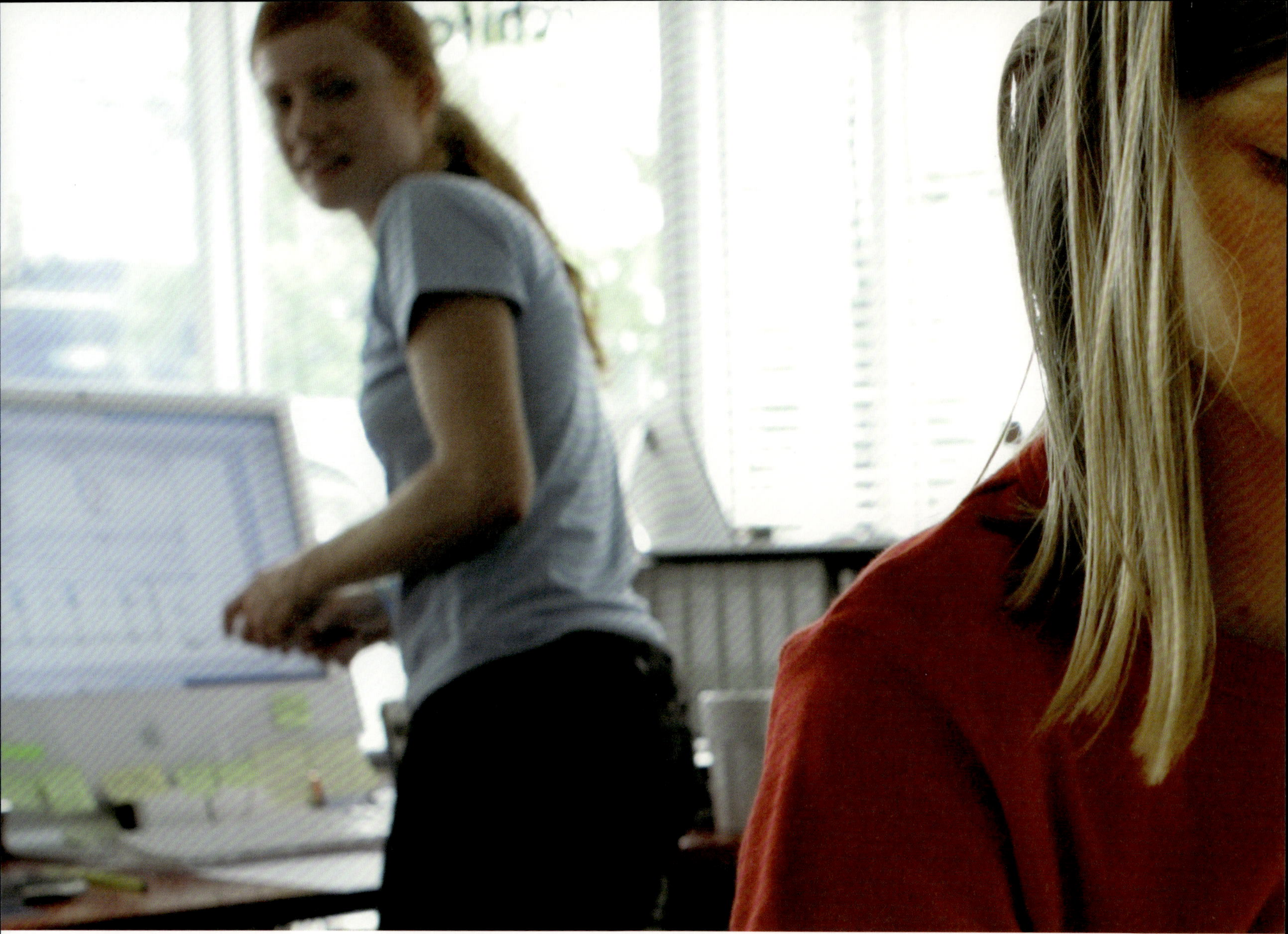

Die Internationalisierung von Wilkhahn in Süd-, West- und Mitteleuropa begann schon in den 50er Jahren. In den vergangenen Jahrzehnten wurden sukzessive Vertriebsbüros und Tochtergesellschaften in Madrid, Paris, Bern, Wien, Rotterdam, London, Antwerpen und Warschau gegründet. Sie sorgen gemeinsam mit ausgewählten Fachhandelspartnern dafür, dass hohe Beratungsqualität, perfekte Logistik und kundenorientierte Serviceleistungen in allen Ländern der Region gewährleistet sind.

The internationalization of Wilkhahn in Southern, Western and Central Europe already began in the 1950s. In the recent decades, sales offices and subsidiary companies have been gradually established in Madrid, Paris, Bern, Vienna, Rotterdam, London, Antwerp and Warsaw. Together with selected specialist partners, they make sure that the high quality of consulting services, perfect logistics and customer-oriented services are guaranteed in all countries of the region.

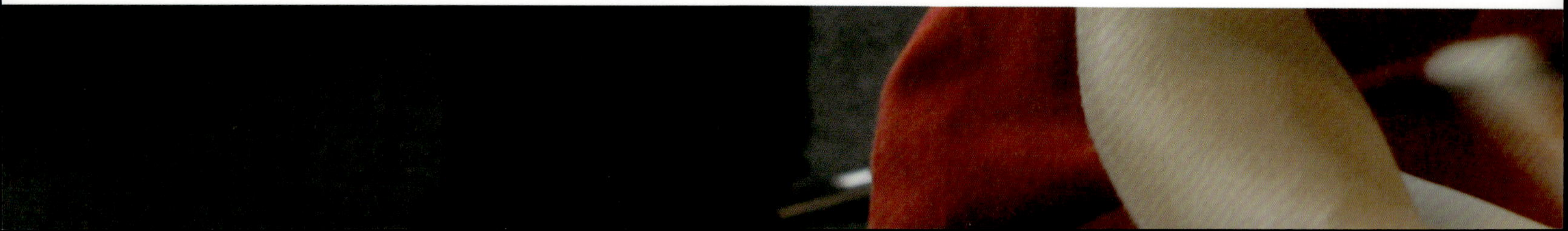

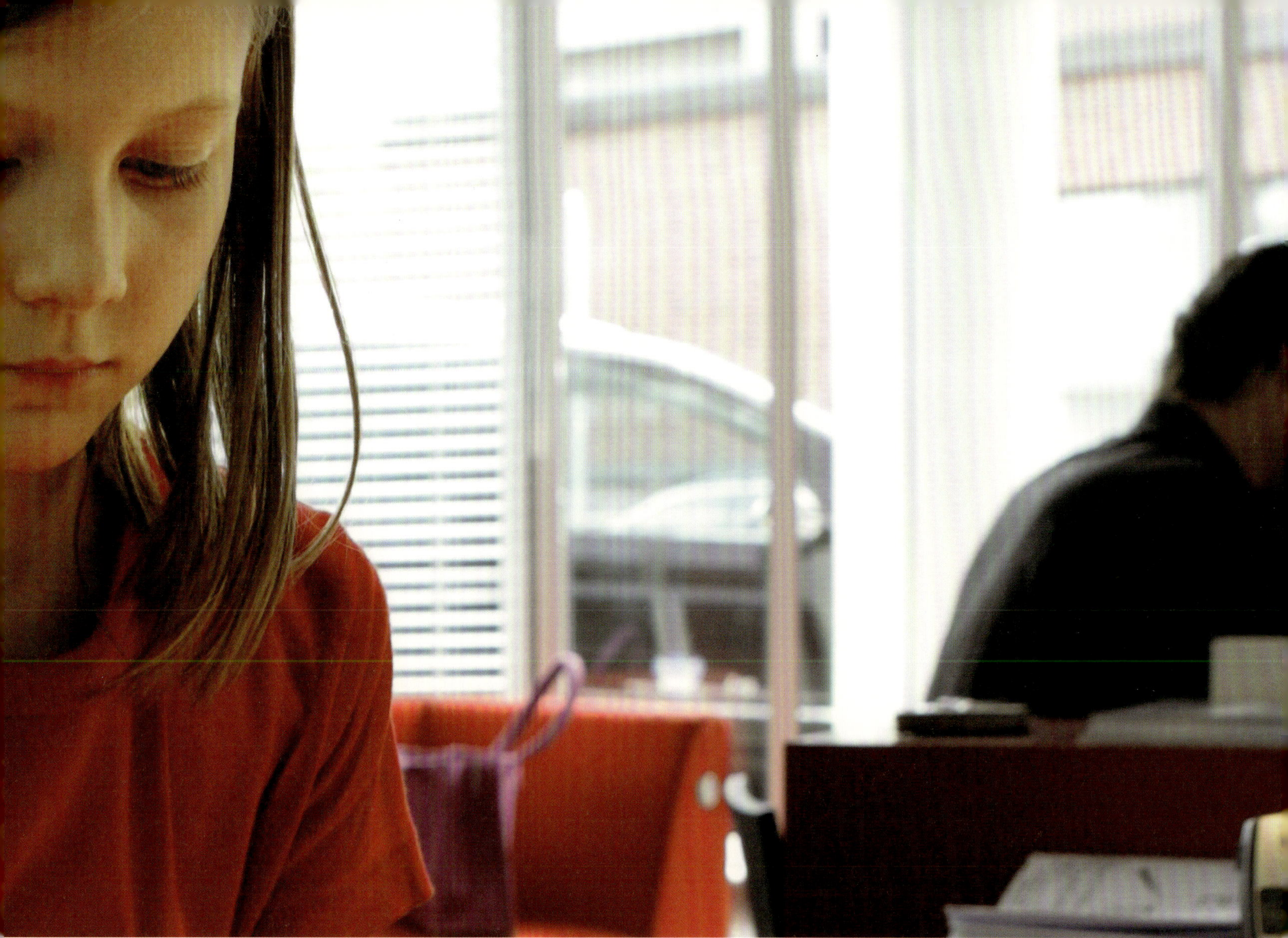

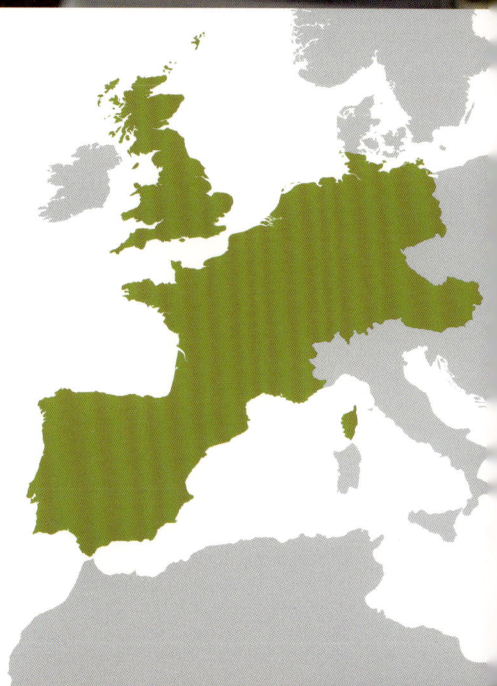

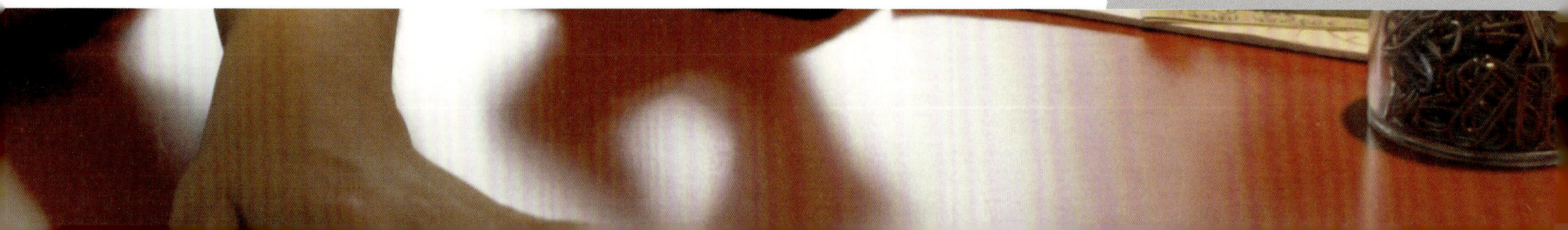

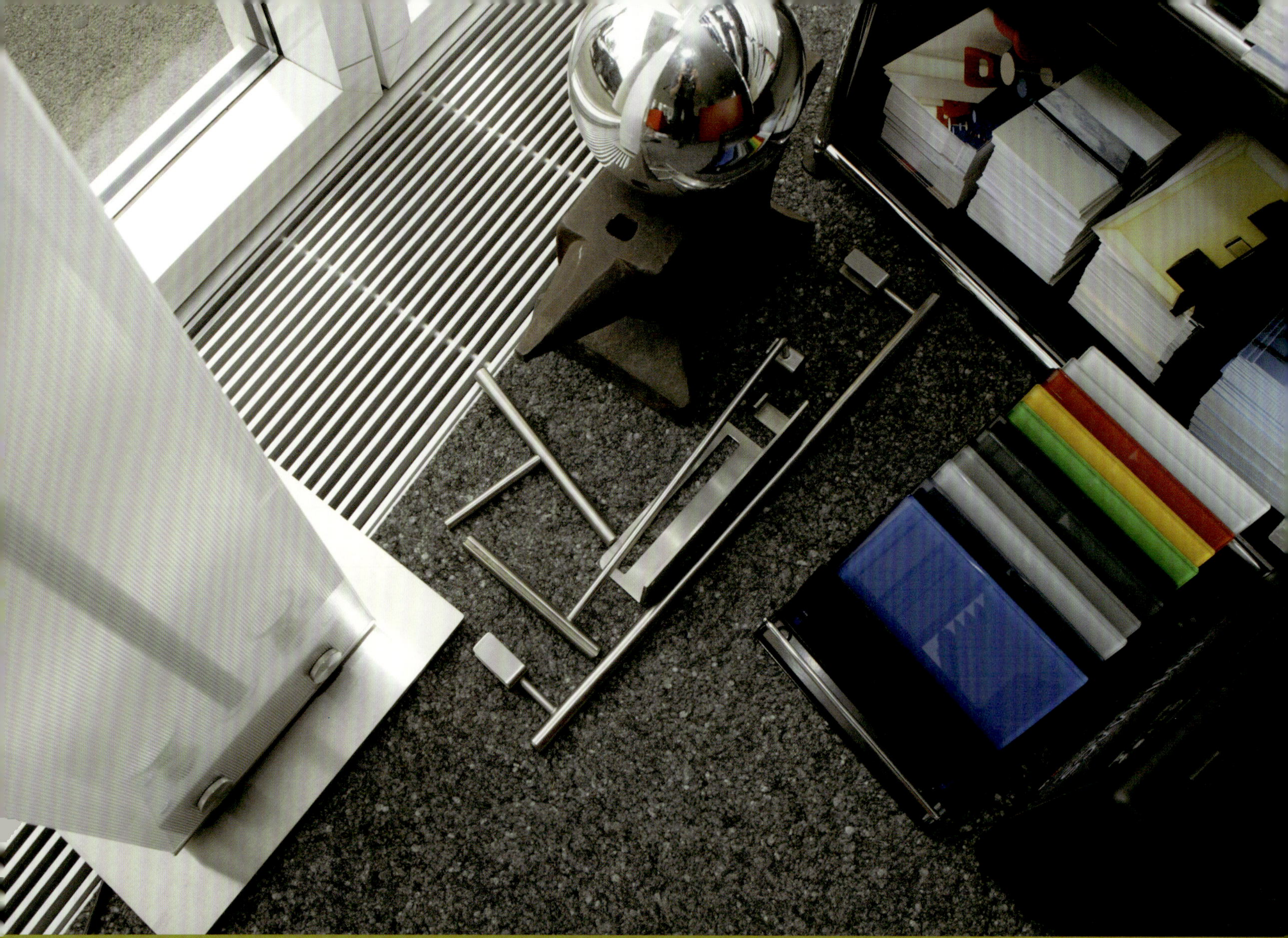

Lassoudry Archiformes GmbH, Basel

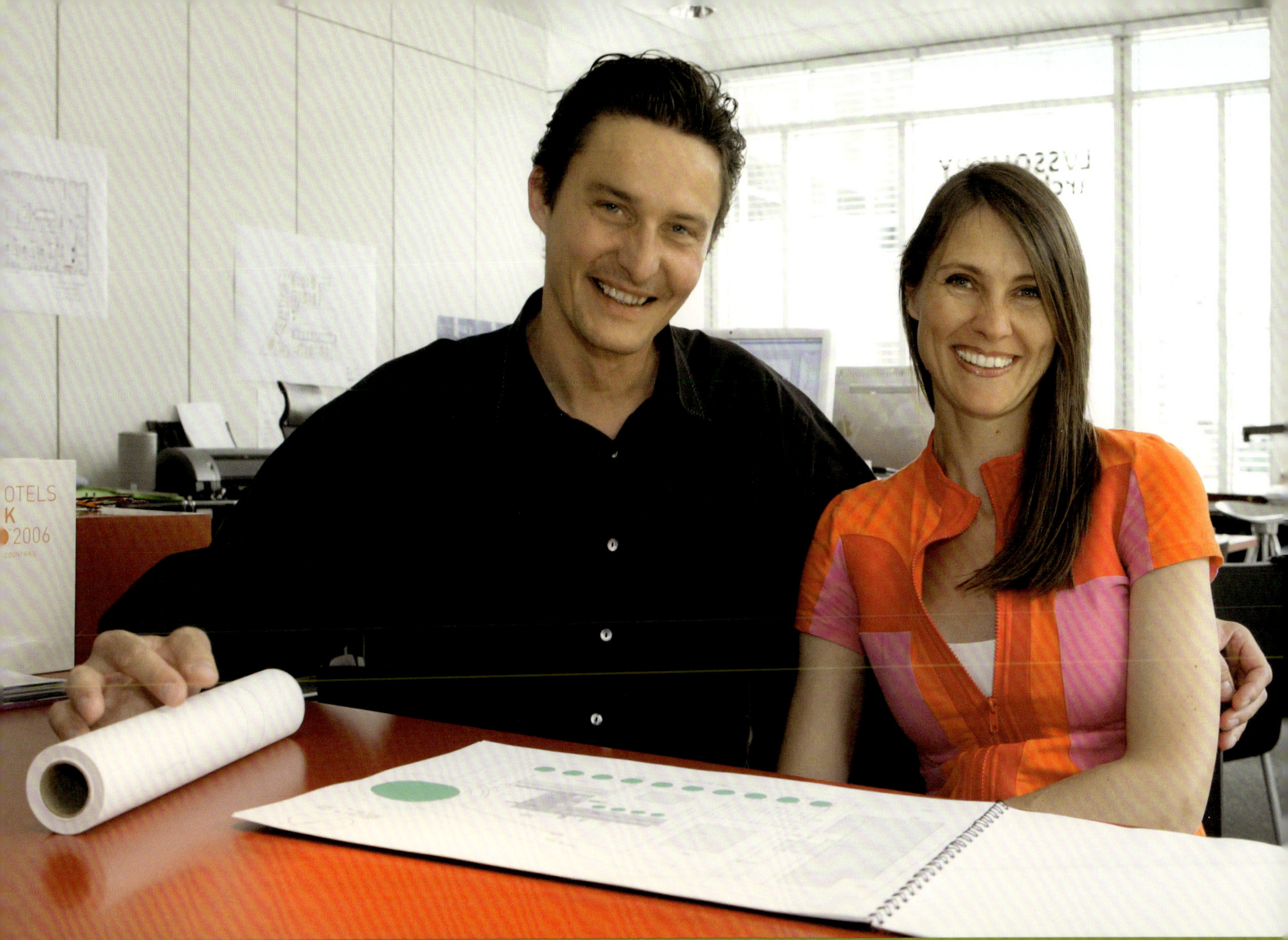

John Ritschl
„Wilkhahn ist eine Firma, die ethische Grundsätze nach innen wie nach aussen, ohne ökonomische Erfolge zu behindern, in der Praxis lebt und umsetzt. Die Langlebigkeit der Wilkhahn-Produkte spiegelt sich in den treuen menschlichen Beziehungen zwischen den Mitarbeitern und Partnern wider."

"Wilkhahn is a company that lives out and realizes ethical principles internally and externally in practice without hindering economic success. The longevity of Wilkhahn products is reflected in the faithful and personal relationships between employees and partners."

Samantha Ritschl-Lassoudry
„Wilkhahn ist ein sehr professionelles und emotionales Unternehmen, das sich – wie jedes gute Design oder jeder gute Text auch – auf den zweiten Blick erschließt und deshalb immer im Gedächtnis bleibt."

"Wilkhahn is a very professional and emotional company which, like every good design or text, is understood at second glance and therefore never forgotten."

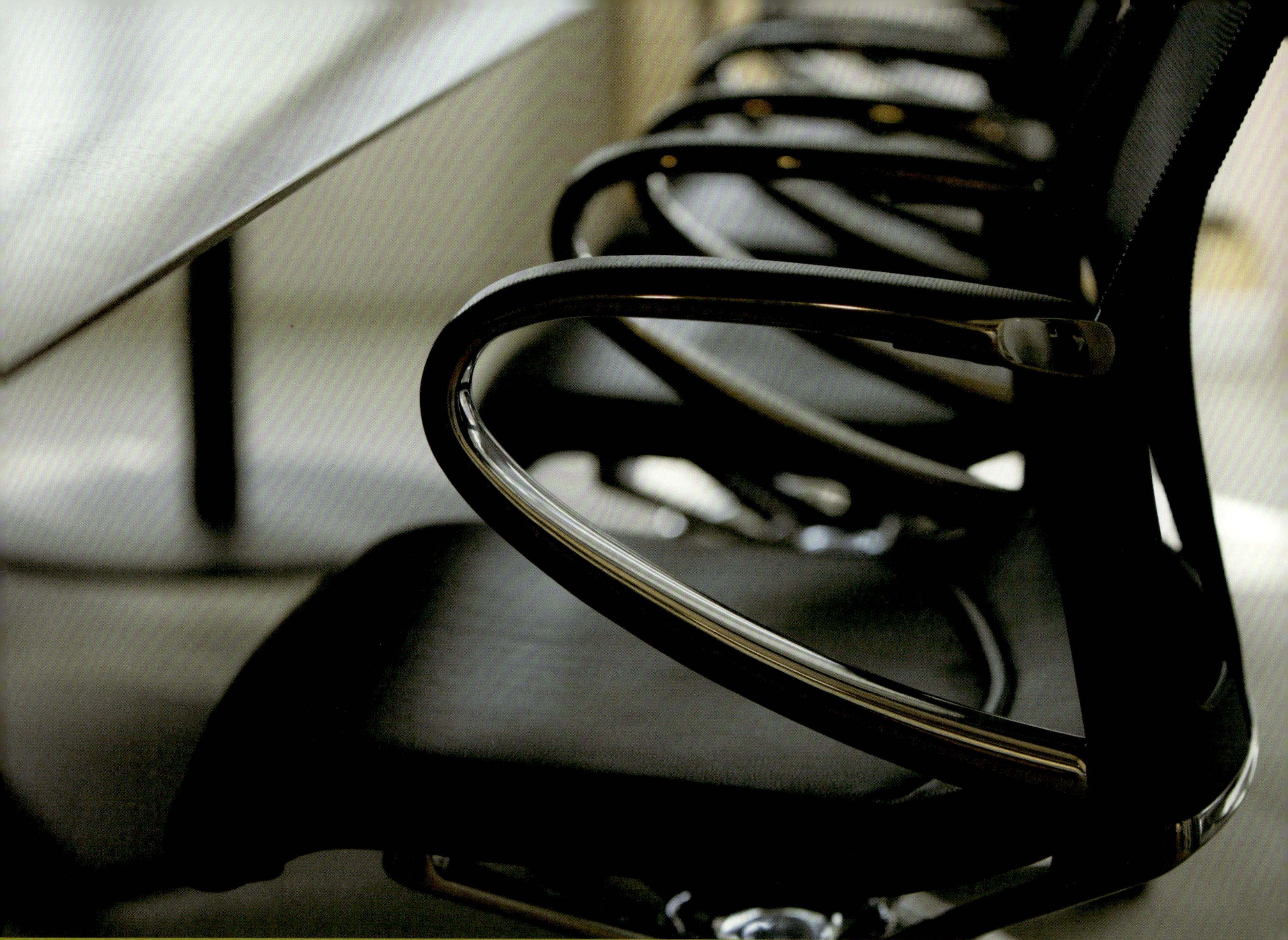

Rainer Hitz
„Wilkhahn hat den längsten Produktlebenszyklus aller mir bekannten Bürostuhlhersteller. So hat der Lebenszyklus der FS-Linie meine ganze Unternehmenstätigkeit begleitet (1985–2007). Design, Form und Funktion sind immer im Einklang mit dem Produkt. Diese Konstante in der Gestaltungshandschrift wie auch die langjährigen, treuen menschlichen Verbindungen stehen für die Marke Wilkhahn in der Schweiz."

"Wilkhahn has the longest product life cycle of all office chair manufacturers that I know. For example, the life cycle of the FS-Line accompanied me for the duration of my entire company activity (1985–2007). Design, form and function are always in harmony with the product. This constant in the design language as well as the long-standing, trusting personal relationships epitomize the Wilkhahn brand in Switzerland."

Aermo, Dietikon
Conference chairs: Modus Medium range

Multi-purpose chair: Aline range; cabinet: Uno range

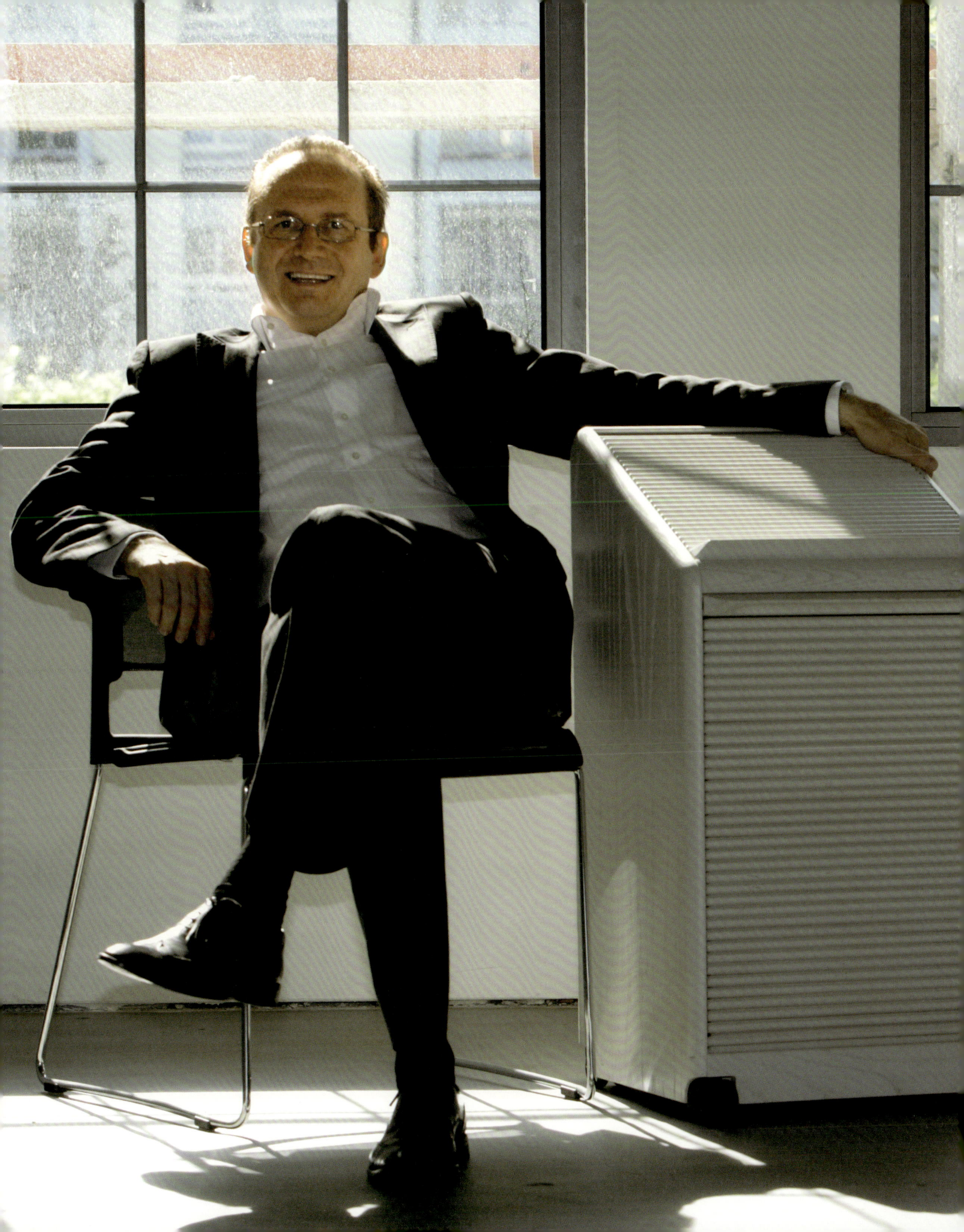

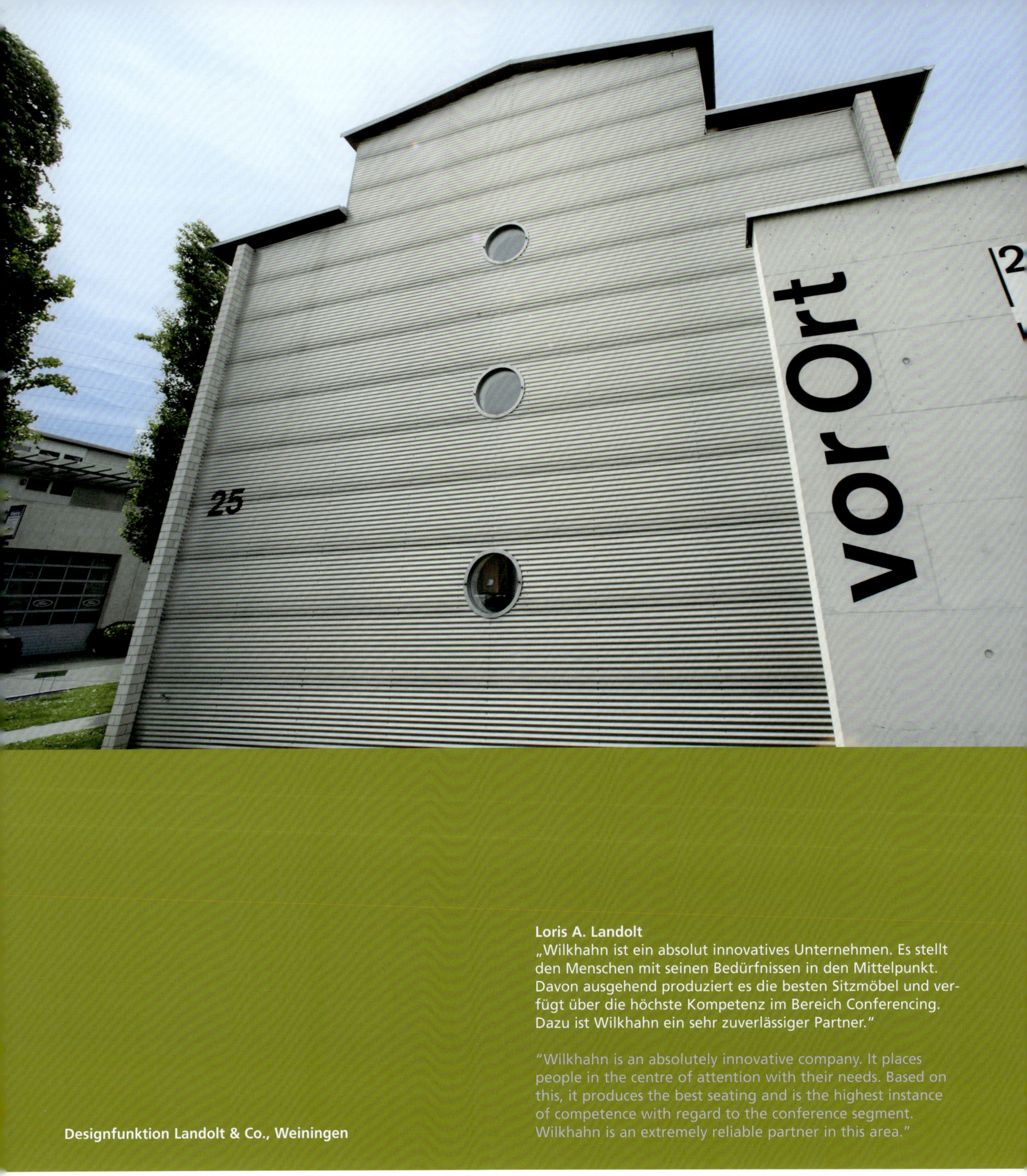

Loris A. Landolt
„Wilkhahn ist ein absolut innovatives Unternehmen. Es stellt den Menschen mit seinen Bedürfnissen in den Mittelpunkt. Davon ausgehend produziert es die besten Sitzmöbel und verfügt über die höchste Kompetenz im Bereich Conferencing. Dazu ist Wilkhahn ein sehr zuverlässiger Partner."

"Wilkhahn is an absolutely innovative company. It places people in the centre of attention with their needs. Based on this, it produces the best seating and is the highest instance of competence with regard to the conference segment. Wilkhahn is an extremely reliable partner in this area."

Designfunktion Landolt & Co., Weiningen

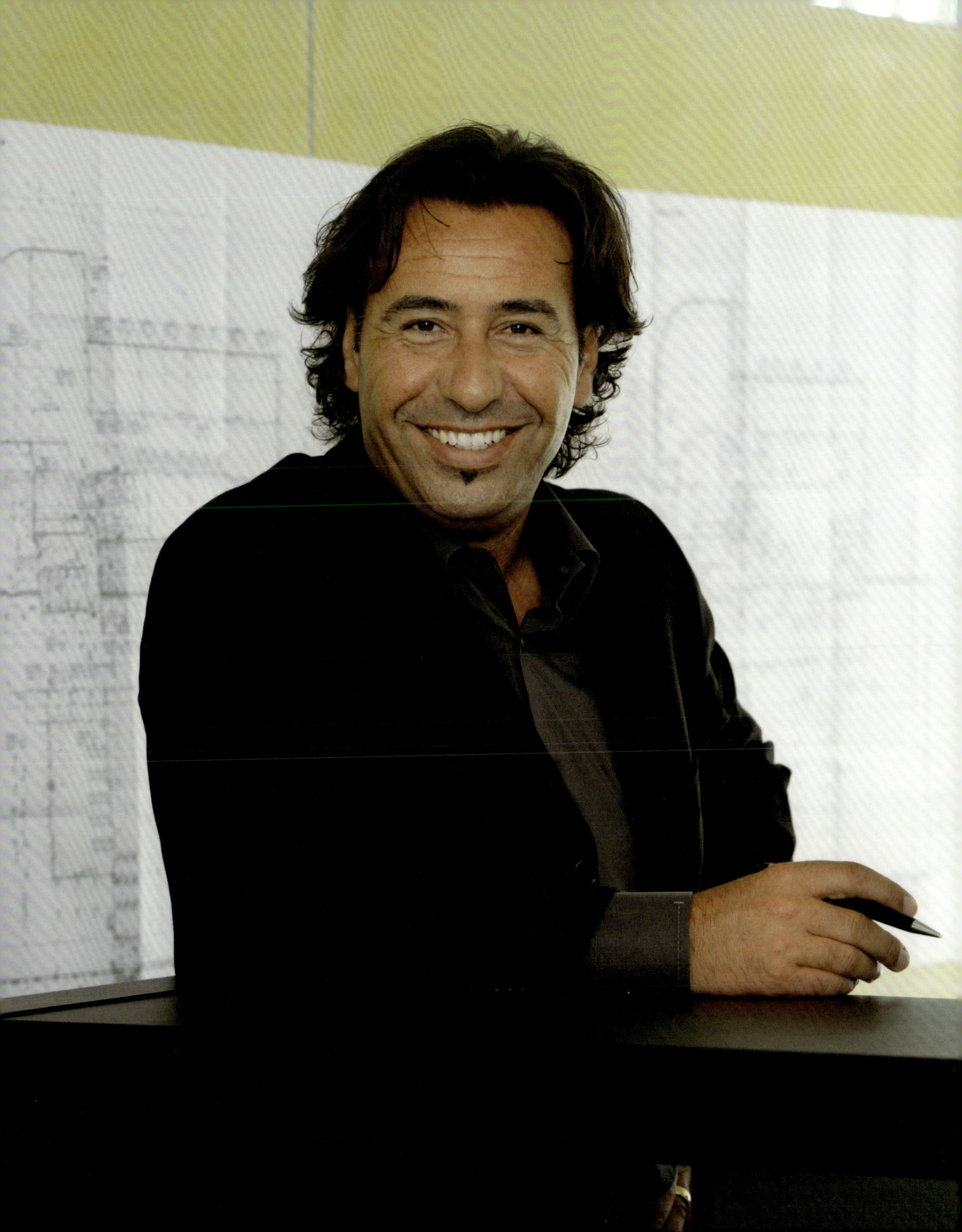

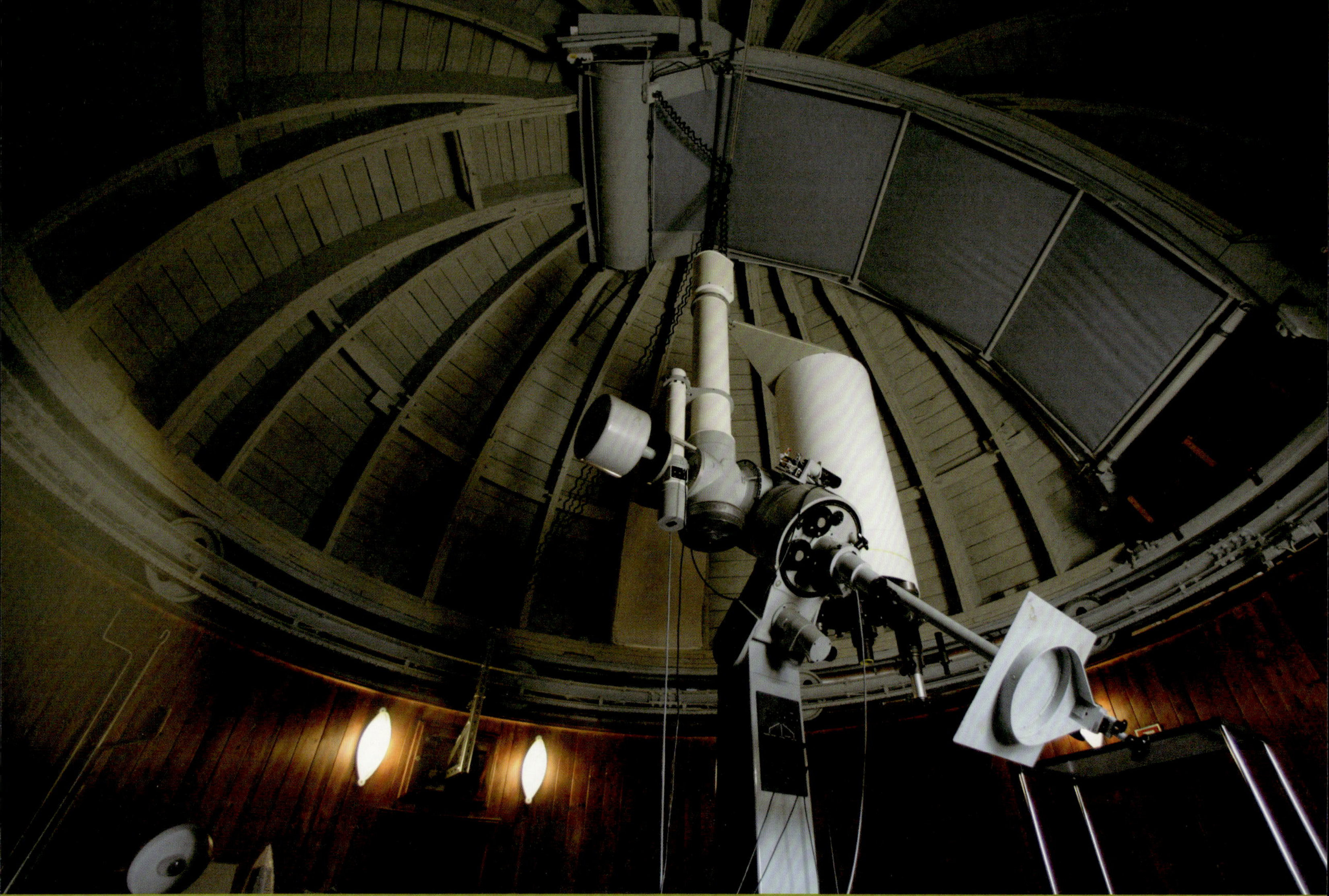

ETH Swiss Federal Institute of Technology, Zurich

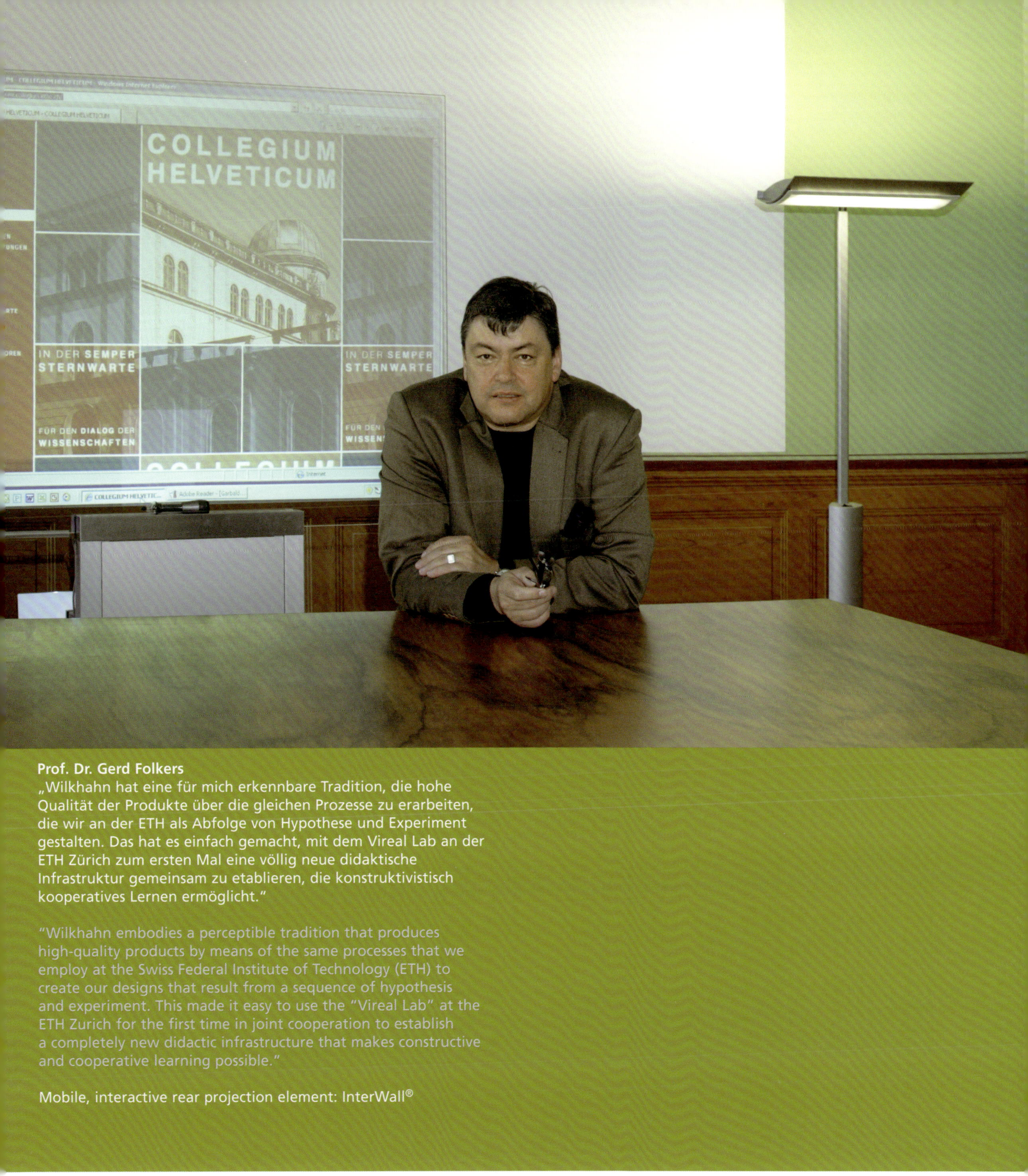

Prof. Dr. Gerd Folkers
„Wilkhahn hat eine für mich erkennbare Tradition, die hohe Qualität der Produkte über die gleichen Prozesse zu erarbeiten, die wir an der ETH als Abfolge von Hypothese und Experiment gestalten. Das hat es einfach gemacht, mit dem Vireal Lab an der ETH Zürich zum ersten Mal eine völlig neue didaktische Infrastruktur gemeinsam zu etablieren, die konstruktivistisch kooperatives Lernen ermöglicht."

"Wilkhahn embodies a perceptible tradition that produces high-quality products by means of the same processes that we employ at the Swiss Federal Institute of Technology (ETH) to create our designs that result from a sequence of hypothesis and experiment. This made it easy to use the "Vireal Lab" at the ETH Zurich for the first time in joint cooperation to establish a completely new didactic infrastructure that makes constructive and cooperative learning possible."

Mobile, interactive rear projection element: InterWall®

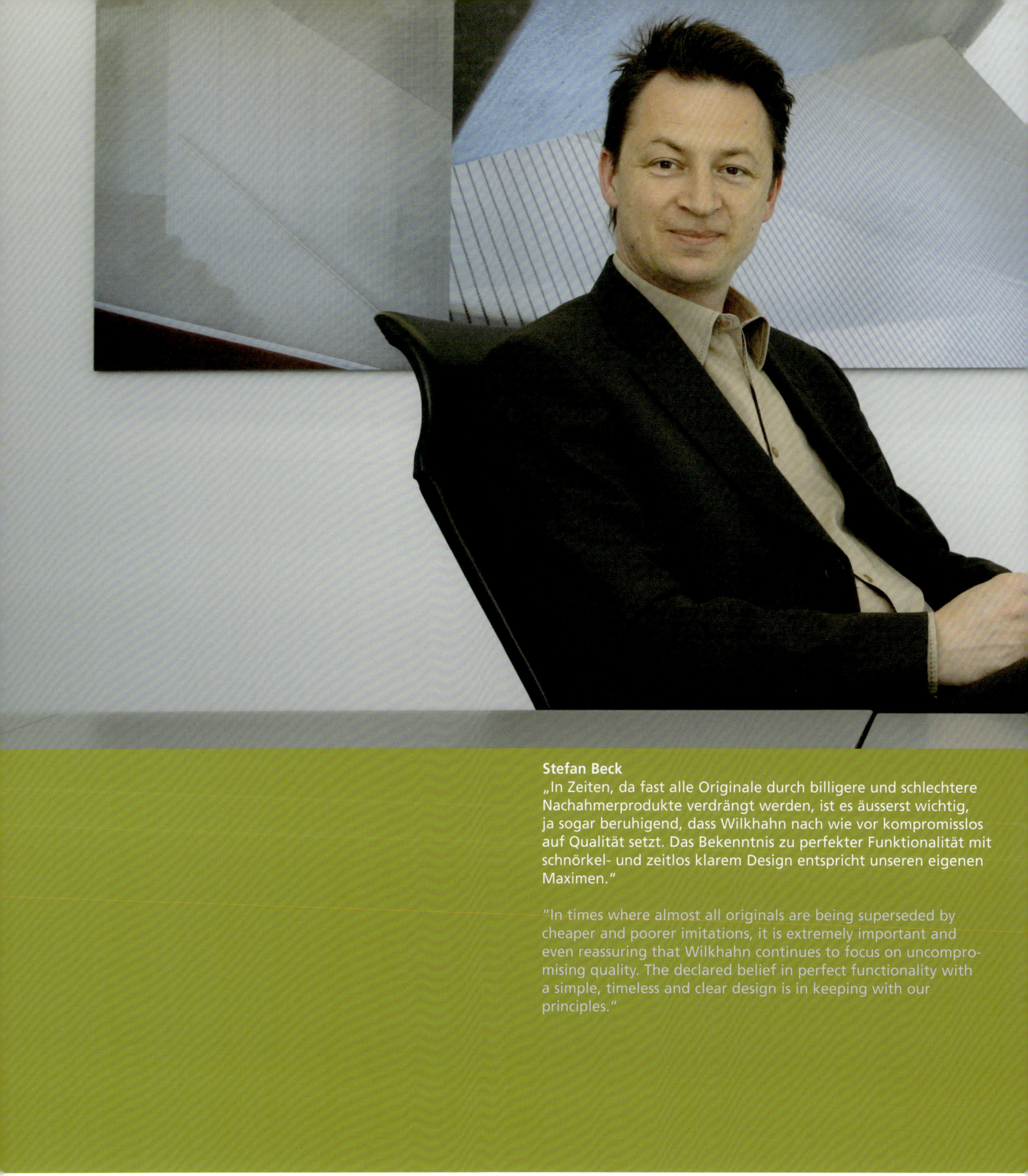

Stefan Beck
„In Zeiten, da fast alle Originale durch billigere und schlechtere Nachahmerprodukte verdrängt werden, ist es äusserst wichtig, ja sogar beruhigend, dass Wilkhahn nach wie vor kompromisslos auf Qualität setzt. Das Bekenntnis zu perfekter Funktionalität mit schnörkel- und zeitlos klarem Design entspricht unseren eigenen Maximen."

"In times where almost all originals are being superseded by cheaper and poorer imitations, it is extremely important and even reassuring that Wilkhahn continues to focus on uncompromising quality. The declared belief in perfect functionality with a simple, timeless and clear design is in keeping with our principles."

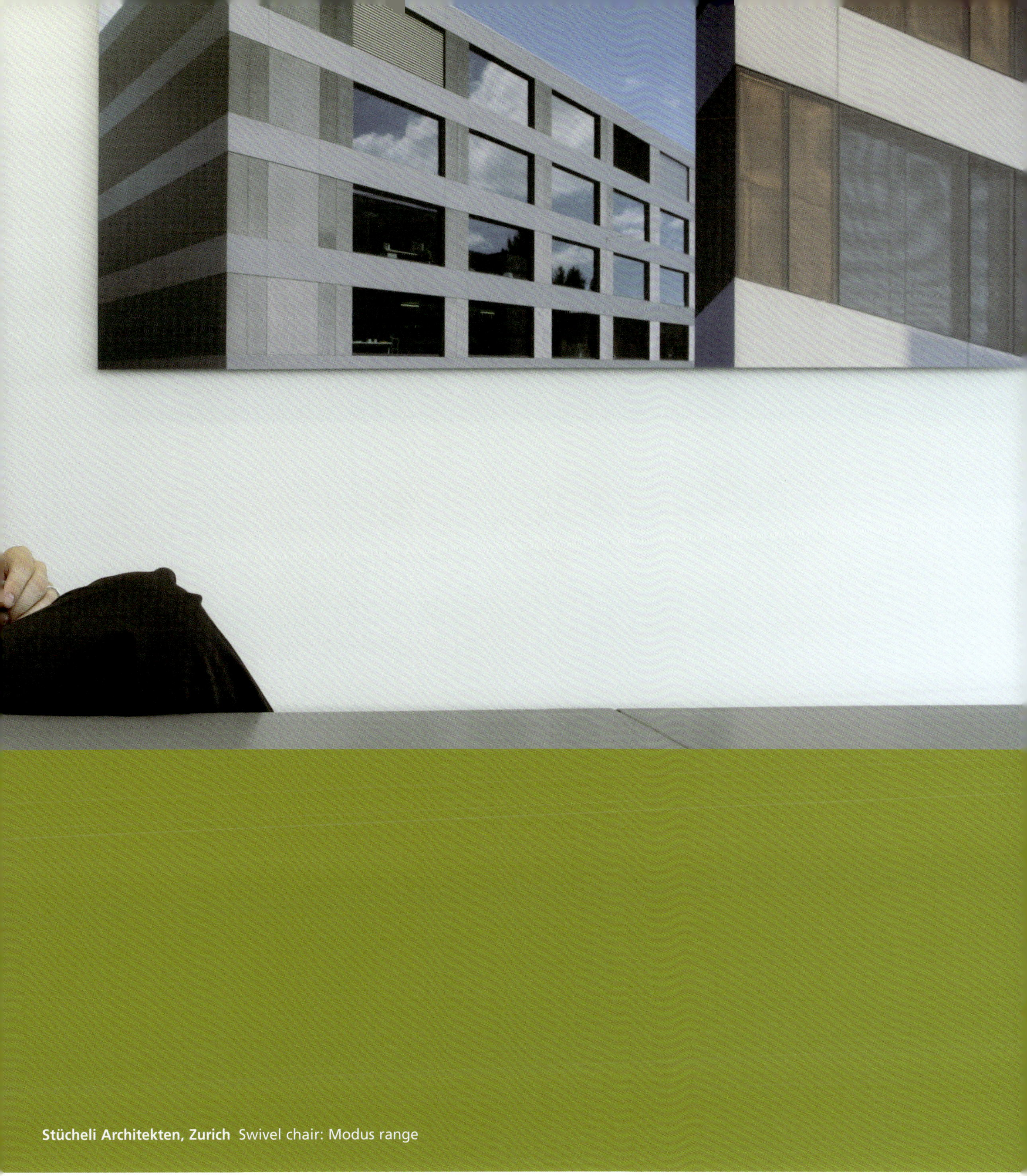

Stücheli Architekten, Zurich Swivel chair: Modus range

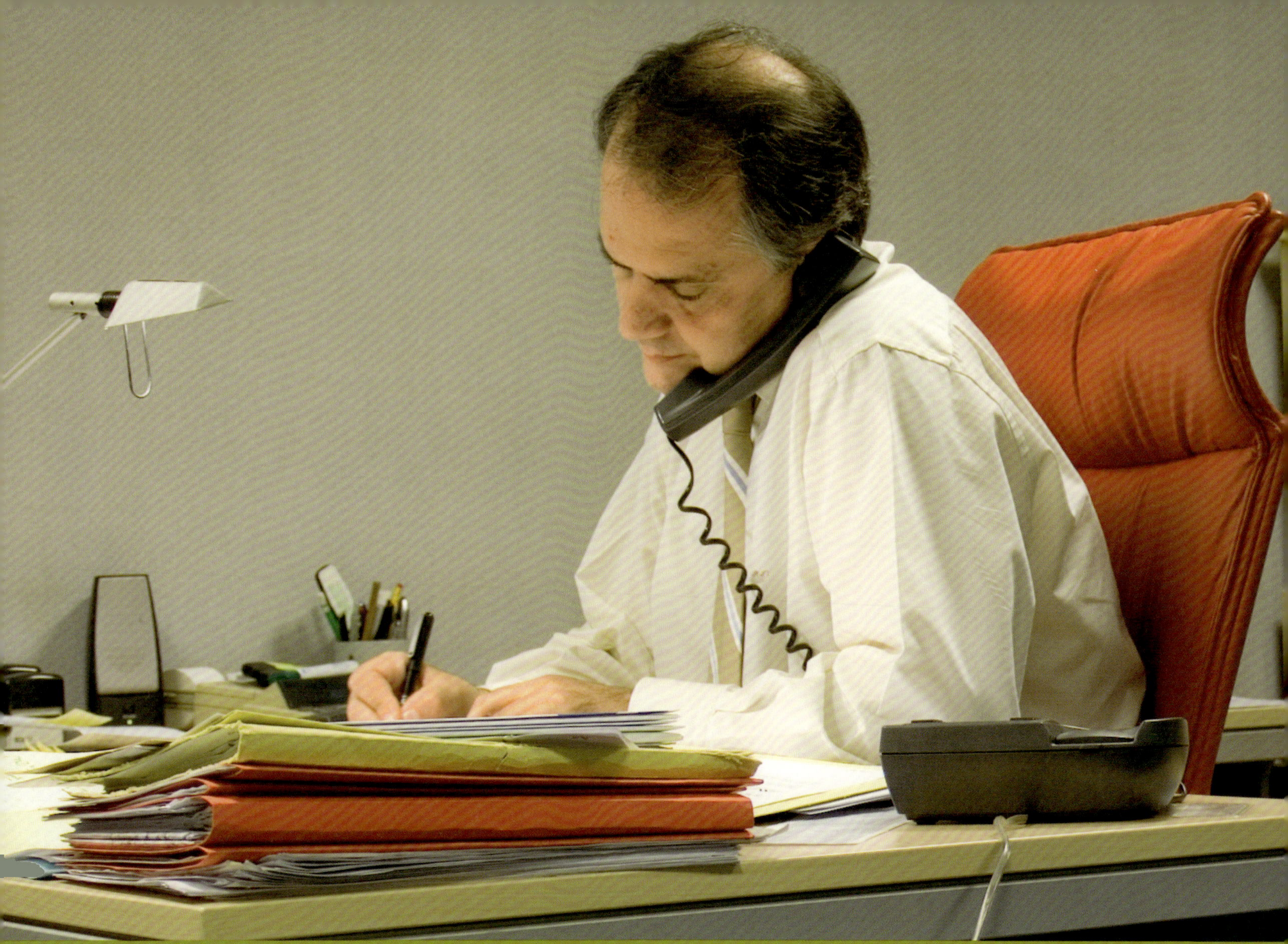

Sergio Tello, President
"La principal diferencia en un producto Wilkhahn con otro de la competencia es su cuidado diseño, calidad y acabados, de las partes visibles pero también de las que no se ven a simple vista; mecanismos bajo las sillas, ingeniería de las mesas, uniones inteligentes, etc."

"The main difference between a Wilkhahn product and one of the competitors is its careful, mature design and the quality and finishes of its visible parts, but also those which one does not see at first glance. For example, mechanisms beneath the chair, superb engineering of the tables or intelligent connections and joints."

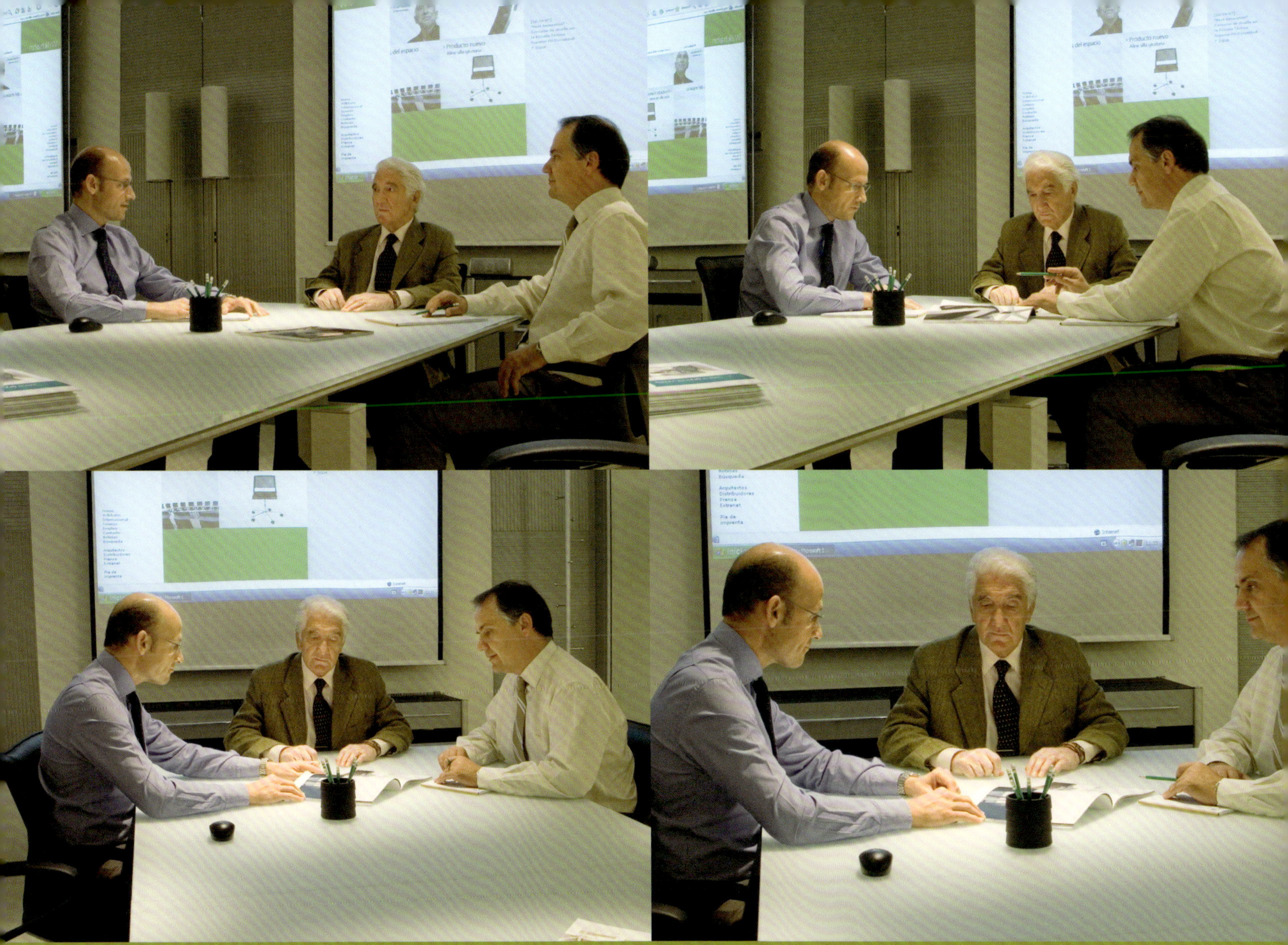

Burodecor S.A. Madrid Swivel chairs: Modus Executive range

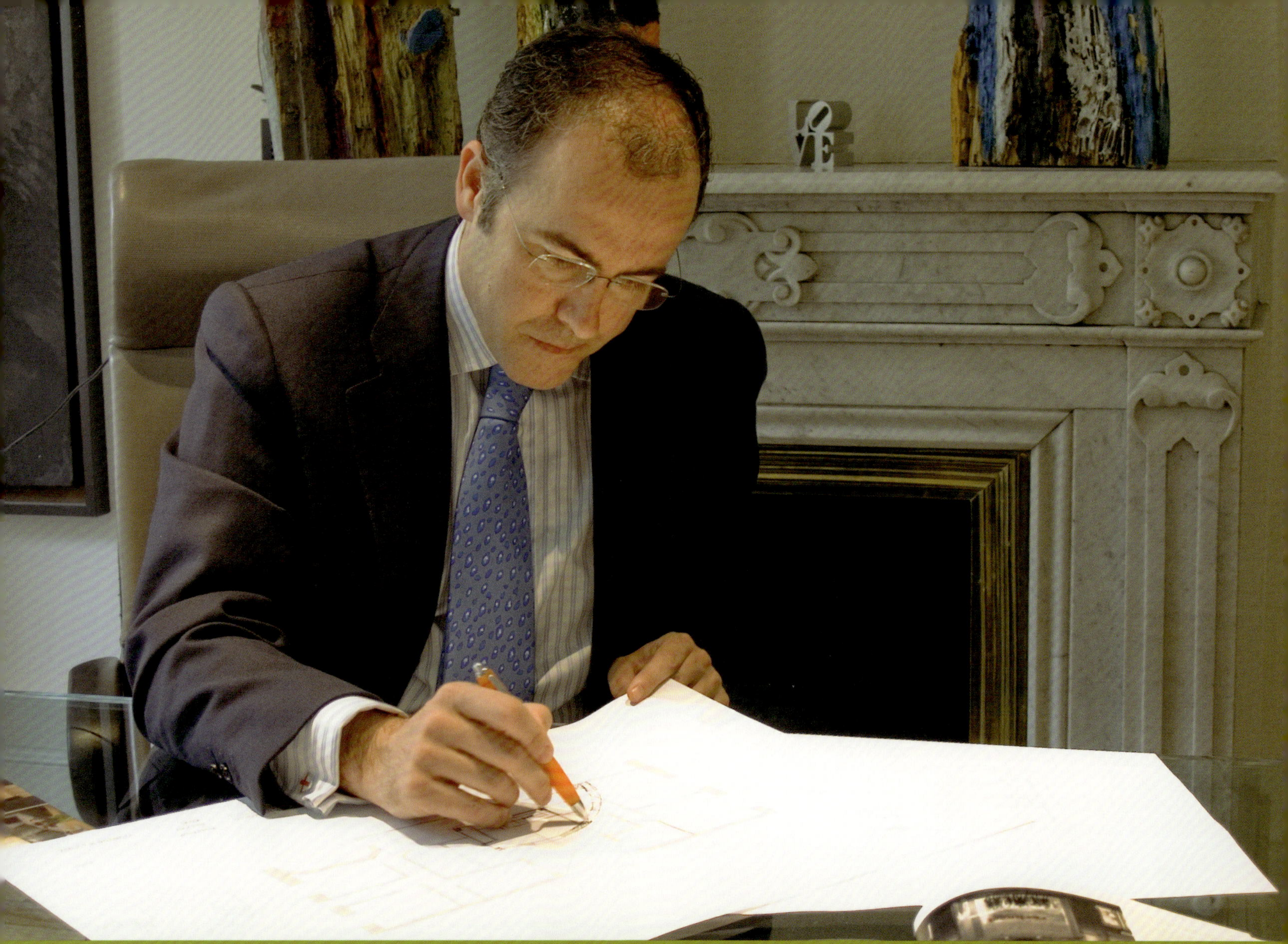

Andrés Dancausa, Architect, Madrid
"Mi primer proyecto de Arquitectura lo realicé sentado en una silla de Wilkhahn, así que me gusta que mis amigos y compañeros Arquitectos hagan lo mismo y diseñen también su primer proyecto como Arquitectos sentados con Wilkhahn. La última silla que regalé a un Arquitecto amigo mío fue la Solist F."

"I realized my first architectural project sitting on a Wilkhahn chair. Therefore I would appreciate my friends and colleague architects doing the same, designing their first project 'as an architect sitting on Wilkhahn'. The last chair I offered to an architect friend of mine was a Solis F."

Swivel chair: Modus Medium range

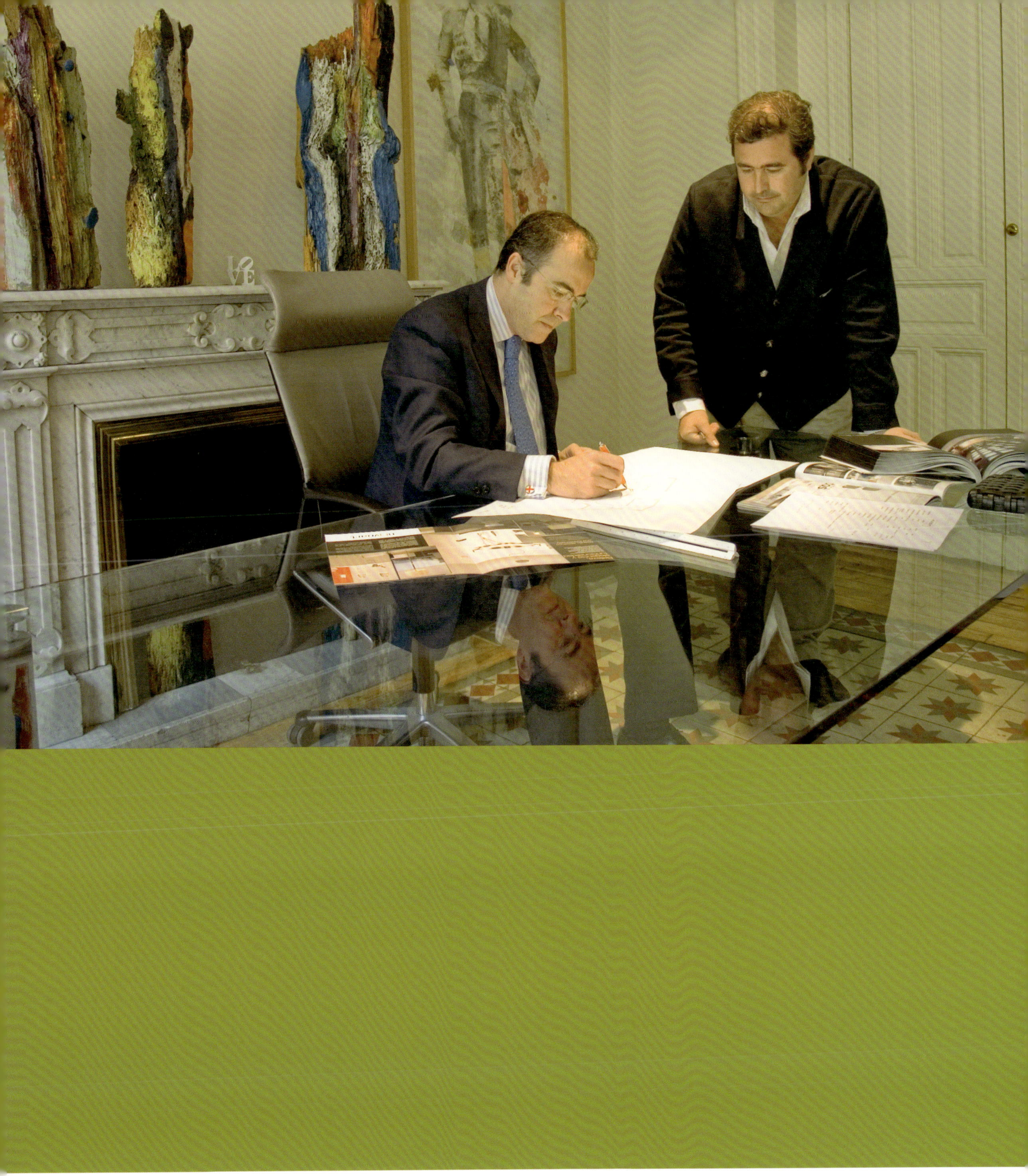

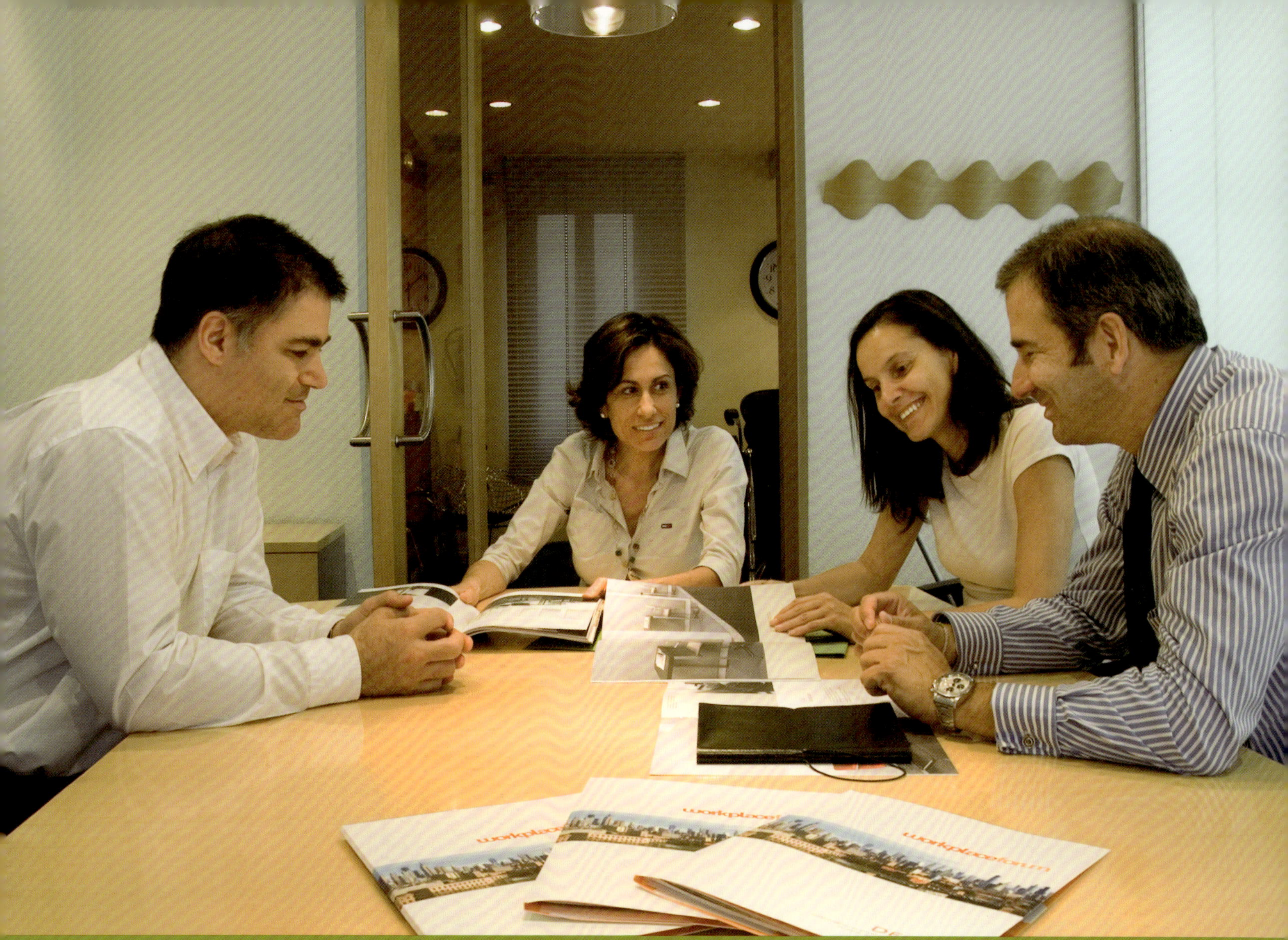

Juan Carredano and Elvira Muñoz
"Para nosotros Wilkhahn es la excelencia en la calidad. Tanto la ingeniería de sus productos como el diseño de los mismos permite que sea una gran referencia para nuestros proyectos e instalaciones. Además, gracias a su 'diseño atemporal', los muebles de Wilkhahn envejecen muy bien y siempre están de moda."

"In our opinion, Wilkhahn stands for 'excellence in quality'. The engineering of its products as well as its design give us great references for our projects and installations. Furthermore, thanks to its 'timeless design' Wilkhahn furniture 'ages' very well and is simply 'up to date'."

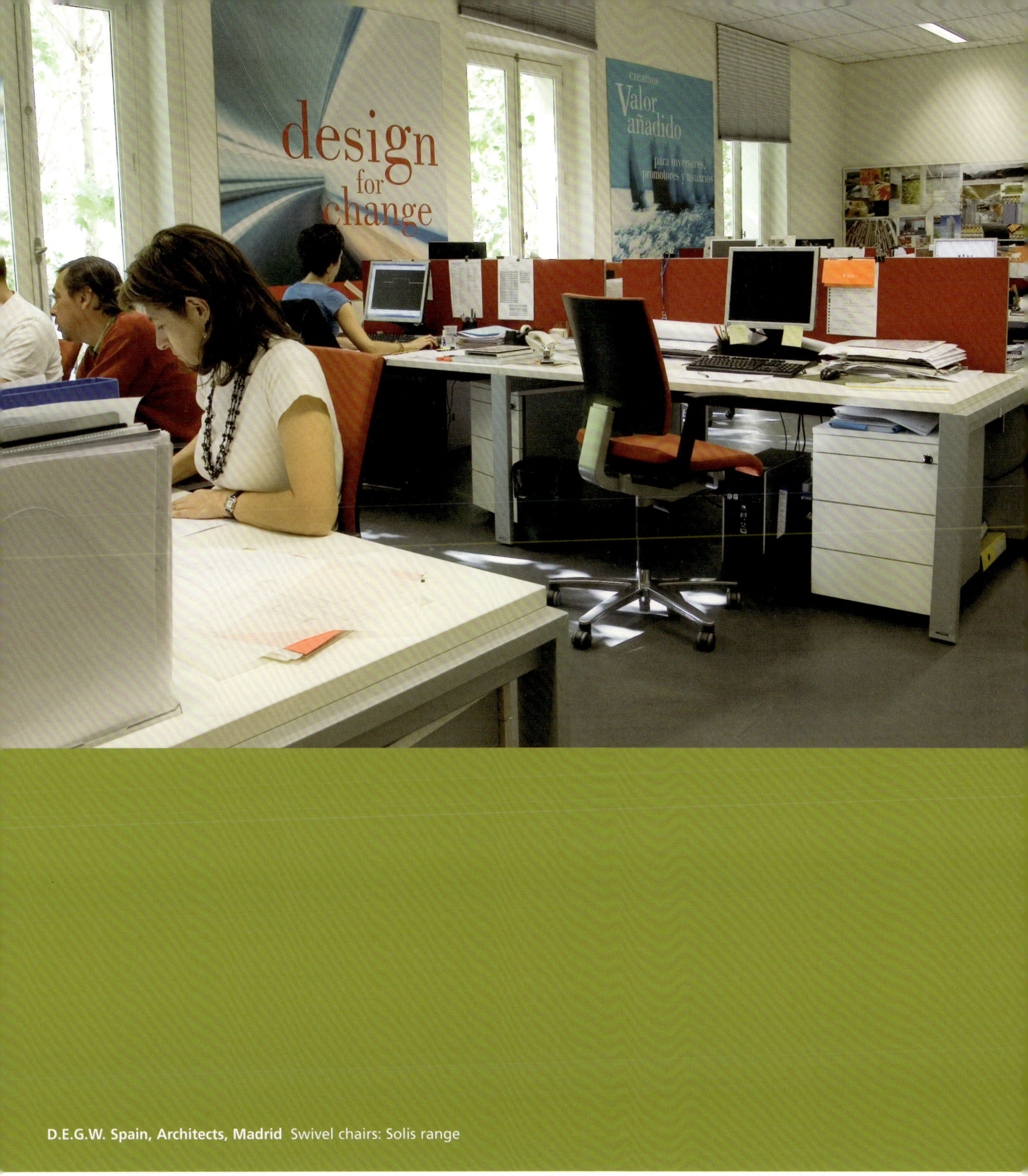

D.E.G.W. Spain, Architects, Madrid Swivel chairs: Solis range

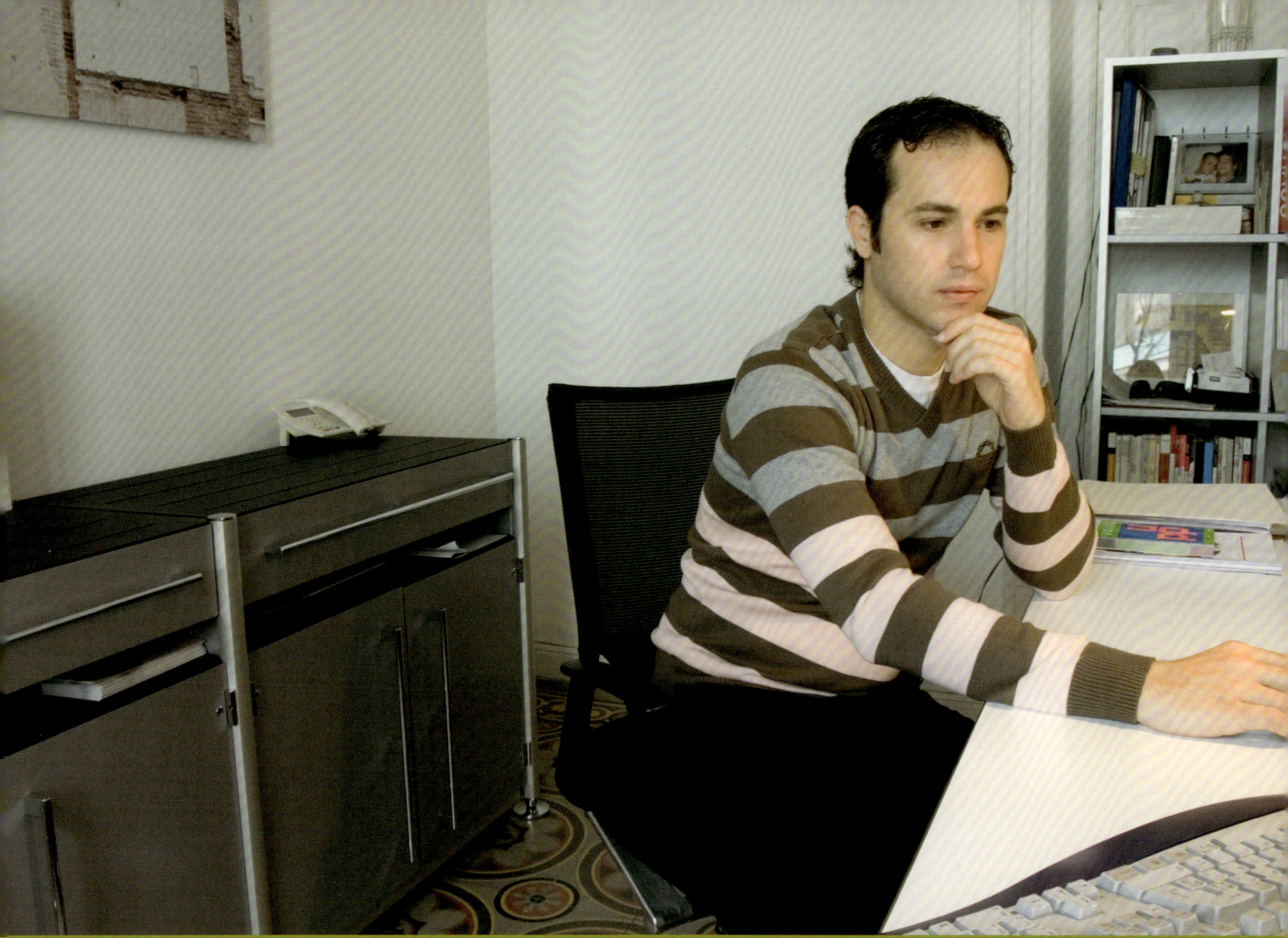

Jose Manuel Fernandez Rodriguez
"… La seguridad de un buen asiento, la garantía de un buen diseño …"

"… The security of a good chair – the guarantee of a good design …"

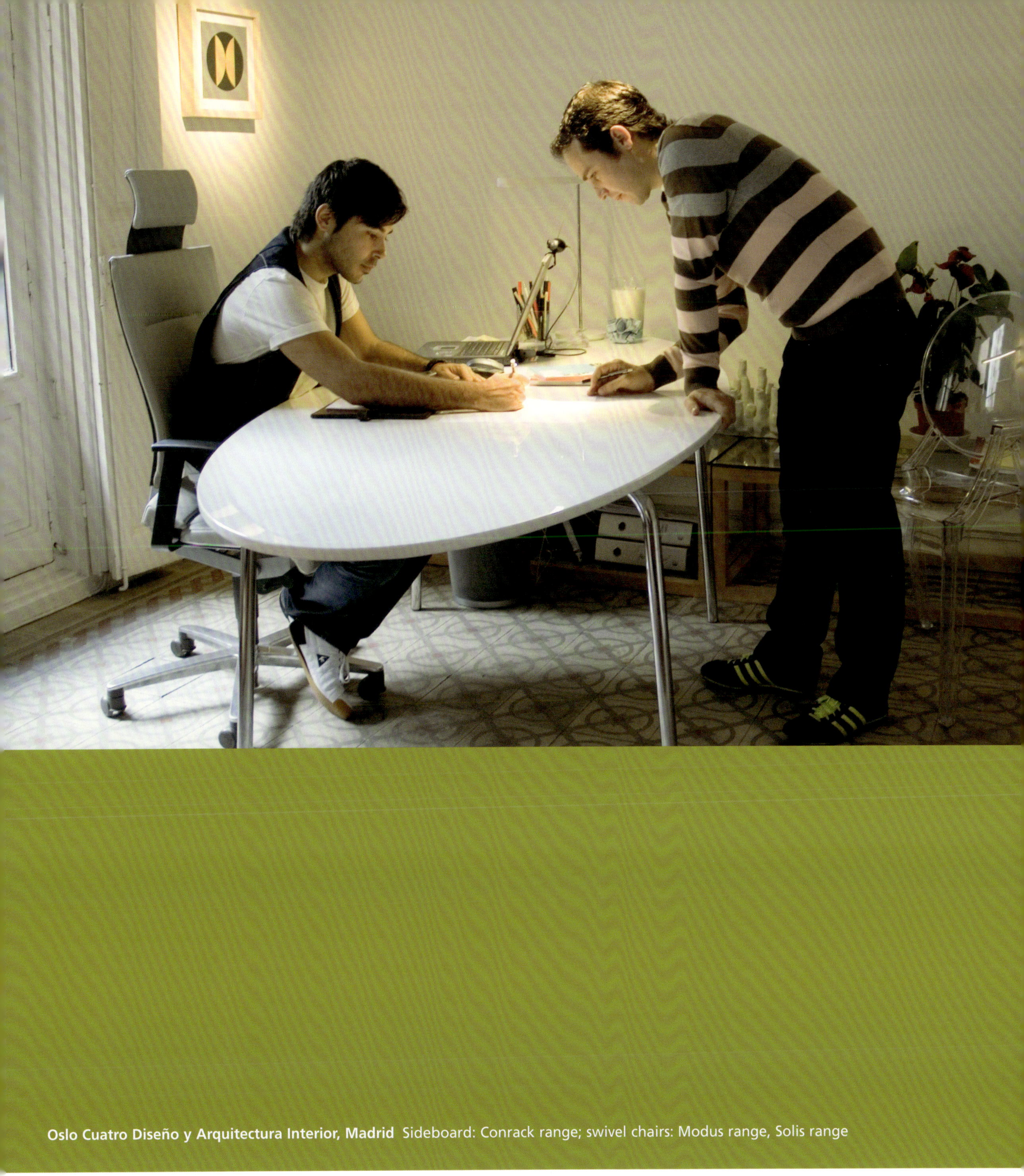

Oslo Cuatro Diseño y Arquitectura Interior, Madrid Sideboard: Conrack range; swivel chairs: Modus range, Solis range

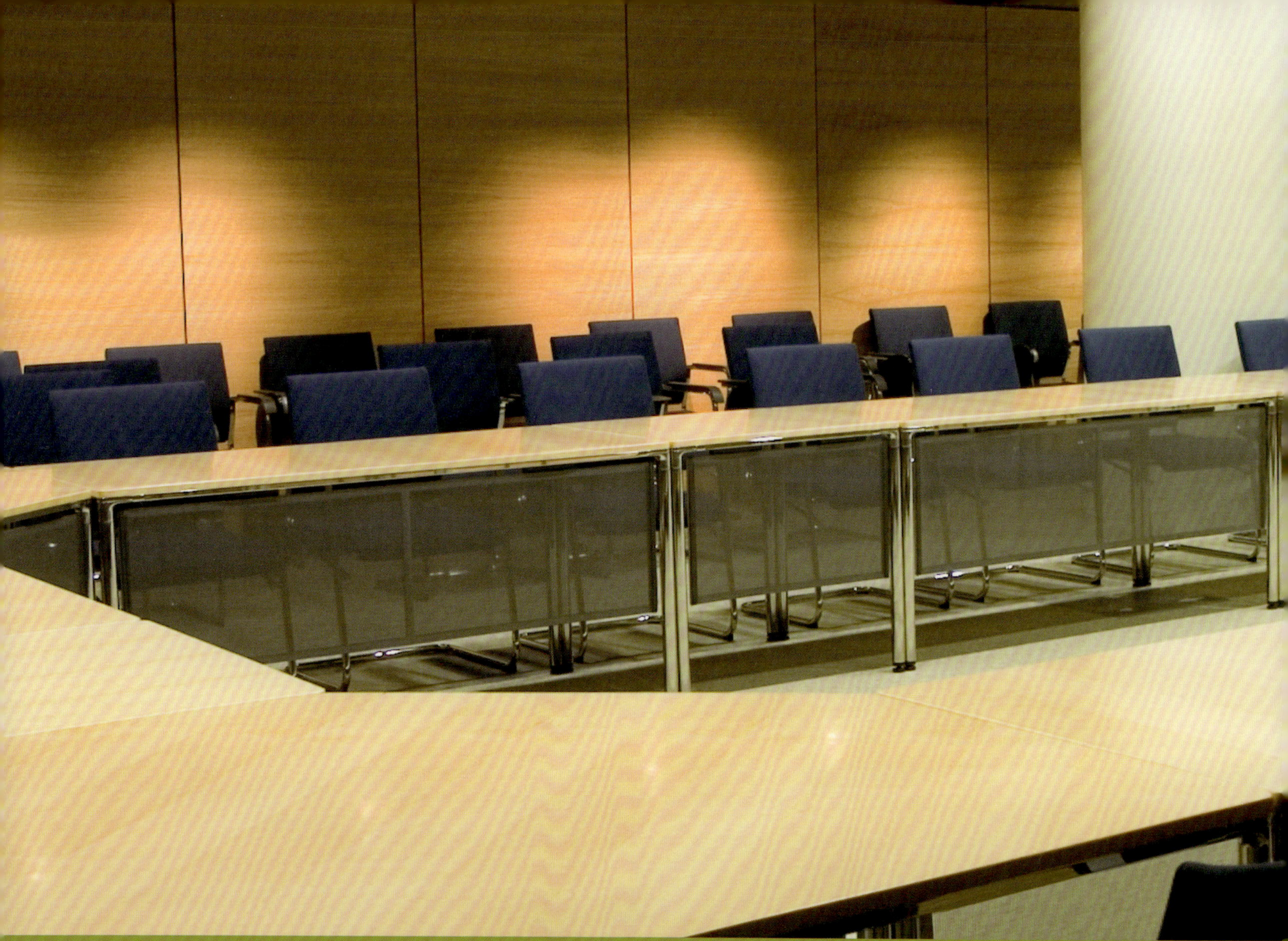

Luis Tortosa
"Primero forma, primero función. Esta principal característica de los muebles diseñados por Wilkhahn fue la razón para su elección como suministrador del mobiliario para nuestras instalaciones."

"First form, first function. This was the principal reason for choosing Wilkhahn as the main supplier of our equipment."

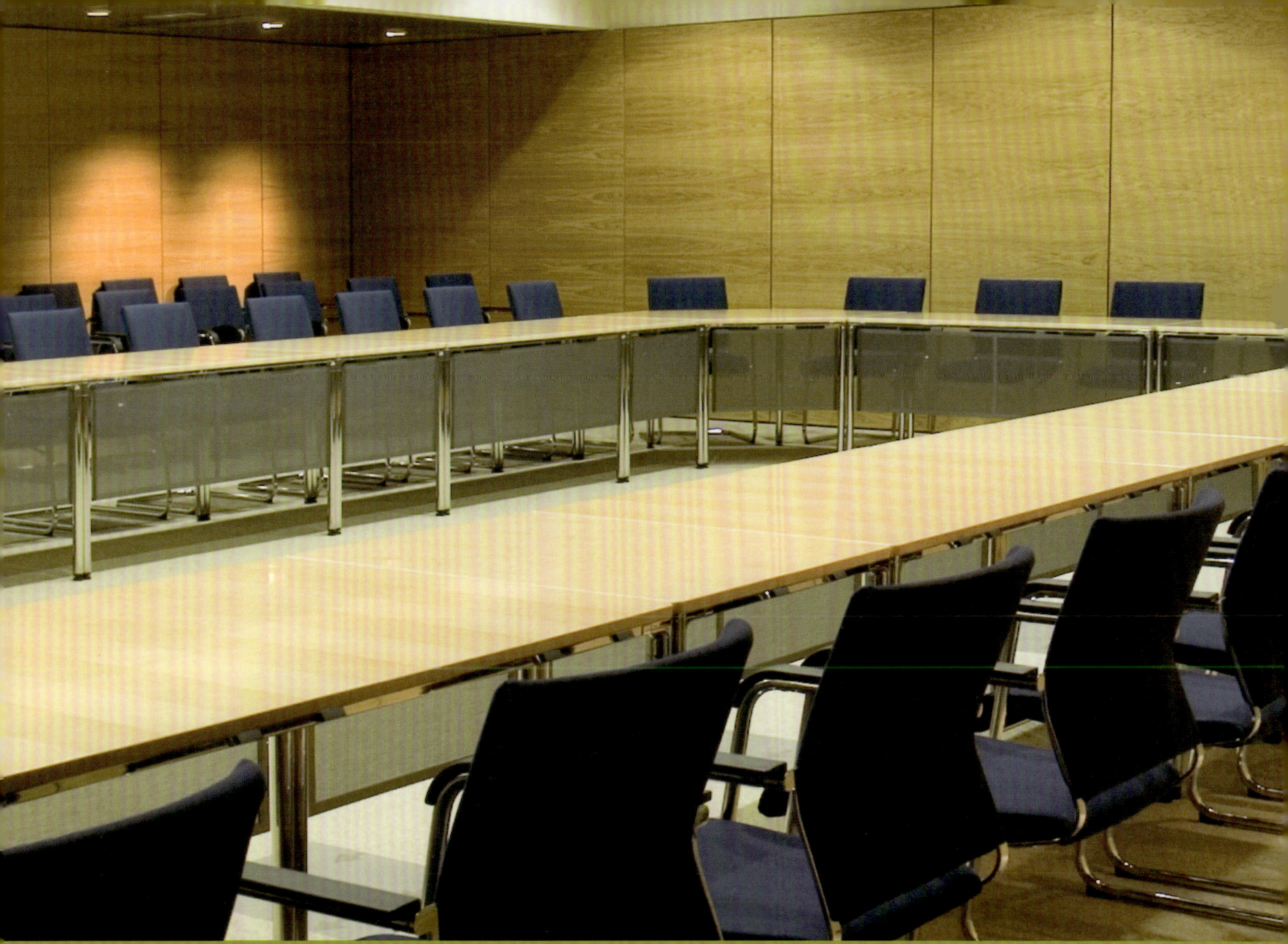

Banco de España, Madrid Table system: Contas range; cantilever chairs: Sito range

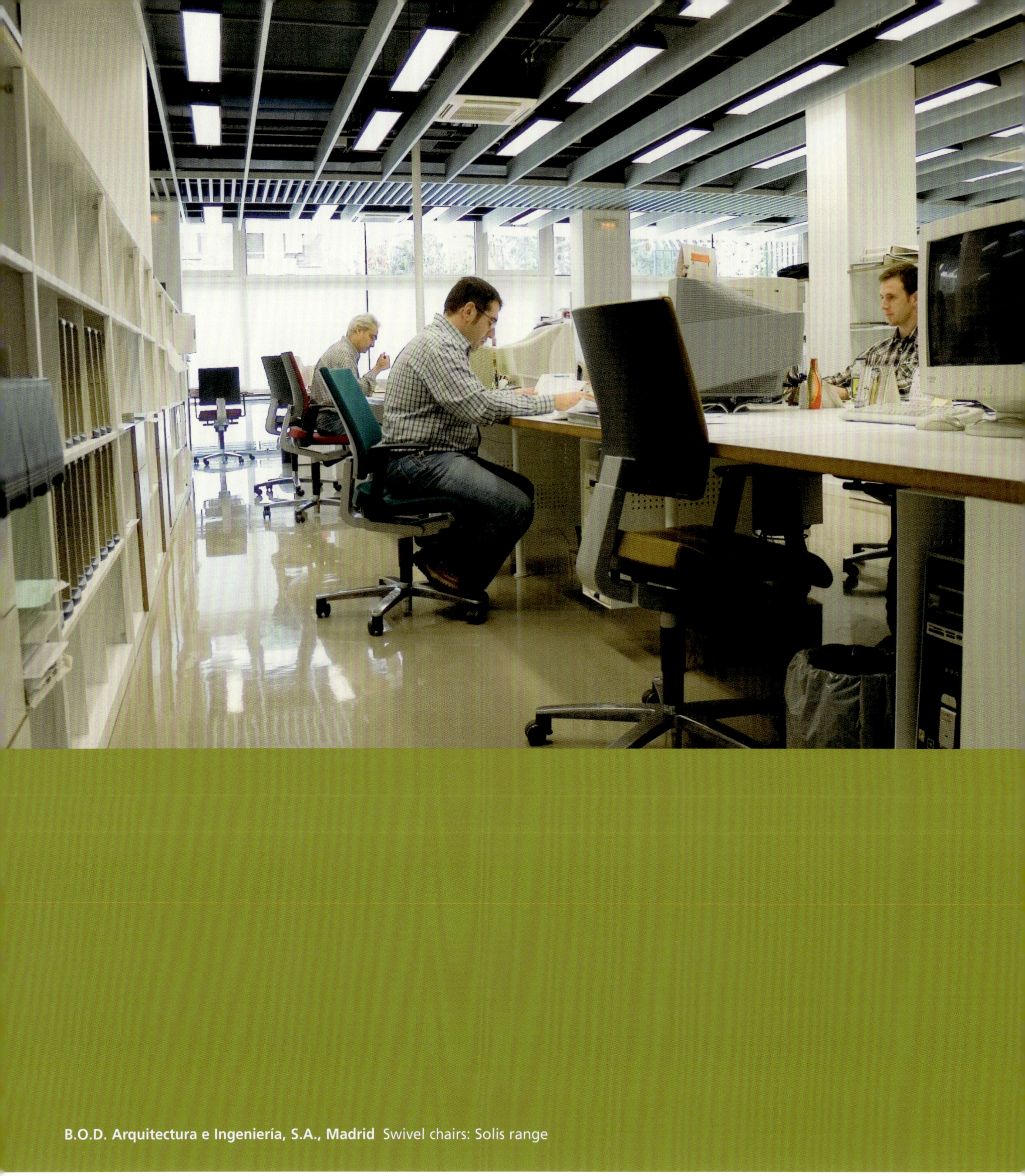

B.O.D. Arquitectura e Ingeniería, S.A., Madrid Swivel chairs: Solis range

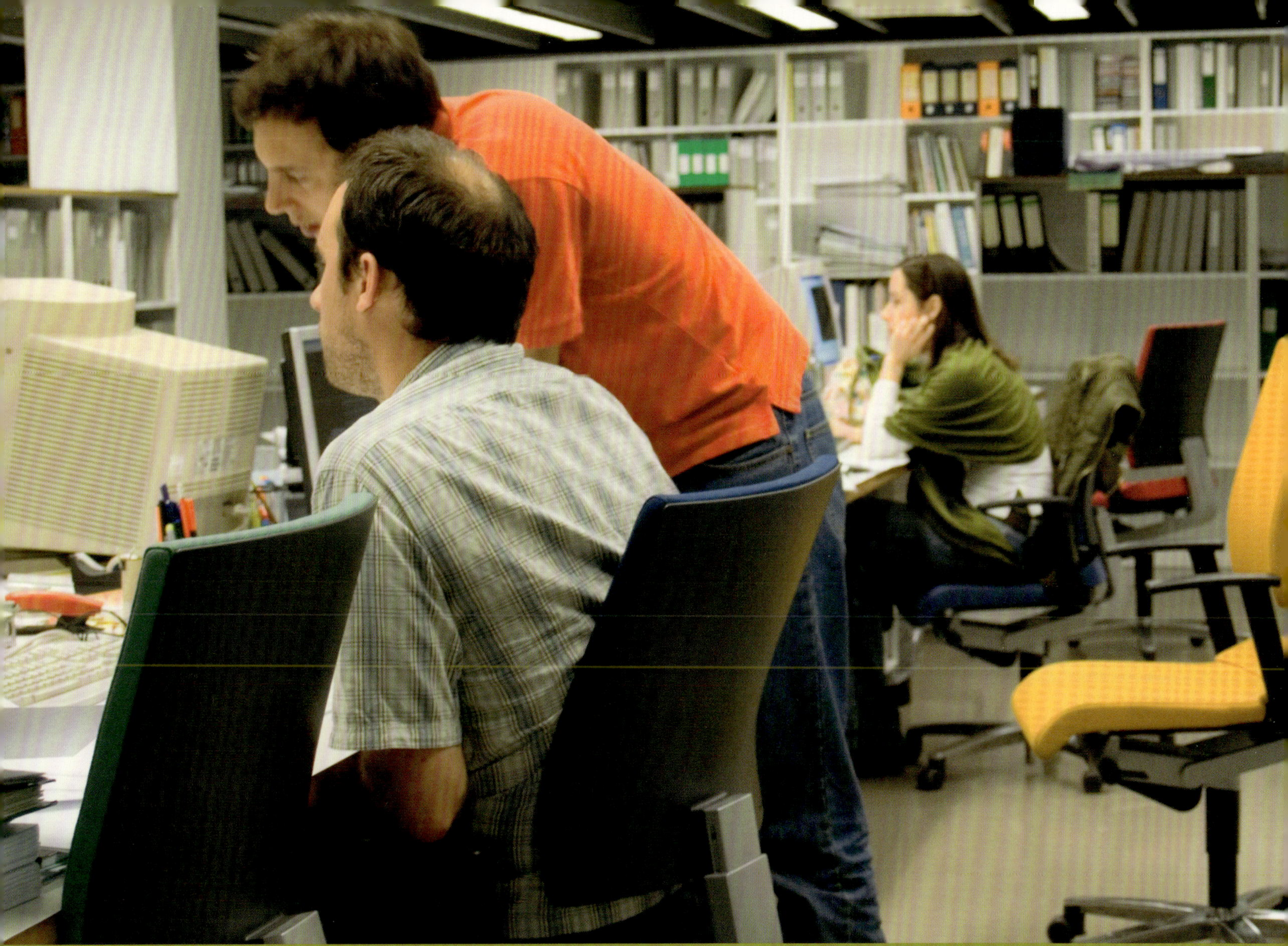

Gorka Ascasíbar Allona

" El mobiliario de una oficina es tan importante para el proyecto de arquitectura como la distribución, los acabados o la iluminación. Las personas se relacionan con el edificio a través de las sillas y las mesas que utilizan diariamente. El mobiliario se convierte entonces en el elemento de contacto entre las personas y la arquitectura. Wilkhahn comparte esta filosofía en cada mueble que diseña."

"Office furniture is as important for an architectural project as its layout, finishes or lighting. The persons are 'connected' with their building through their chairs and tables which they use daily. Therefore furniture becomes a link between the staff and architecture. Wilkhahn shares this philosophy every time it designs furniture."

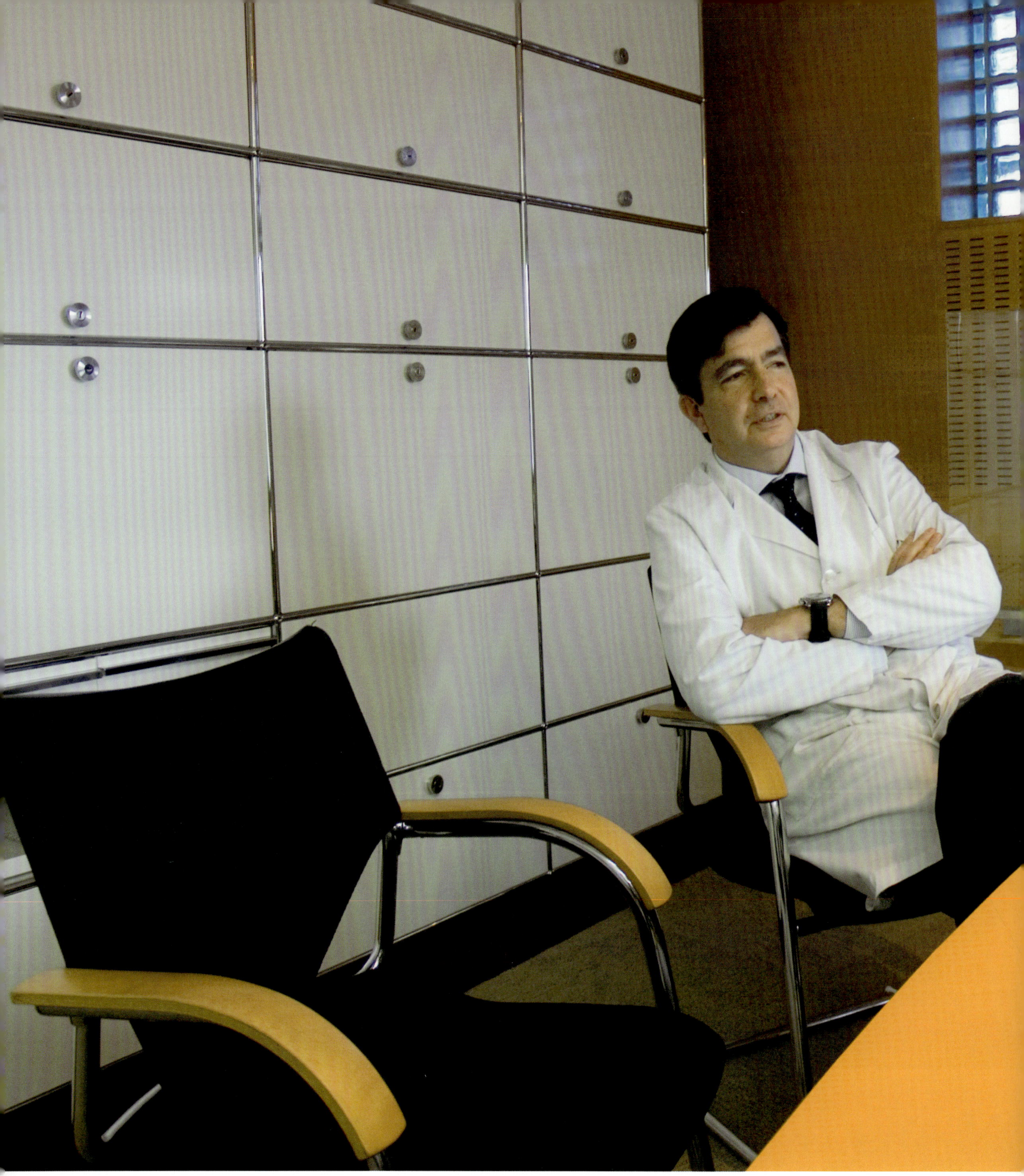

Carlos Lamela, Architect, Madrid
"Para un arquitecto Wilkhahn representa la perfección en la simbiosis del hombre y sus herramientas ergonómicas para el trabajo de oficina: sillas, mesas, escritorios. Este exacto acoplamiento es fruto de un siglo de investigación, calidad en la ejecución y practicidad. Todo ello sin olvidar el esfuerzo que Wilkhahn está haciendo en los últimos años en que sus productos se elaboren con las técnicas y procesos más respetuosos posibles con nuestro entorno."

"For an architect, Wilkhahn is the epitome of the perfect symbiosis between humans and their ergonomic tools for the office: the chairs, tables and desks. This perfect combination is the result of 100 years of research and quality in the realization of design. Not to forget the efforts Wilkhahn has made in the last years to bring the development of its products into line with the environment through technology and processes in the most responsible way possible."

Cantilever chairs: Modus range

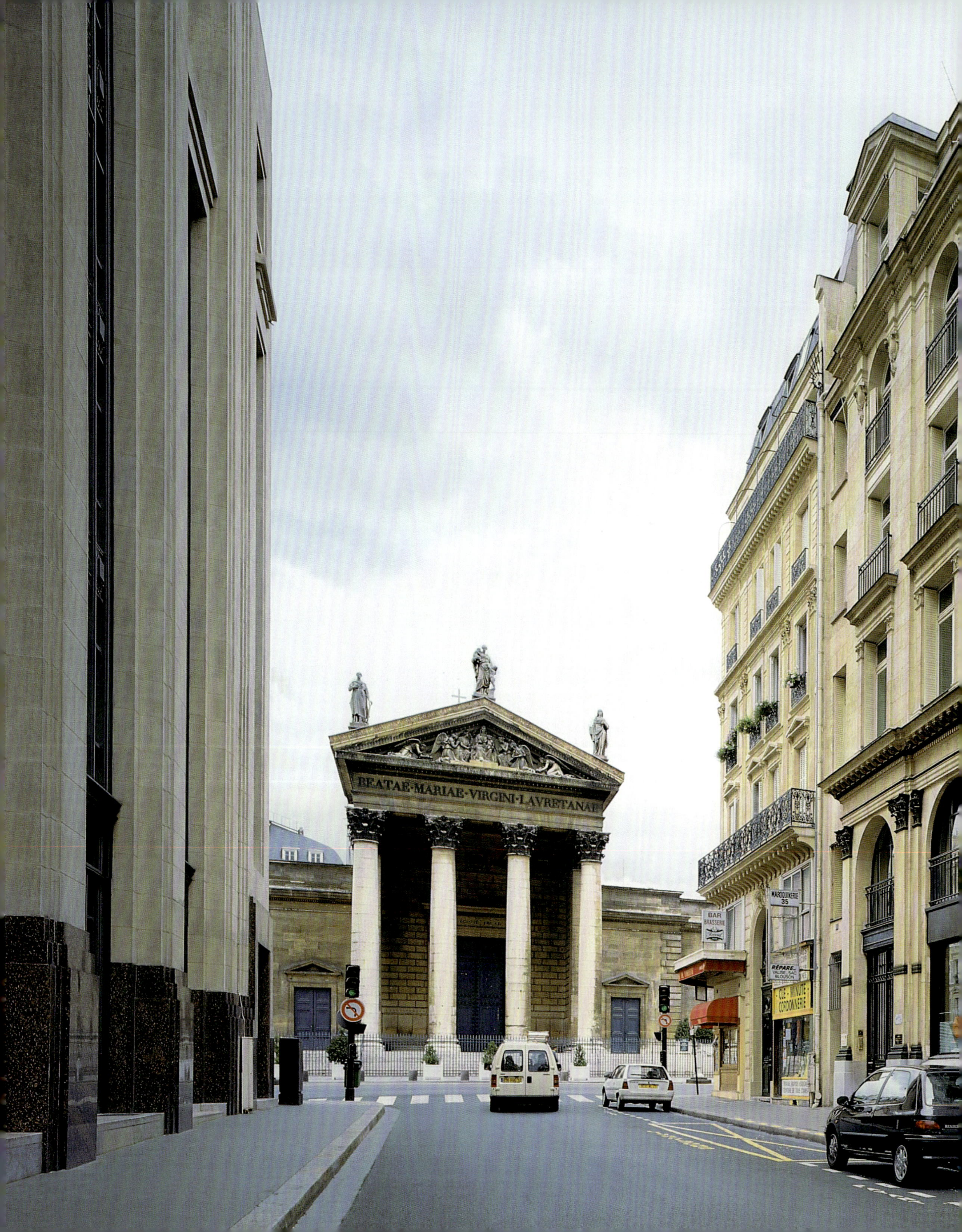
BEATAE·MARIAE·VIRGINI·LAVRETANAE
BAR BRASSERIE
MAROQUINERIE 35
RÉPARE VALISE. SAC BLOUSON
CLE - MINUTE CORDONNERIE

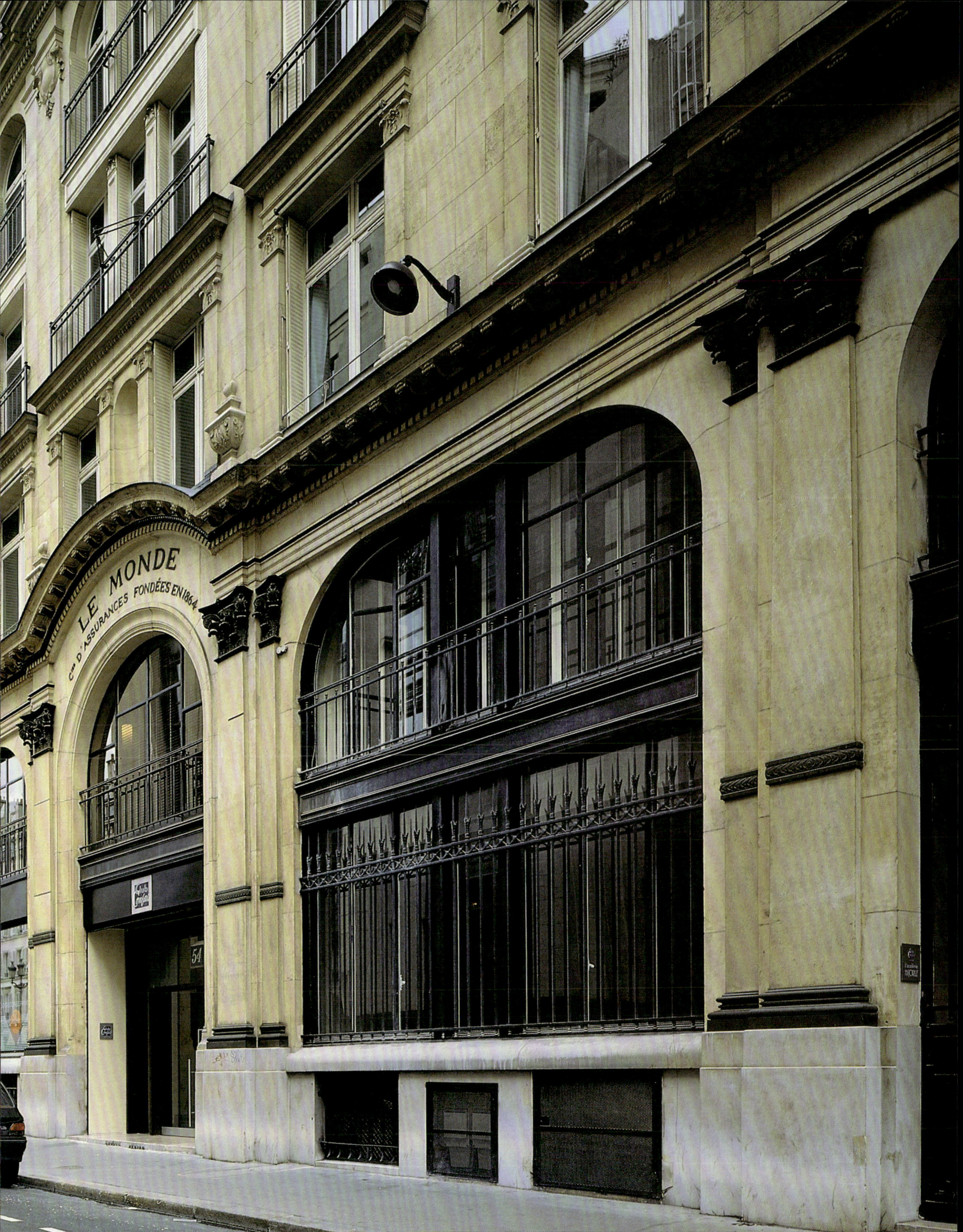
LE MONDE
Cie D'ASSURANCES FONDÉES EN 1864
54

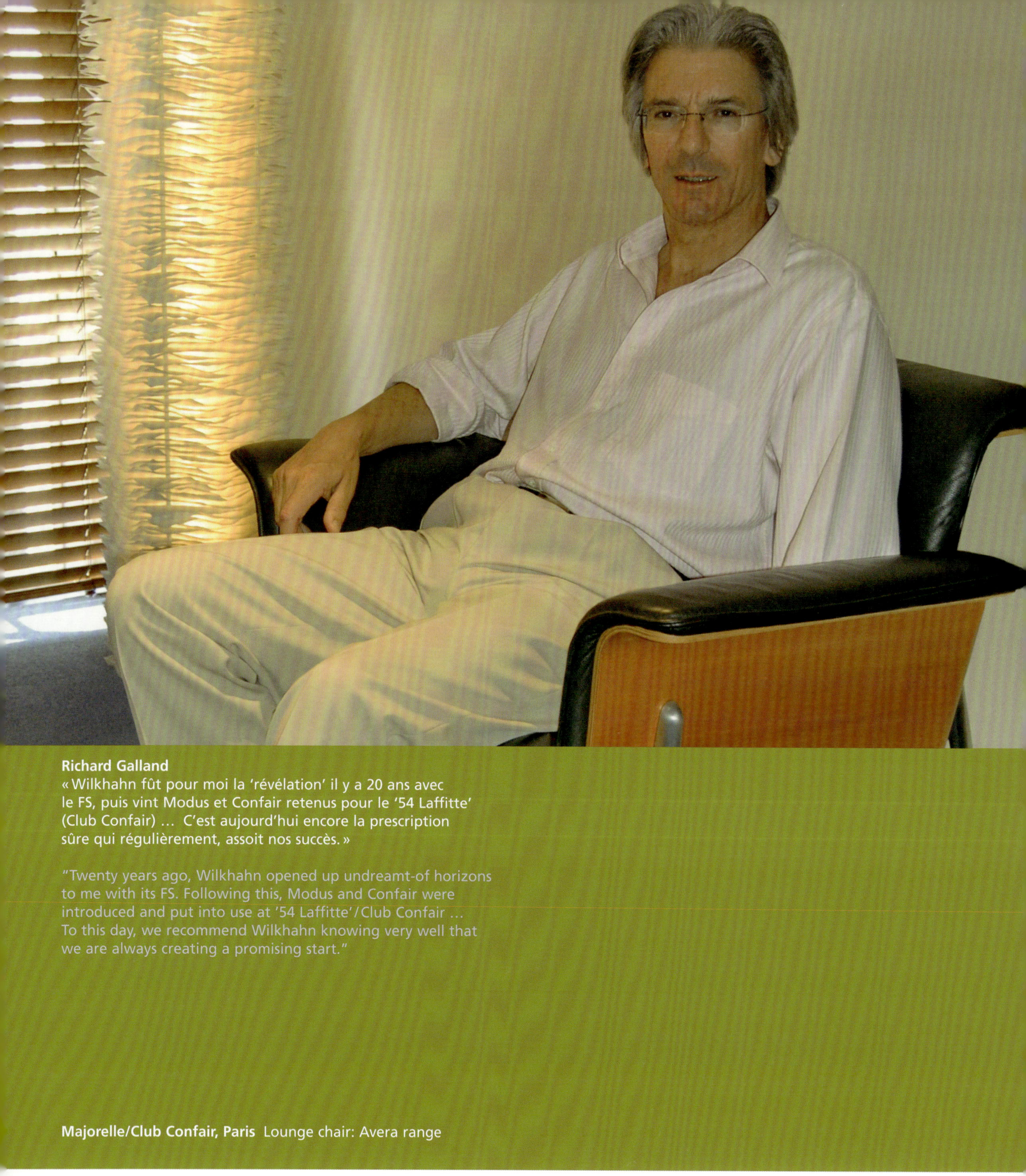

Richard Galland
« Wilkhahn fût pour moi la 'révélation' il y a 20 ans avec le FS, puis vint Modus et Confair retenus pour le '54 Laffitte' (Club Confair) … C'est aujourd'hui encore la prescription sûre qui régulièrement, assoit nos succès. »

"Twenty years ago, Wilkhahn opened up undreamt-of horizons to me with its FS. Following this, Modus and Confair were introduced and put into use at '54 Laffitte'/Club Confair … To this day, we recommend Wilkhahn knowing very well that we are always creating a promising start."

Majorelle/Club Confair, Paris Lounge chair: Avera range

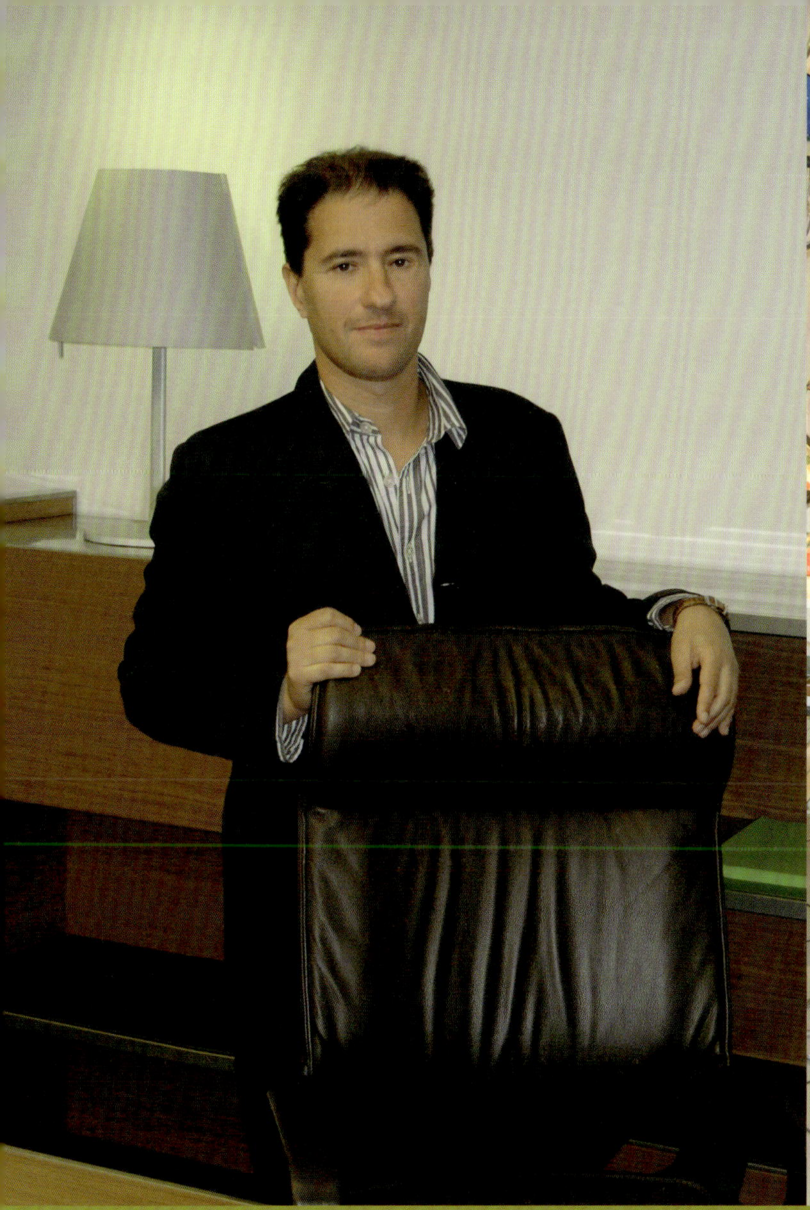

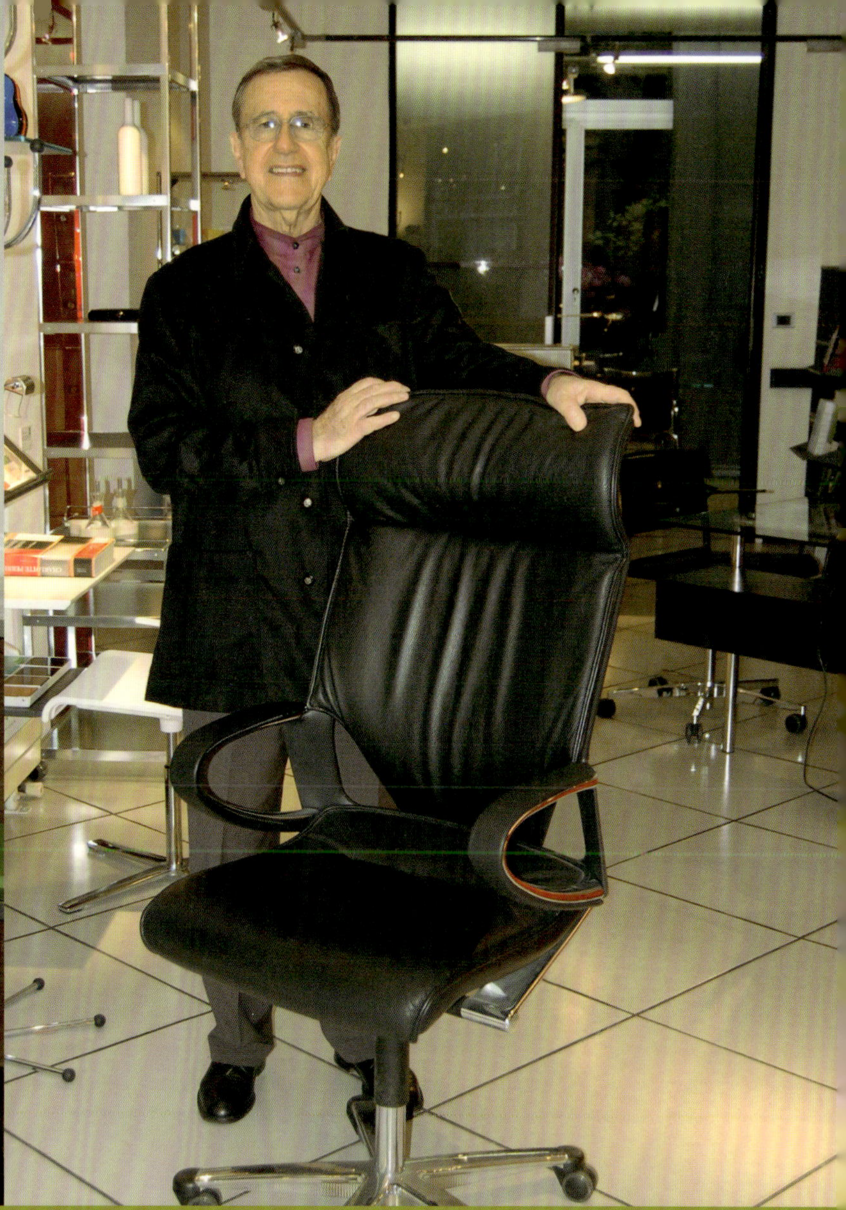

Paul Silvera
« Je suis trop jeune distributeur Wilkhahn (4 ans) mais je suis attaché au personnel de Wilkhahn et je trouve que Wilkhahn c'est un design fonctionnel et sans concession. »

"I have been in the business as a dealer for only four years now so I am not completely in the picture as far as the time-related perspective goes. I have a very good relationship to the Wilkhahn team though and I am highly impressed with the functional, uncompromising design."

SILVERA, Paris Swivel chair: Modus Executive range

Pierre Perrigault
« Ce voyage de Wilkhahn dans un siècle de création a été le catalyseur de belles aventures humaines et professionnelles auxquelles nous avons eu l'honneur de participer. »

"Wilkhahn's journey of development in the last 100 years of creative design became a catalytic converter for many exciting and wonderful adventures in a human as well as a business sense. We are honoured to have taken part in it."

MFI, Paris Swivel chair: Modus Executive range

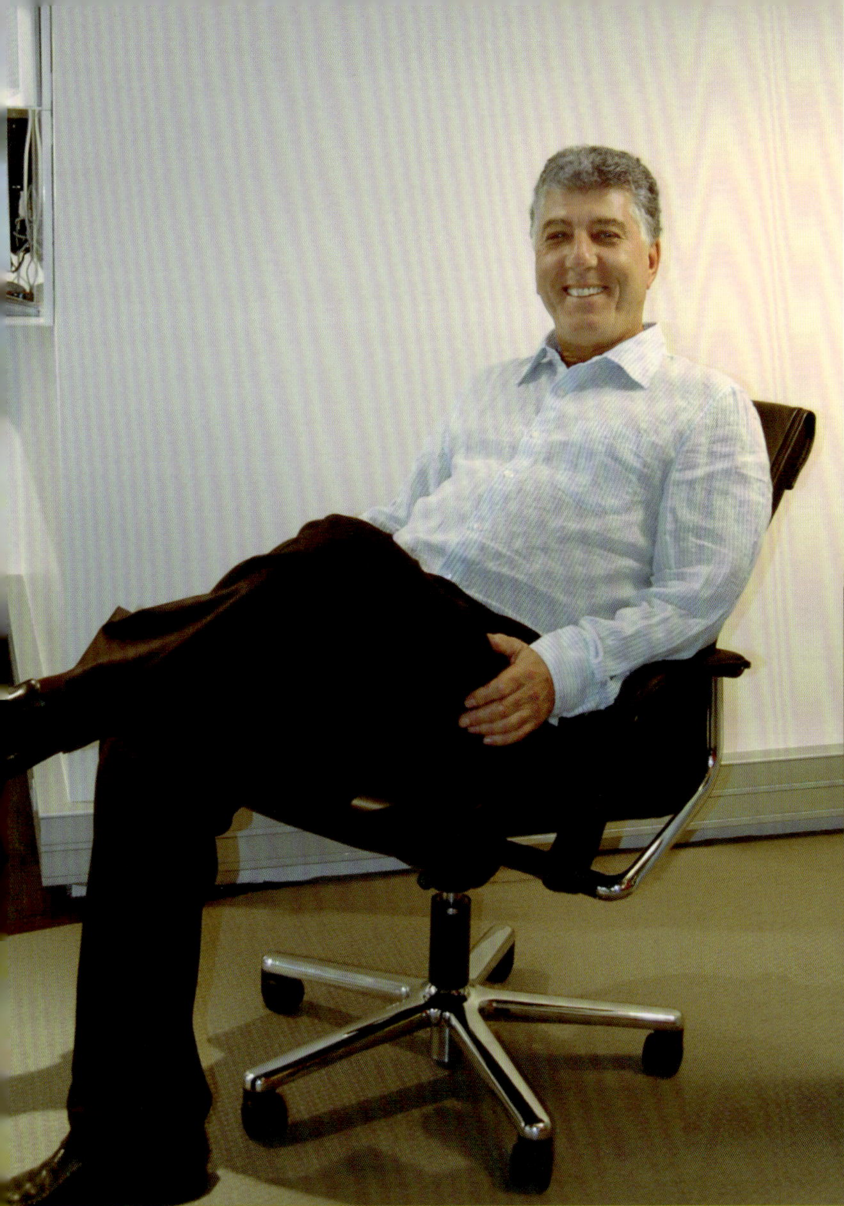

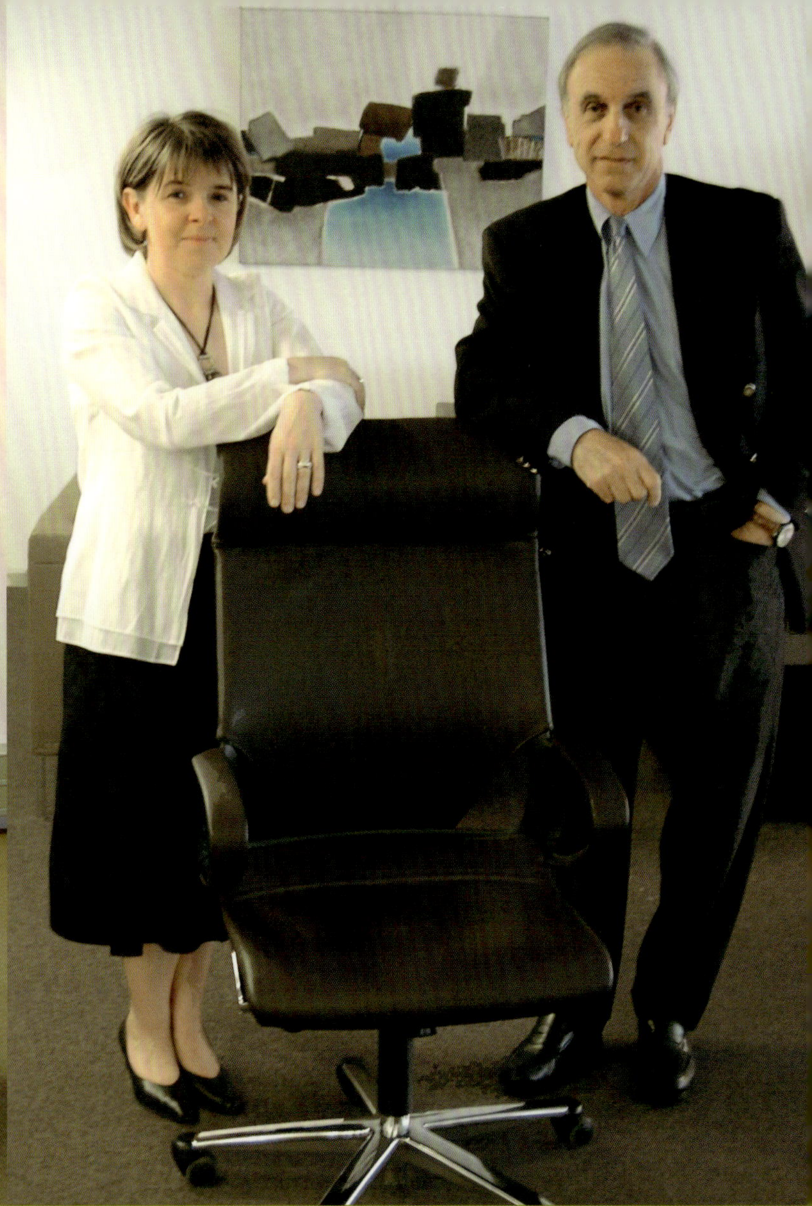

Bernd Tordjman
« Créativité, intemporalité, qualité les trois atouts qui font de Wilkhahn l'indispensable partenaire du présent et du futur. »

"Creativity, timelessness, quality – these three characteristics make Wilkhahn an indispensable partner – today and tomorrow."

Catherine Guyot, Georges Schulmann
« Forum est heureux et fier d'être associé aux 100 ans de Wilkhahn, partenaire essentiel du passé, totalement tourné vers l'avenir ! »

"It is a pleasure and an honour for Forum to participate in the 100th anniversary of Wilkhahn, a partner that is both very close as well as uncompromising and future-oriented."

Siltec, Paris Swivel chair: FS range

Forum Diffusion, Paris Swivel chair: Modus Medium range

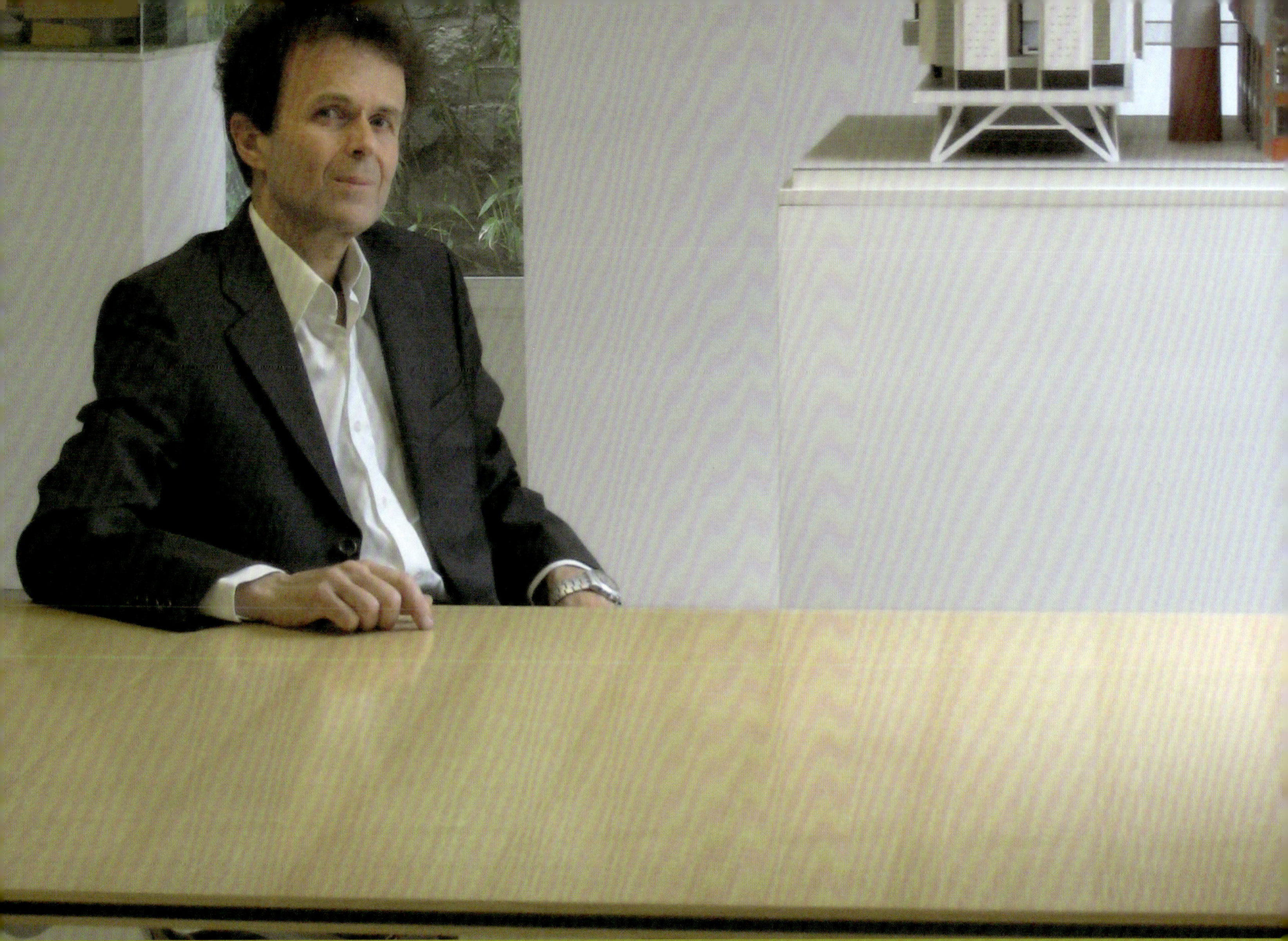

Frédéric Borel, Architect
« Les meubles Wilkhahn traversent le temps en mariant ergonomie, confort et élégance.
Leur design épuré, sans surenchère esthétique, allie pureté des lignes, équilibre et fiabilité. »

"Wilkhahn furniture will survive time by combining ergonomics, elegance and comfort.
Its minimalist, understated design is always clearly balanced and consistent."

Agence Borel, Paris Swivel chair: FS range; table: Confair folding table

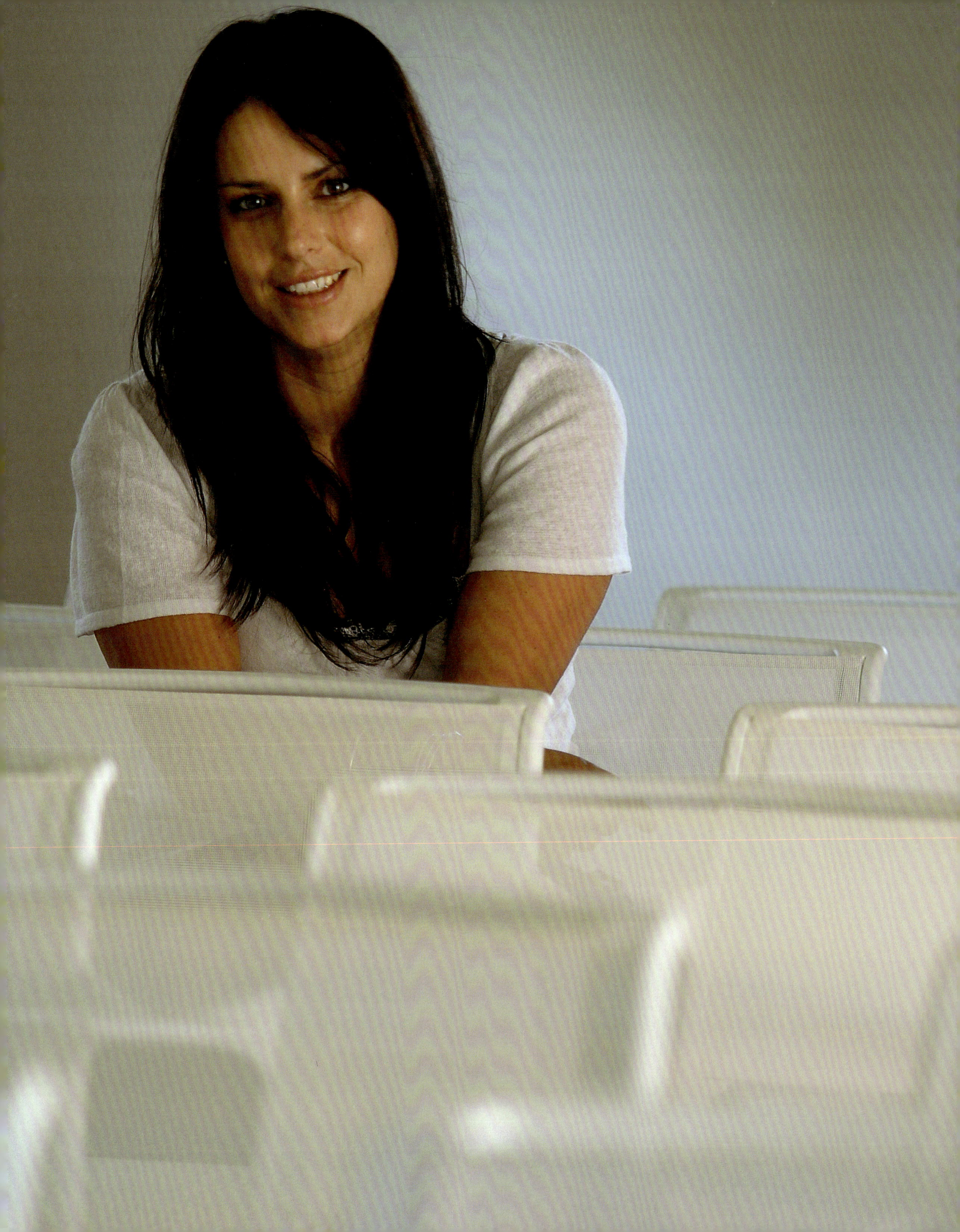

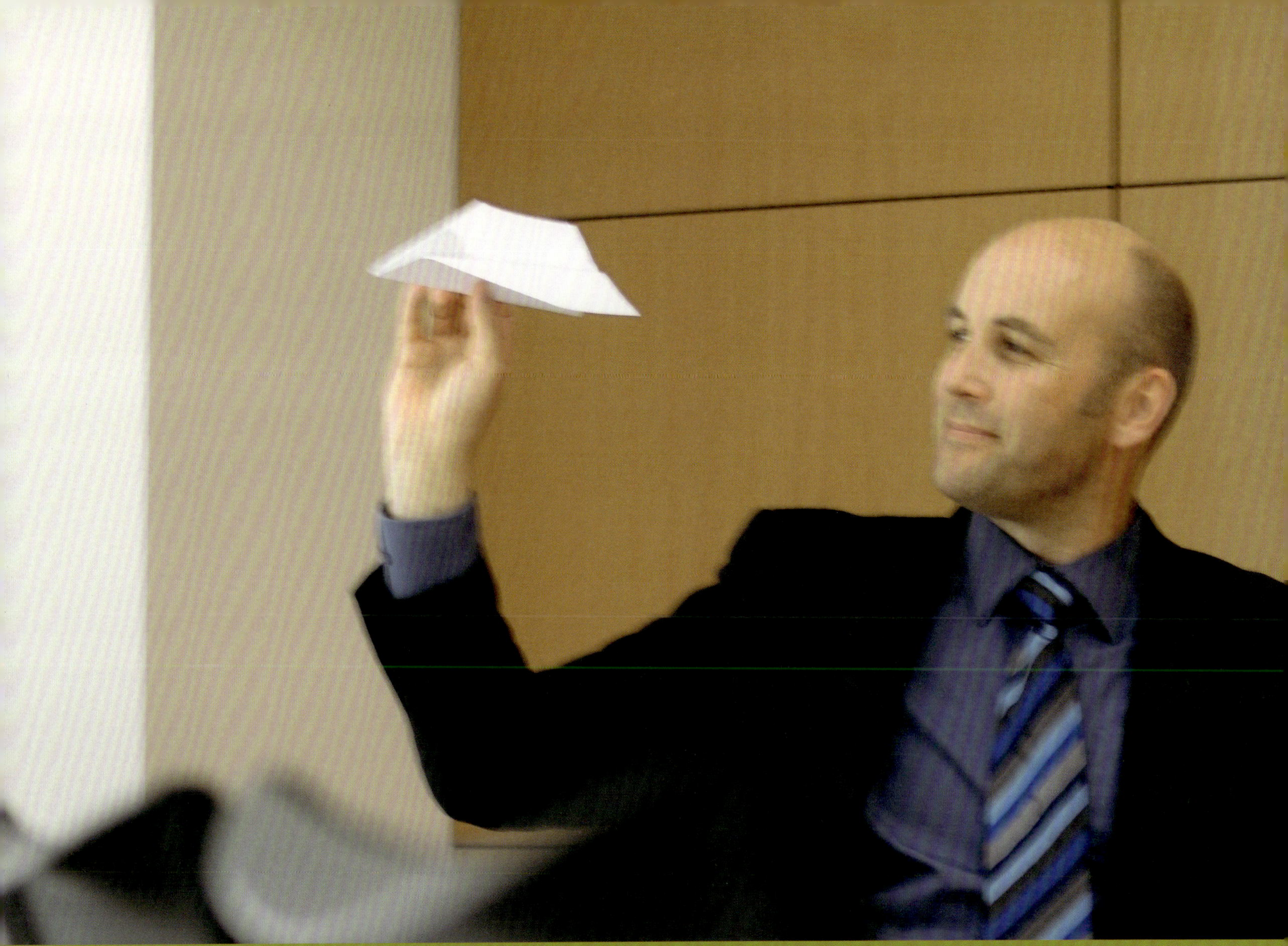

Karl Chisman, Facility Manager
"Wilkhahn design means for me the creation of a comfortable working environment encouraging both productivity and efficiency. The most important aspect of Wilkhahn products is the combination of attractive designs manufactured in qualitative and lasting materials."

KPMG, Luxembourg
Cantilever chairs: Modus range

Sofie Waldeyer, Managing Director
„Mooi, ik hou van minimalistisch design, dat er eenvoudig uitziet maar zeer complex is. Het is gedetailleerd en ergonomisch, meer dan alleen maar een stoel. En Functionaliteit! Heel veel verschillende vormen die zeer éénvoudig aan te passen zijn."

"I love the minimalist design, which seems simple but is complex. It is detailed and ergonomic. It is more than a chair. And functionality! There are so many different shapes and they are very easy to adapt."

Hof ter Musschelen, Conference Center, Brussels
Multi-purpose chair: Aline range

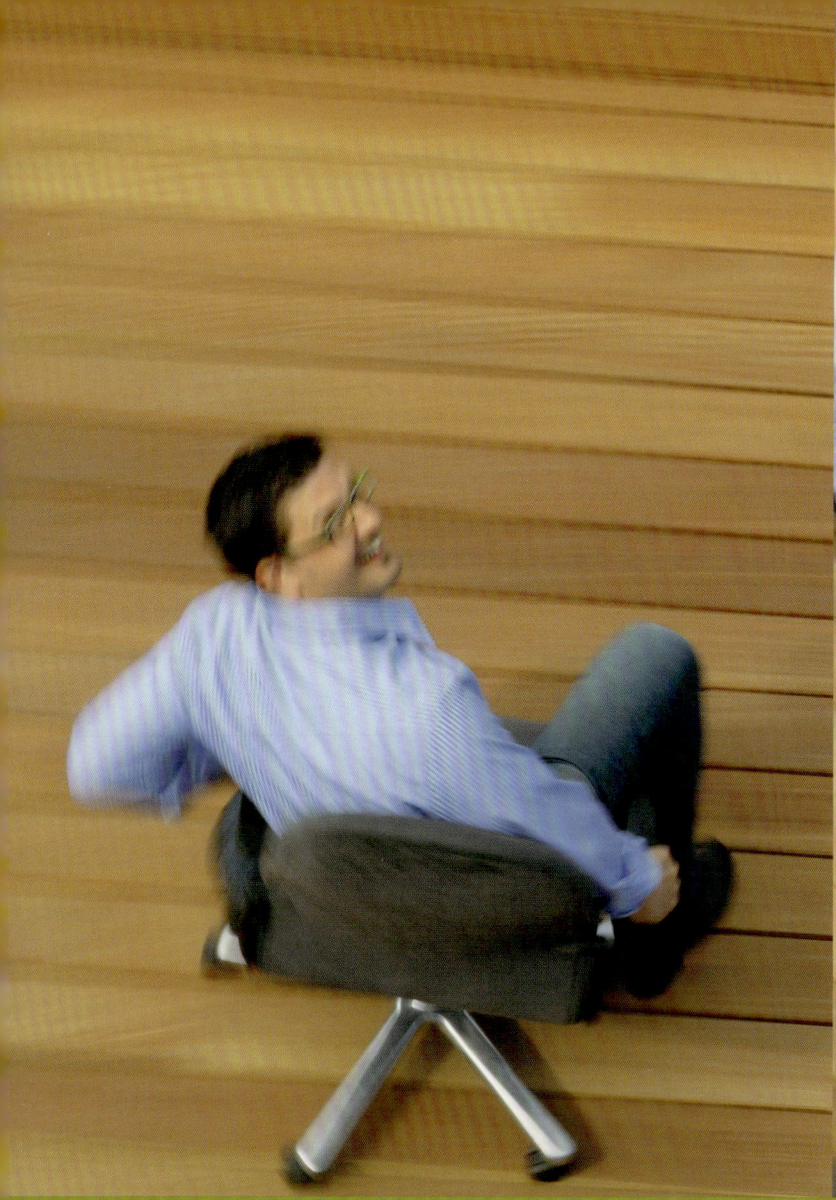

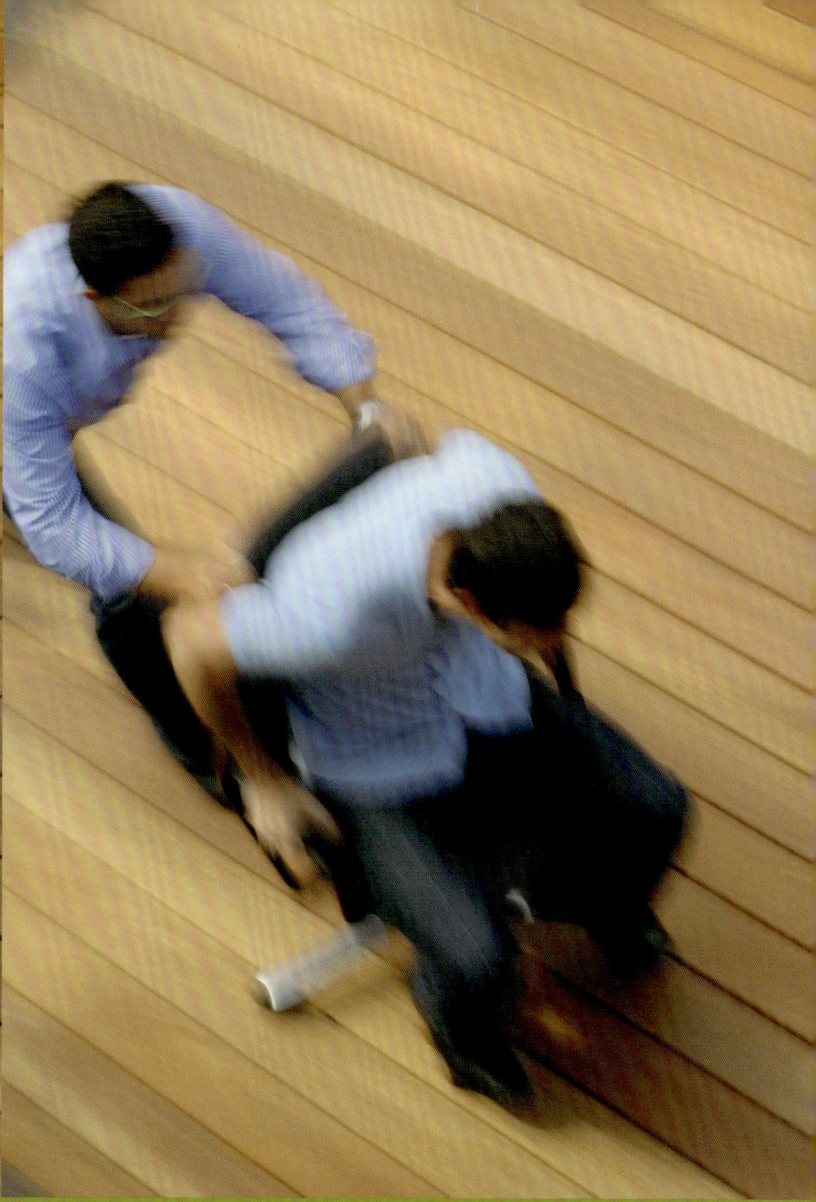

Nico van Daele, Head of Facilities
„Kwaliteit en design staan op de voorgrond"

"Quality and design are in the foreground."

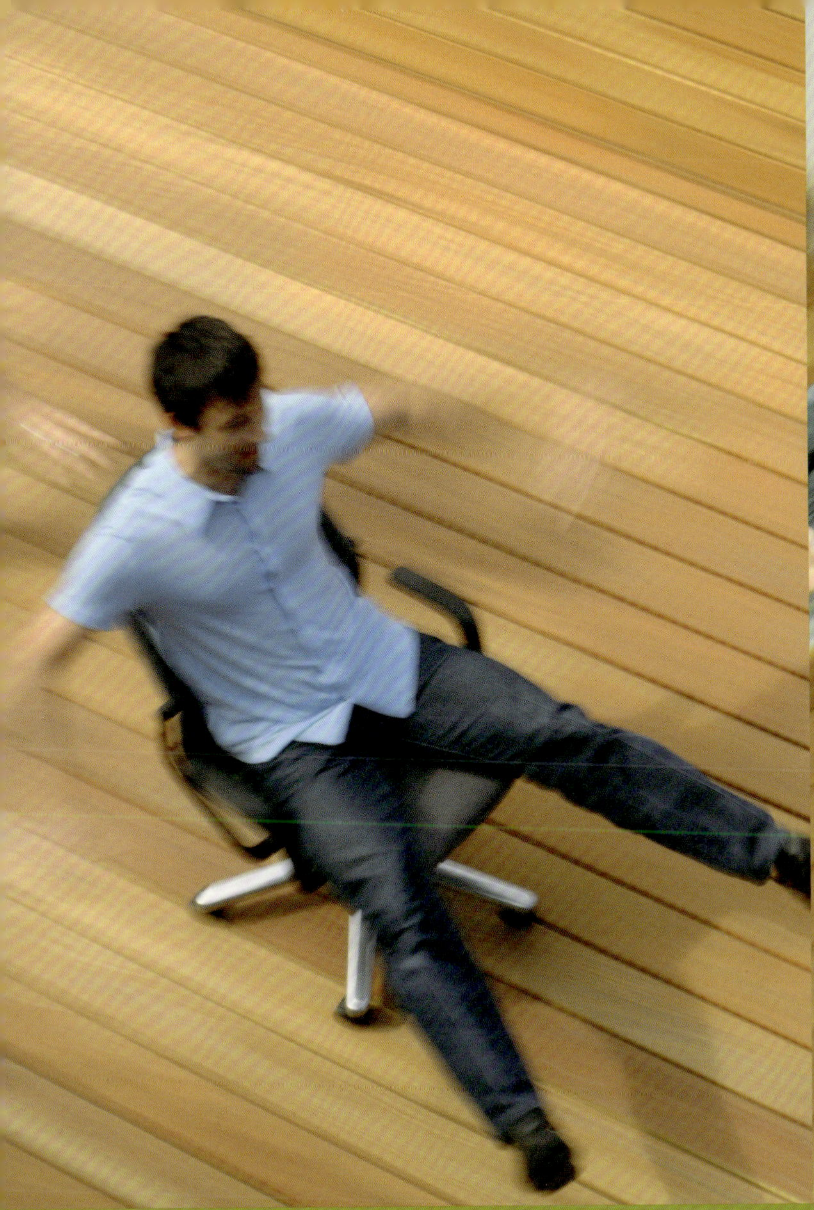

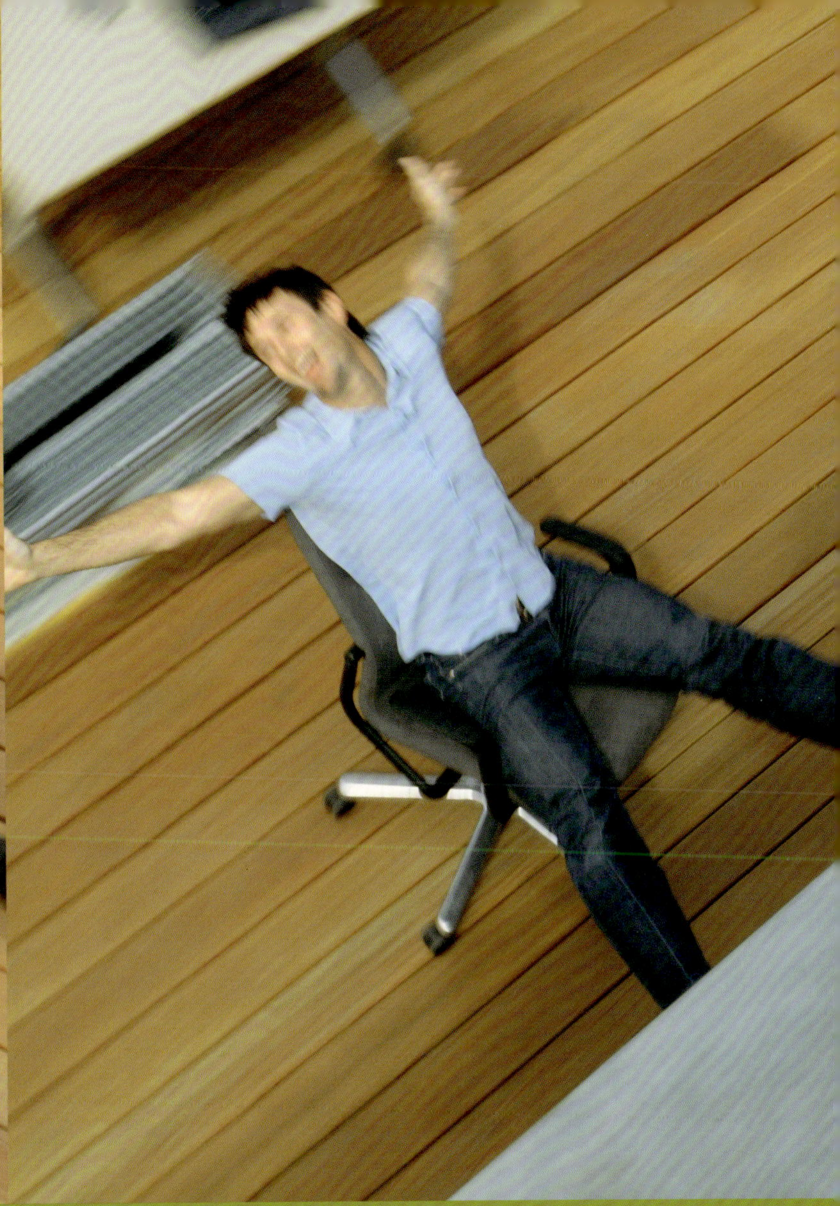

Michael Laeremans, Head Innovations
„Wilkhahn design? Het past! Het is functioneel!"

"Wilkhahn design? It fits! It's functional!"

De Lijn, Mechelen Swivel chairs: FS range

Marcel Eeckhout
„Minimalistisch design. Transparantie. Kwaliteit."

"Minimalist design. Transparency. Quality."

MCS – Enterprise facility management software, Antwerp Swivel chair: Modus Executive range

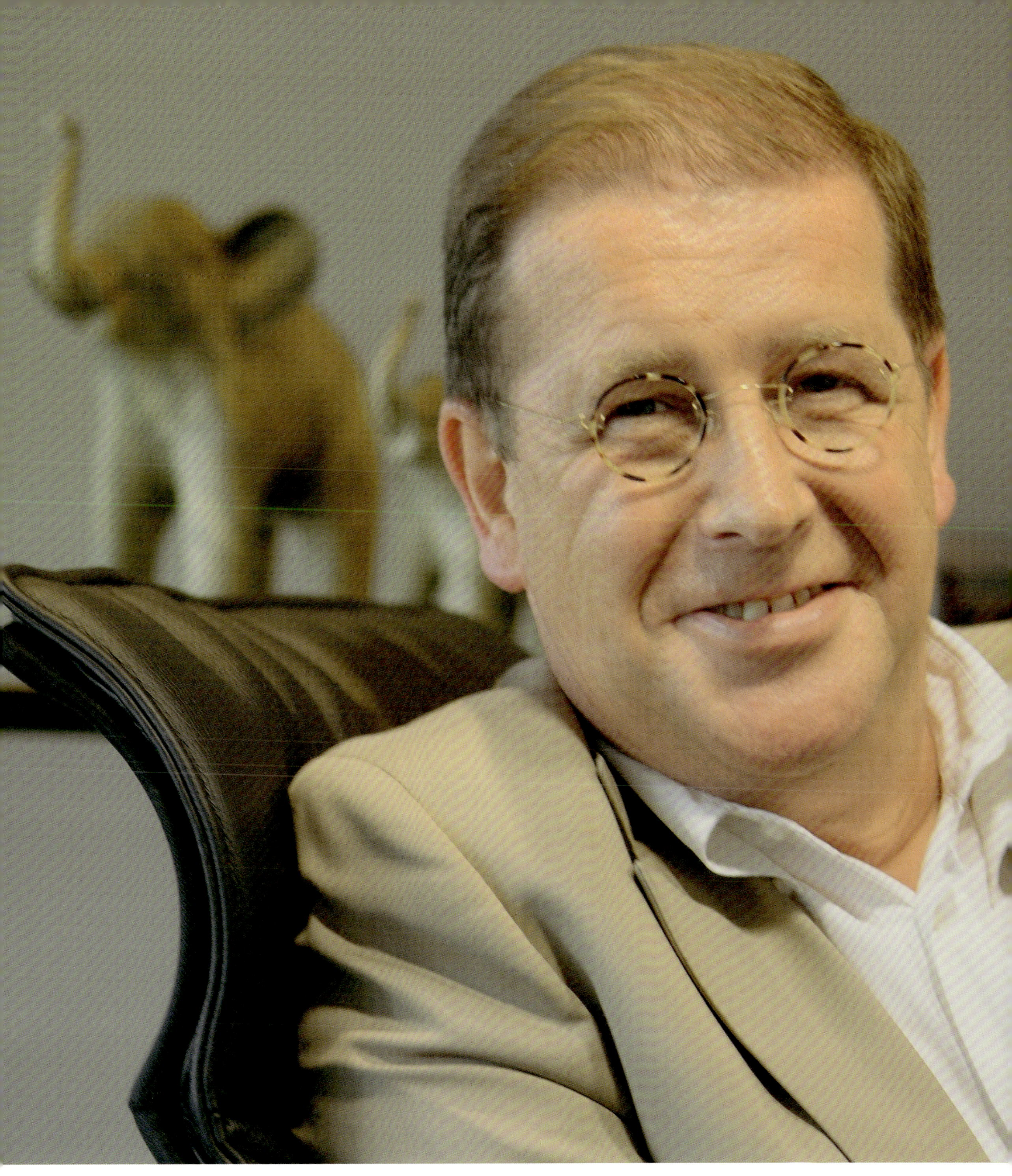

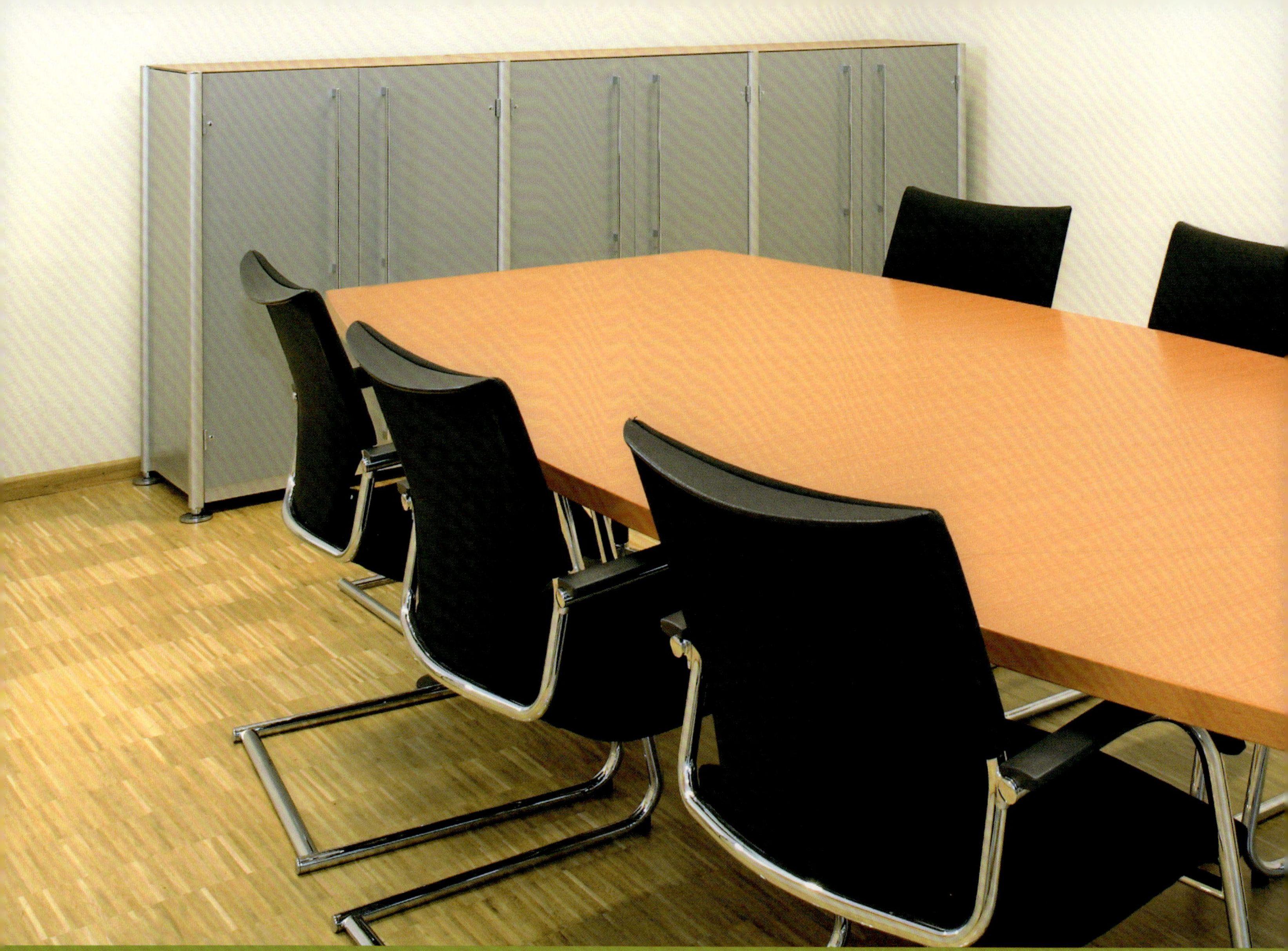

Mirjam Spooren, Facility Manager
„Meer dan 1300 kast-eenheden in verschillende configuraties werden in minder dan 7 weken geproduceerd voor ons hoofdkantoor in Amsterdam. Geen enkele unit moesten we weigeren wegens schade. Speciale hoogwaardige board-room tafels voorzien van de nodige electrificatie en aluminium inleg werden in dezelfde periode gemaakt in een zeer speciale fineersoort (himalaya teak). Deze tafel wordt maandelijks omgebouwd wegens een andere toepassing van de ruimte. Op facilitair vlak bereikt Wilkhahn het hoogste niveau. Hoogwaardige stapelbare sledestoelen hebben hun plaats gevonden rond de tafel."

"More than 1,300 storage units in several different configurations were produced in less than seven weeks for our HQ in Amsterdam. Not a single unit had to be rejected due to damage. Special high-quality boardroom tables with the required electrical connections and aluminium inlays were produced in the same period in an unusual veneer (Himalaya teak). Each month, this table is re-configurated due to a different function of the room. From a facility point of view, Wilkhahn achieves the top level. High-standard stackable cantilever chairs have found their place around the table."

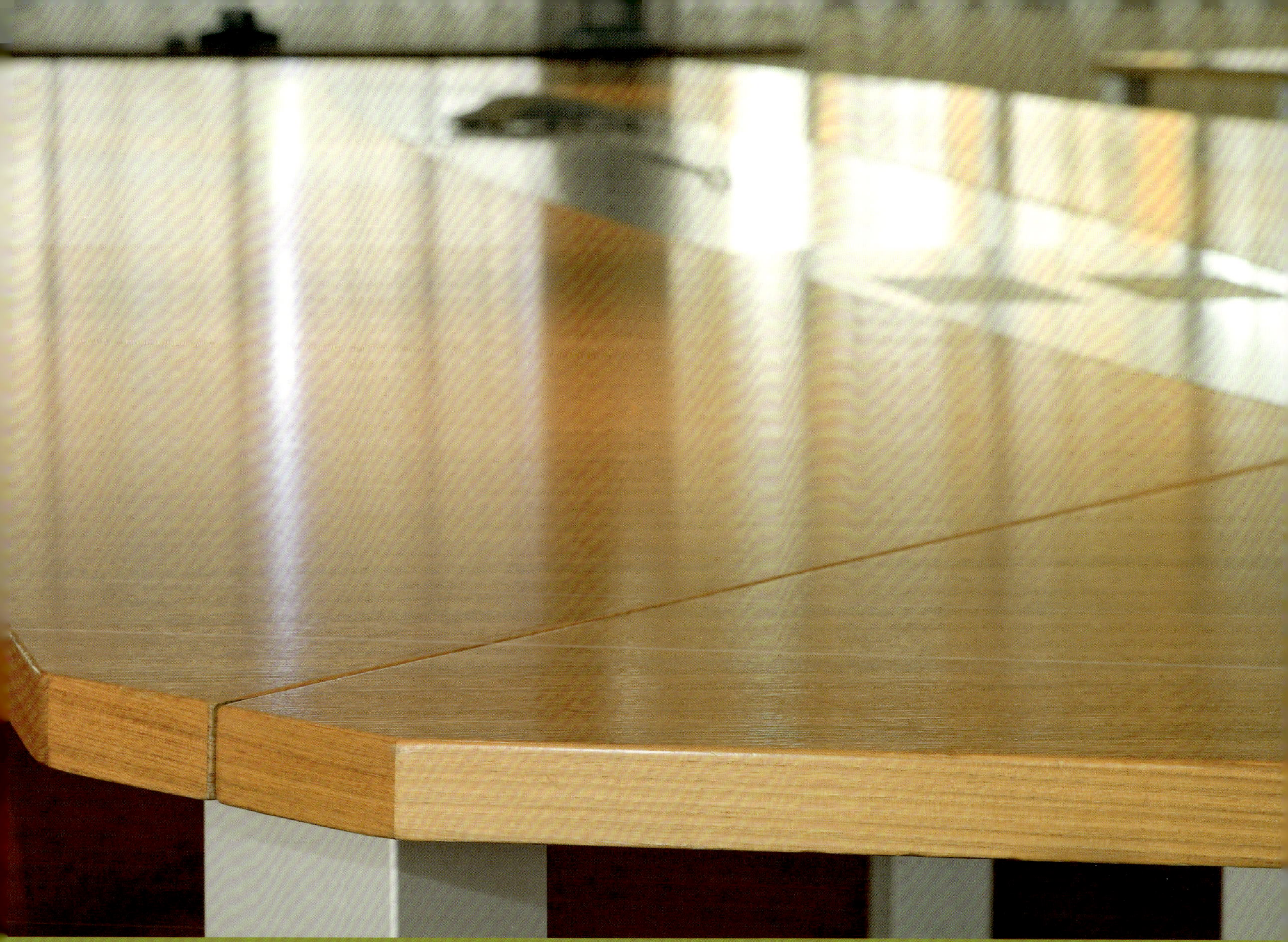

Loyens & Loeff, Rotterdam Table configuration: Logon range, Palette range; stackable cantilever chairs: Modus range

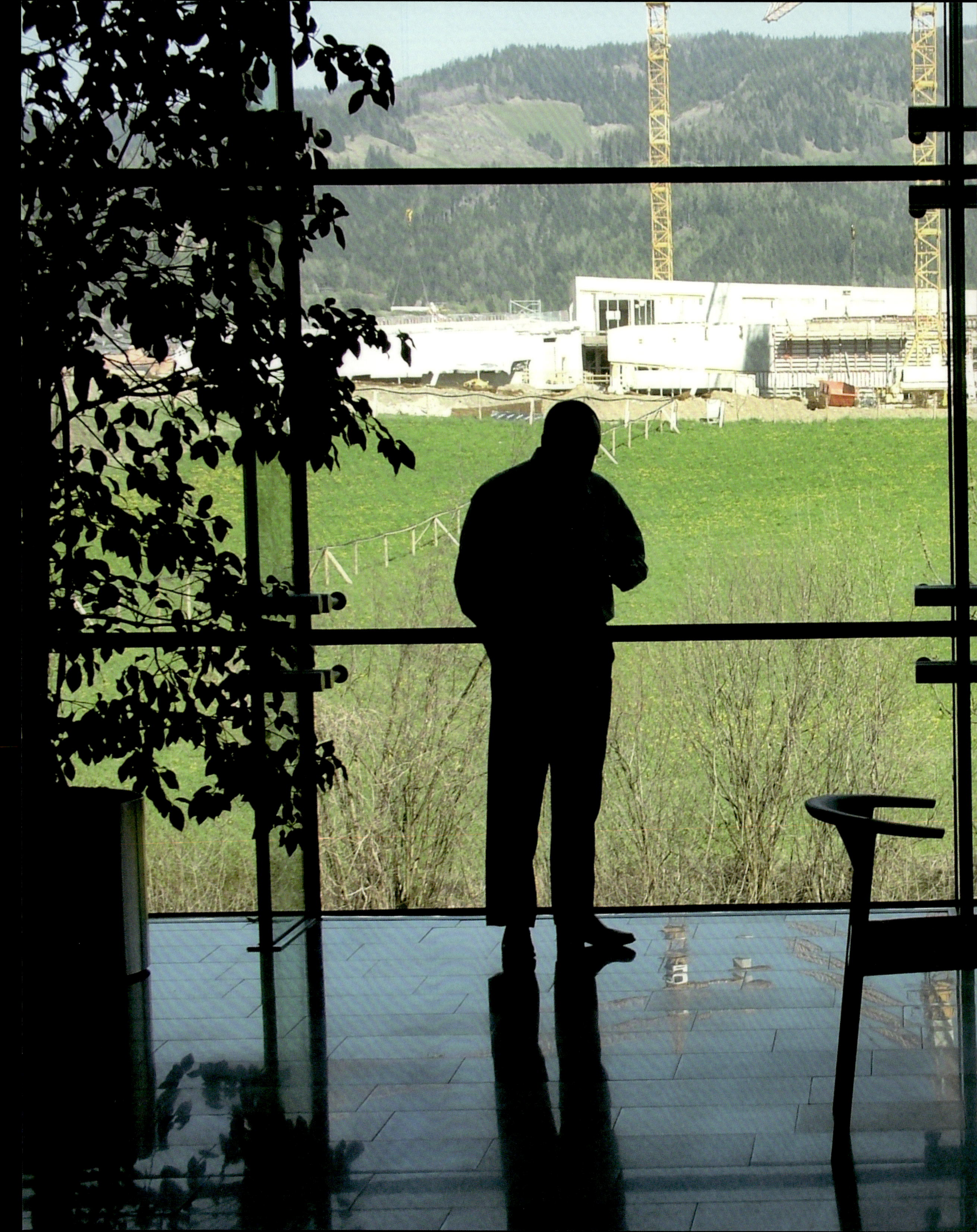

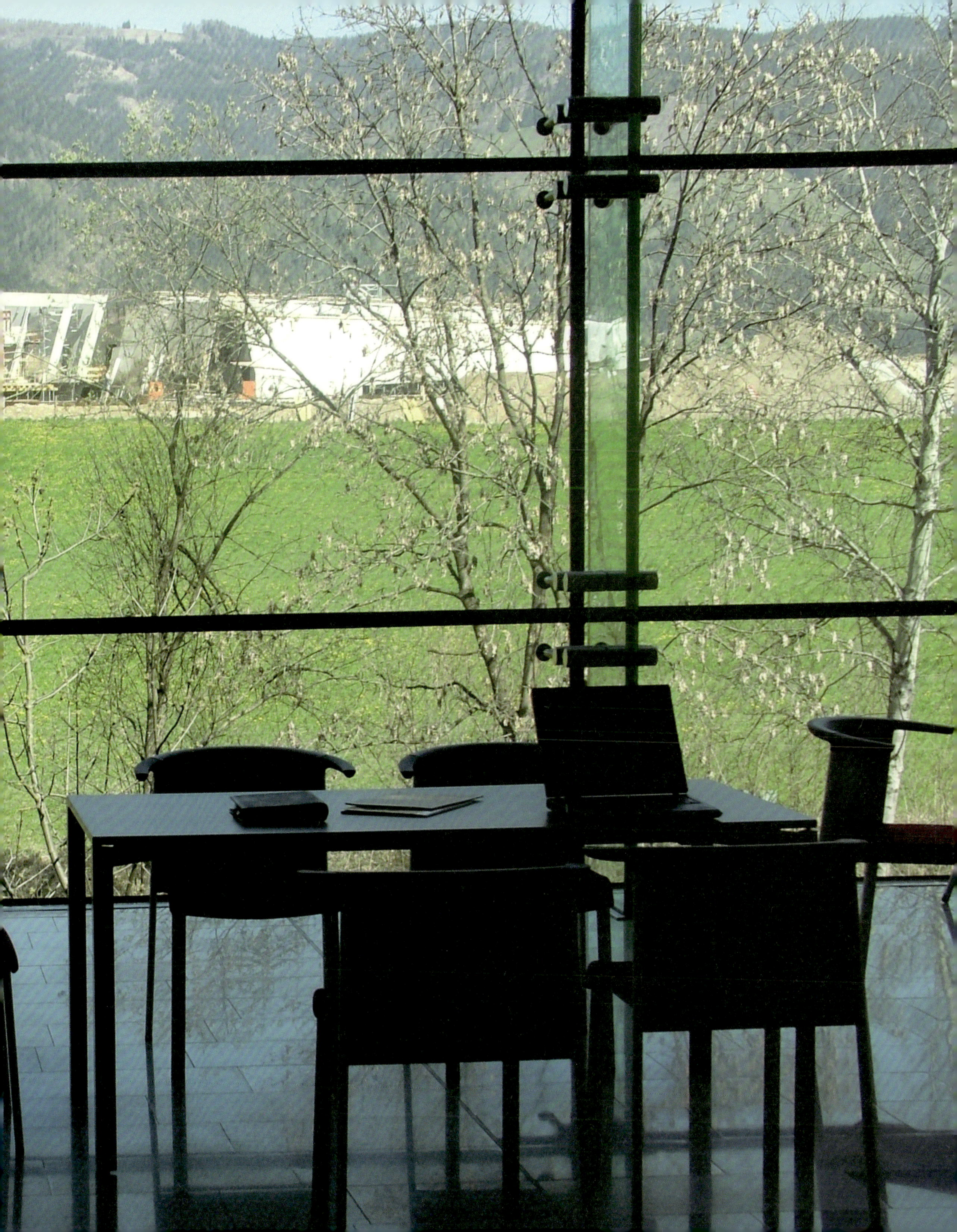

Rainer Ogrinigg, Director
„Das Besondere für unsere Seminargäste ist die individuelle Betreuung. Wir besprechen vorher alles gut mit den Trainern und Veranstaltern. Wir organisieren sogar die Anreise, geben dem Trainer quasi ein VIP-Treatment, wie beispielsweise ein größeres Zimmer mit eigenem Arbeitsbereich. Bei den Abläufen vor Ort zeigen wir dann auch noch immer eine große Flexibilität. Und unser Speisenangebot ist gehobene Haubengastronomie. ‚Tagen wie die Fürsten' ist deshalb unser Motto.
Weil sich große Firmen und Konzerne ihre eigenen Akademien einrichten, um dort das mittlere Management zu trainieren, gehen wir eher auf das gehobene Management zu, und das natürlich im kleinen Kreis. Das alte Gemäuer, moderne Akzente und eine wunderbare Landschaft sind ideale Voraussetzungen für erfolgreiche Tagungen, Konferenzen und Seminare. Am wichtigsten aber sind die zu 100% persönlich ausgewählte Hardware, zu der auch das Mobiliar zählt, und die Software, sprich: das Team, das sich umsichtig um die Gäste kümmert."

"The special aspect for our seminar guests is that they are taken care of individually. We discuss all the details beforehand with the trainers and organizers. We even organize the journey there and give the trainers VIP treatment so to speak, such as a larger room with their own personal working space. We also demonstrate great flexibility at the location with regard to the order of events. And selection of dishes is real gourmet cuisine. "Conduct your seminar like the lords" is therefore our motto.
Due to the fact that big companies and groups establish their own academies to train those in middle management, we cater for higher management instead and this, naturally in a small circle. The old walls, a modern touch and beautiful scenery are ideal conditions for successful meetings, conferences and seminars. The most important thing however is the hardware, all of which is personally selected and to which the furnishings belong, as well as the software which is the team that looks after the guests attentively."

Seminarhotel Schloss Gabelhofen, Obersteiermark Seminar chairs: Confair range

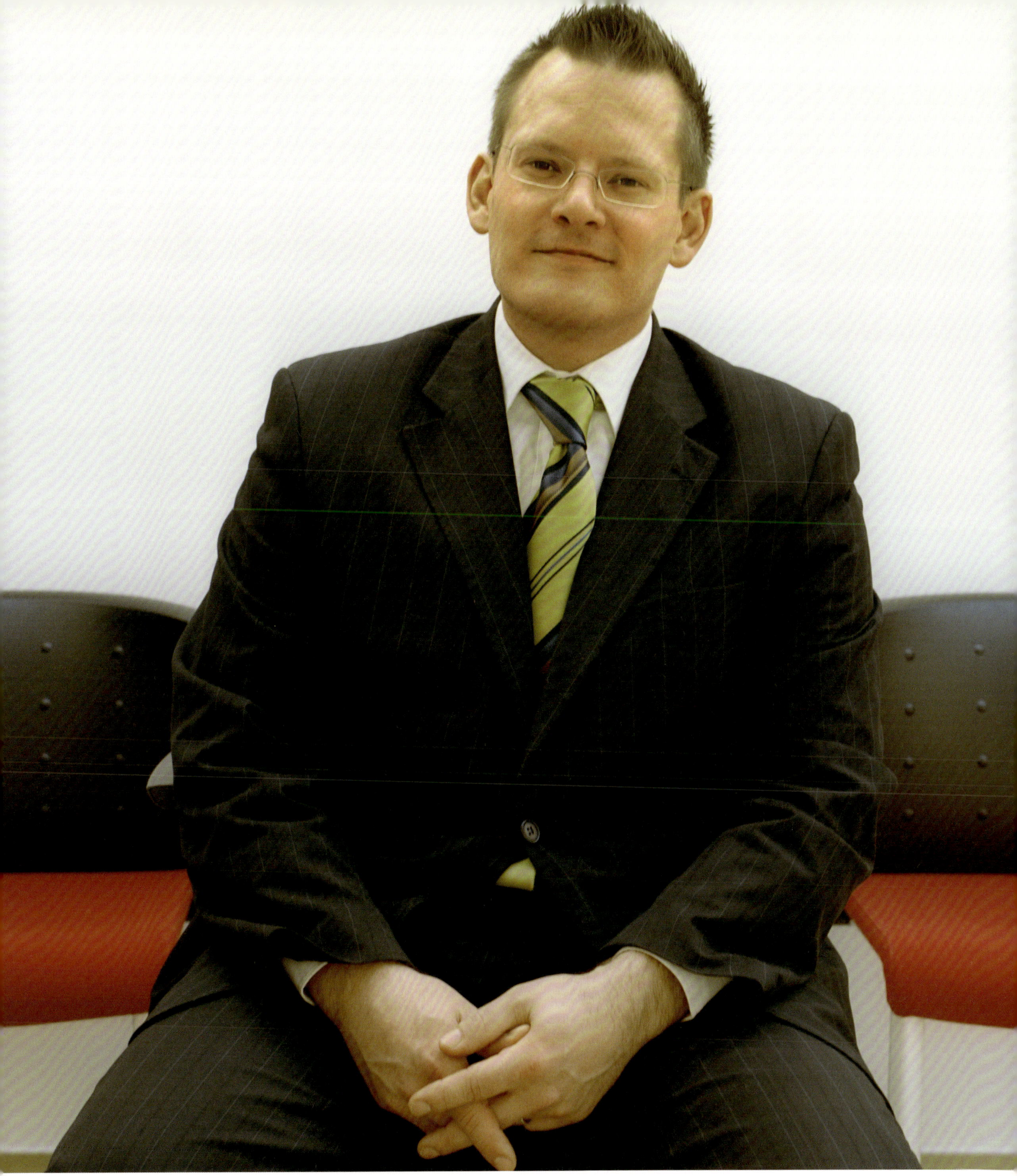

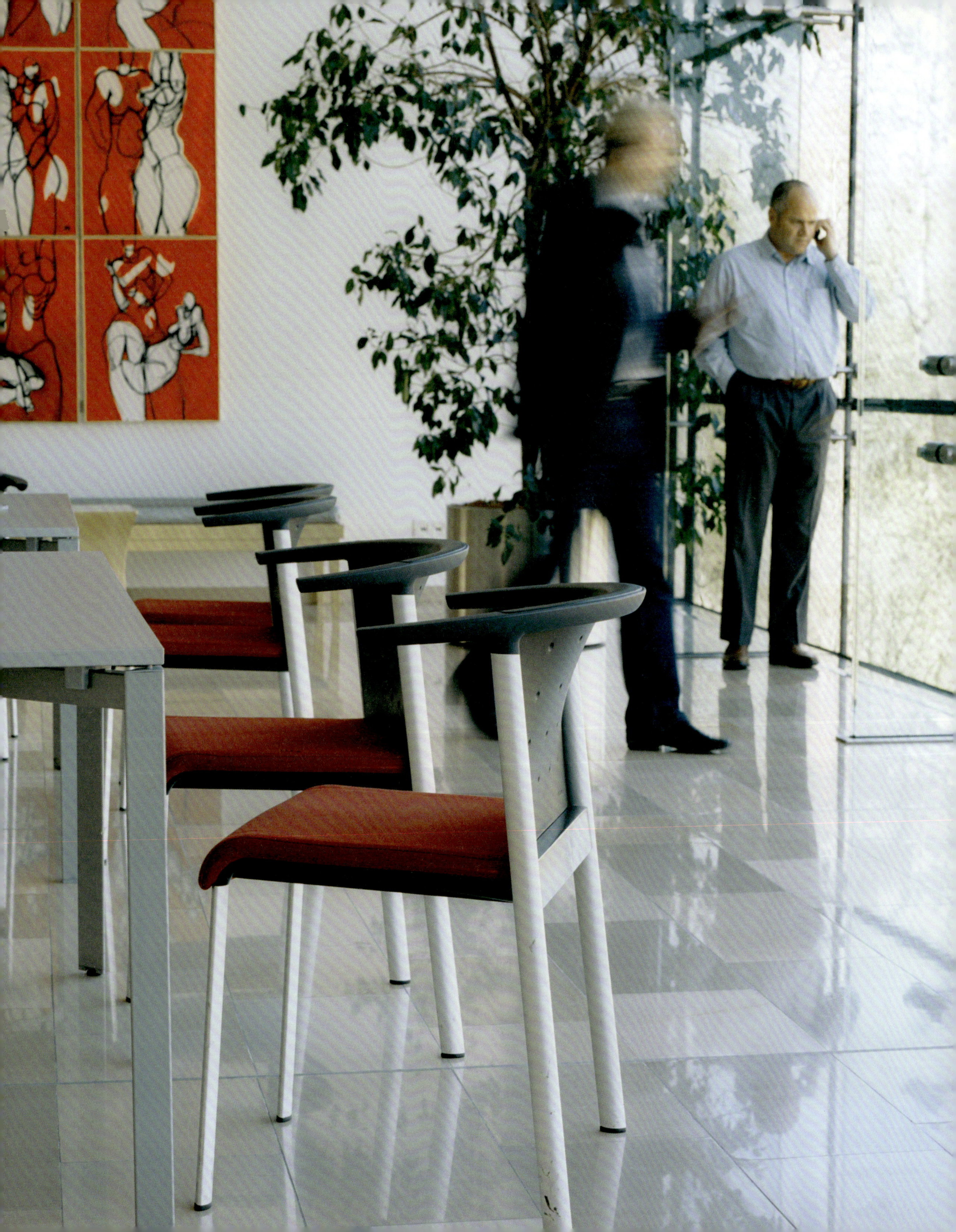

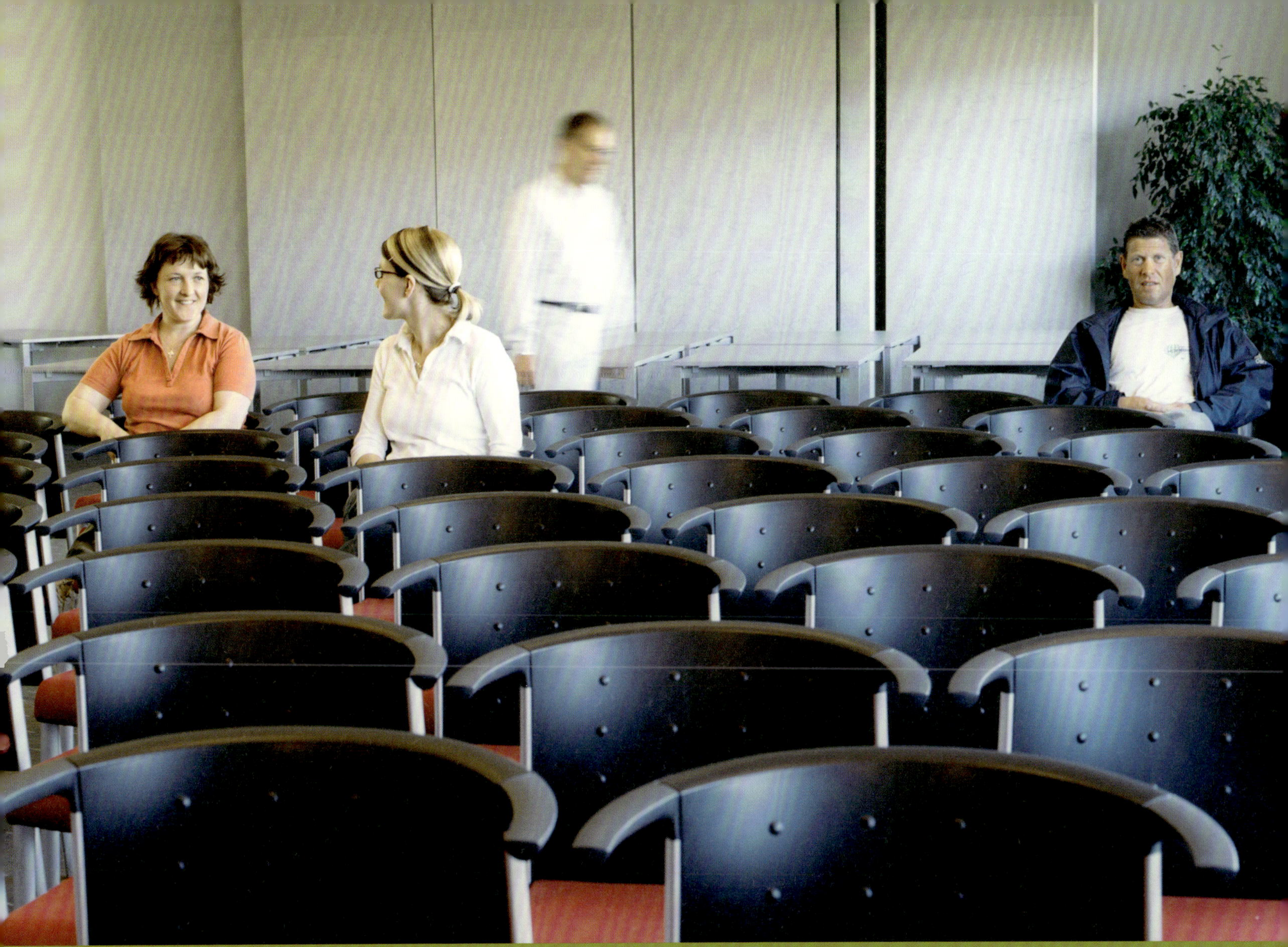

Seminarhotel Schloss Gabelhofen, Obersteiermark Seminar chairs: Confair range

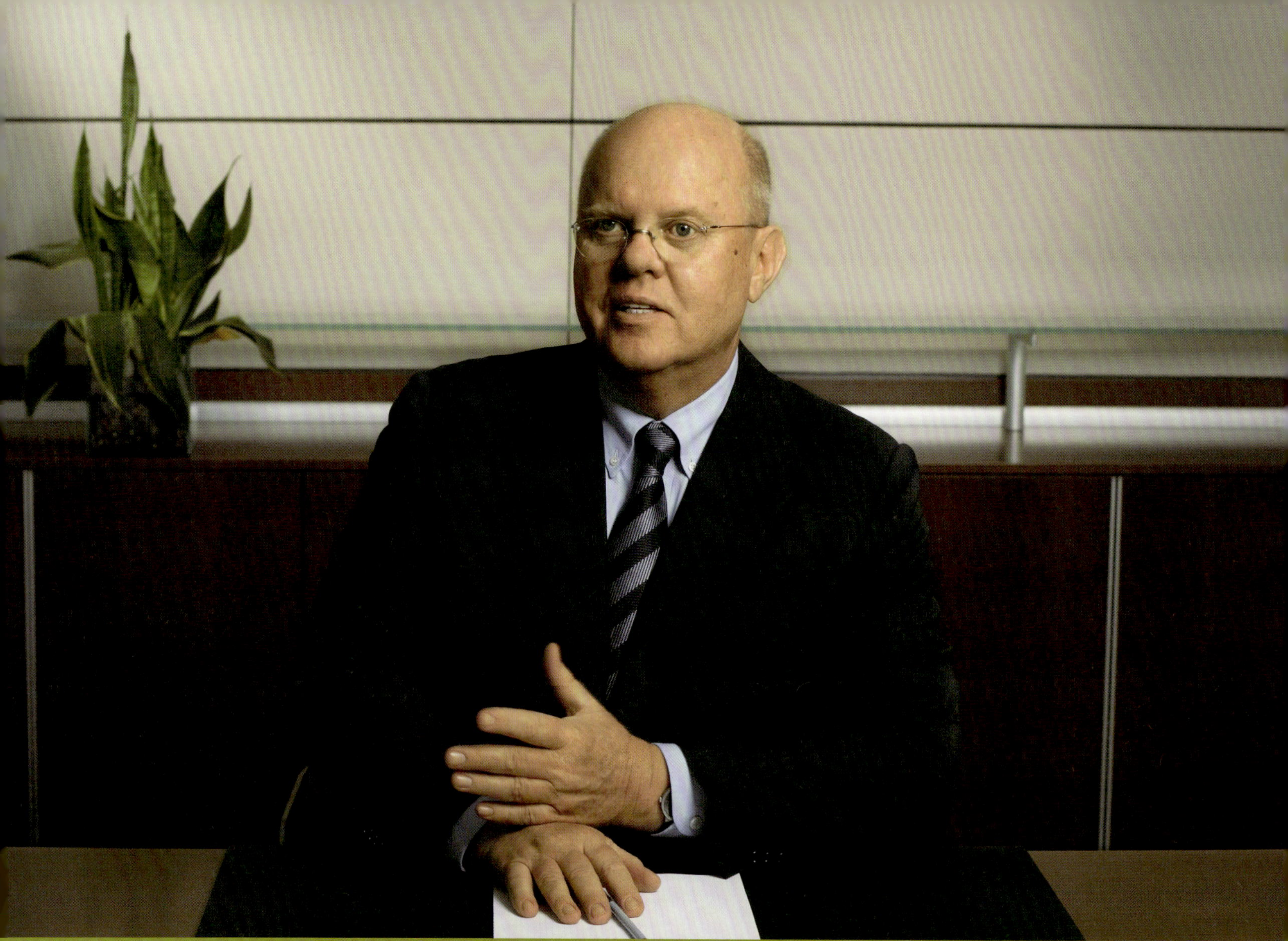

Manfred Bene, President of the Supervisory Board
„Die Beziehung zwischen Wilkhahn und Bene begann mit der Begegnung zweier Menschen mit ähnlicher (Wert-)Haltung: von Fritz Hahne und mir. Das war Ende der 1970er Jahre, als ich Sätze wie ‚Zuerst prägen wir unser Umfeld, dann prägt das Umfeld uns' zu meinen eigenen machte – und zum Leitsatz des Unternehmens Bene. Eine ausgeprägte Unternehmenskultur zu entwickeln, zu pflegen und eine Vorreiterrolle in der Branche einzunehmen, war damals wie heute jeweils unser Anspruch.
Bene und Wilkhahn, in den 1980er Jahren auch in Person des langjährigen Vorsitzenden der Geschäftsführung Theodor Diener, trafen und treffen einander in vielen gemeinsamen Interessen: das Interesse an Menschen, an der Herstellung von Qualitätsprodukten, am Verständnis von Design als kultureller Disziplin im Sinne einer Kritik am puren Funktionalismus. Uns verbindet, Räume als Bühnen für Menschen zu sehen."

"The relationship between Wilkhahn and Bene began when two people met that had similar approaches (values): Fritz Hahne and myself. This was at the end of the 1970s when I made a habit of saying things like "First we shape our surroundings and then our surroundings shape us" and also making this Bene's motto. Our aspiration then and now was to develop and cultivate a distinctive corporate culture and to position ourselves as forerunners in the business.
Bene and Wilkhahn, as well as long-standing Managing Director Theodor Diener in the 1980s, had and still have many common interests: interest in people, in the production of quality products, in our understanding of design as a cultural discipline within the context of criticism of pure functionalism. What links us is to see spaces as a stage for people."

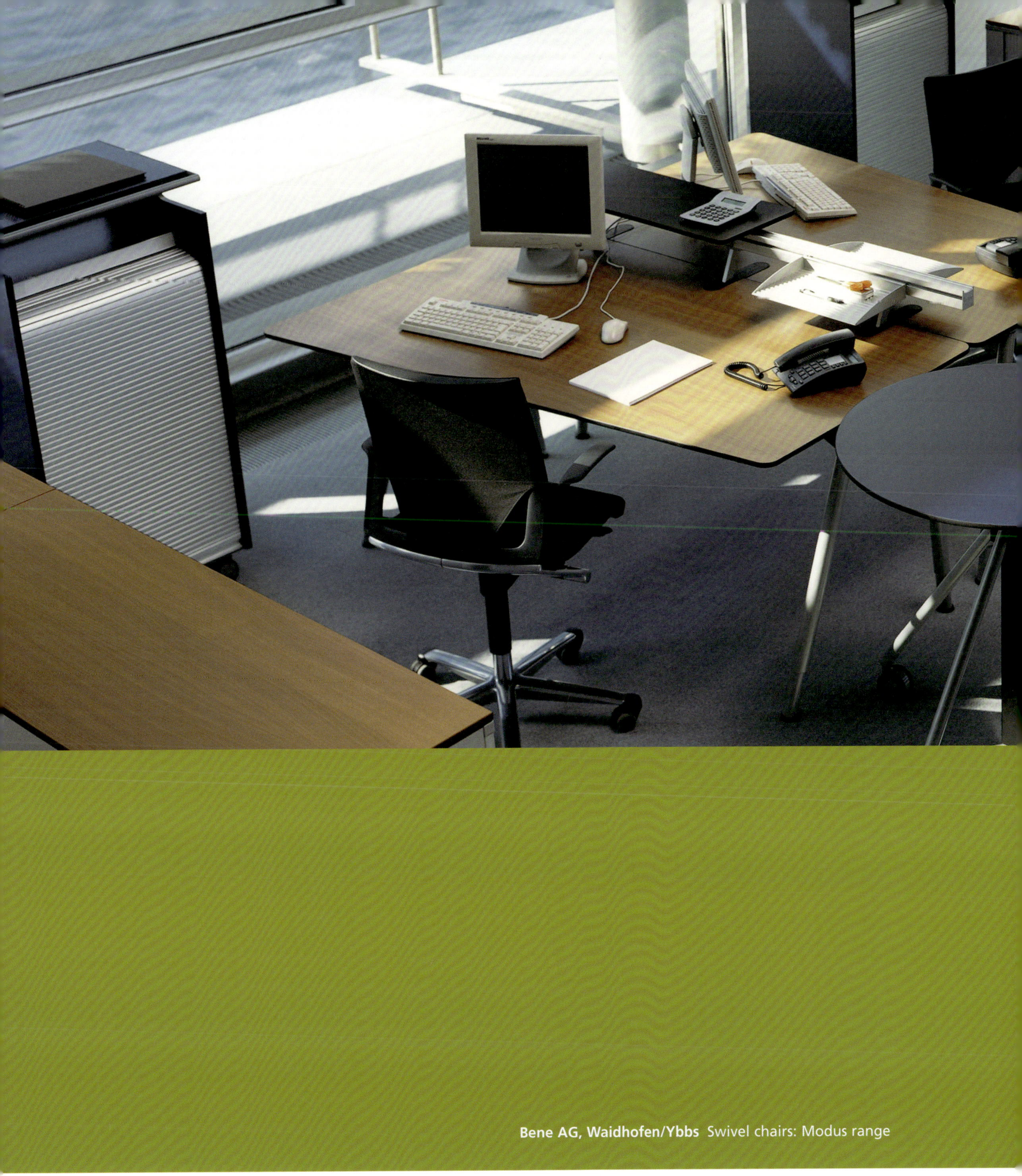

Bene AG, Waidhofen/Ybbs Swivel chairs: Modus range

Mag. Bernhard Grubhofer, Office Manager
„Wir haben 330 Angestellte und ein paar freie Dienstnehmer. Desksharing gibt es nur für Außendienstler. Eine allzeit funktionierende IT, kurze Kommunikationswege und eine gute Ausstattung unterstützen einen reibungslosen Ablauf ungemein. Wir haben uns eine hochwertige Ausstattung gegeben, um alle Mitarbeiter in allen Arbeitsbereichen positiv zu unterstützen. Dabei achten wir natürlich auf die Ergonomie. Wir haben aber auch einen Fitnessbereich zur Verfügung. Bei Wilkhahn begeistern mich die Funktionalität, die Qualität und das repräsentative Erscheinungsbild. Diese Möbel sind einfach leicht veränderbar und flexibel und zeigen ein gutes Preis-Leistungs-Verhältnis."

"We have 330 employees and a few free-lance collaborators. Desksharing is only for those who work in the field. Good functioning IT, short communication channels and suitable facilities are invaluable in supporting a smooth workflow. We have provided high-quality furnishings to support all employees in all work areas positively. We naturally focus on ergonomics. A fitness area is also available. I am impressed by Wilkhahn's functionality, the quality and the prestigious appearance. This furniture is simply easy to adjust and is flexible as well as providing good cost effectiveness."

Raiffeisen Capital Management, Vienna Lounge chairs: Confair range

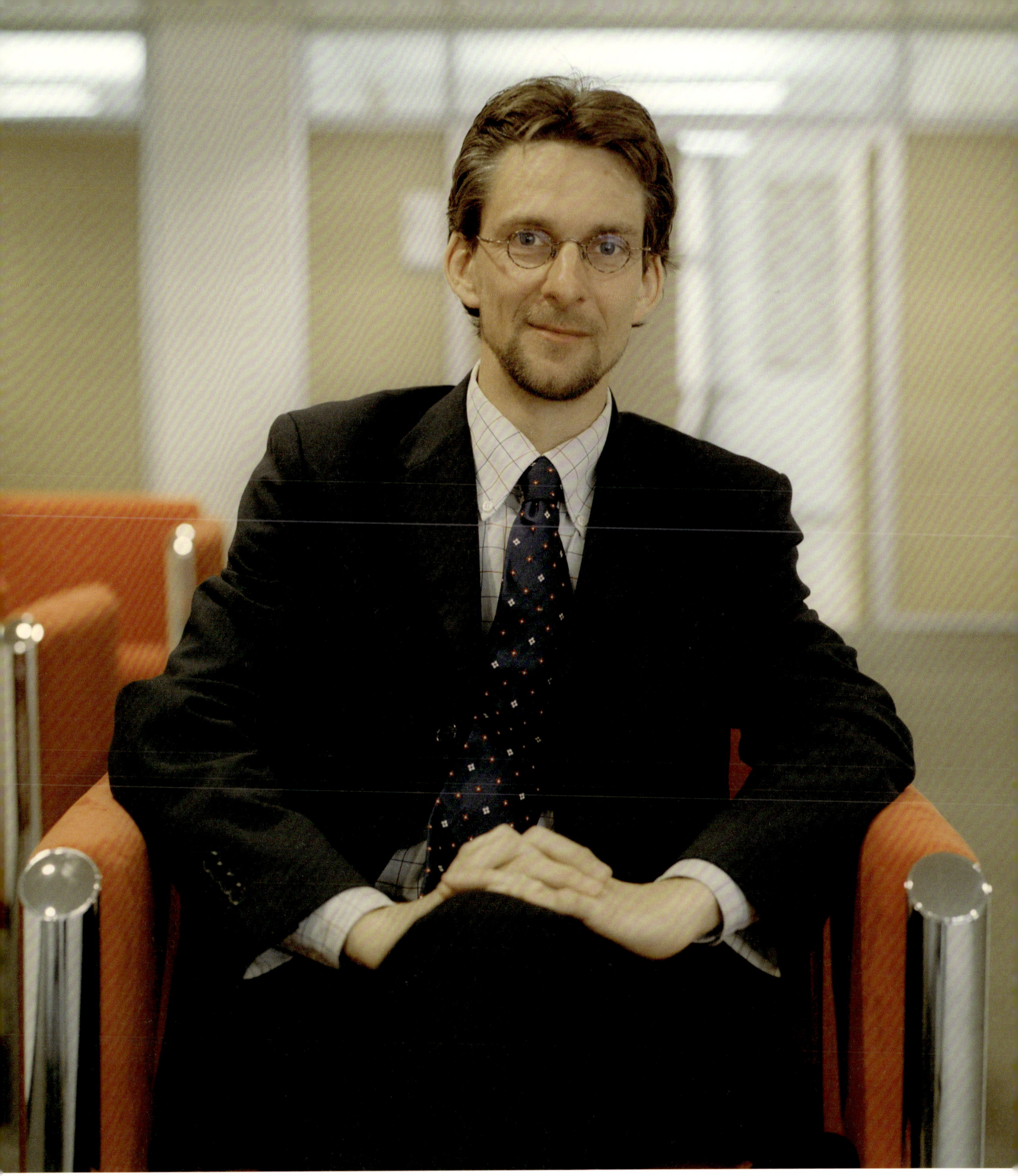

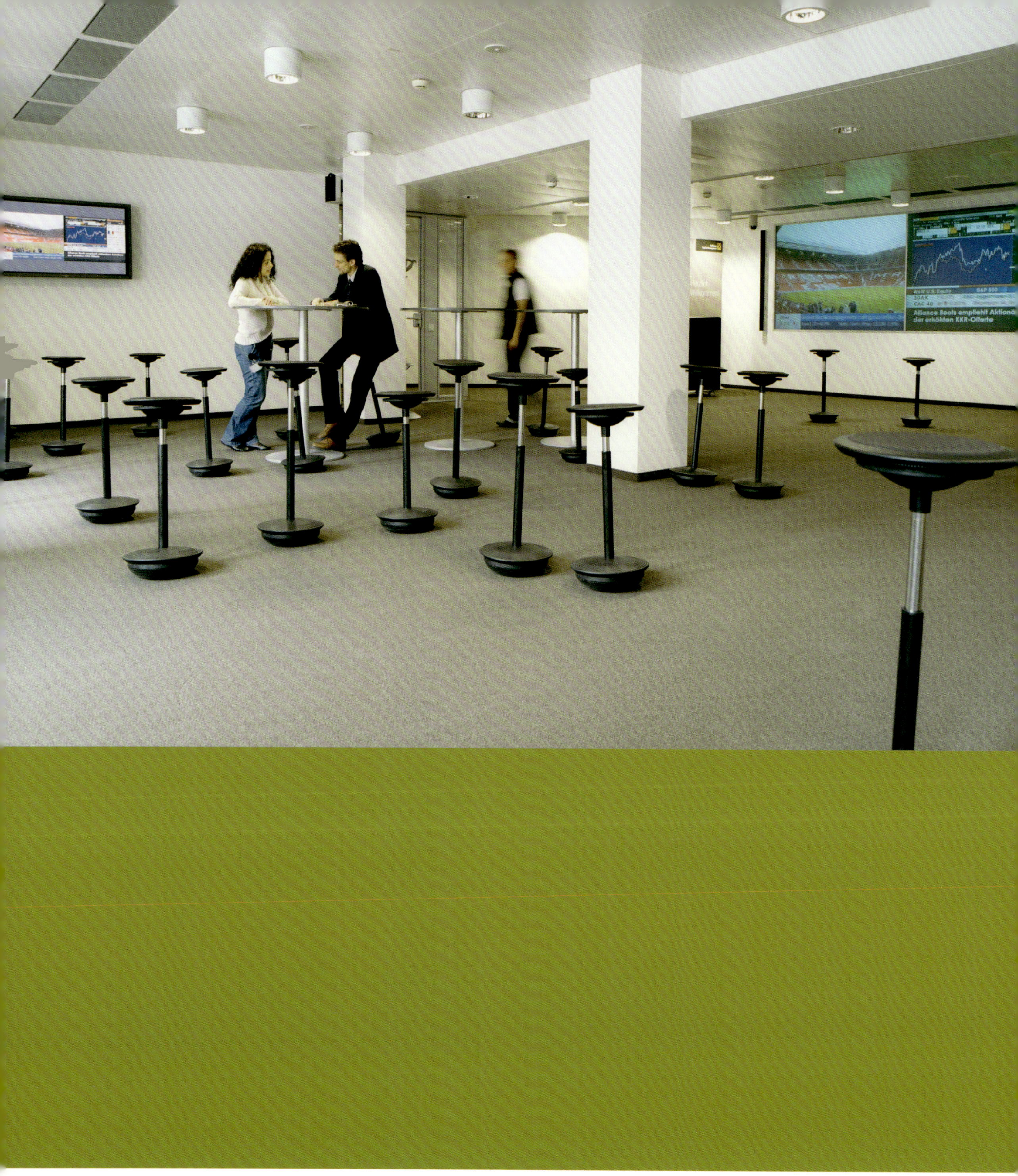
SDAX
CAC 40
Alliance Boots empfiehlt
der erhöhten KKR-Offerte

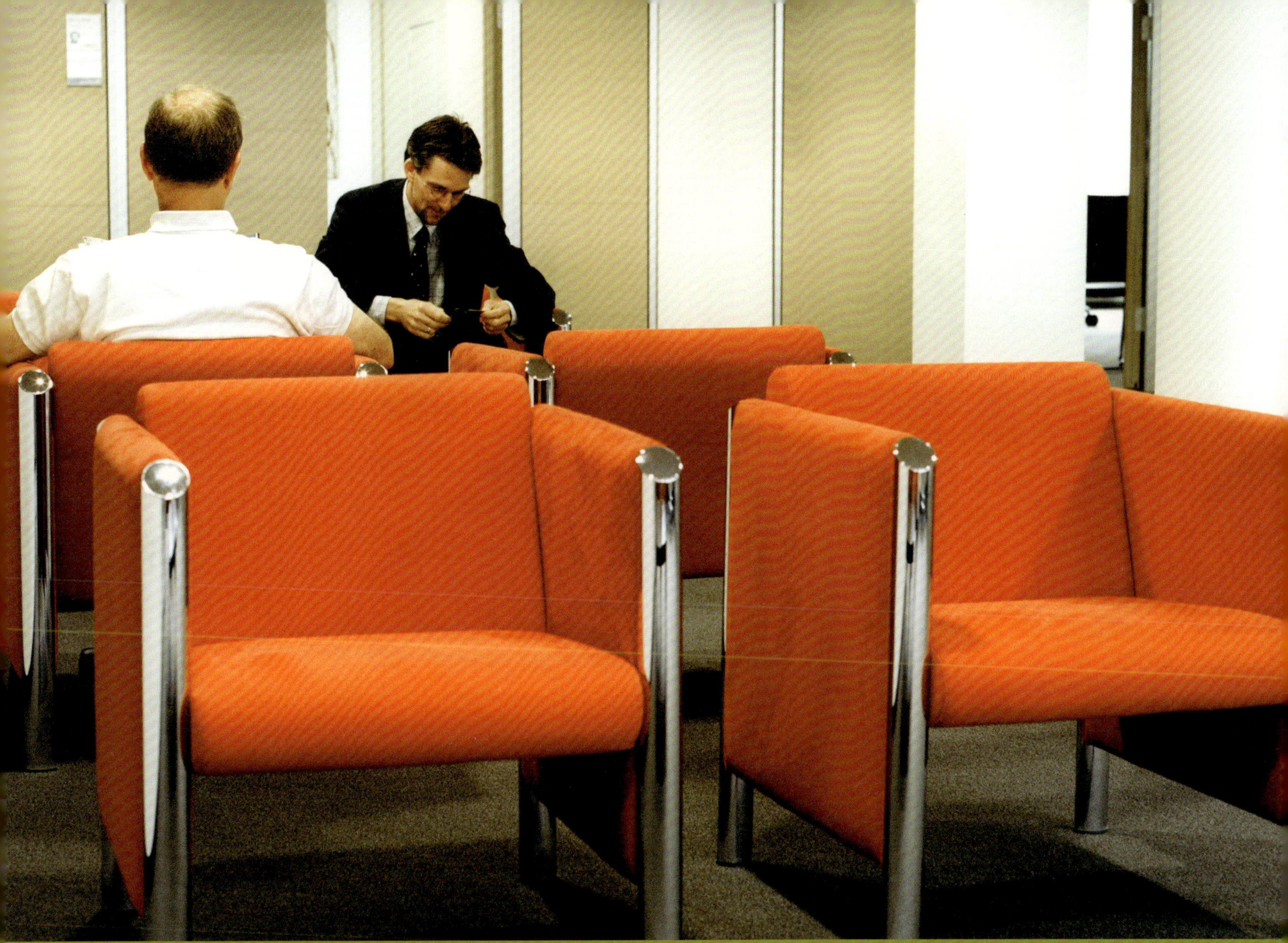

Raiffeisen Capital Management, Vienna Dynamic stools: Stitz range; lounge chairs: Cubis range

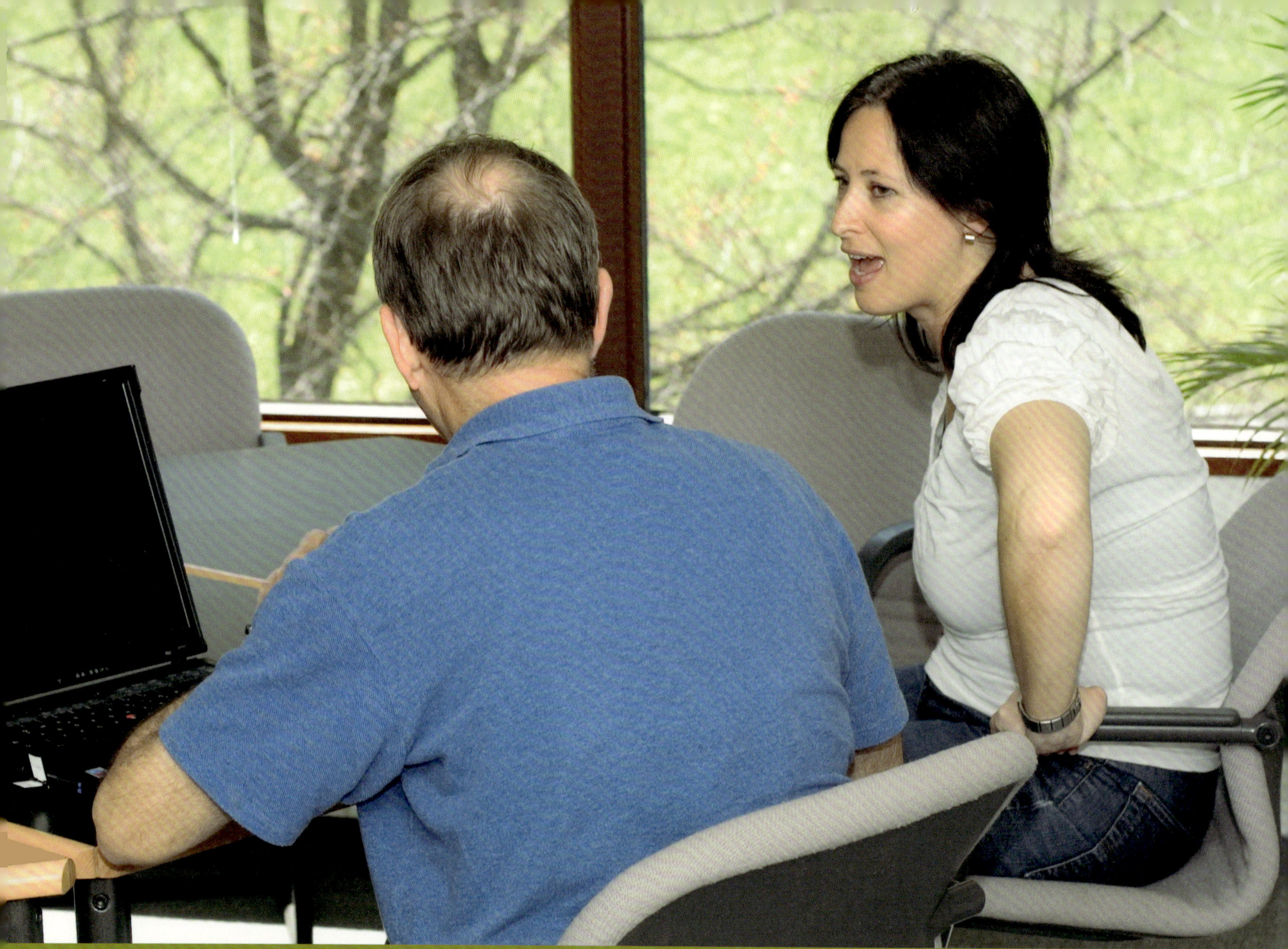

Susanne Erhart, Assistant to the President

„SOS steht für ‚Societas Socialis', also für soziale Gemeinschaft. Wir haben hier eine besondere Art von Menschen im Haus, die einen tieferen Sinn darin sehen, so zu arbeiten, dass dabei etwas herauskommt. Im Gegensatz zu ordinären Büros, deren oberstes Bestreben es ist, immer reicher zu werden. Trotzdem sind wir genauso wie vorgenannte auf einen florierenden Arbeitsablauf angewiesen und brauchen Maschinen und Möbel. 92 treue Begleiter fühlen sich in diesem Hause sehr wohl! Ich glaube, dass wir längere und häufigere Meetings haben als Kolleginnen und Kollegen in der freien Wirtschaft. Schließlich geht es darum, wie wir den Kindern dieser Welt einen geraden und guten Weg bahnen können."

"SOS stands for "Societas Socialis", i.e. for a social community. We have special types of people in the house that see a deeper purpose in working so that there are results. This stands in contradiction to normal offices whose efforts are primarily directed at making even more money. Nevertheless, we are just as dependent on a flourishing workflow as these persons, and we require machines and furniture. Ninety-two faithful companions feel very comfortable in this house! I believe we have longer and more frequent meetings than colleagues in the free market economy. Finally, it is about how we can pave an honest and good way for the children of this world."

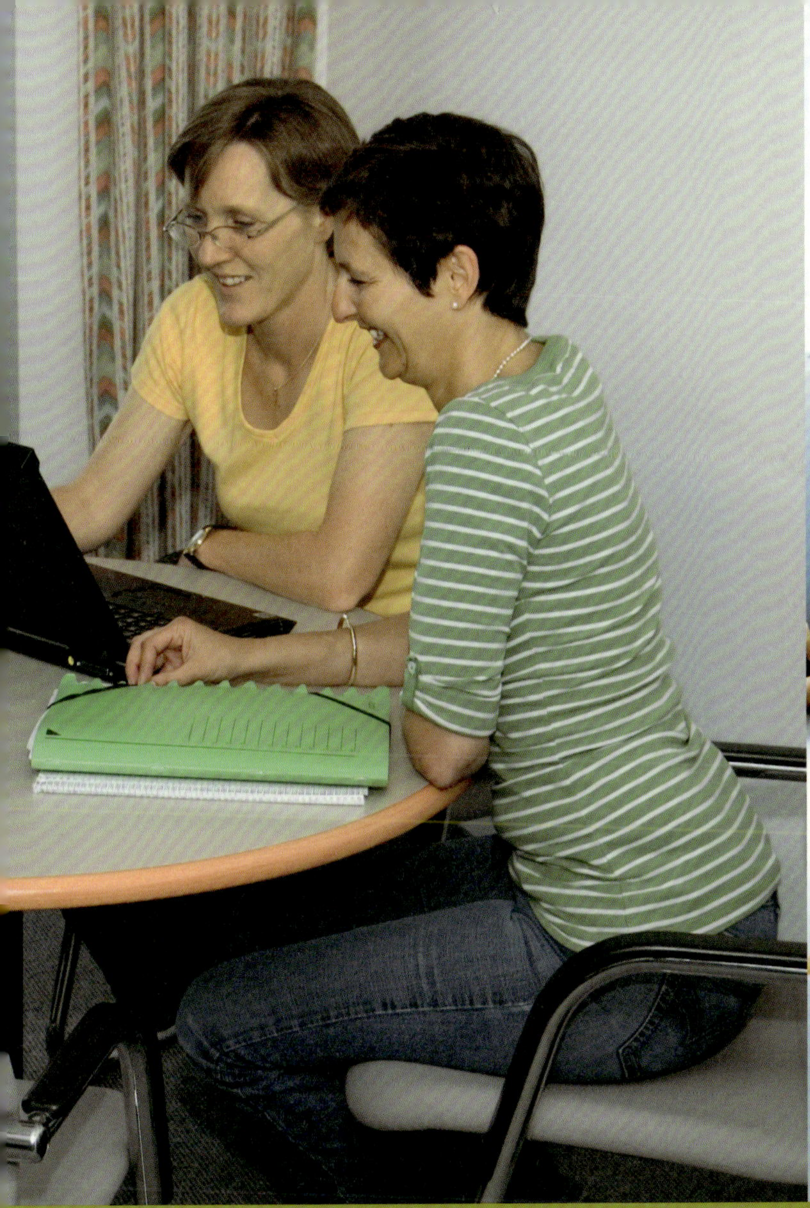

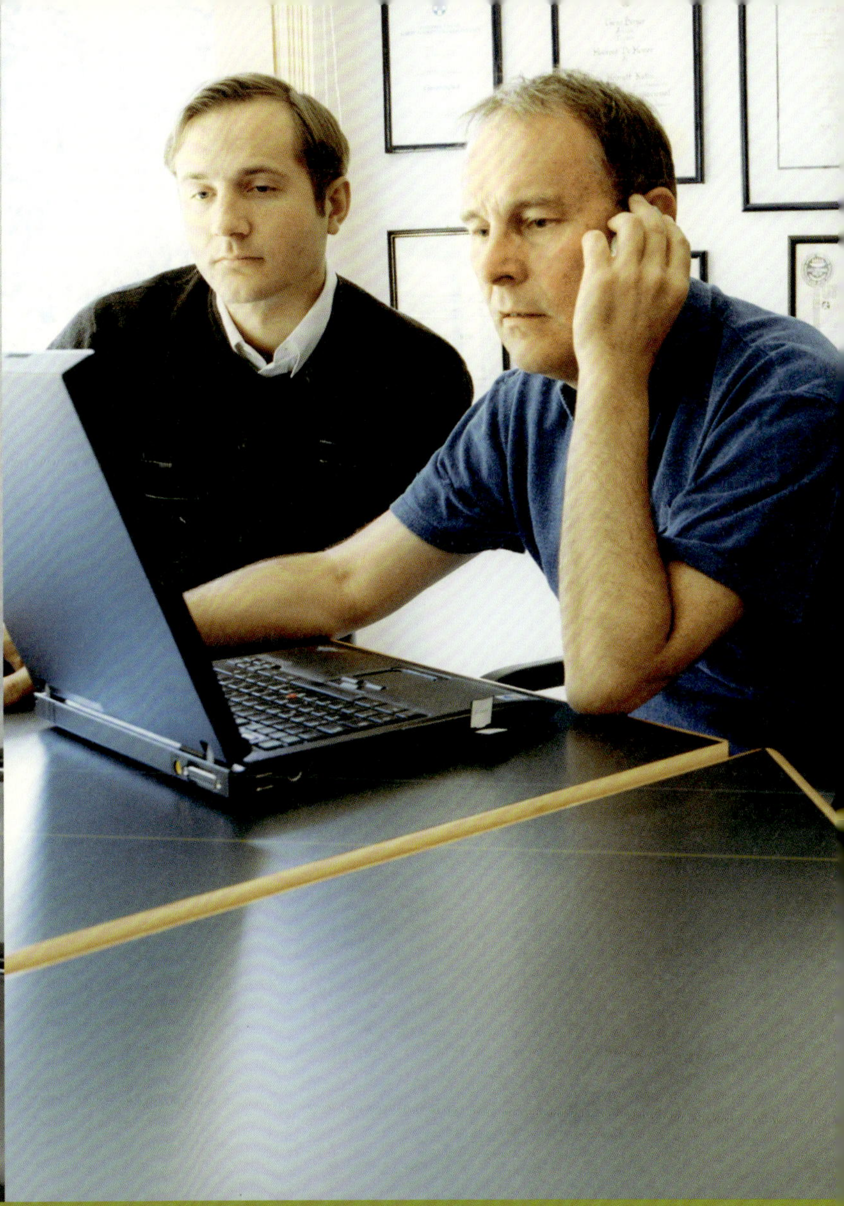

SOS Kinderdorf International, Innsbruck Cantilever chairs: FS range

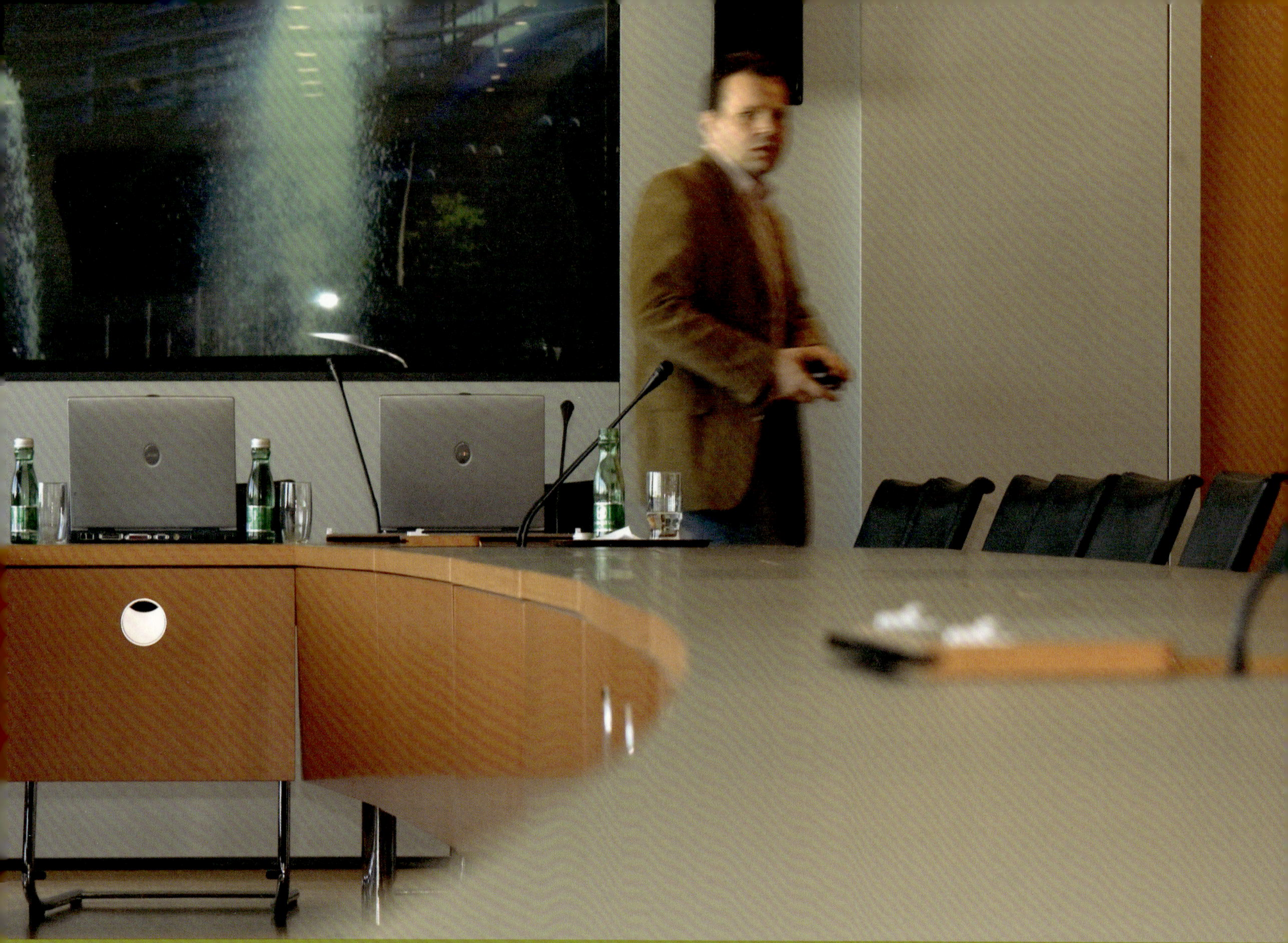

Dr. Walter Leiss, Club Director
„Für die Einrichtung des Konferenzraums gab es mehrere Angebote, dabei war für uns das Gesamtkonzept und speziell die Perfektion der Medienintegration entscheidend. Auch die Stühle verbinden das Nützliche mit dem Angenehmen, denn man sitzt sehr angenehm darauf. Weil für uns die Gesamtheit wichtig ist, gibt es keine besonders exponierten Plätze. Es gibt zwar eine Hierarchie, aber wenn konferiert wird, sind alle Diskussionen auf einer gleichrangigen Ebene."

"There were several tenders for the furnishing of the conference interior; the overall concept and particularly the perfection of the media integration were crucial for us. The chairs also combine functionality with pleasure as they are very comfortable to sit on. As the concept is important to us in its entirety, there are no exposed places in particular. Although there is a hierarchy, discussions are carried out at a level of equal status when we have our conferences."

Austrian People's Party (ÖVP) Lower Austria, St. Pölten
Table configuration: Palette range; cantilever chairs: Modus range

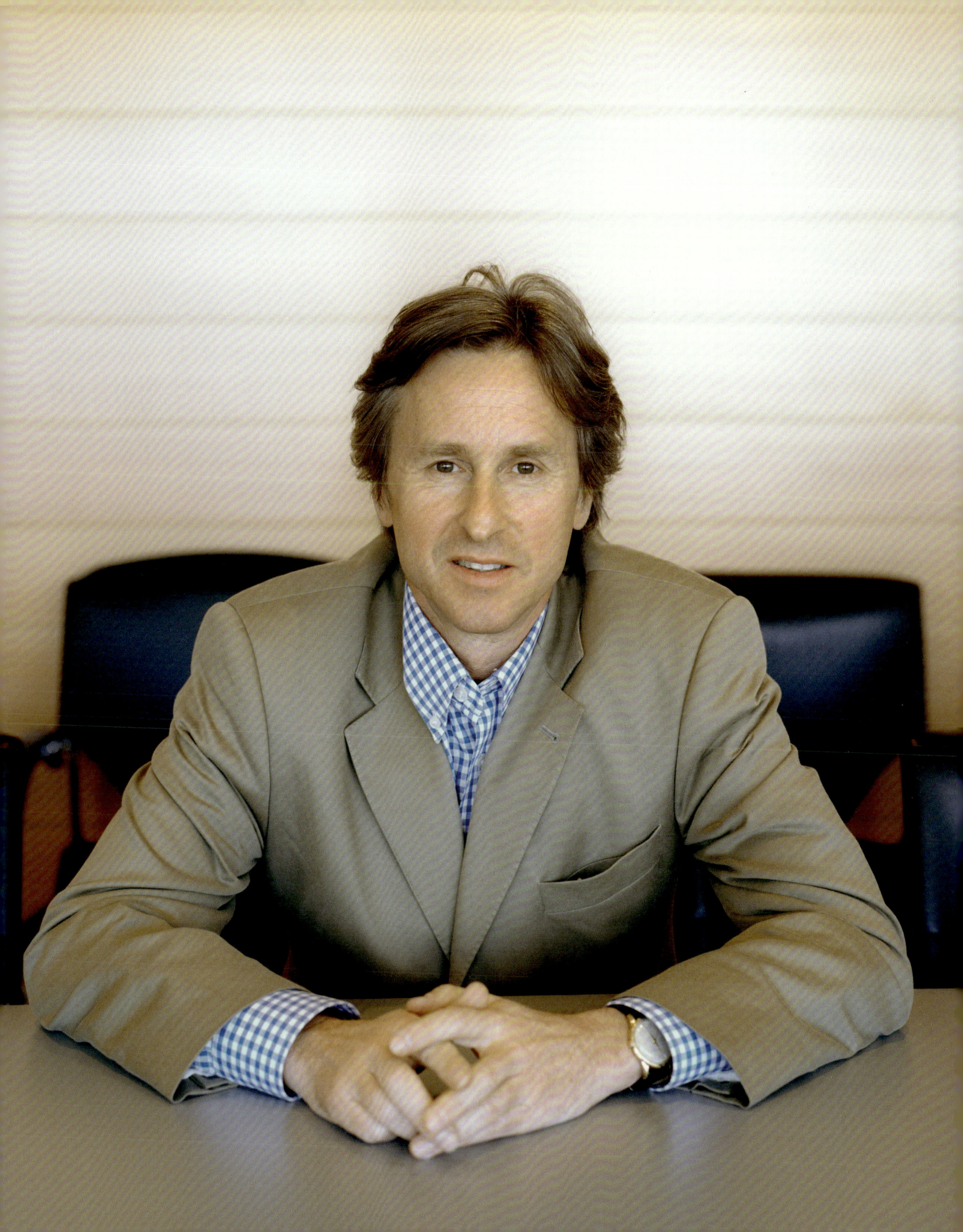

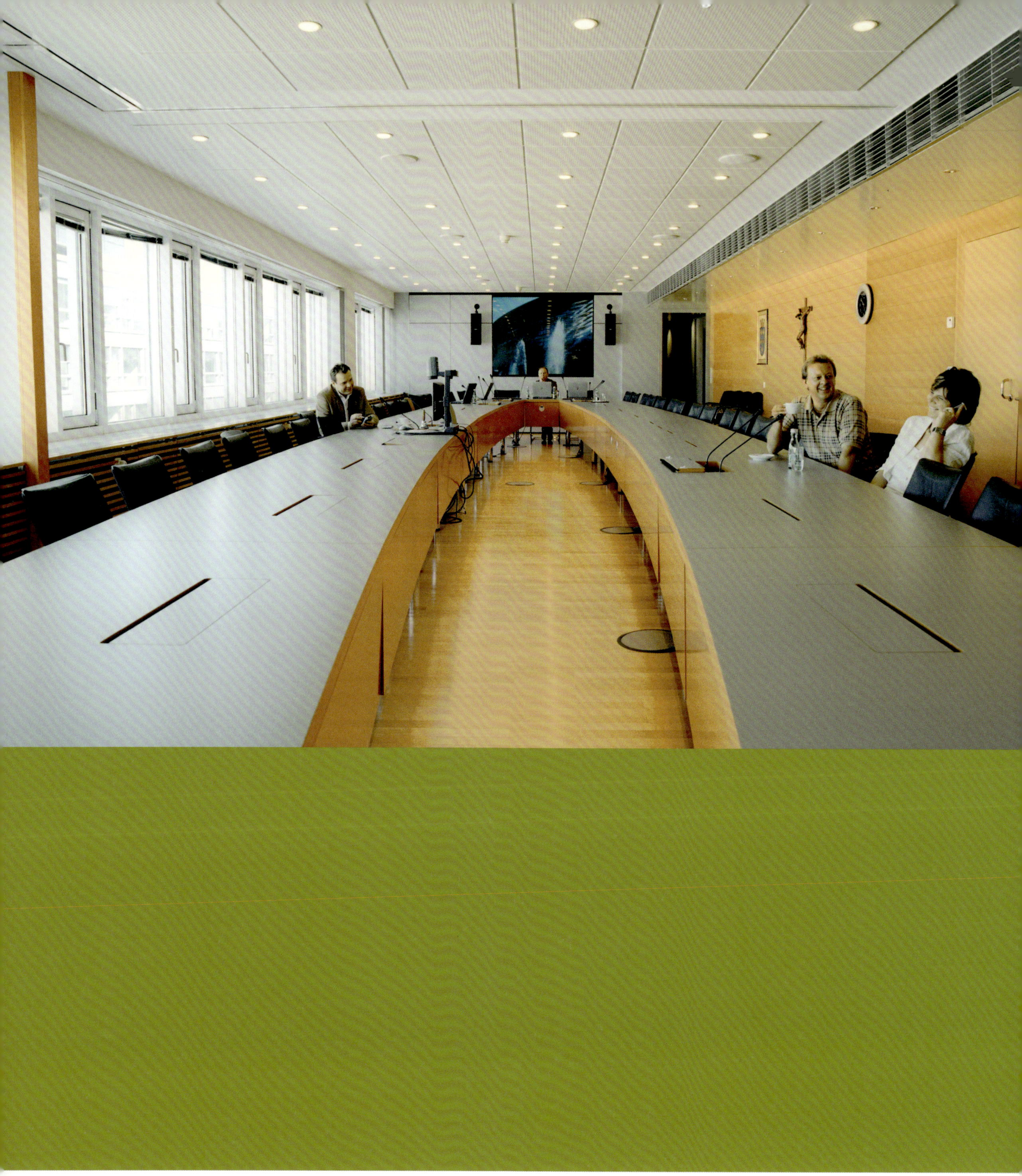

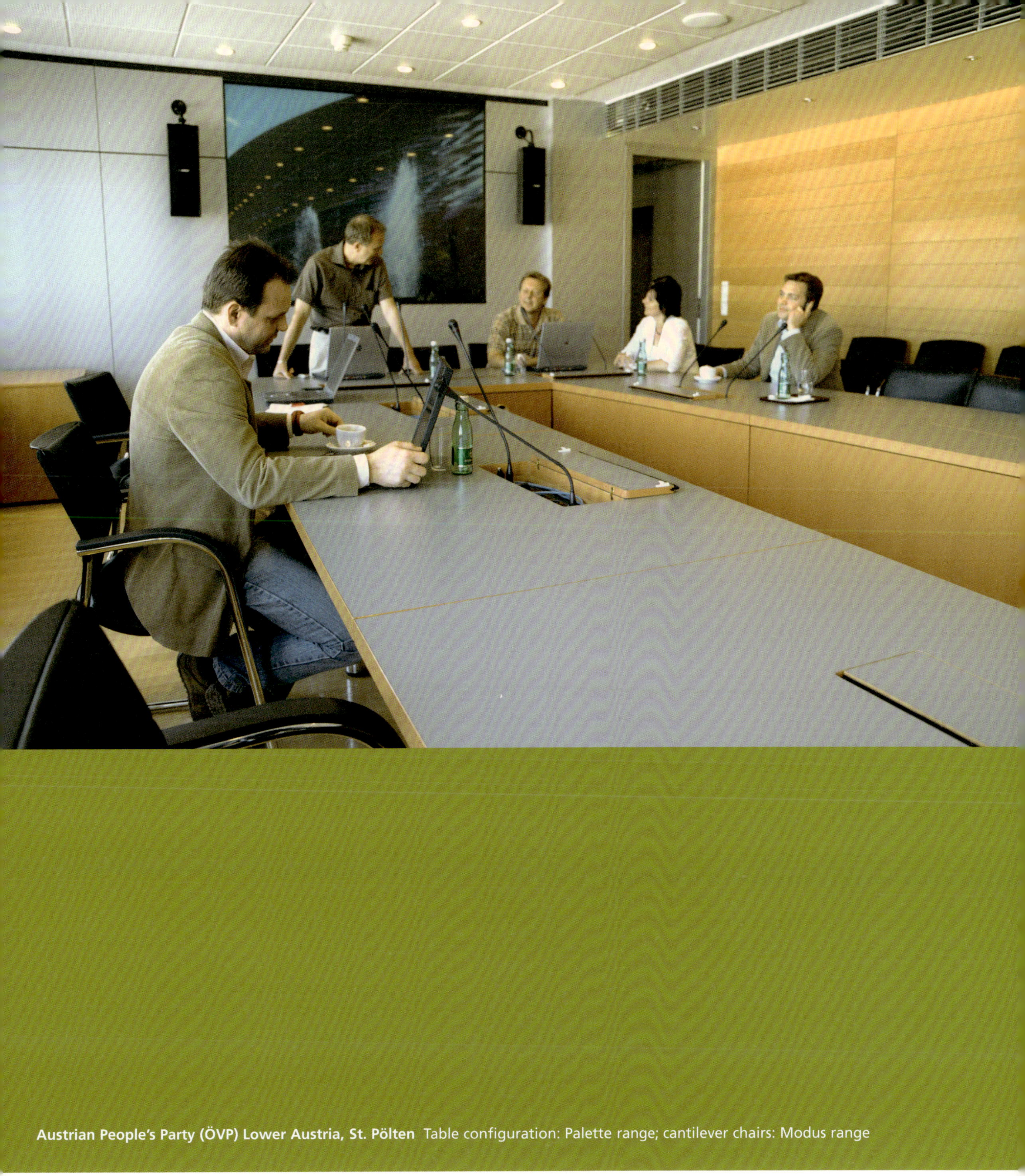

Austrian People's Party (ÖVP) Lower Austria, St. Pölten Table configuration: Palette range; cantilever chairs: Modus range

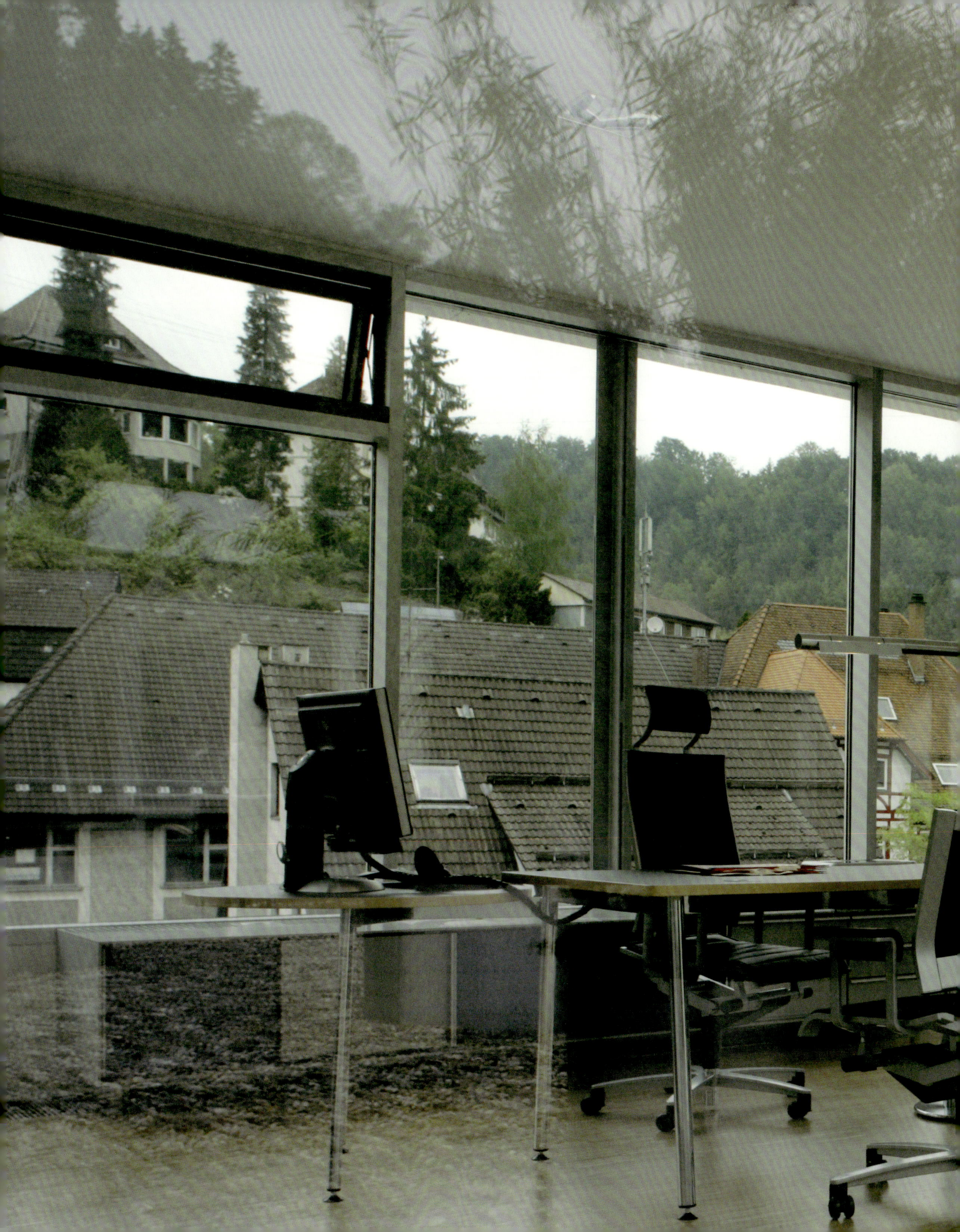

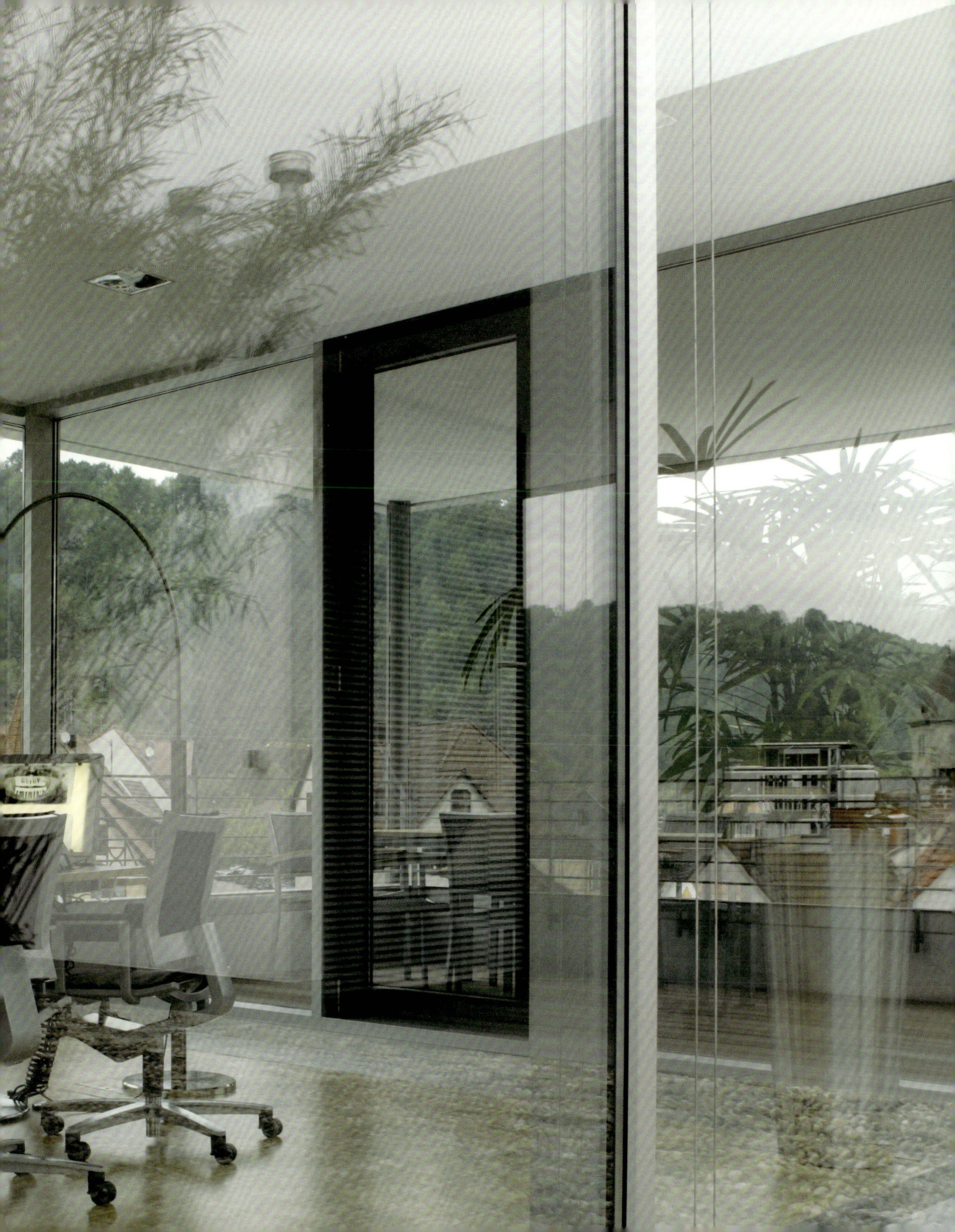

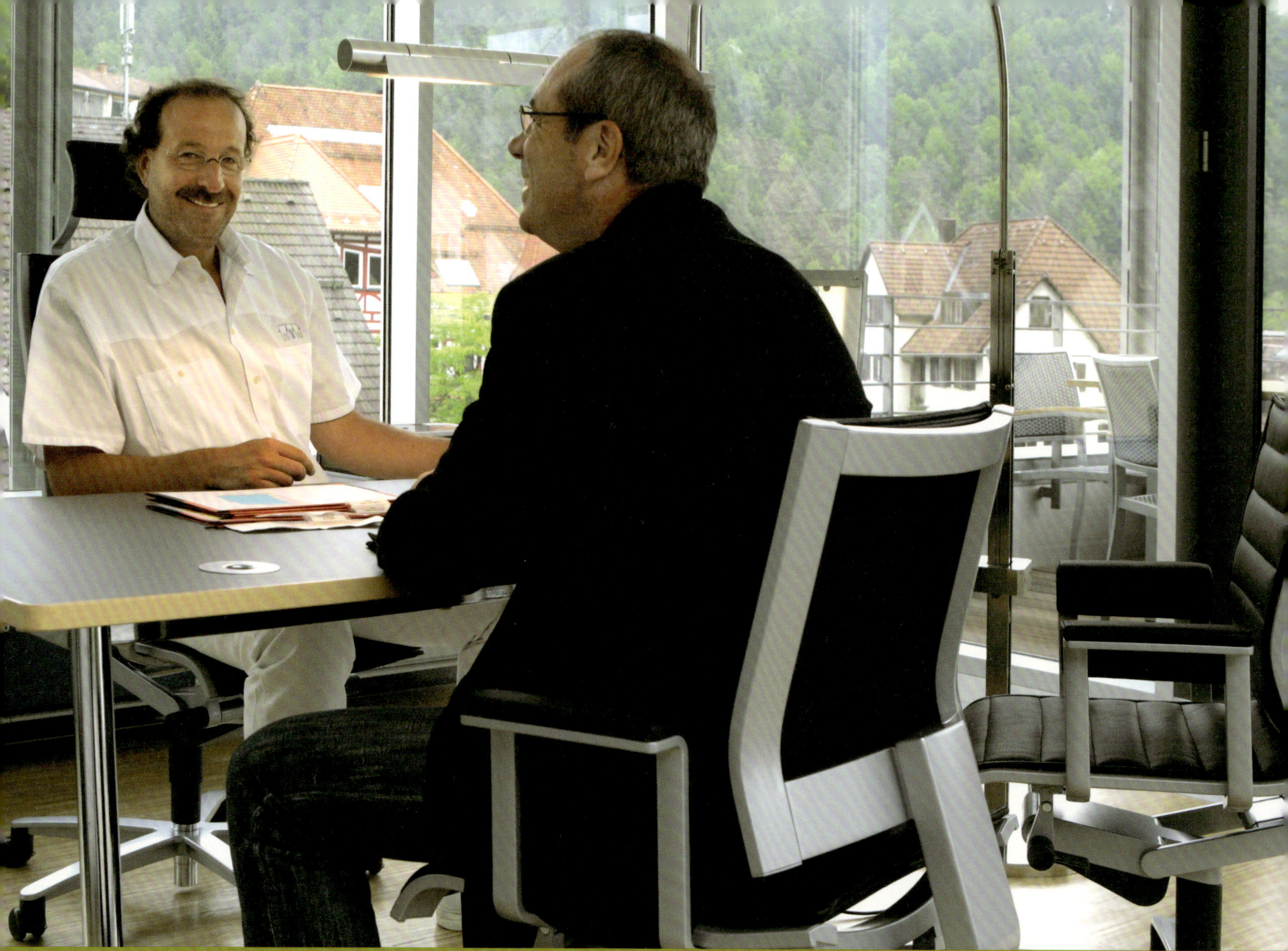

Dental Clinic, Oberndorf Swivel chair: Solis F range

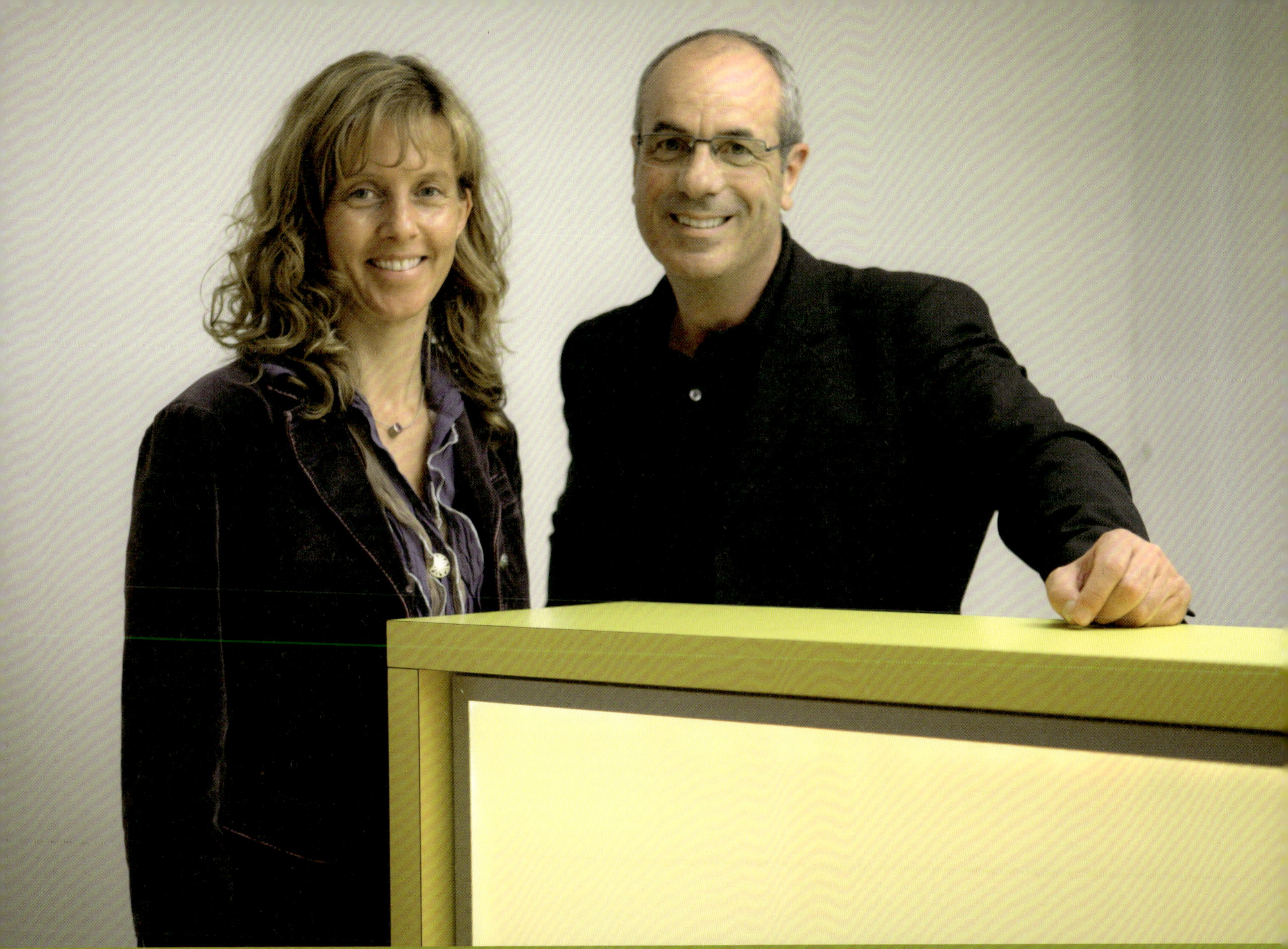

Dietrich Hopf, Bettina Pfäffle
„Wir freuen uns, für Bauherren mit ausgeprägter Individualität zu arbeiten, deren Ansprüche das Angebot einer standardisierten Fertighausproduktion übersteigen und die in ihre Wirtschaftlichkeitsbetrachtungen auch Qualitätsaspekte und Nachhaltigkeit mit einbeziehen. Schon im Studium haben wir die Wurzeln der auf das Bauhaus zurückgehenden modernen Architektur lieben gelernt. Deshalb beziehen wir uns direkt auf diese Tradition der ‚Neuen Sachlichkeit' und übersetzen sie in die Zukunft. Wir schätzen kompromisslose Qualität, perfekte Funktionalität und schnörkellose Gestaltung."

"We enjoy working for clients with distinctive individuality whose demands extend beyond the offer of a standard pre-fabricated house and who, in their analysis of economic efficiency, also include aspects concerning quality and sustainability. During our studies, we already came to appreciate the roots of modern architecture which go back to Bauhaus. We therefore apply this tradition of "New Functionalism" direct and translate it into the future. We also therefore highly value uncompromising quality, perfect functionality and simple design."

Architekten Hopf & Pfäffle, Oberndorf

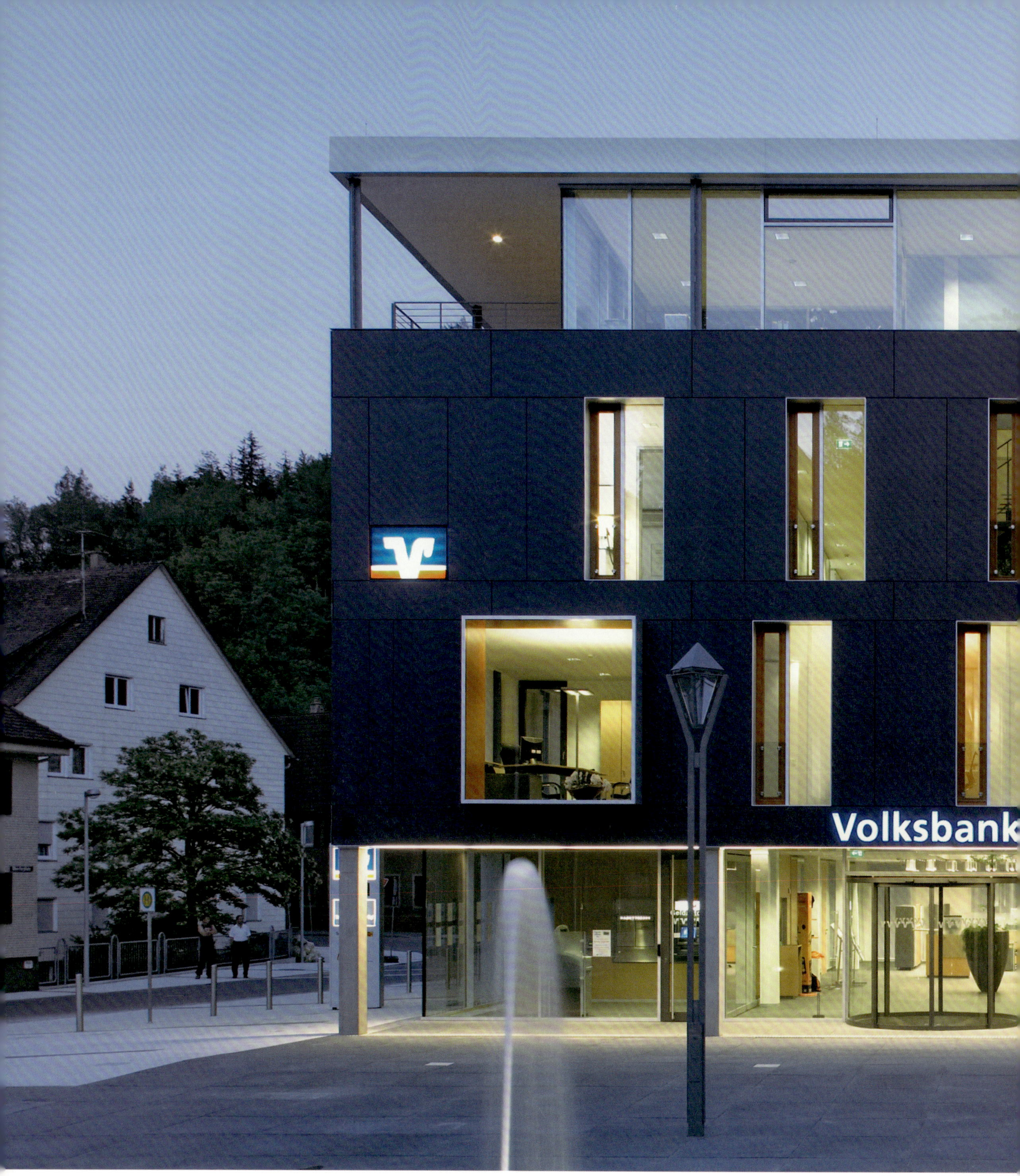
Volksbank

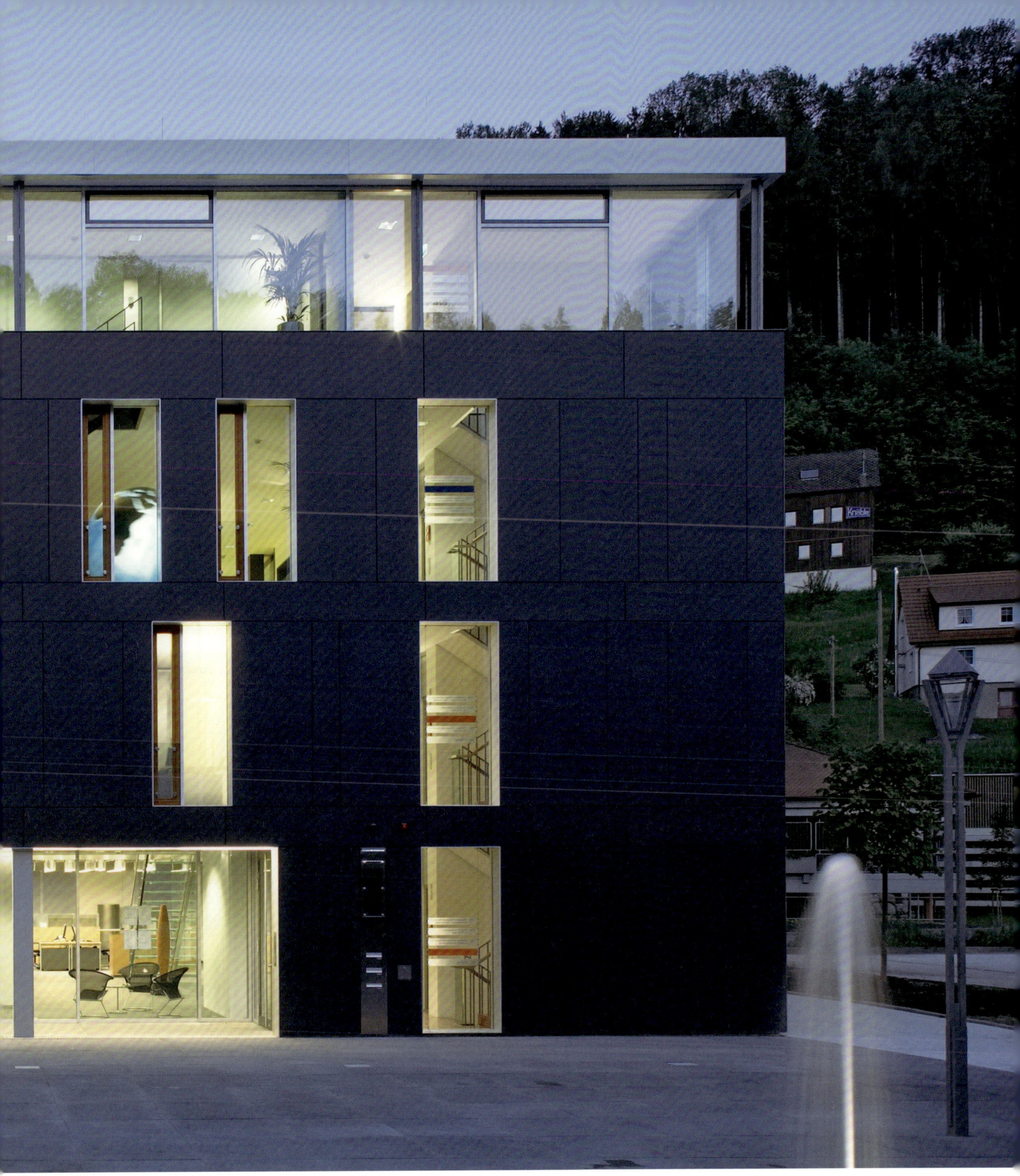

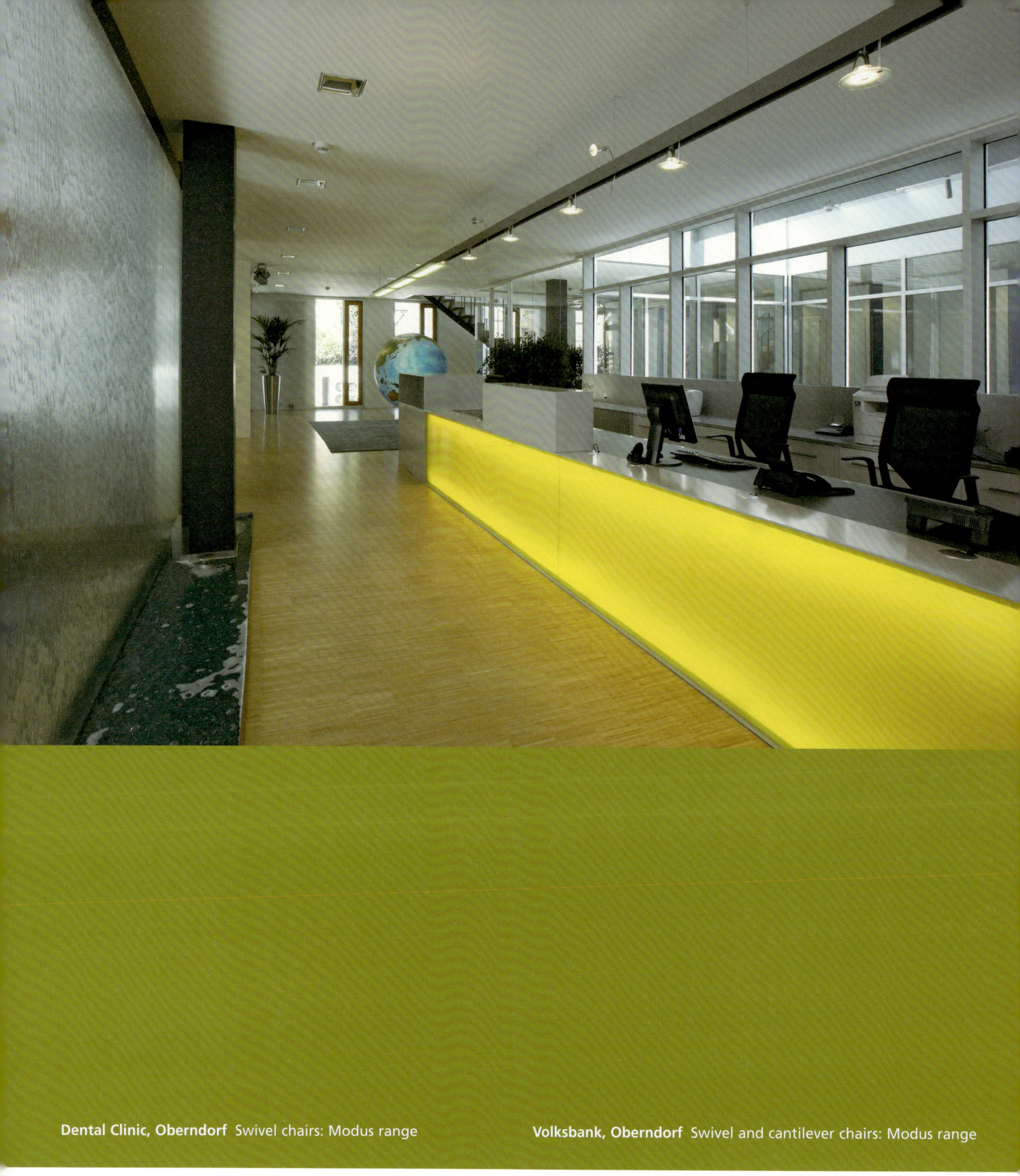

Dental Clinic, Oberndorf Swivel chairs: Modus range

Volksbank, Oberndorf Swivel and cantilever chairs: Modus range

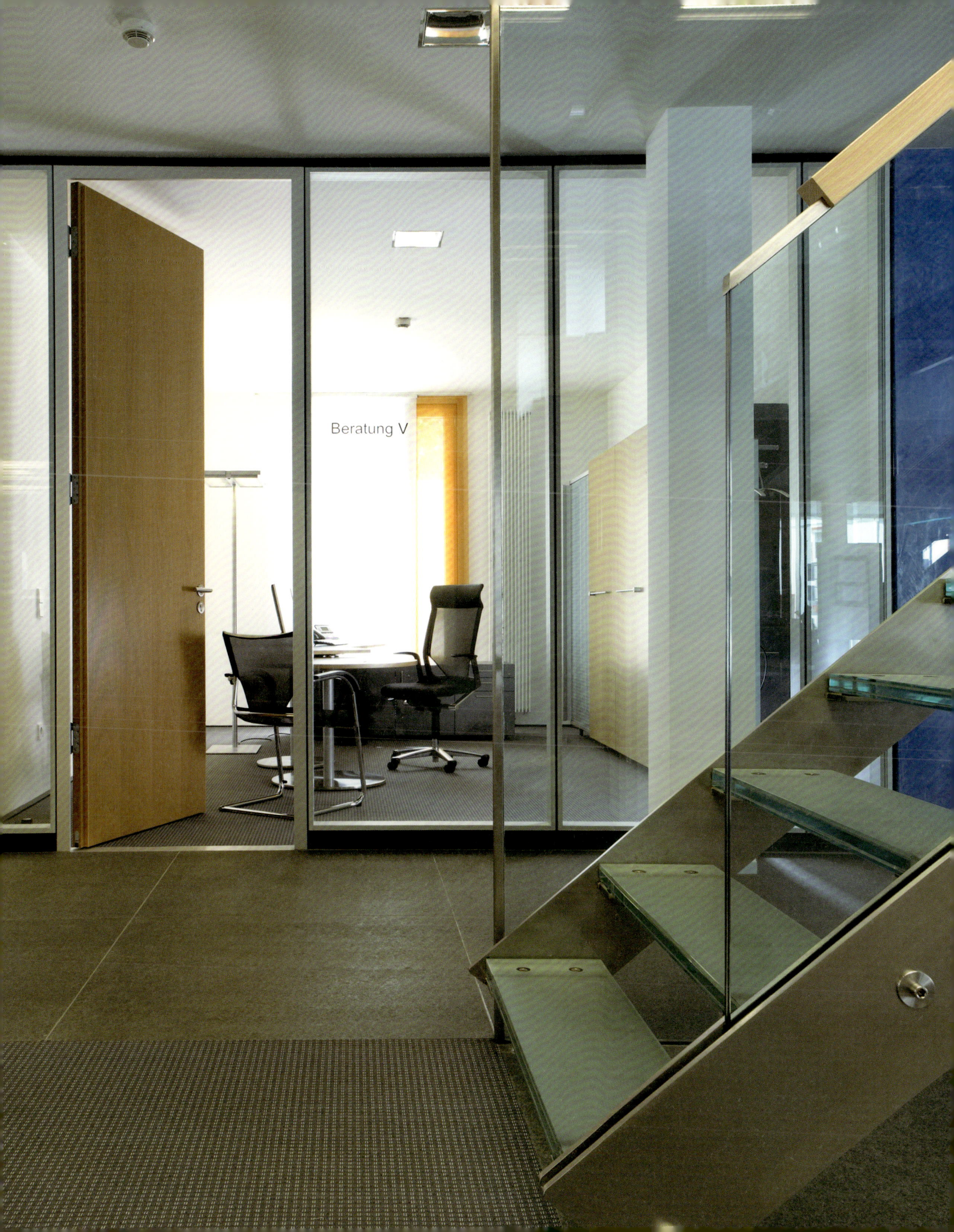
Beratung V

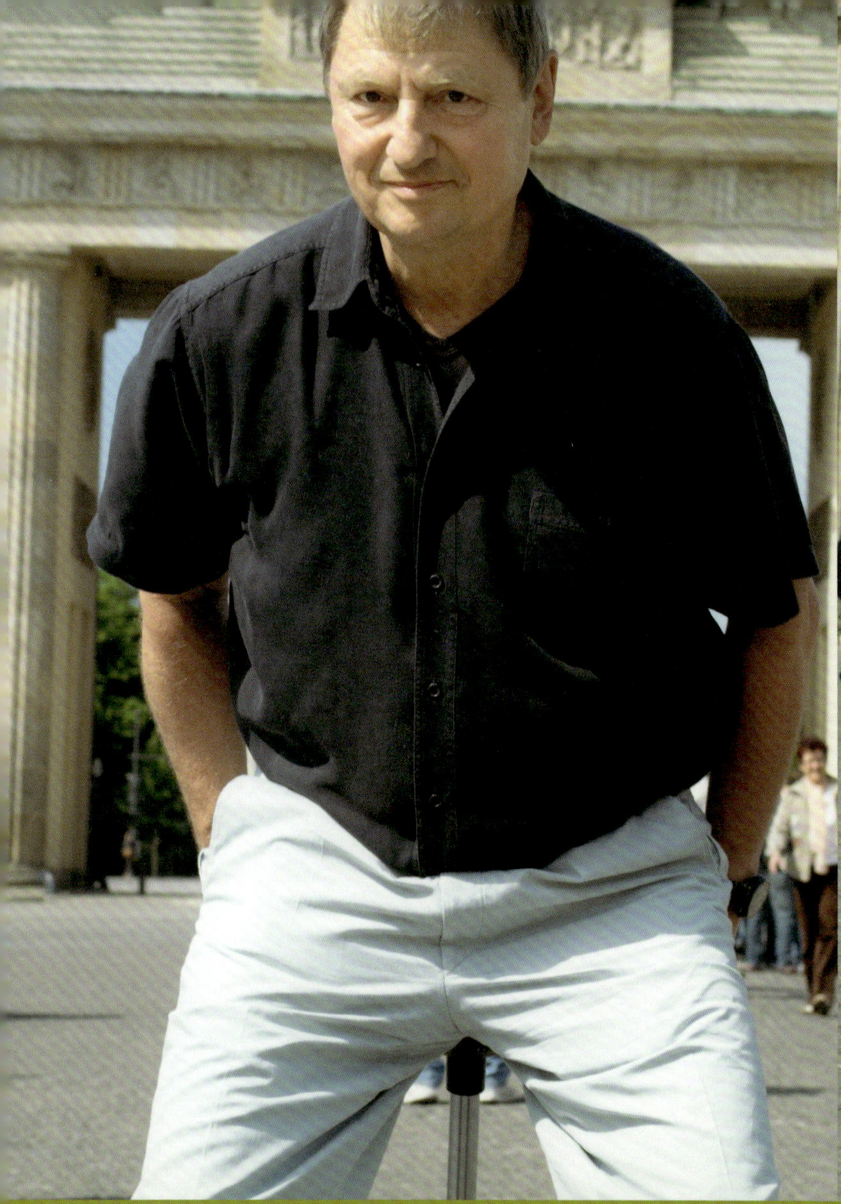

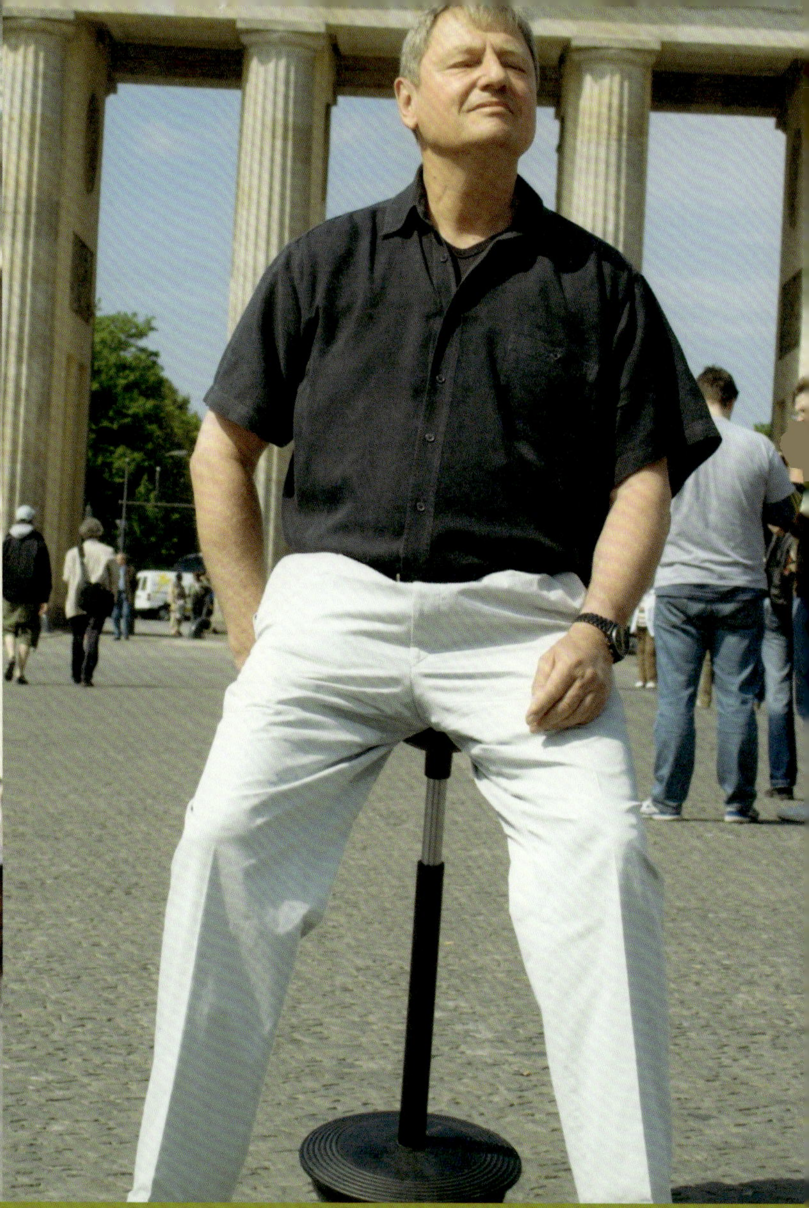

Gerhard Hootz, Berlin
„Wenn ich an die Zusammenarbeit mit Wilkhahn denke, wird mir das Herz warm! Menschen, die „menschlich" sind! Ein Partner, der ehrliche, saubere Produkte entwickelt und anbietet. Ich bin stolz, dazuzugehören und das Unternehmen von 1961 bis 1999 mit dem Einrichtungshaus Deha aktiv begleitet zu haben."

"My heart becomes warm when I think of the collaboration with Wilkhahn! People who are "human"! A partner that develops and sells honest, clean products. I am proud to belong and to have supported the company actively from 1961 to 1999 with the Deha furnishing house."

Dynamic stool: Stitz range

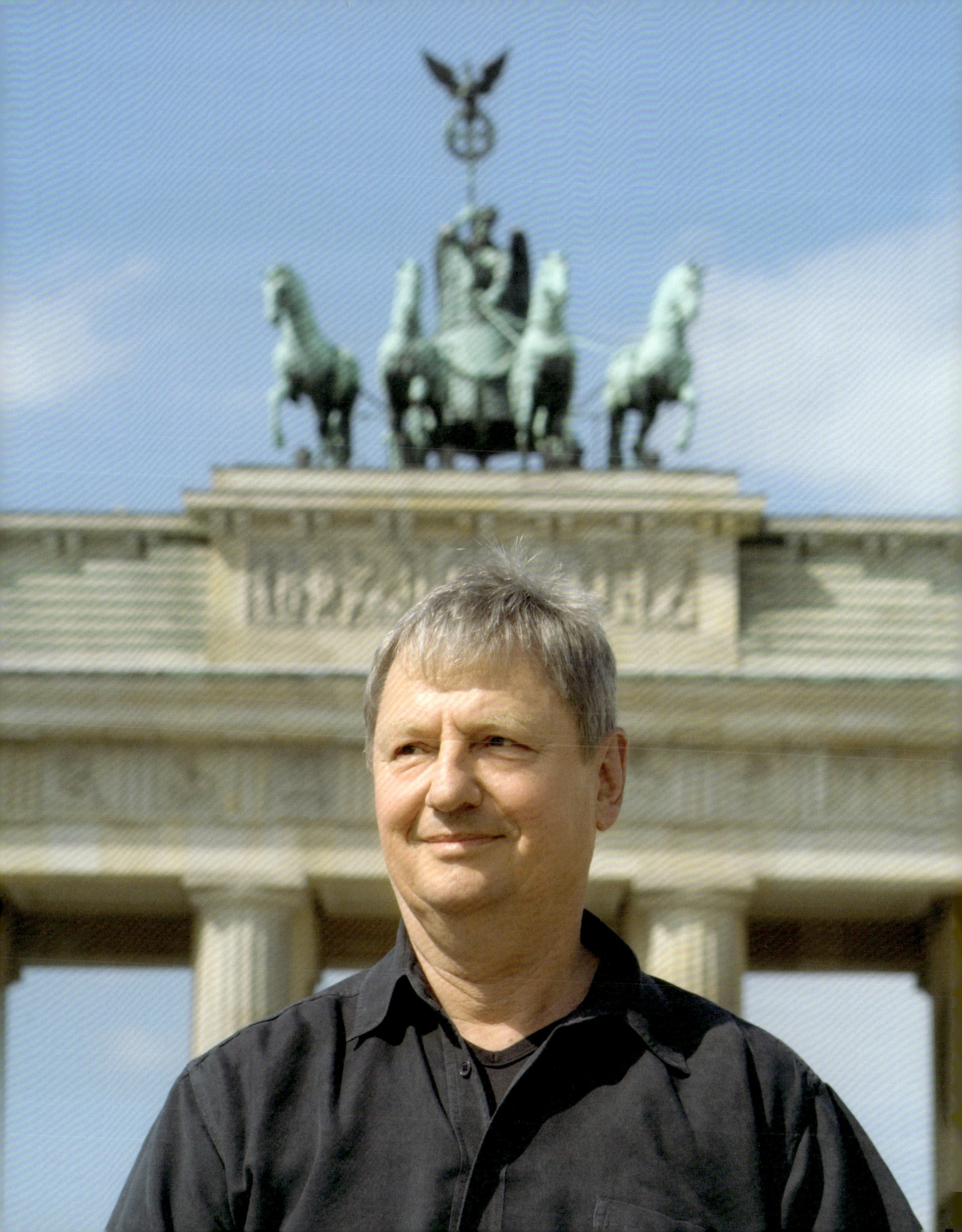

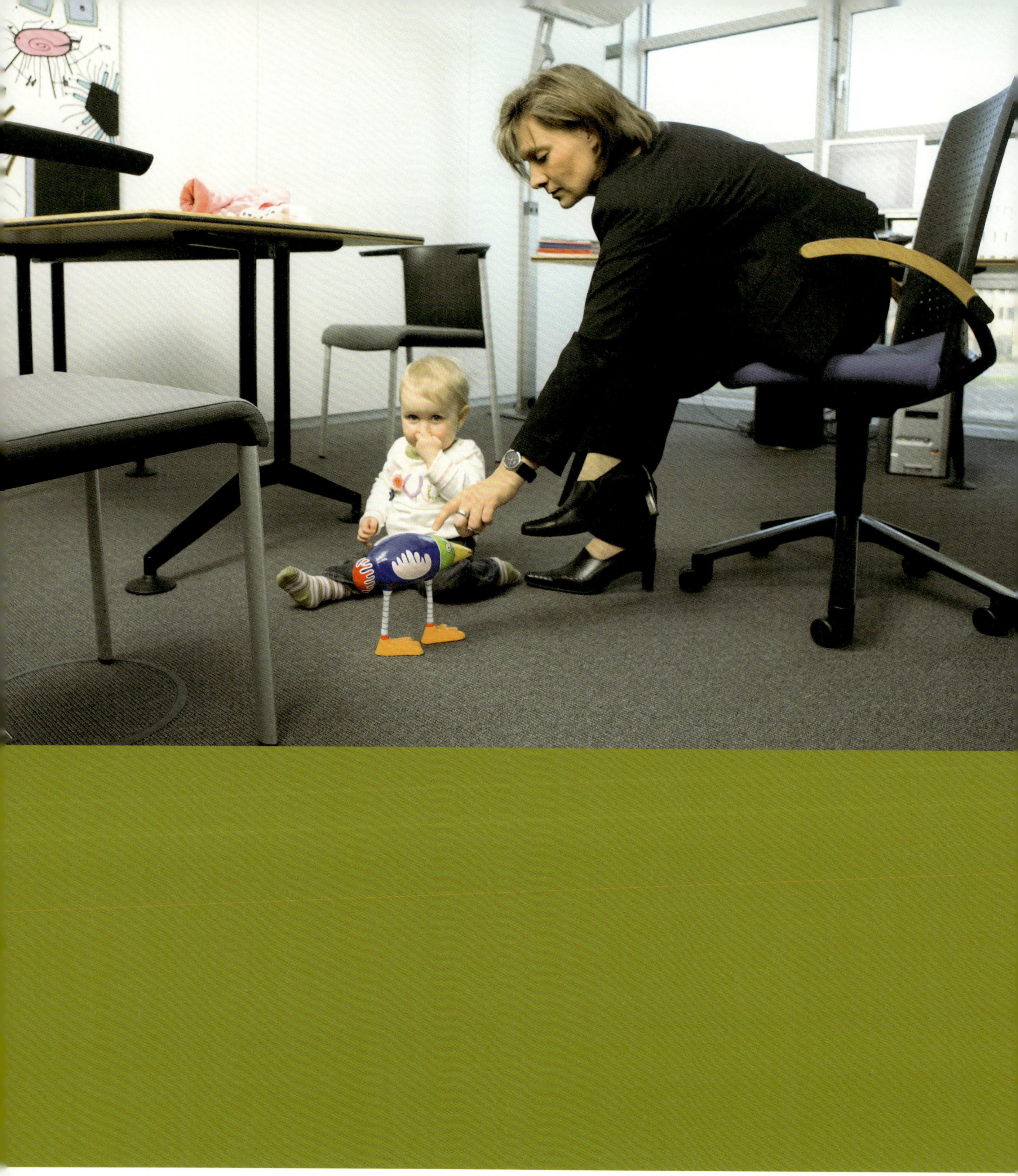

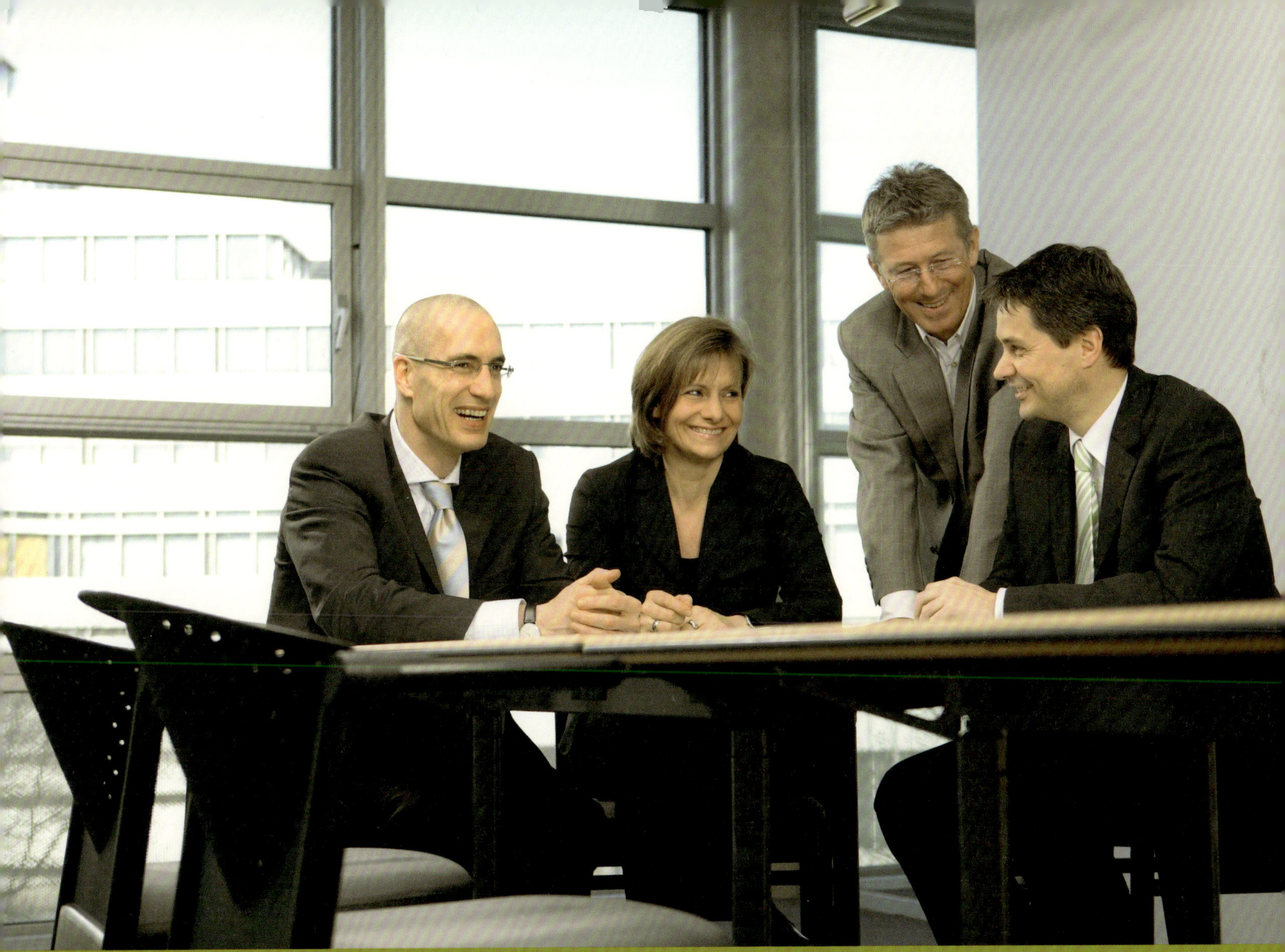

imug, Hanover Tables: Logon range, Confair folding table; swivel chair: Picto range; seminar chairs: Confair range, Kendo range

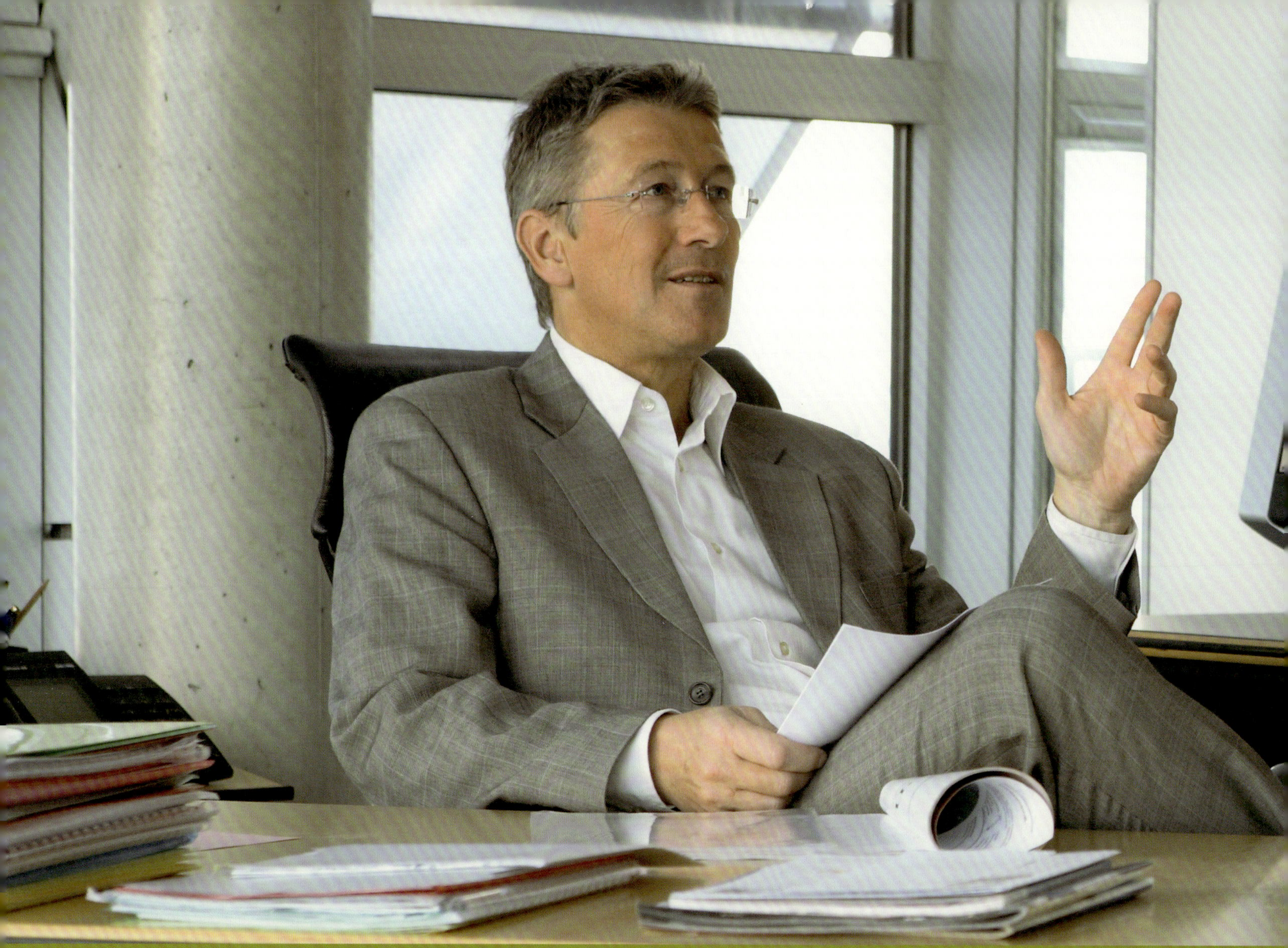

Dr. Ingo Schoenheit, Managing Partner
„imug ist eine Abkürzung und steht für Institut für Markt-Umwelt-Gesellschaft. imug ist ein Spin-off des renommierten Marketing-Lehrstuhls von Frau Prof. Dr. Dr. h.c. Ursula Hansen an der Leibniz Universität Hannover. Imug steht heute für eine konsequente Kunden- und Gesellschaftsorientierung des Marketings. Unser Analyse- und Beratungsansatz ist nachhaltig. Für uns gehören Kunden-orientierung, Führungskompetenz, Mitarbeiterorientierung und gesellschaftliche Verantwortung zusammen. Als Marketingexperten sehen wir unsere Aufgabe darin, Win-Win-Situationen mit den ‚Mitteln des Marktes' herbeizuführen. Qualität, Leistung und Fairness sollen überzeugen."

"imug is an abbreviation that stands for the Institute for Market, Environment and Society. It is a spin off of the renowned marketing chair of Prof. Dr. Dr. h.c. Ursula Hansen at the Leibniz University, Hanover. Imug today stands for consistent orientation towards customers and society with regard to marketing. Our analysis and consulting approach is sustainable. In our opinion, customer focus, executive competence, employee focus and social responsibility are linked together. As marketing experts, we see our task as one of winning-winning situations which one brings about using the 'resources of the market'. Quality, performance and fairness should be convincing."

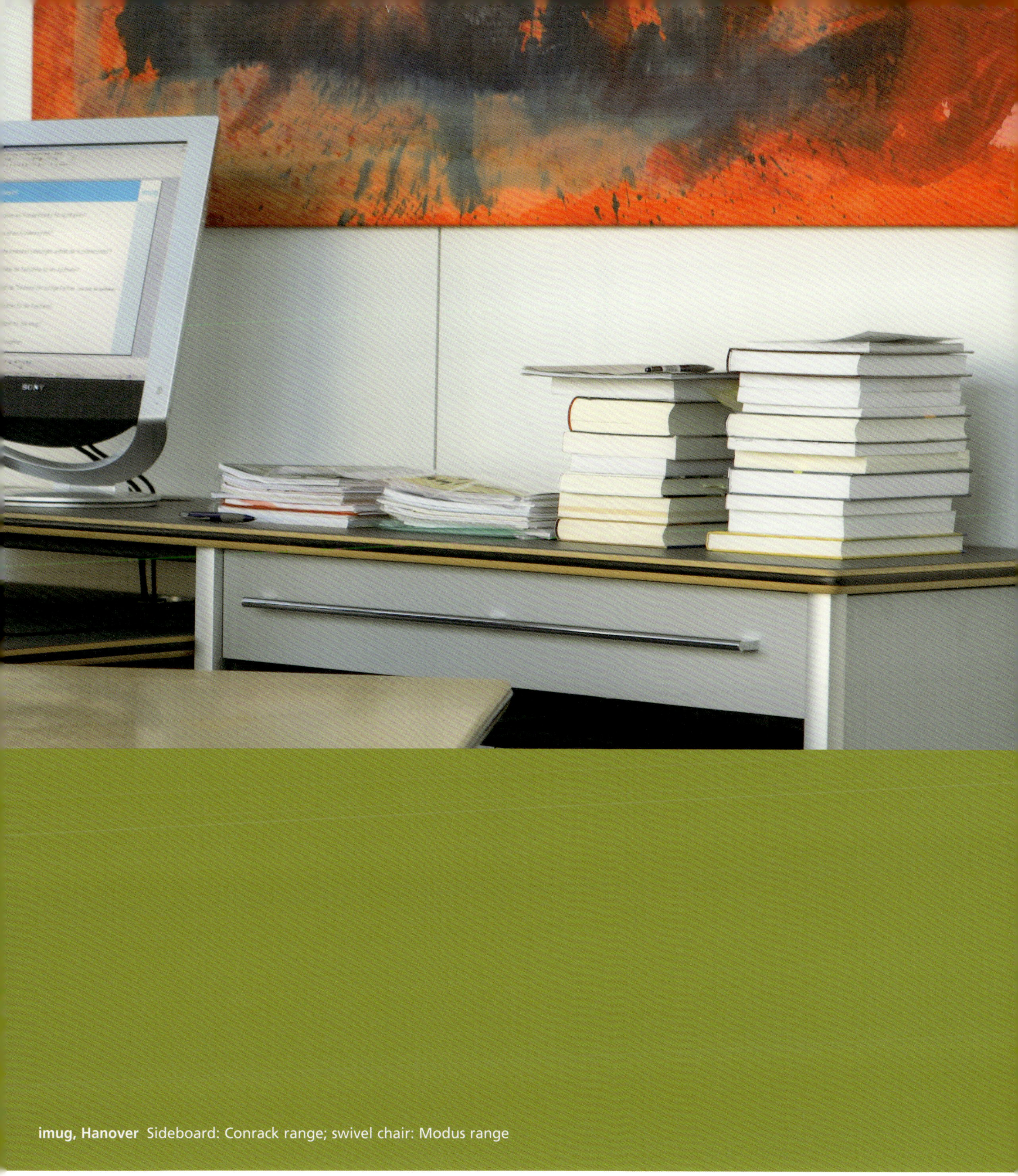

imug, Hanover Sideboard: Conrack range; swivel chair: Modus range

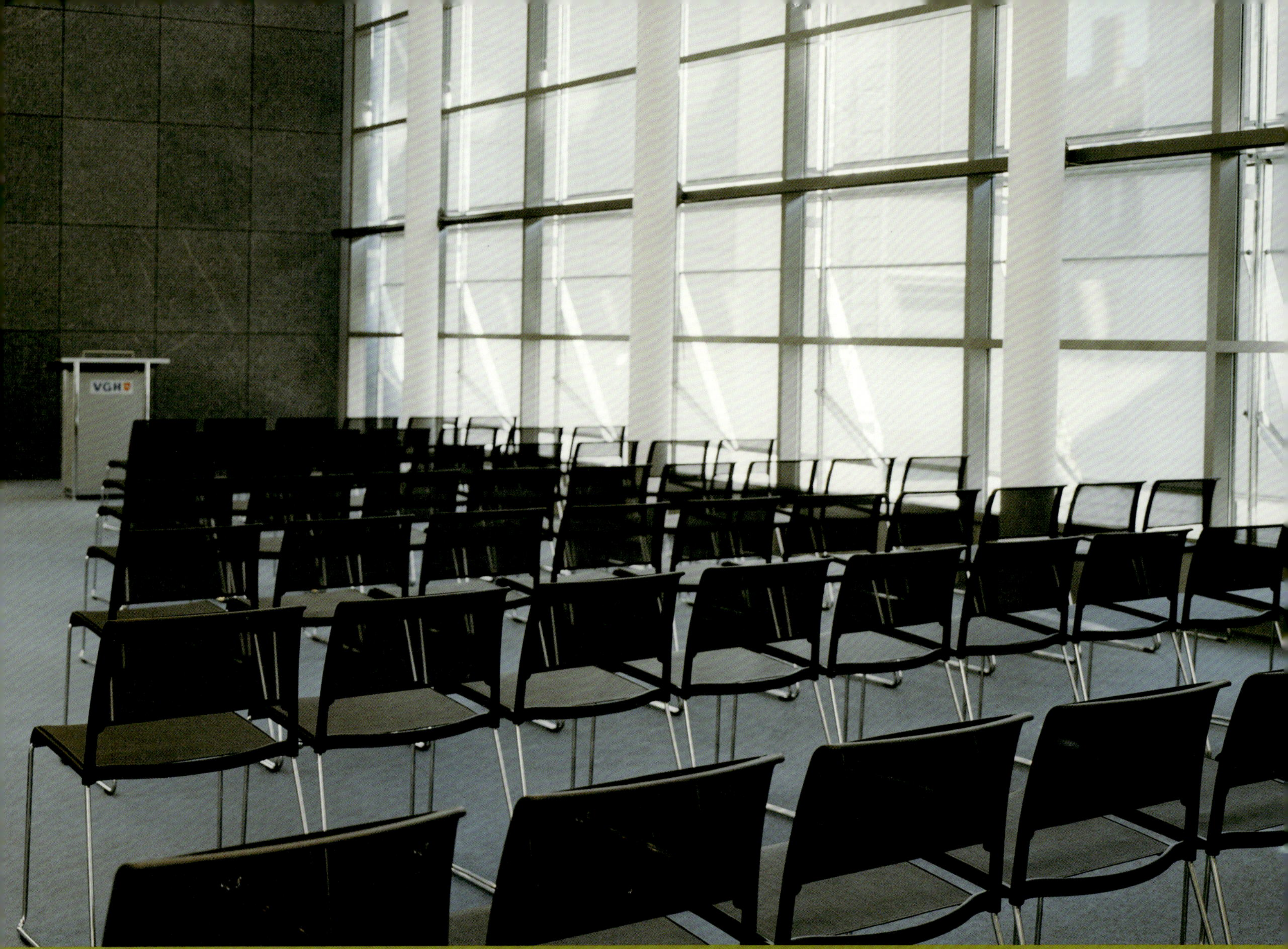

Jochen Herdecke, Member of the Board
„Als Versicherer steht für uns der faire Umgang mit Kunden und Mitarbeitern an oberster Stelle. Offenheit, Klarheit und Berechenbarkeit sind unsere Prinzipien. Davon profitieren unsere Kunden, unsere Mitarbeiter, das Unternehmen und die Menschen in unserer Geschäftsregion. Genau wie Wilkhahn verfolgen wir diese Prinzipien schon lange und erfolgreich in Niedersachsen. Wilkhahn steht für uns für Langlebigkeit, Innovation und zeitloses Design. Herzlichen Glückwunsch zum Firmenjubiläum."

"As an insurance company, fair treatment of customers and employees is of utmost importance to us. Openness, clarity and calculability are our principles that benefit our customers, employees, the company and the people in our region of business. Just like Wilkhahn, we have been following these principles successfully in Lower Saxony for a long time. Wilkhahn stands for longevity, innovation and timeless design. Congratulations on the occasion of your company anniversary."

VGH Versicherungen, Hanover Lectern: Conrack range; multi-purpose chairs: Aline range

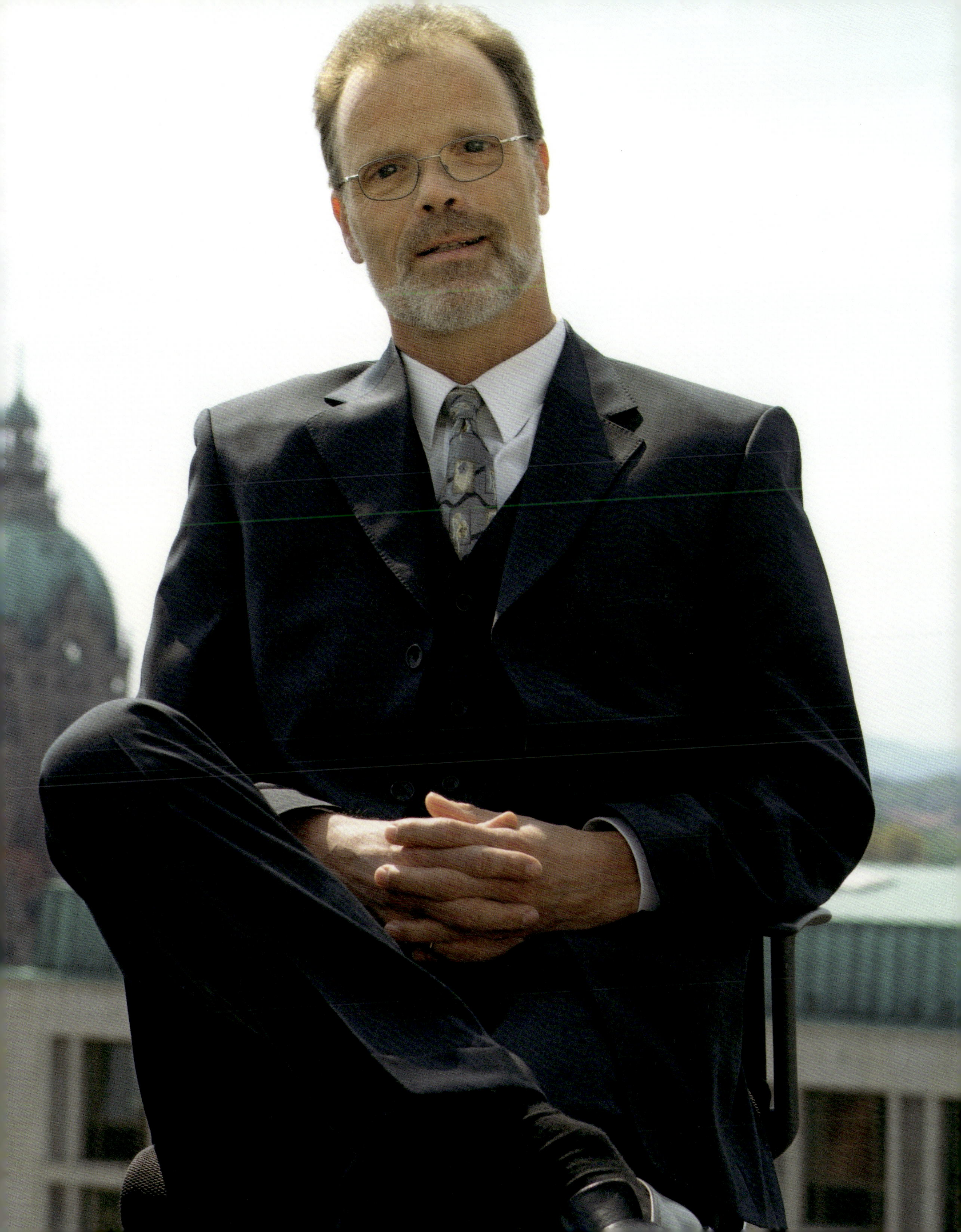

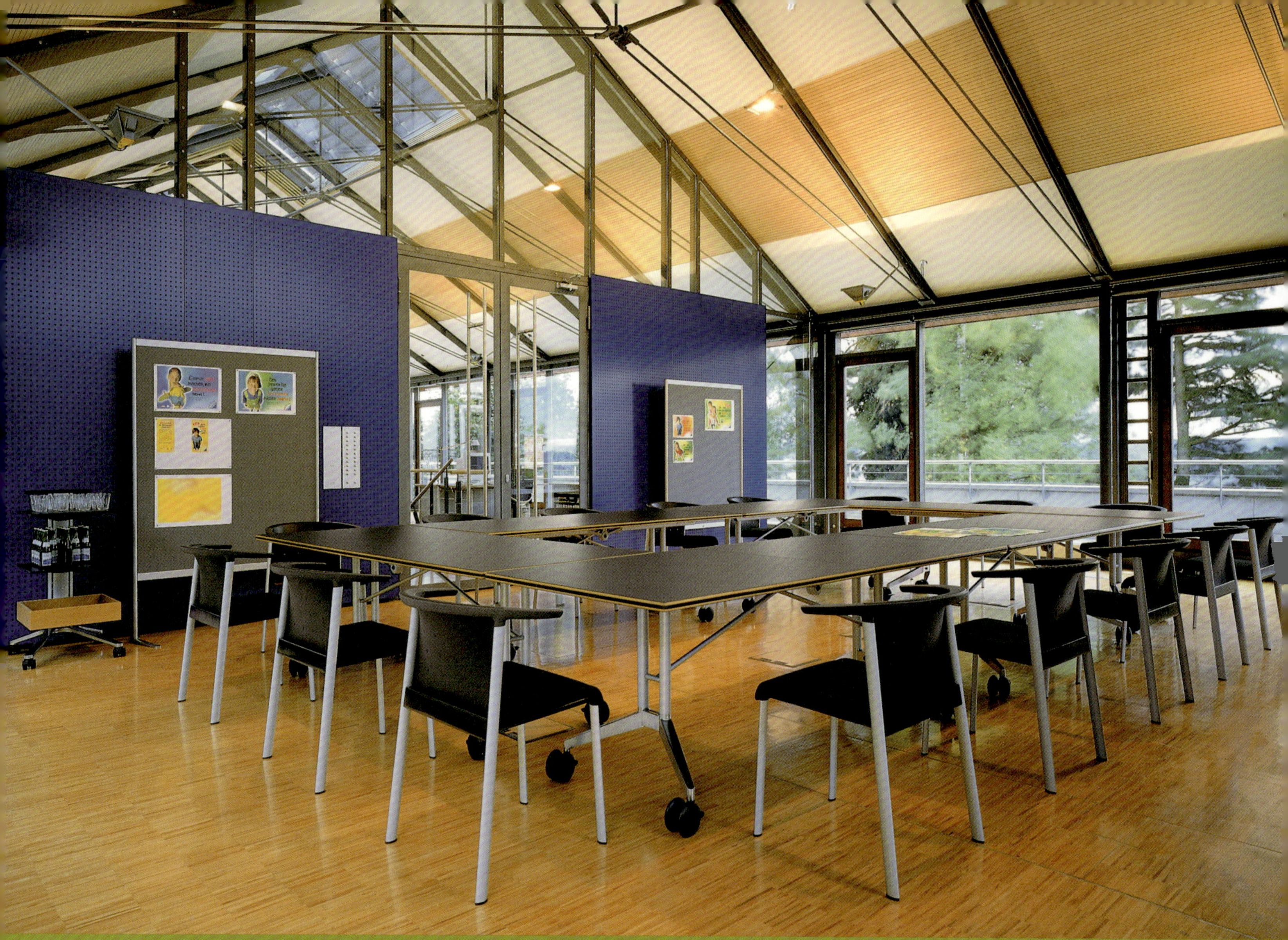

Jean-Claude Parent, Managing Partner
„Das Confair-Programm passte sowohl von seiner Ästhetik als auch Funktion am besten zu unserer Architektur und zu unserer Auffassung von kreativem Arbeiten. Darüber hinaus haben wir uns ganz bewusst für die Haltung und Konsequenz von Wilkhahn entschieden. Wir fanden, dass die Marke damit recht gut zu uns passte."

"The Confair range best suited our architecture and was in keeping with our idea of creative work with regard to its aesthetics and function. Furthermore, we chose Wilkhahn, very aware of the company's approach and consistency. We felt that the brand fitted well to us."

Schindler, Parent & Cie. GmbH, Meersburg Table configuration and seminar chairs: Confair range

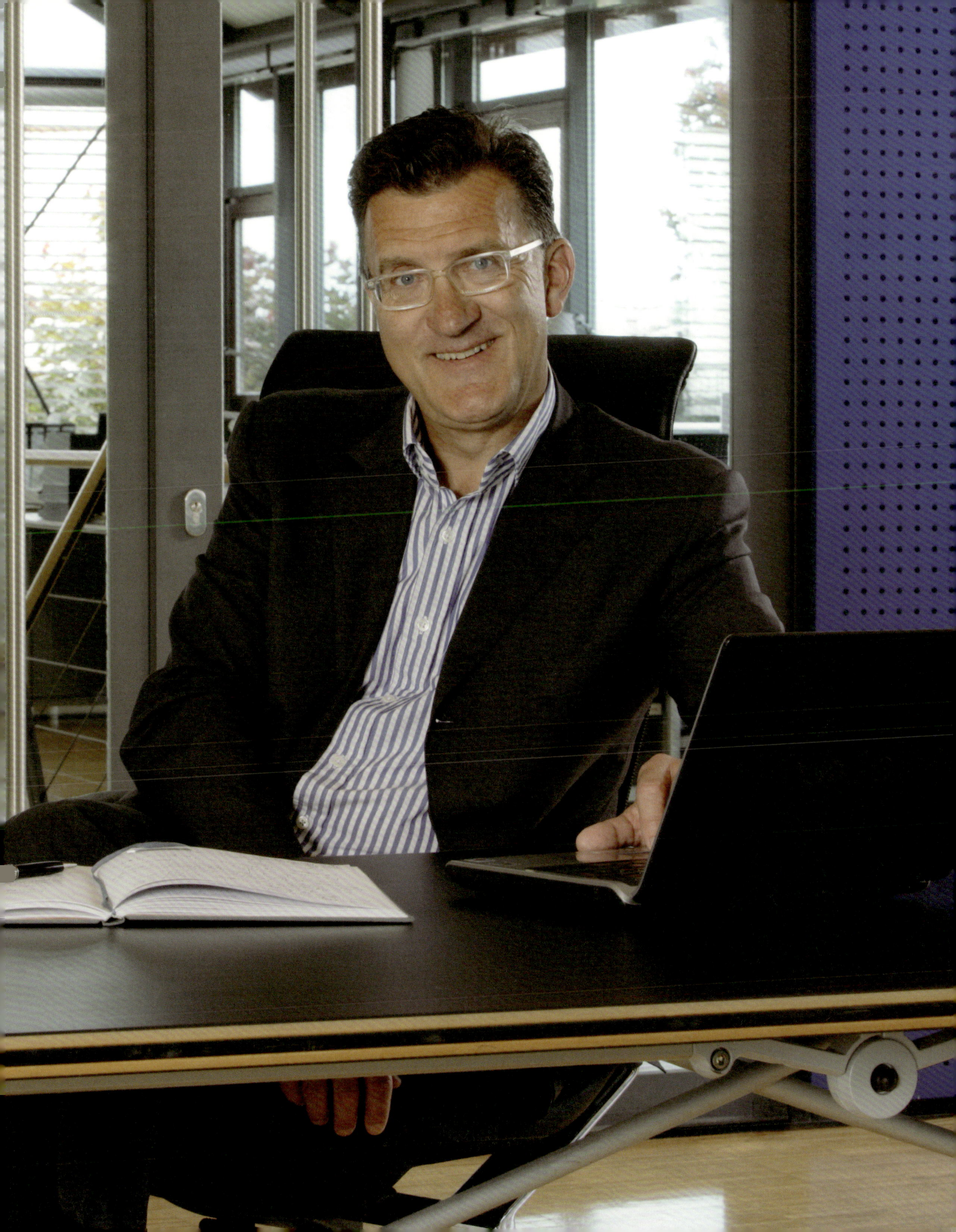

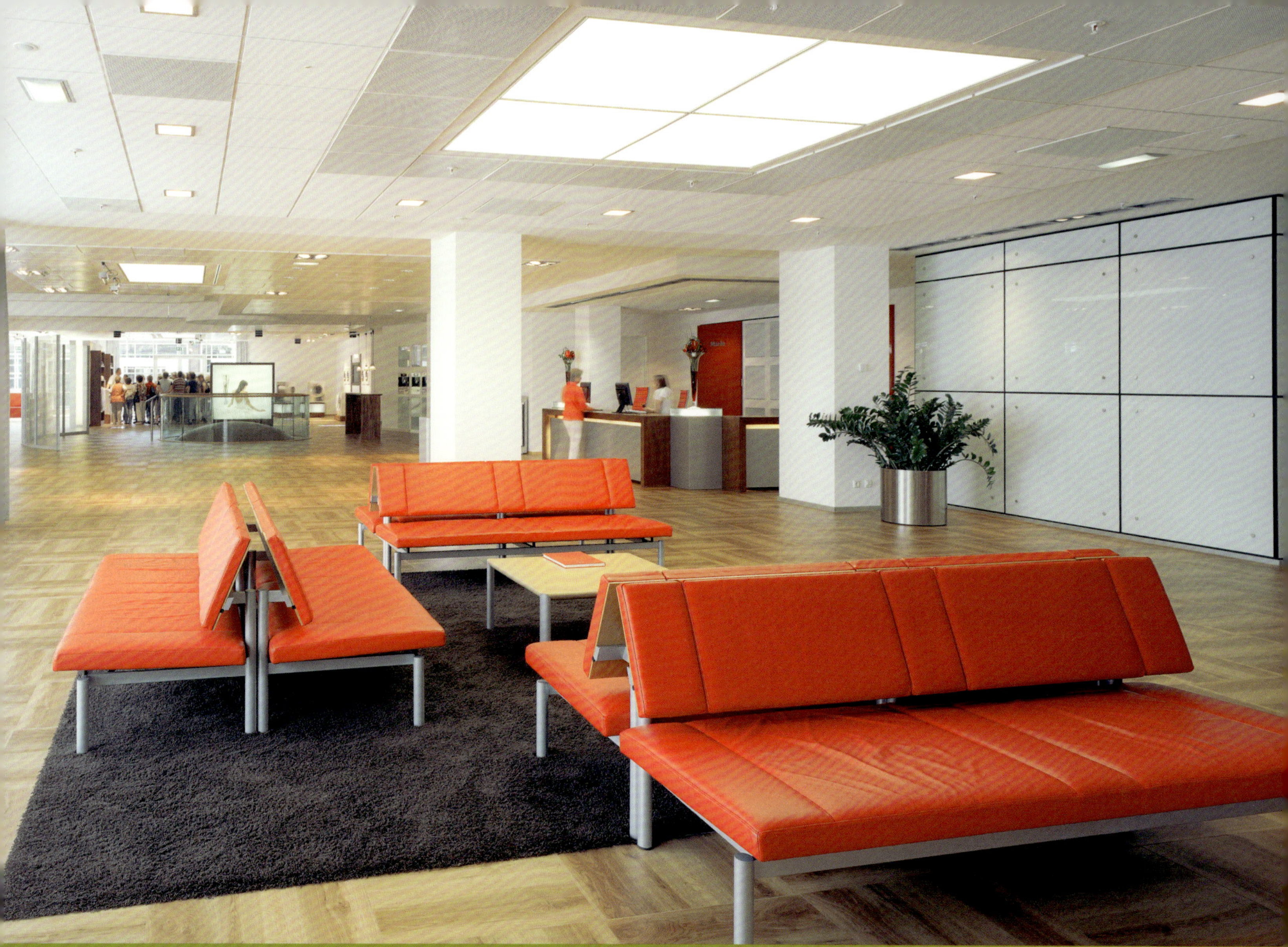

Dr. Markus Miele, Managing Partner
„Miele hat auch in Zukunft auf dem globalen Markt große Chancen, denn Qualität überzeugt. Miele ist als Marke ein Gütesiegel. Unser Firmenmotto lautet ‚Immer besser'. Ohne jede der heute so beliebten Wortspielereien ist die Aussage klar: Wir wollen selbst immer besser werden und wir wollen besser sein als alle anderen. Eine ehrgeizige, sogar etwas provozierende Leitlinie. Sie gilt seit Gründung des Unternehmens 1899, ist also genauso langlebig wie unsere Hausgeräte. Dabei immer aktuell – wie die Geräte. Es gibt immer einen Markt für die Spitze, einen Markt für das Bessere, auch wenn es teurer ist. Wir versuchen, der Zeit vorauszudenken. Wir wollen Trends setzen, statt Trends hinterherzulaufen. Unser Wachstum kommt aus der Technologieführerschaft und unserer Innovationsfähigkeit. Das ist unsere Auslegung von Qualität."

"Miele also has great future opportunities on the global market because quality is convincing. The Miele brand is a stamp of quality itself. Our corporate motto is "Continually better". The statement is clear without any play on words that is so popular nowadays: we want to continue improving and we want to be better than all the others. An ambitious, perhaps even provocative guideline. It has applied since the founding of our company in 1899 and is thus just as long-lasting as our household appliances. But always up-to date – like our appliances. There is always one brand for the top, a market for the better end even though it is more costly. We try to think future-oriented. We want to set new trends instead of running after them. Our growth emerges from the leading role in technology and our capability to be innovative. This is our interpretation of quality."

Miele & Cie. KG, Gütersloh Lounge settee: Cana range; swivel chair: Solis F range

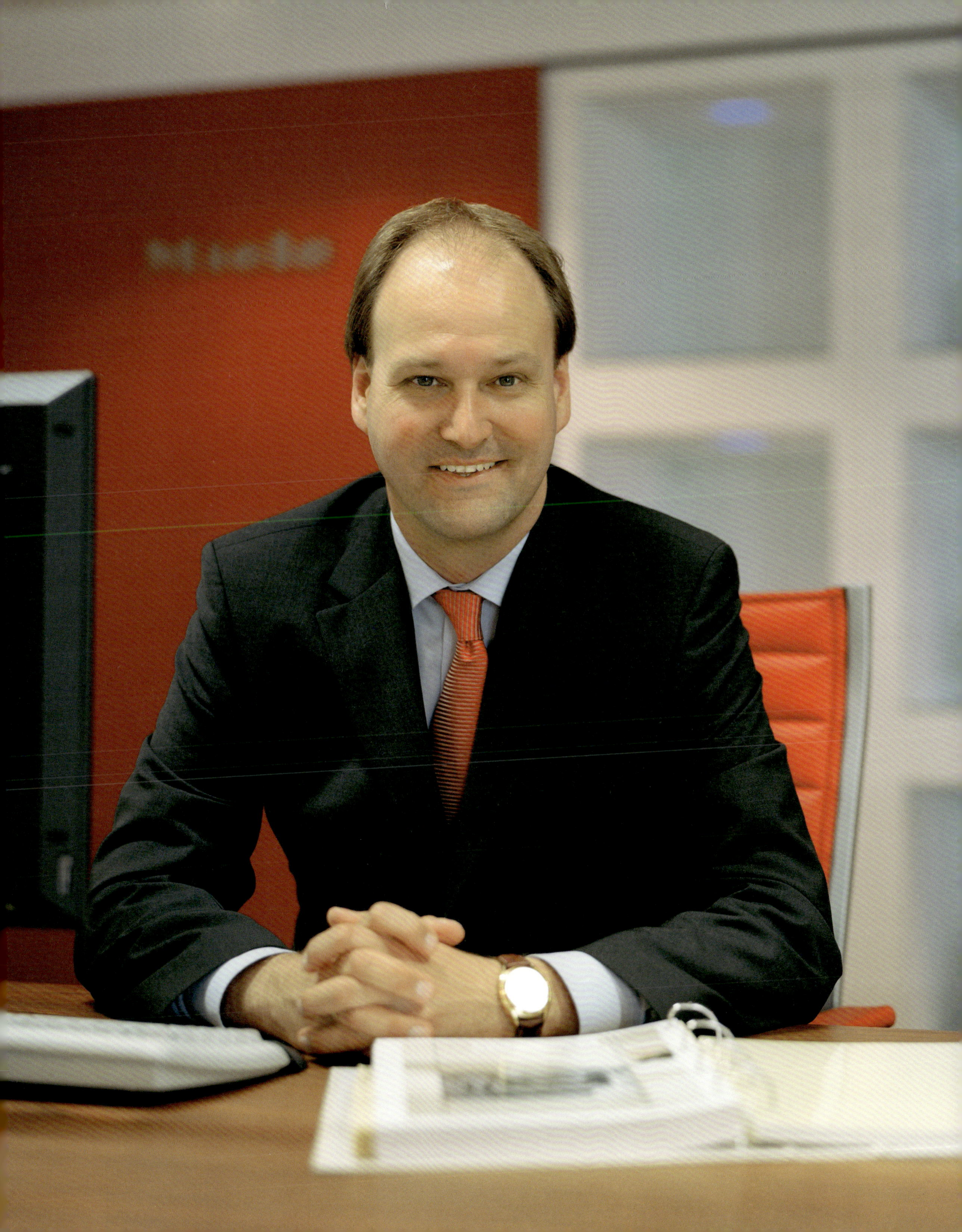

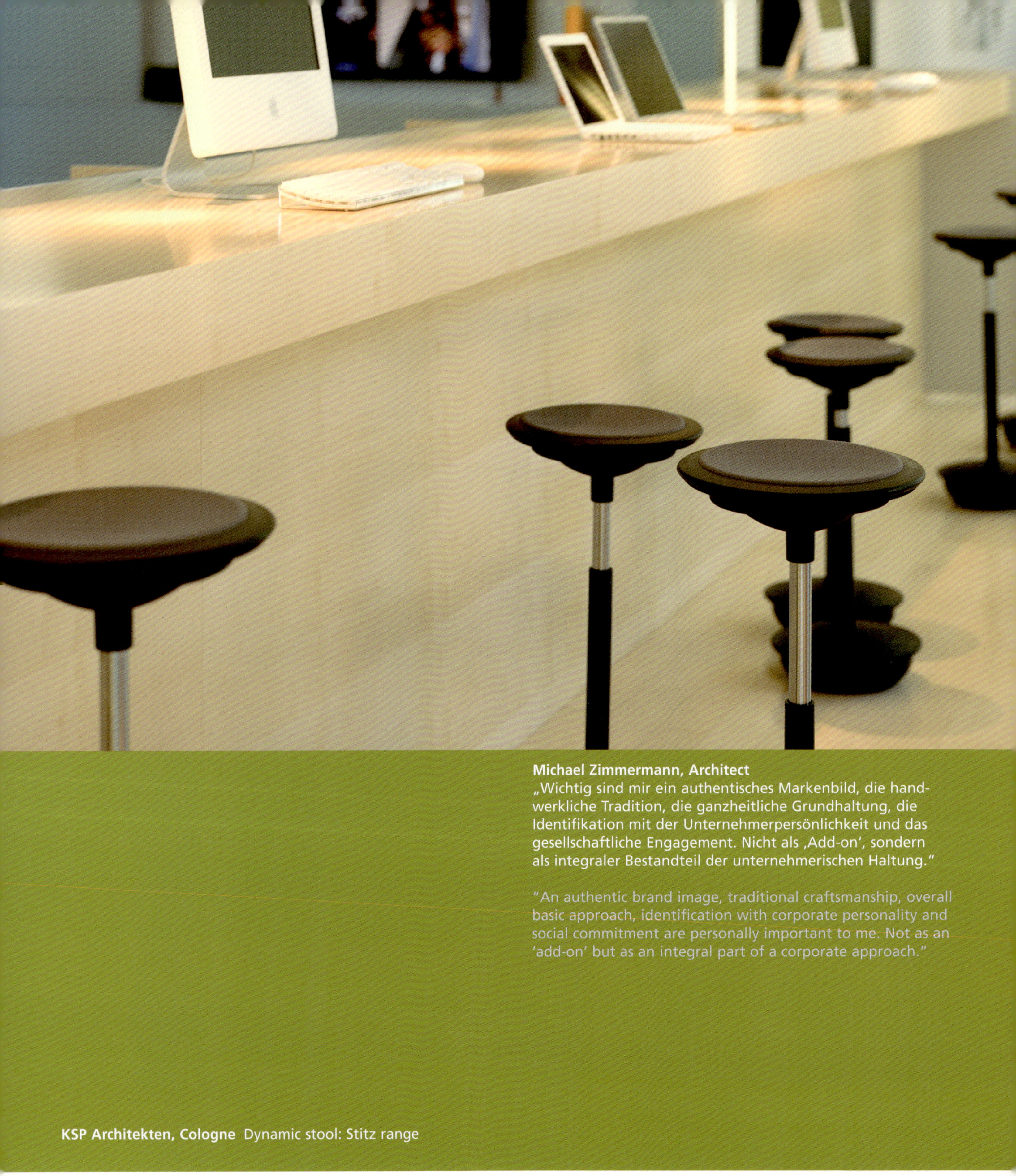

Michael Zimmermann, Architect
„Wichtig sind mir ein authentisches Markenbild, die handwerkliche Tradition, die ganzheitliche Grundhaltung, die Identifikation mit der Unternehmerpersönlichkeit und das gesellschaftliche Engagement. Nicht als ‚Add-on', sondern als integraler Bestandteil der unternehmerischen Haltung."

"An authentic brand image, traditional craftsmanship, overall basic approach, identification with corporate personality and social commitment are personally important to me. Not as an 'add-on' but as an integral part of a corporate approach."

KSP Architekten, Cologne Dynamic stool: Stitz range

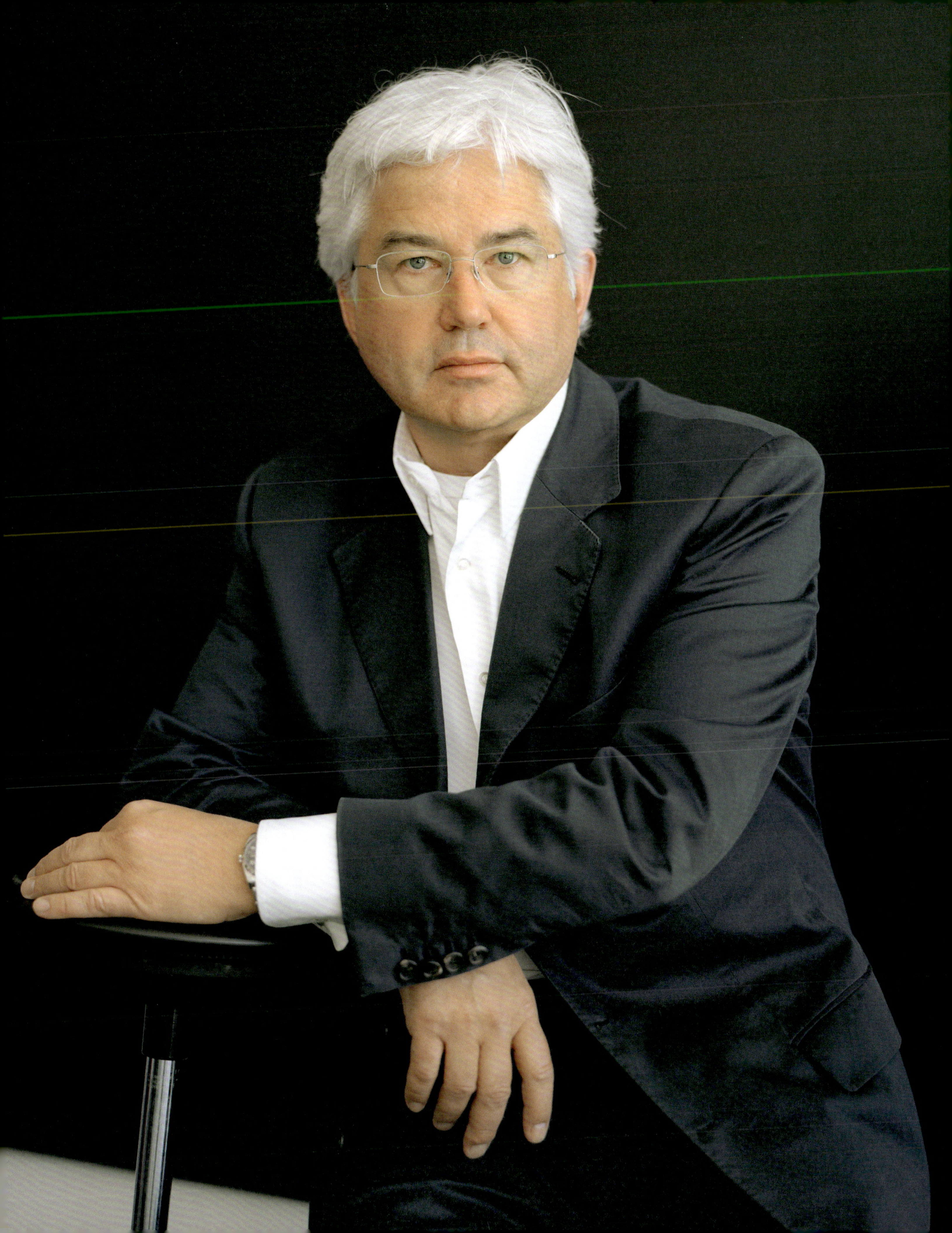

Dietmar Spielmann
„Erfolg braucht Design: Das ist die unternehmerische Erfahrung vieler Jahrzehnte. Die Klarheit, die wir von einem gelungenen Produktdesign erwarten, sollte auch im Kopf jedes Unternehmers herrschen – verbunden mit dem unverzichtbaren Schuss Esprit, mit kreativer Unruhe. Und wo das der Fall ist, da besteht dann meist auch eine ausgeprägte Neigung, sich bei der Einrichtung der Firma für Design zu entscheiden. Wer das tut, weiß, dass er etwas zu sagen hat. Hier hat man als Händler die Chance, zum wahren Partner zu werden, die originelle Verbindung von Rationalität und Ästhetik zu finden, eben richtiges Design. Damit, mit Design for Business, haben wir die besten Erfahrungen gemacht. Was mit dem Werkbund, mit dem Bauhaus und Herstellern wie Wilkhahn begann, ist längst zum Ausdruck unserer kommerziellen Kultur geworden. Er besitzt, abgesehen von allen funktionellen Vorzügen, eine ideelle Wertigkeit. Dass das alles auch seinen Preis hat, ist ein Glück für den Händler, solange die Kunden den Wert zu schätzen wissen, solange es Produkte gibt, die ihren Ansprüchen genügen. Aber dabei können wir uns ja getrost auf Wilkhahn verlassen. 100 Jahre Tradition mit immer neuen Ideen versprechen eine gute Zukunft für das Design made in Germany."

"Success requires design: this is the corporate experience of many decades. The clarity we expect from a successful product design should also be dominant in the mind of every entrepreneur and indisputably linked with a dash of esprit and creative restlessness. And where this is the case, there is usually a very distinctive inclination to decide for design when furnishing the company. Those who do this know that they have something to say within a company. In this case, the dealer has the chance to become a true partner, to find the original combination of rationale and aesthetics, exactly the right design. Our experience has been very satisfactory with design for business. What began with the Werkbund, Bauhaus and manufacturers like Wilkhahn has long become an expression of our commercial culture. Apart from all functional advantages, it has non-material value. That this all has its price is a fortunate circumstance for the dealer as long as the customers are able to appreciate the value, as long as there are products which meet their demands. But this is where we need have no fears about relying on Wilkhahn. One hundred years of tradition with a constant flow of new ideas are a promising future for design made in Germany."

Spielmann's Officehouse, Kronberg

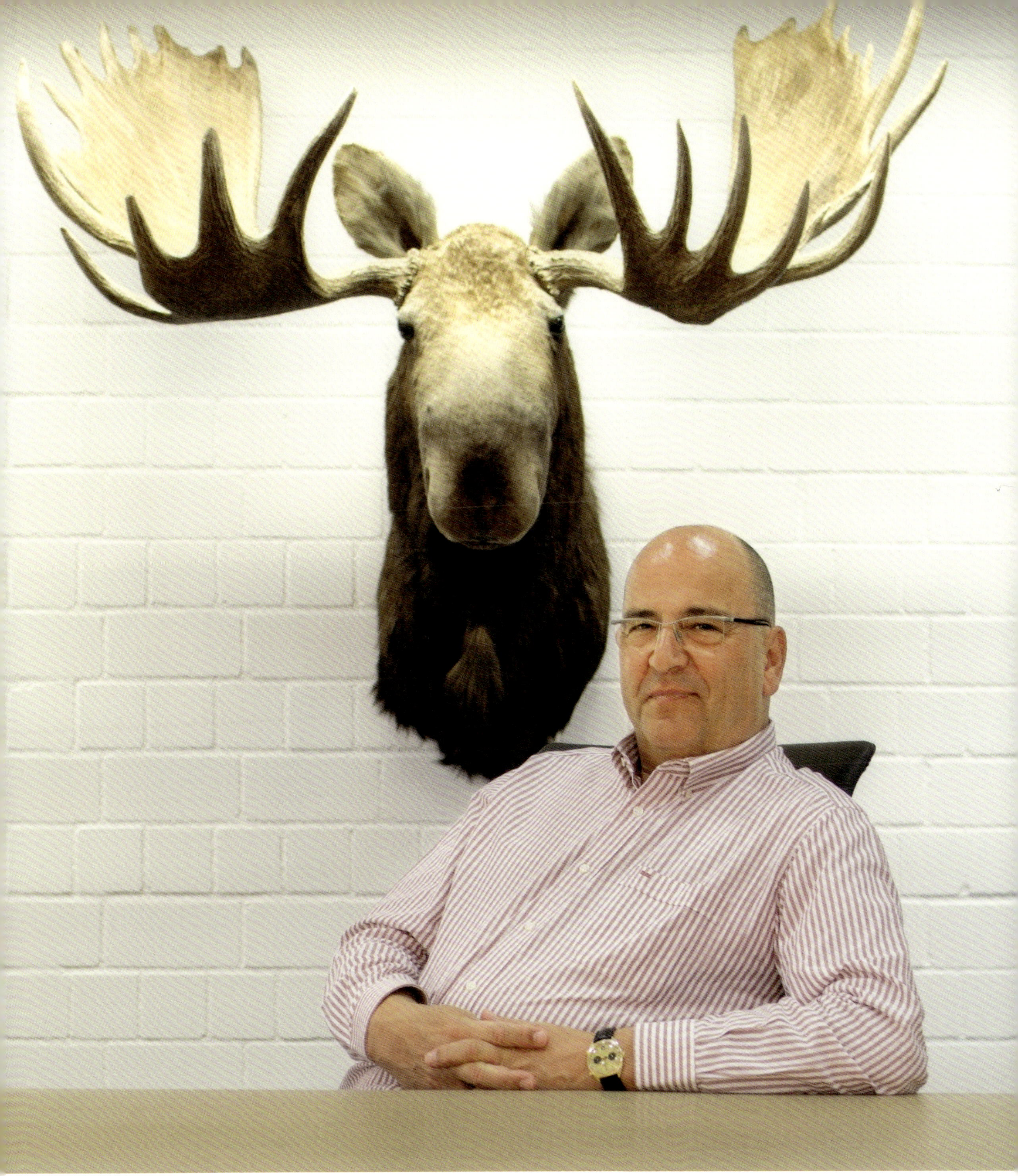

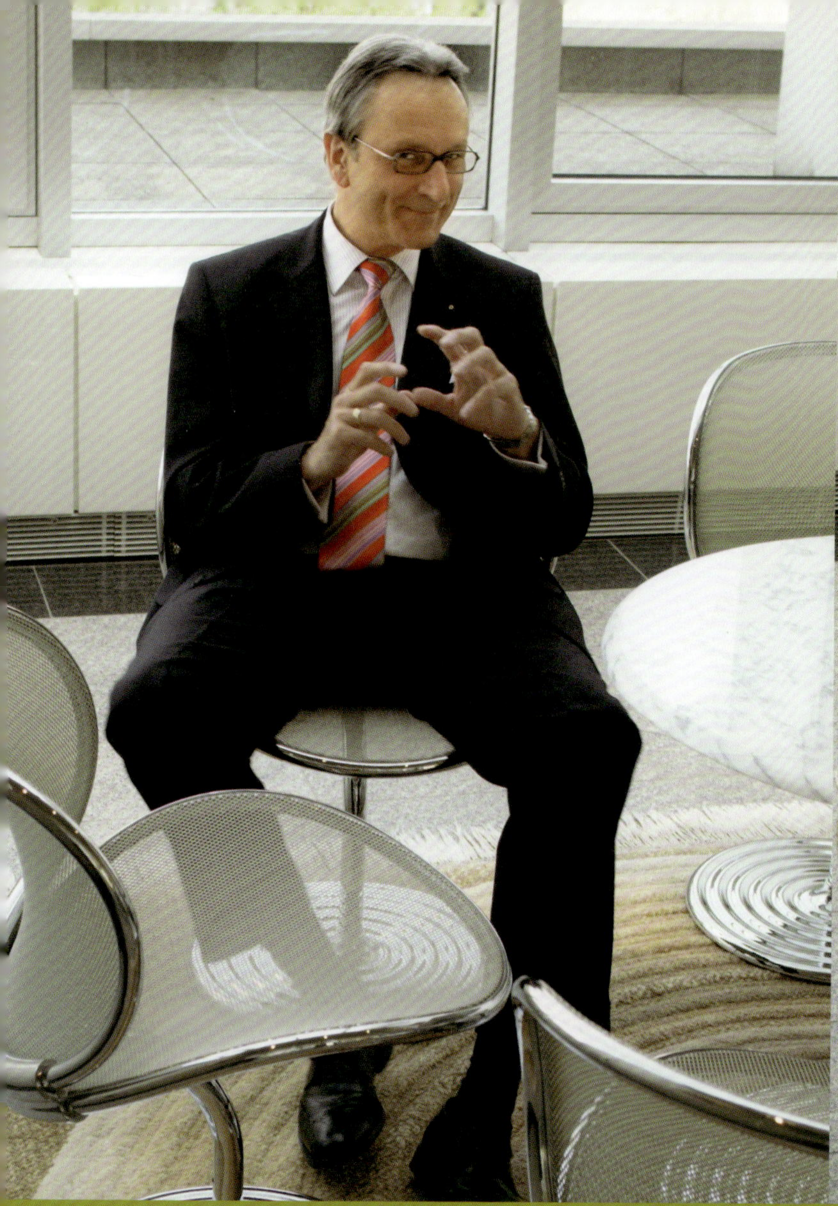

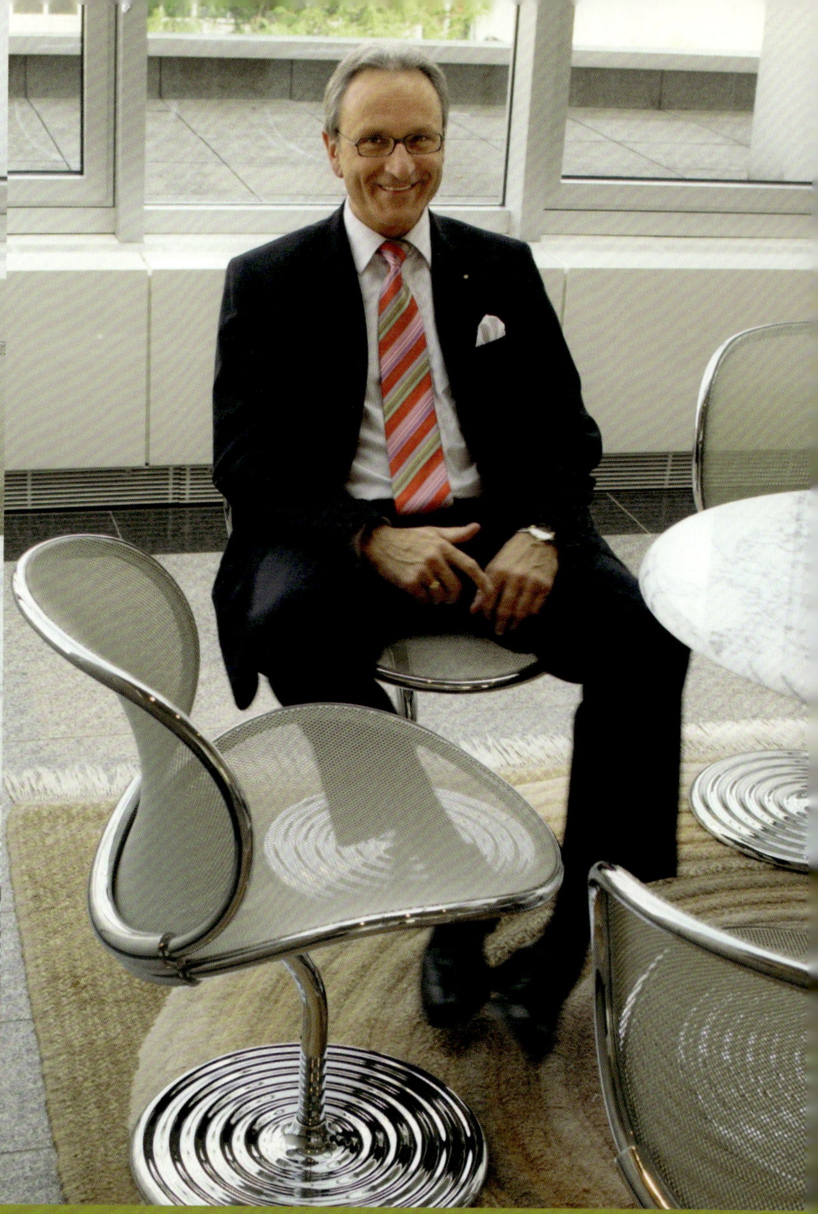

Martin Grapentin, Chairman
„Die Landessparkasse zu Oldenburg gehört zu den größten Sparkassen in Deutschland. 220 Jahre Geschichte sind Ausdruck von Verlässlichkeit, Treue und Verbundenheit mit dem alten Oldenburger Land. Zugleich stehen sie für eine erfolgreiche Kombination von Tradition, Kompetenz und Innovation."

"The Landessparkasse zu Oldenburg is one of the largest savings banks in Germany. A history of 220 years is testimony of reliability, trust and of closeness to the traditional Oldenburg region. At the same time, this stands for a successful mix of tradition, competence and innovation."

Landessparkasse zu Oldenburg, Oldenburg Lounge chair: O-line range

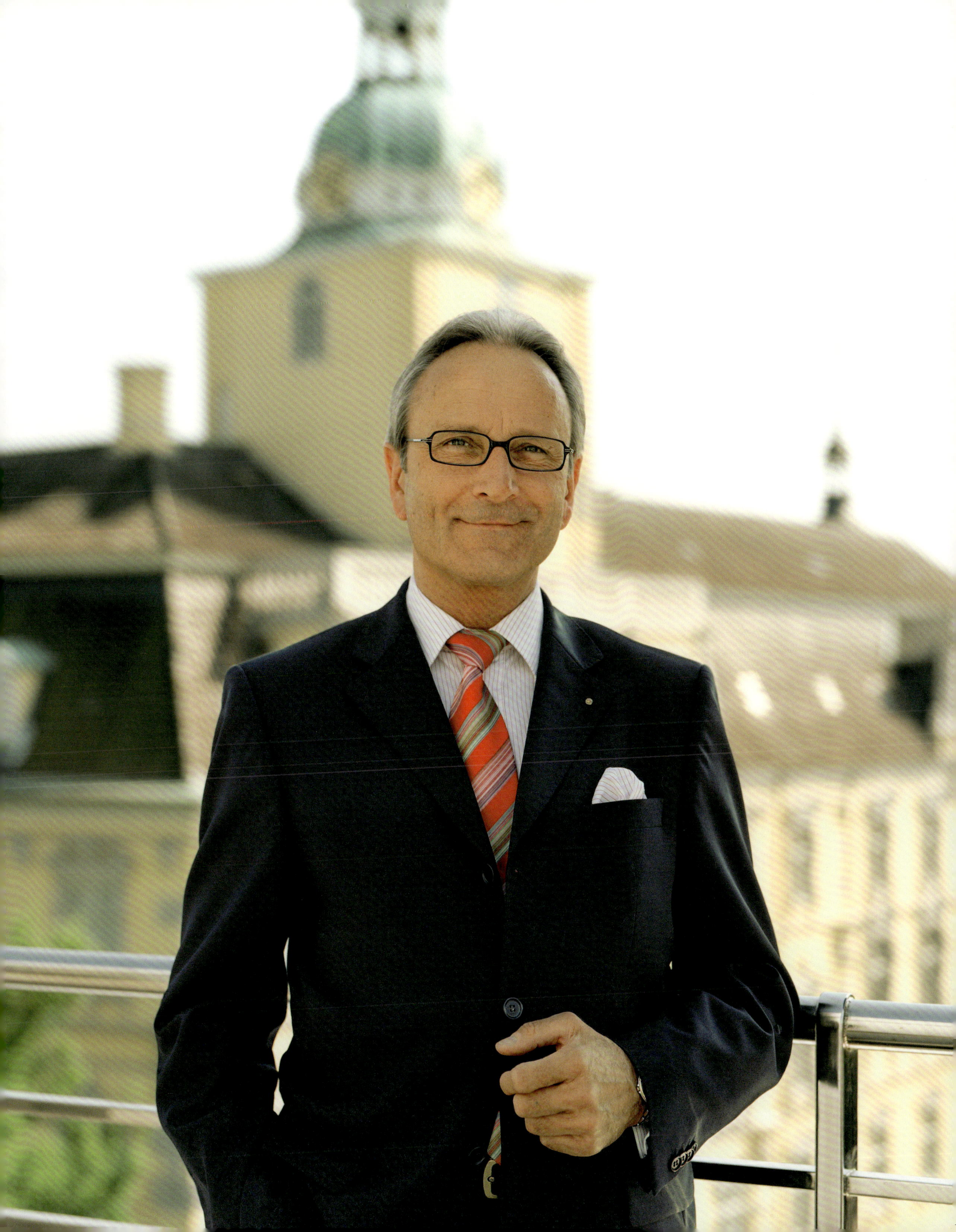

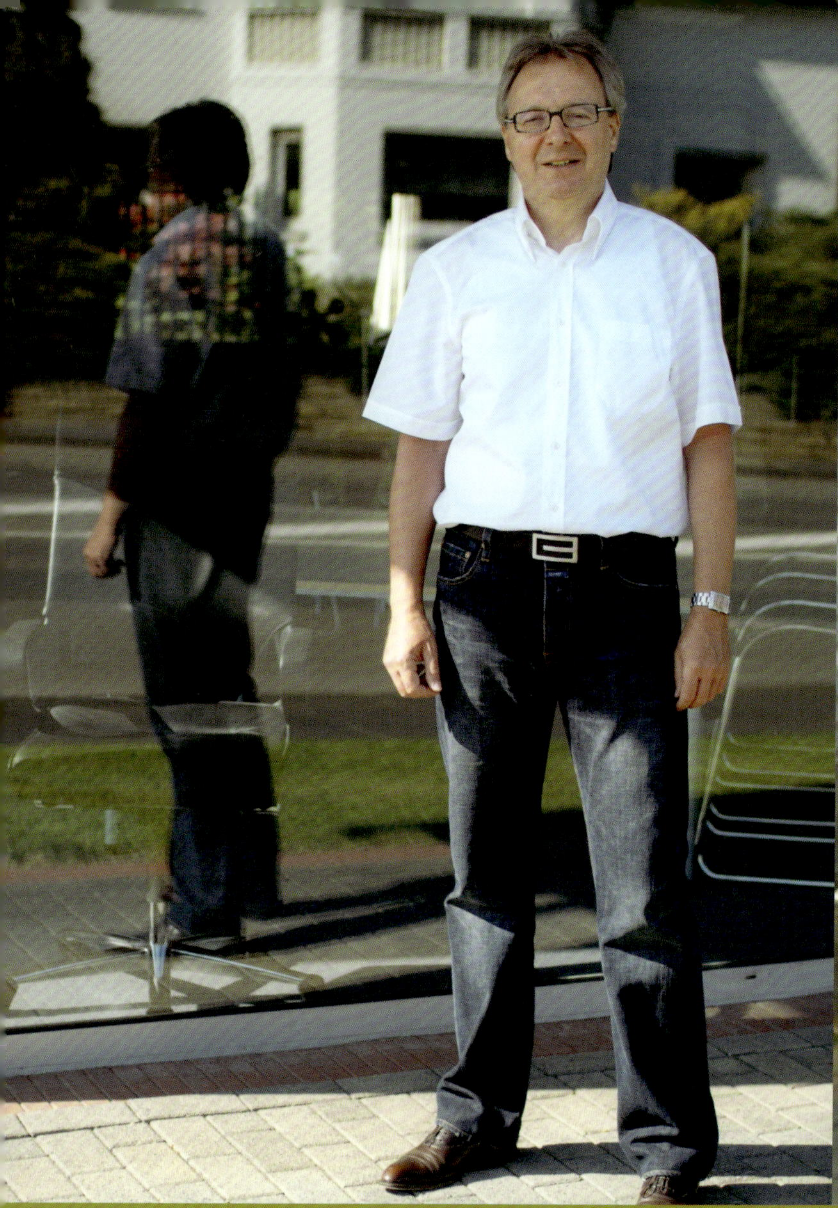

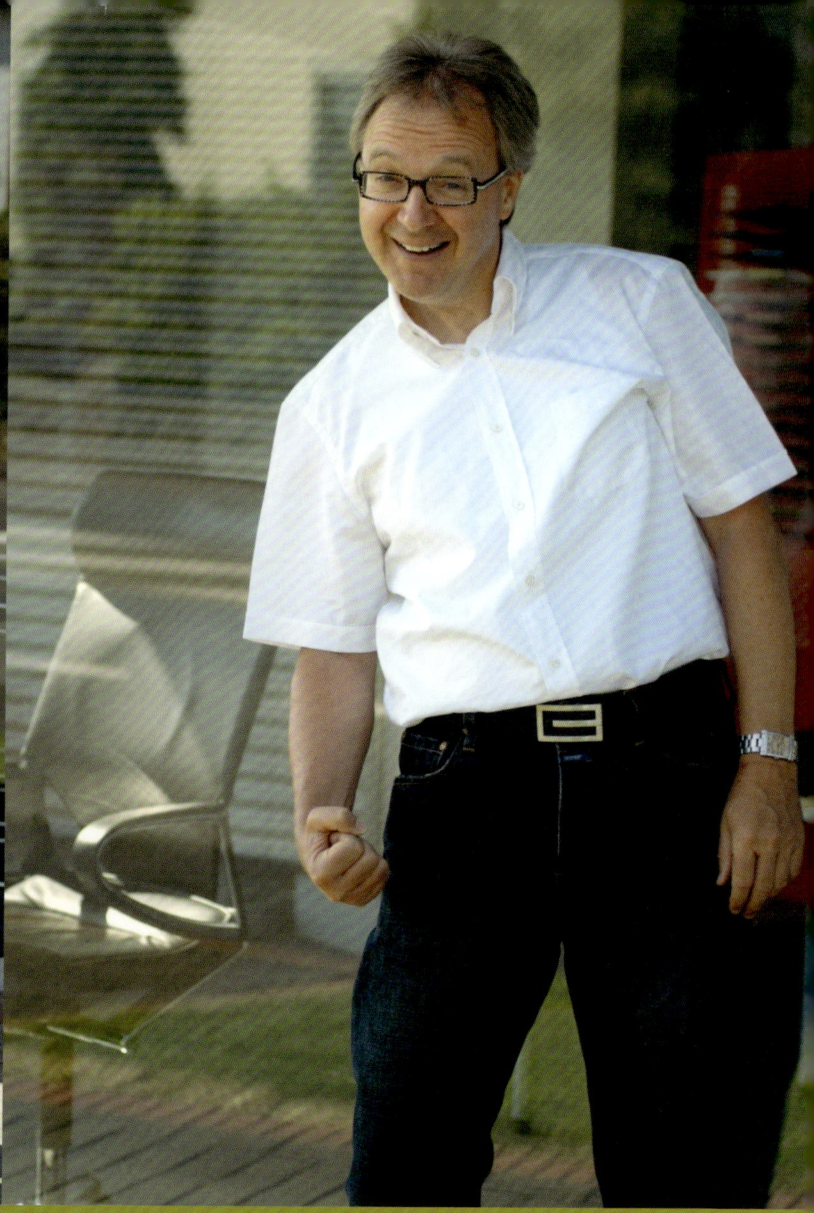

Pro Office, Lemgo

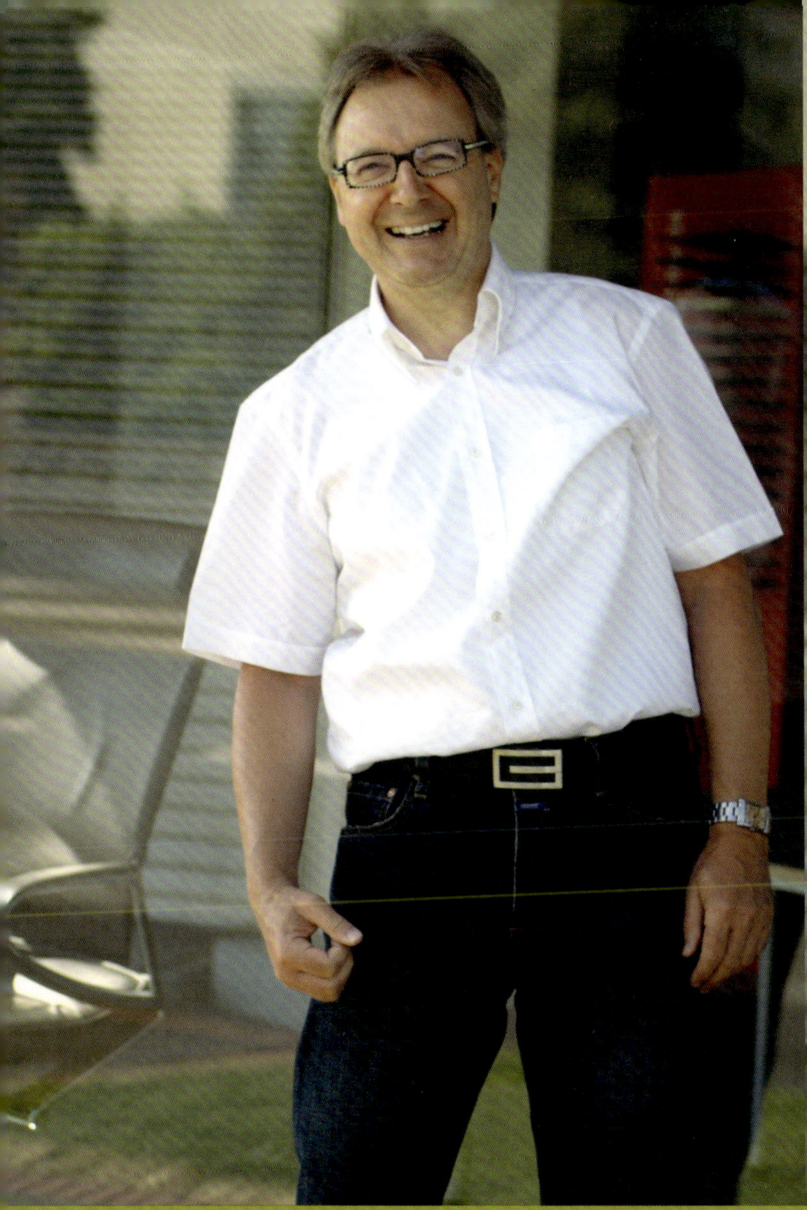

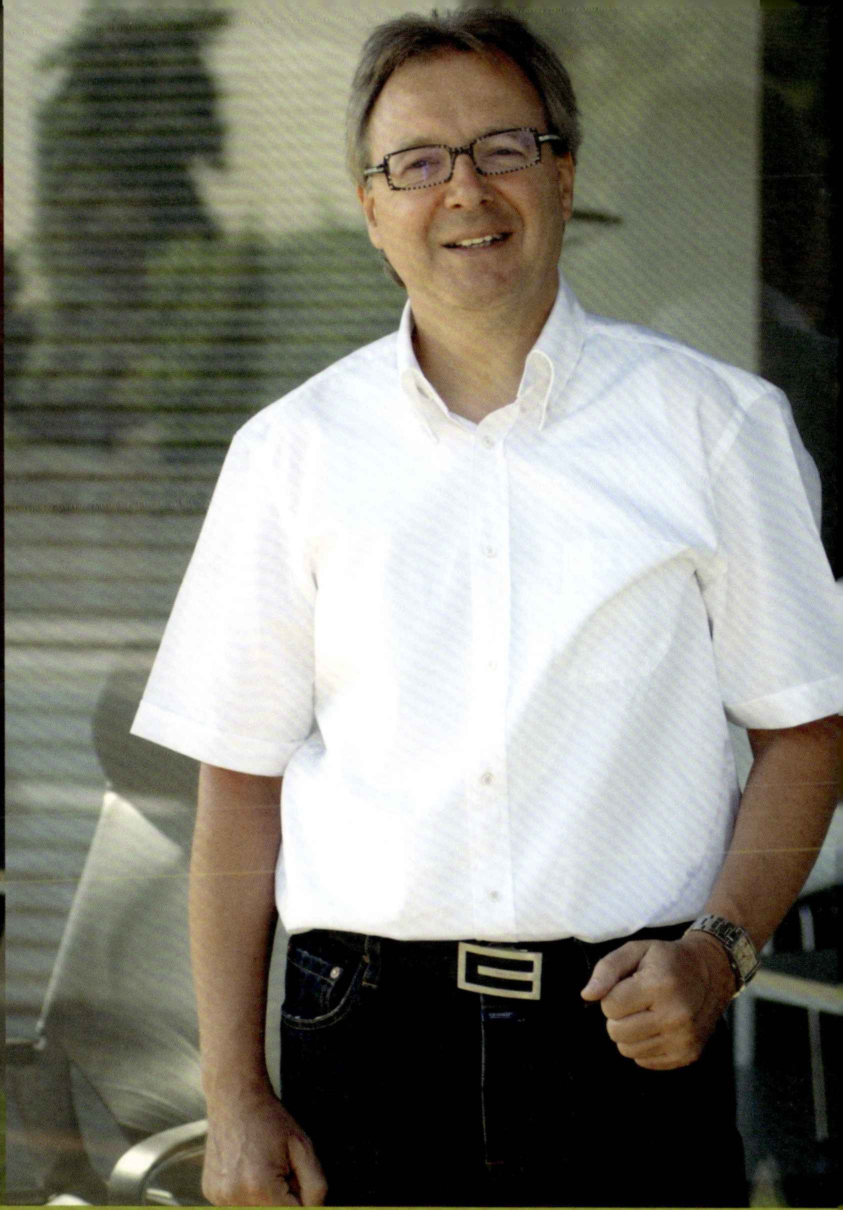

Michael Kahl
„Ich mag an Wilkhahn das gute Design mit ausgeprägter Qualität, das unternehmerische TUN + HANDELN, die gute Zusammenarbeit mit dem Fachhandel und die klare Strategie. Besonders wichtig sind für mich dabei eine konsequente Weiterentwicklung der Produkte, sowohl der vorhandenen Modelle als auch Neuentwicklungen, und der Mut, neue Wege zu gehen, auch abweichend von alltäglichen Entwicklungen. Jungen Designern Chancen zu öffnen und der vertrauensvolle Umgang mit allen beteiligten Menschen einer solchen soliden Partnerschaft – all das gehört bei Wilkhahn dazu."

"What I like about Wilkhahn is the good design with its distinctive quality, the entrepreneurial approach throughout, the excellent cooperation with the specialist shops and the clear strategy. I also find the consistent further development of the products particularly important – both of the present models as well as new developments, and the courage to try out new things even if they differ from everyday developments. Giving young designers opportunities and the trusting contact with all those involved in such a solid partnership is all part of Wilkhahn."

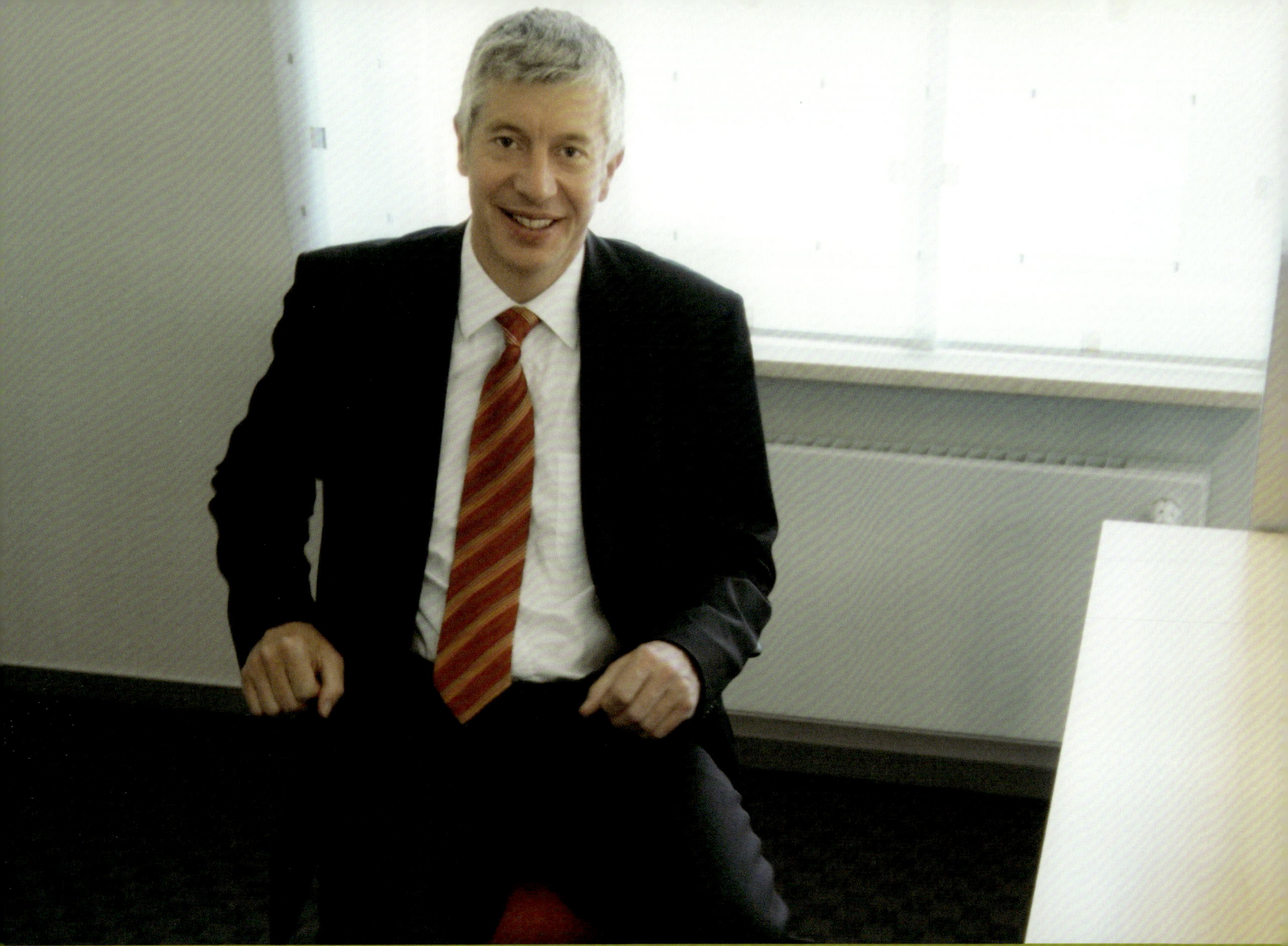

Heinz Feldmann, Vice Chairman
„Der Vertrieb von individuellen und maßgeschneiderten Finanz-Dienstleistungs-Produkten erfordert hervorragende Mitarbeiterinnen und Mitarbeiter, eine leistungsfähige Beratungsunterstützung sowie eine solide und ansprechende Beratungsatmosphäre."

"The selling of individual and customized financial services products requires outstanding staff members, efficient advisory support and a pleasant atmosphere when carrying out advisory services."

Sparkasse LeerWittmund

Samir Ayoub
„Die Qualität und das zeitlose Design der Wilkhahn-Programme sind für mich besonders wichtig. Die Gestaltqualität ist erstklassig, da sie von Bestand ist."

"The quality and timeless design of Wilkhahn ranges are very important for me. The design quality is first rate due to its enduring character."

Objektform, Bonn
Seminar table: Timetable range

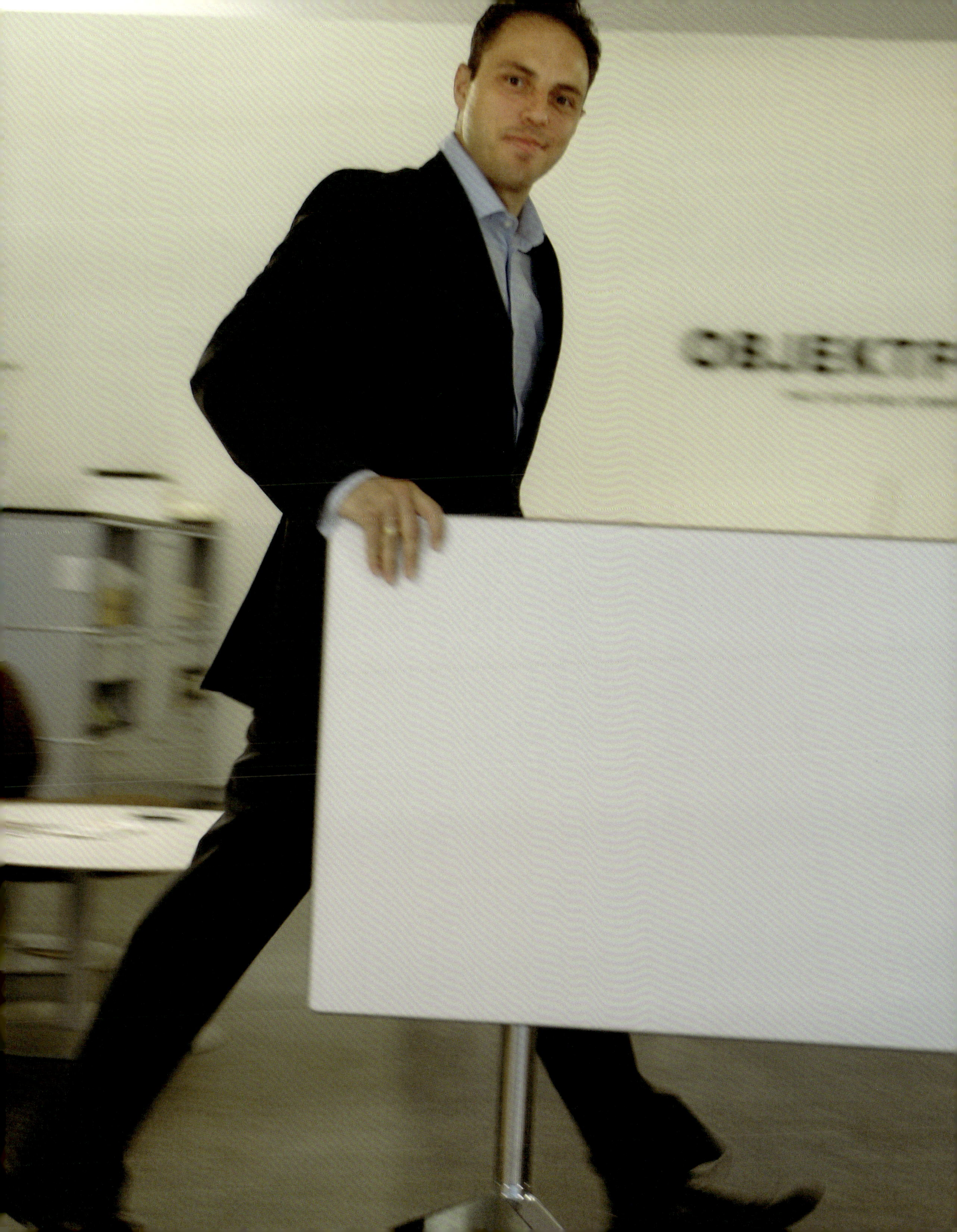

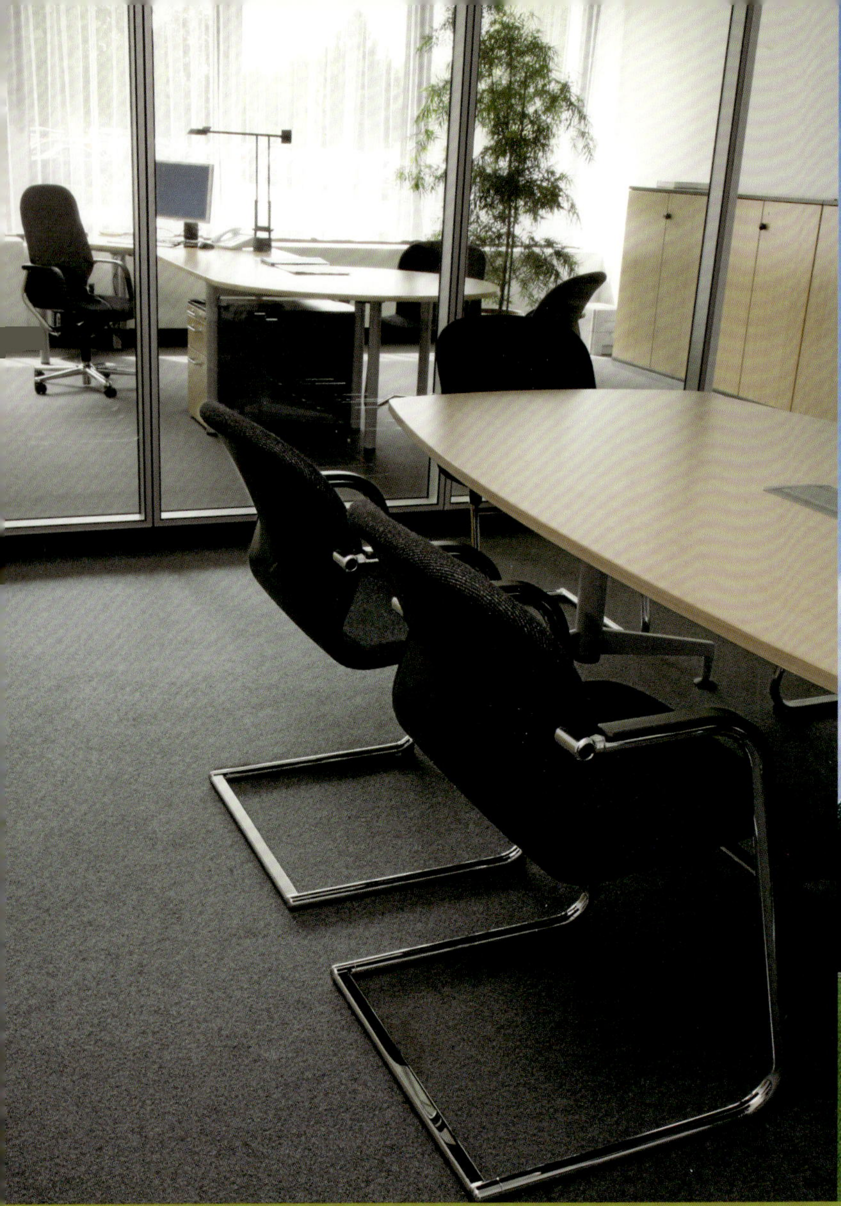

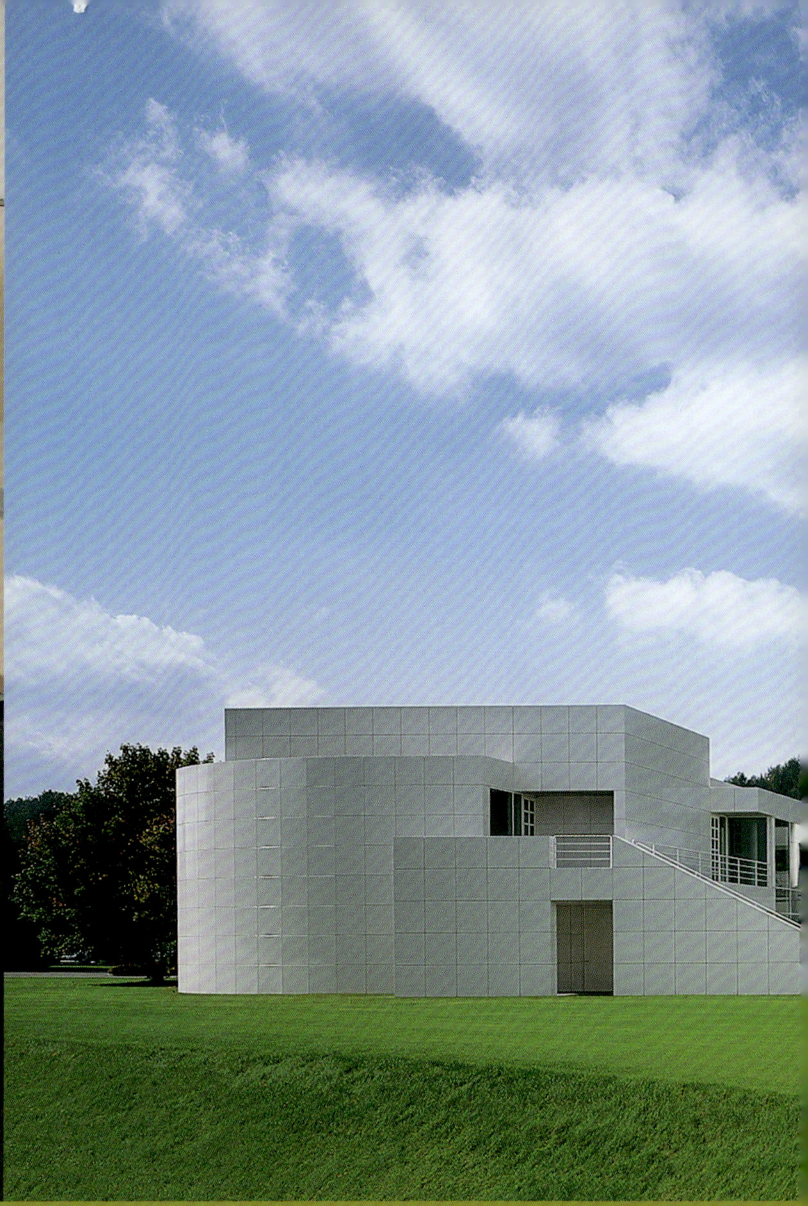

Max Weishaupt GmbH, Schwendi Swivel and cantilever chairs: FS range

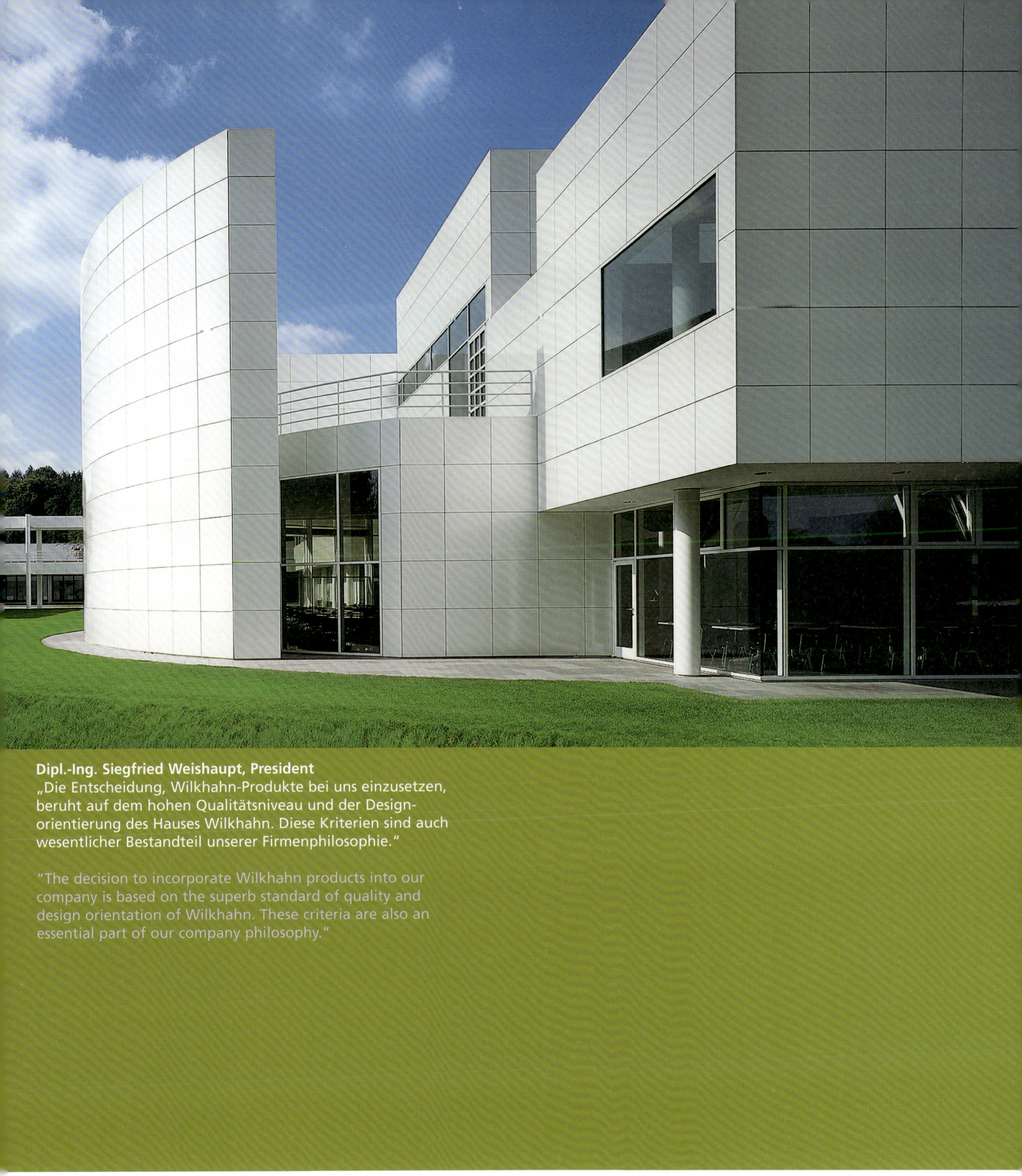

Dipl.-Ing. Siegfried Weishaupt, President
„Die Entscheidung, Wilkhahn-Produkte bei uns einzusetzen, beruht auf dem hohen Qualitätsniveau und der Designorientierung des Hauses Wilkhahn. Diese Kriterien sind auch wesentlicher Bestandteil unserer Firmenphilosophie."

"The decision to incorporate Wilkhahn products into our company is based on the superb standard of quality and design orientation of Wilkhahn. These criteria are also an essential part of our company philosophy."

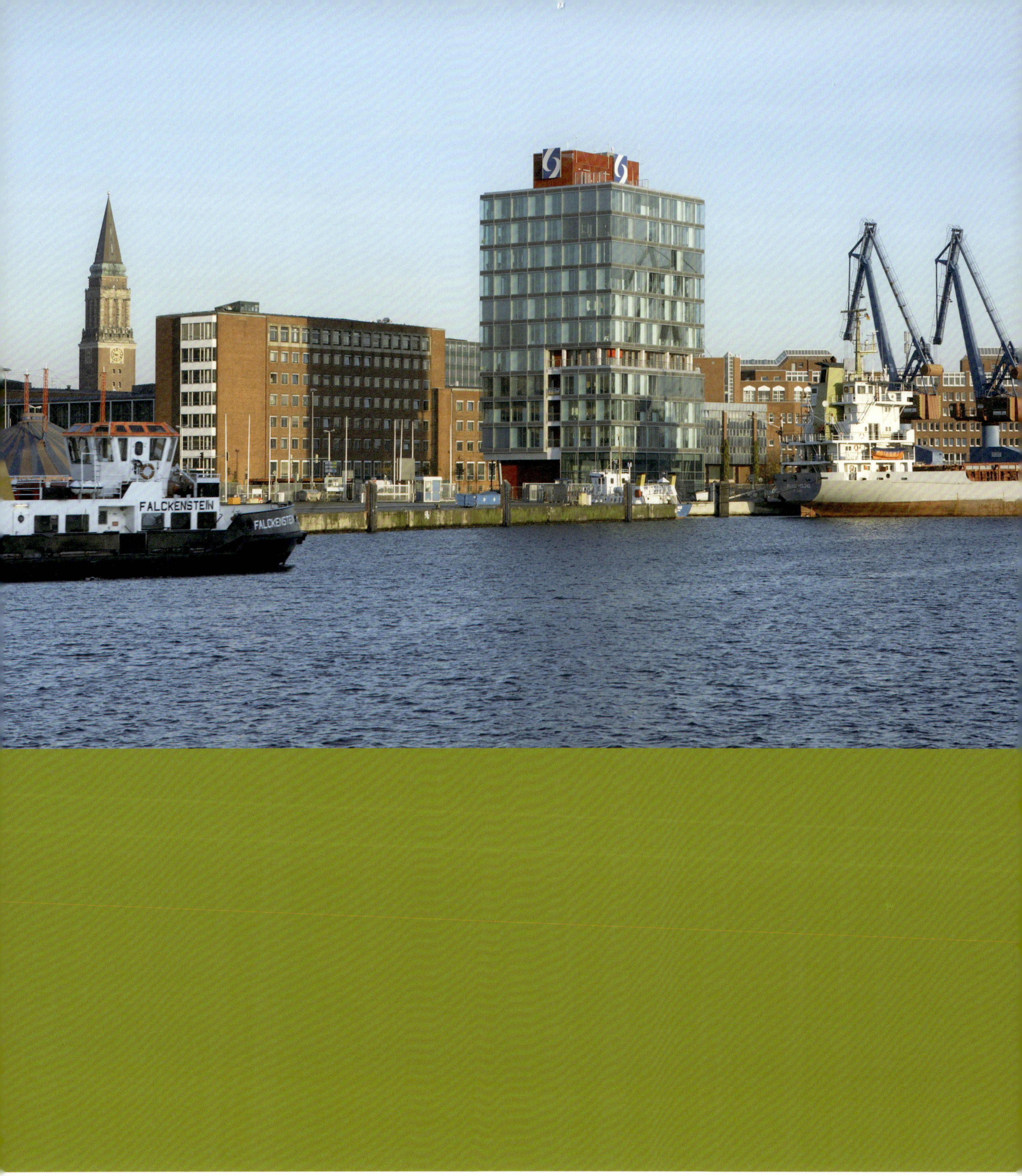
FALCKENSTEIN

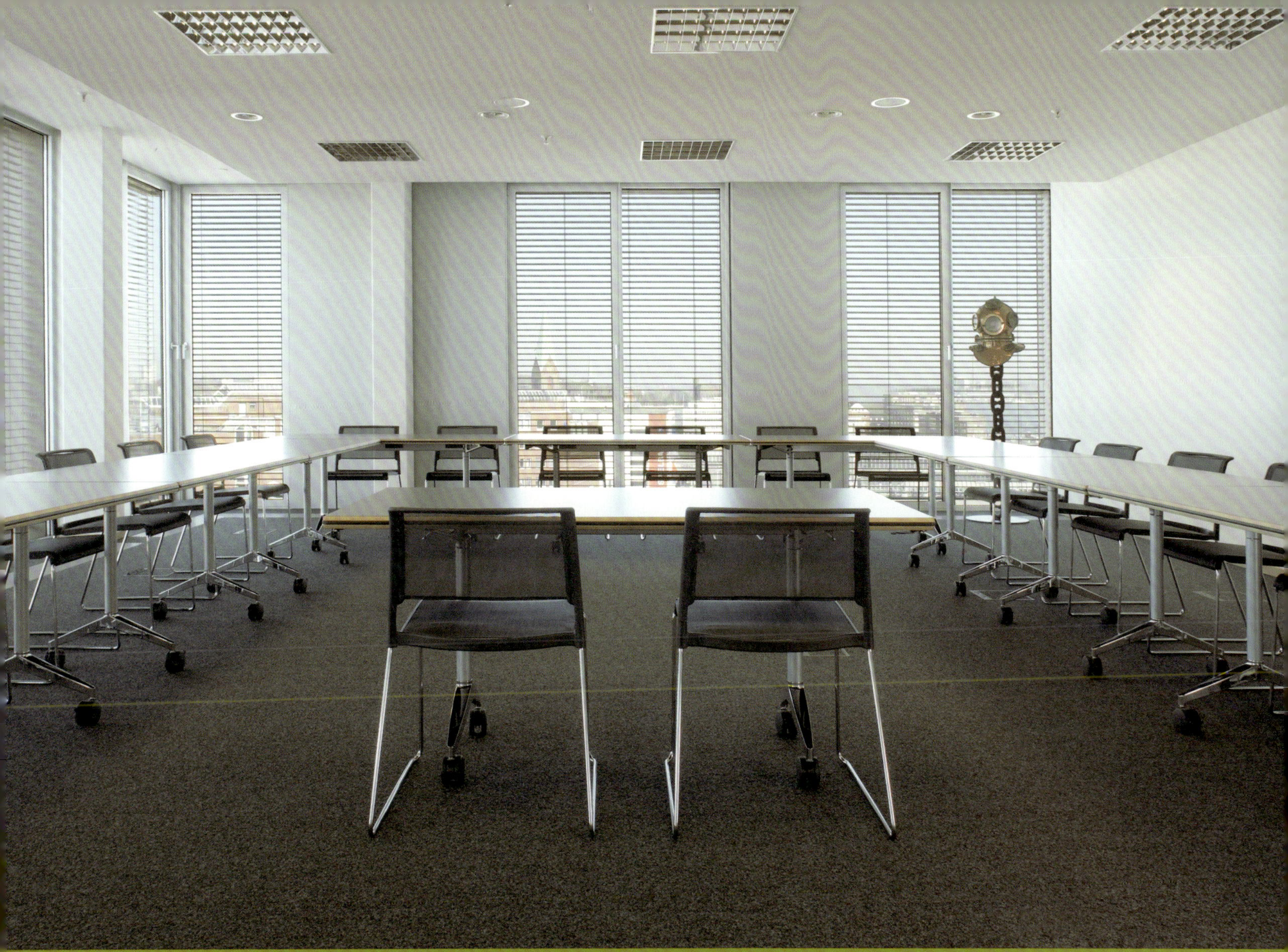

Seehafen Kiel GmbH & Co., Kiel Seminar tables: Timetable range; multi-purpose chairs: Aline range

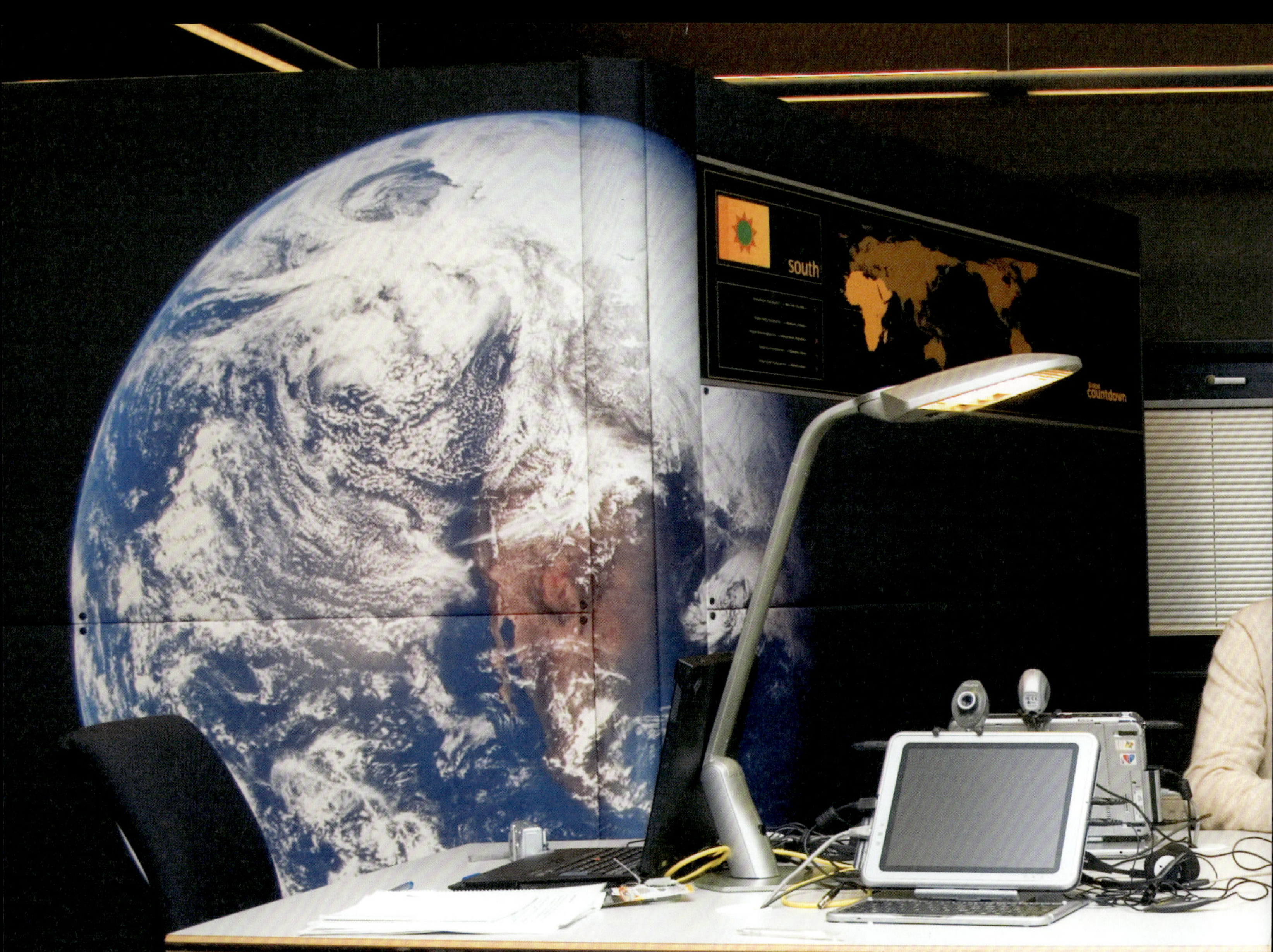

Die Märkte in Nordeuropa und Skandinavien werden vom Wilkhahn-Export betreut. In den vergangenen Jahren konnten hier gemeinsam mit Partnern und Architekturbüros wunderschöne Projekte realisiert werden. Durch die traditionell hohe Affinität der Region zu qualitätsvoller und werthaltiger Gestaltung wird Wilkhahn hier von Architekten und Innenarchitekten besonders geschätzt.

The markets in northern Europe and Scandinavia are looked after by Wilkhahn Export. Together with partners and architects, we have been able to realize some wonderful projects here in the last years. Owing to the traditionally strong affinity in the region for qualitative and sound design, Wilkhahn is very highly regarded here by architects and interior designers.

Telenor, Oslo Table: Confair folding table

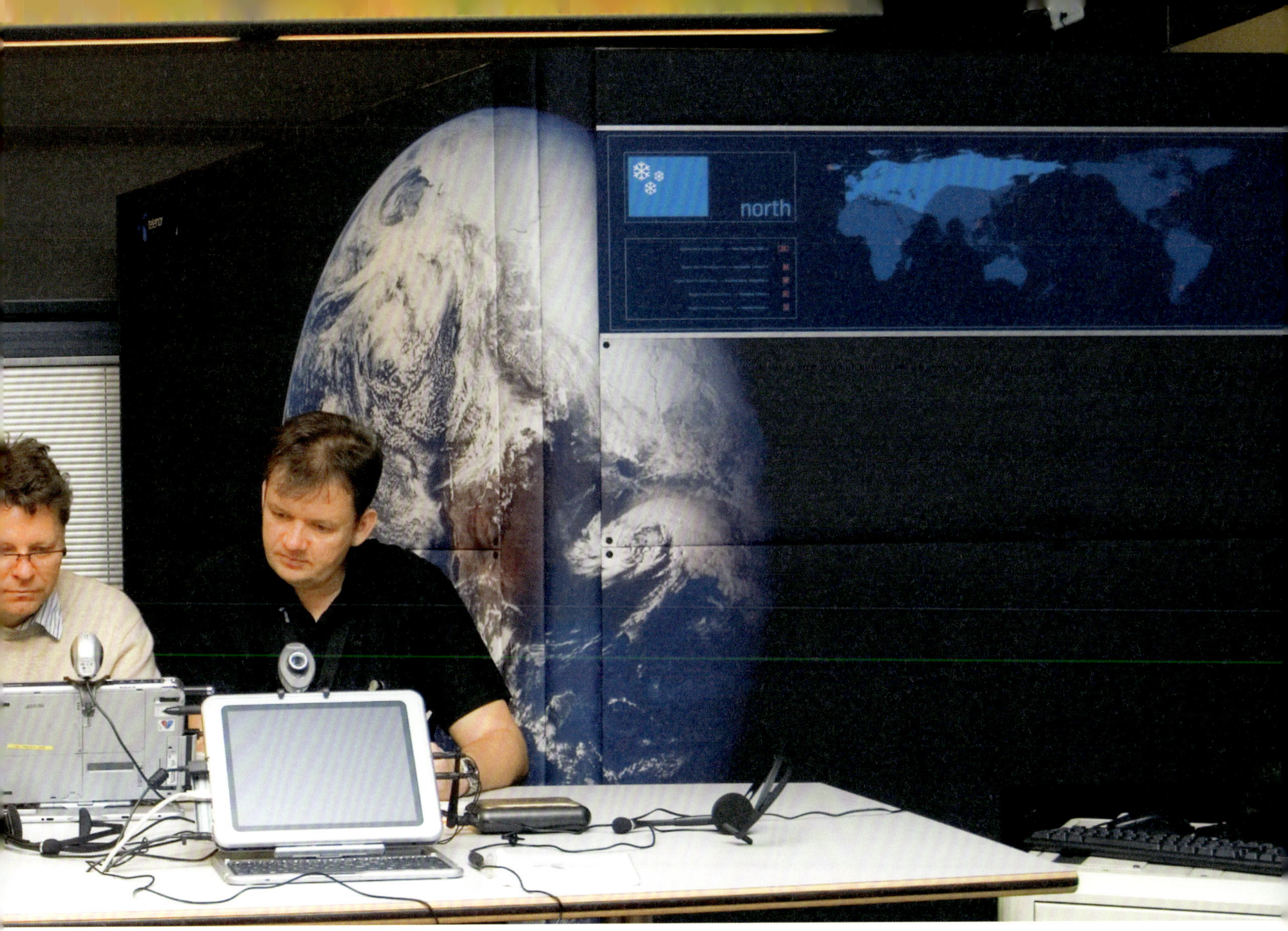
north

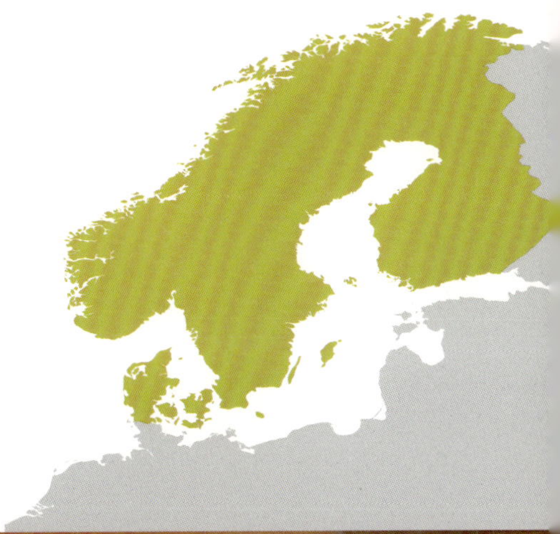

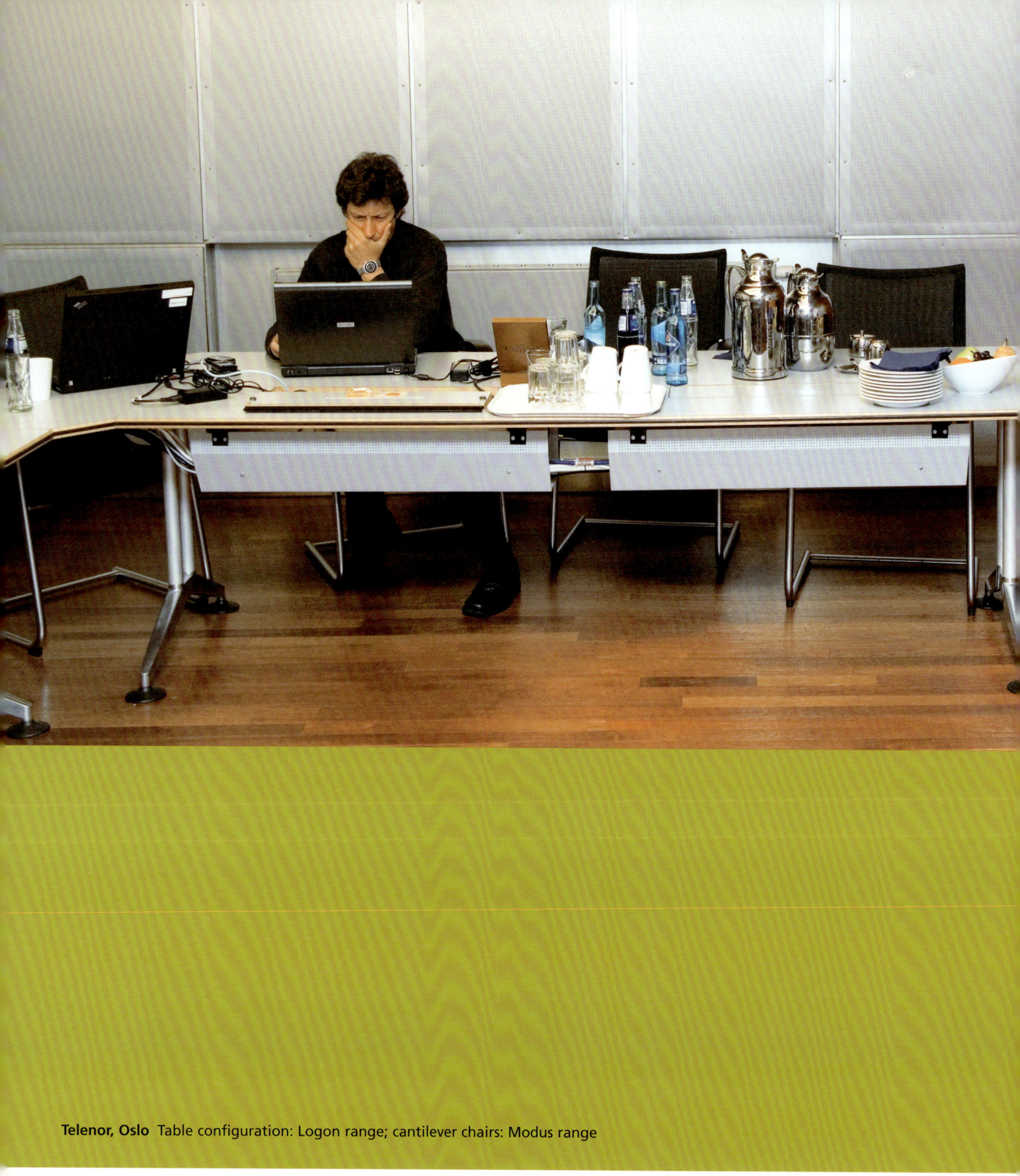

Telenor, Oslo Table configuration: Logon range; cantilever chairs: Modus range

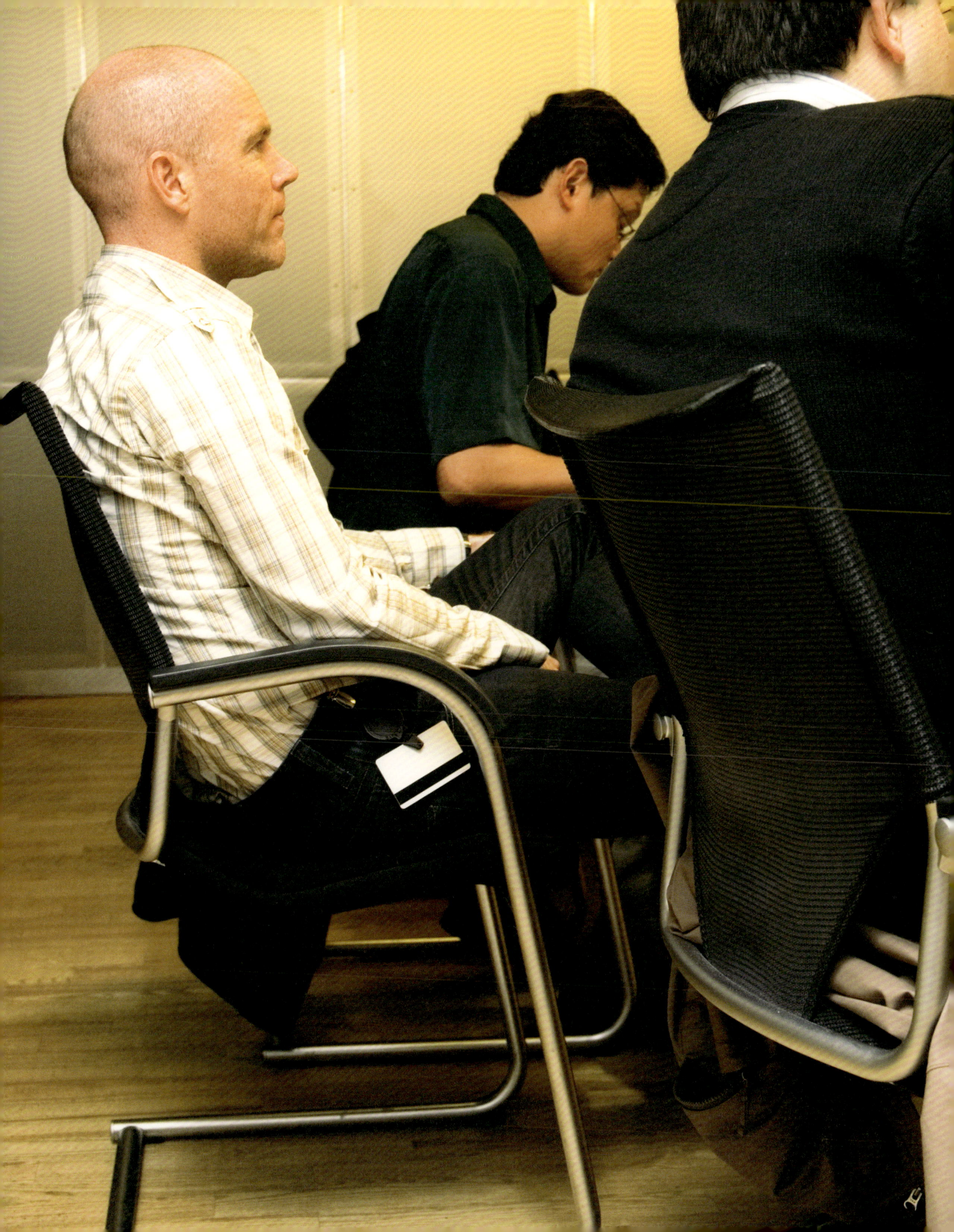

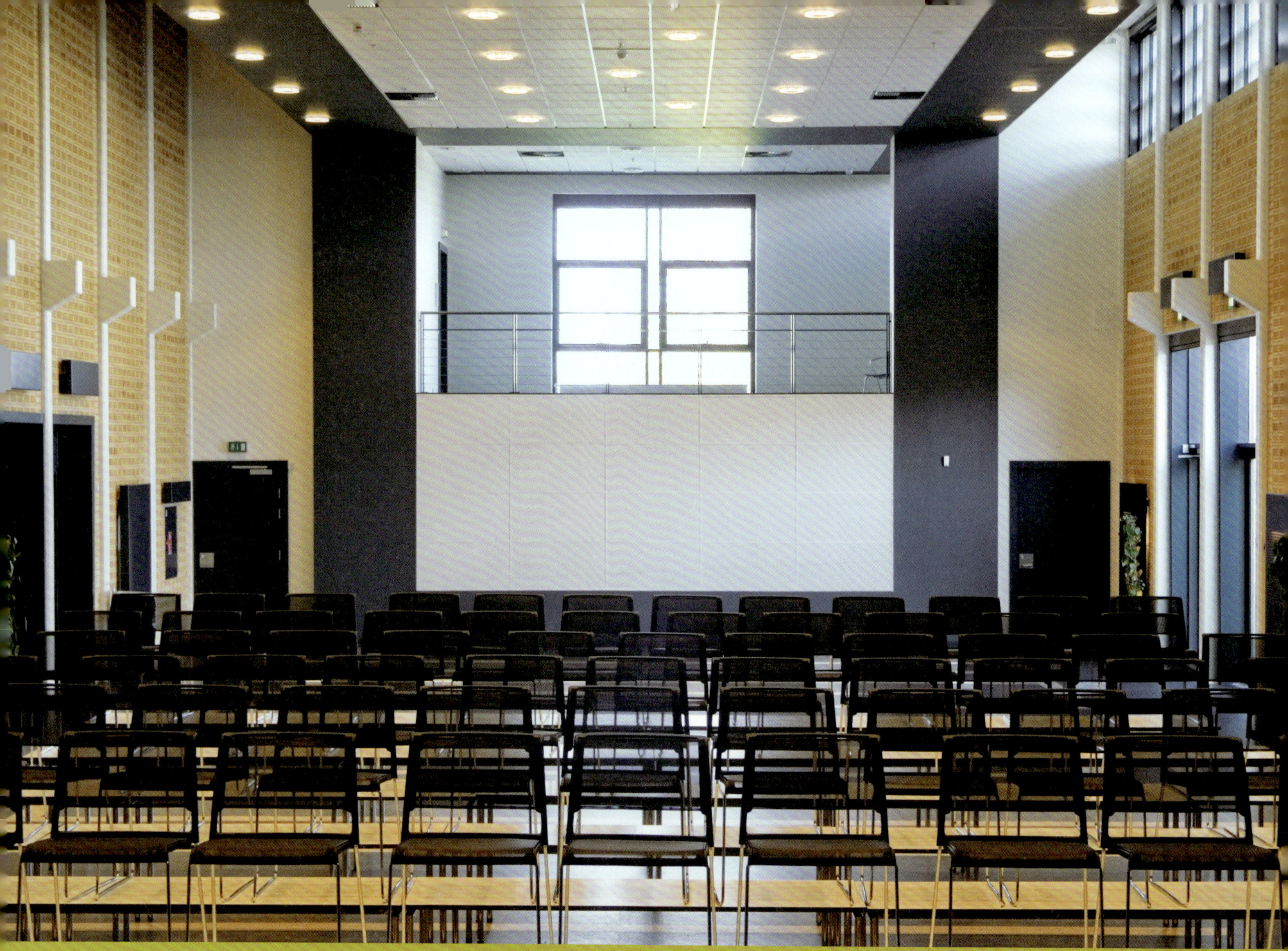

Lars Ramsing, Ramsing + Ramsing
"For Odense Tekniske skole, var det vigtigt at finde et stol med let design, integreret koblingsbeslag og let at stable, da rummet bliver brugt til mange forskellige formål. For Media huset, var det vigtigt at finde en let enkel stol, med et innovativt design udtryk."

"What was important for Odense Technical College was that the chairs have a light design, integrated connectors and are easily stackable because the rooms are used for a lot of different purposes. For the new 'Cantina' at the media centre, it was important to choose a light chair, which also had an innovative design."

OS Odense Technical College, Odense Multi-purpose chairs: Aline range

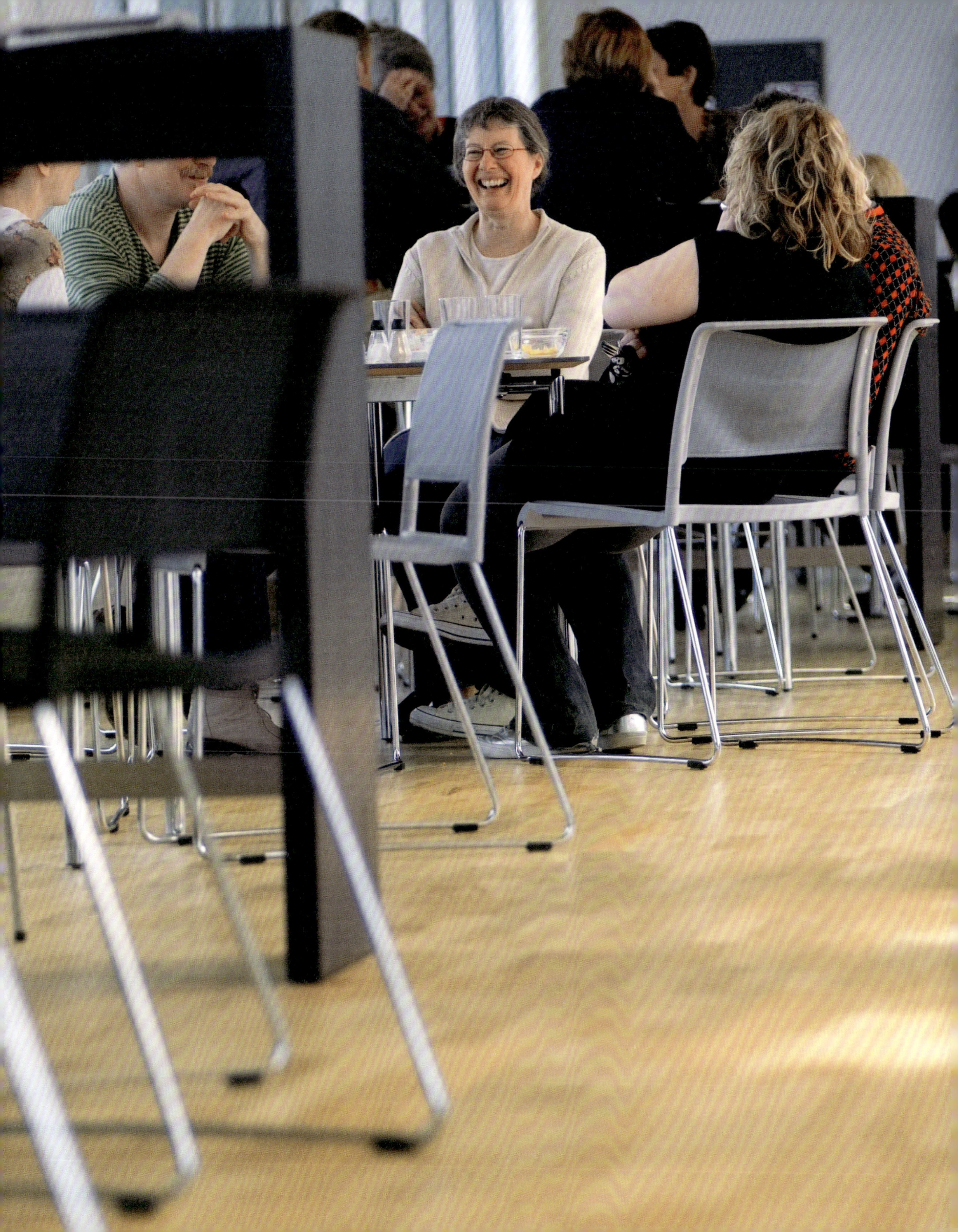

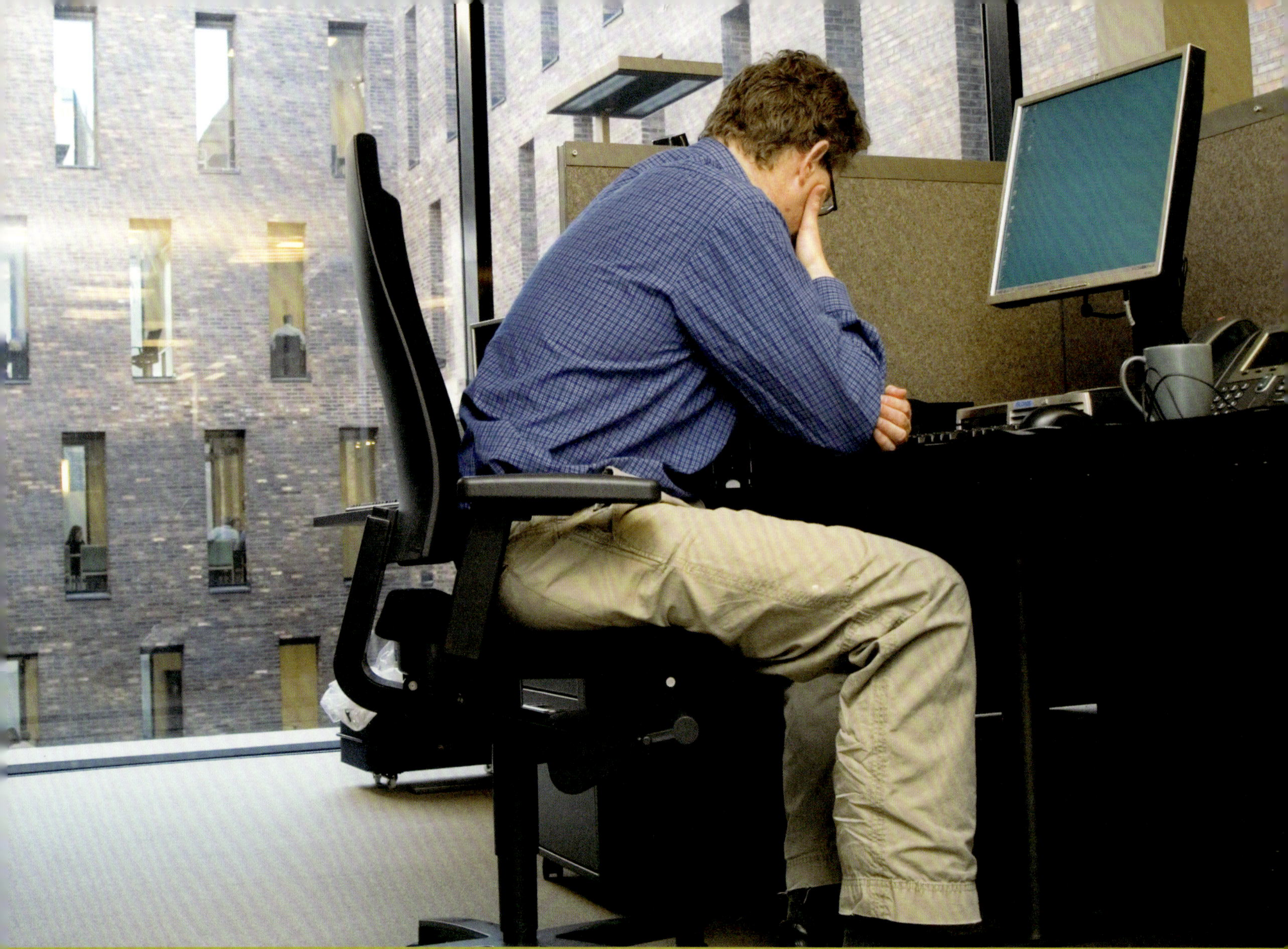

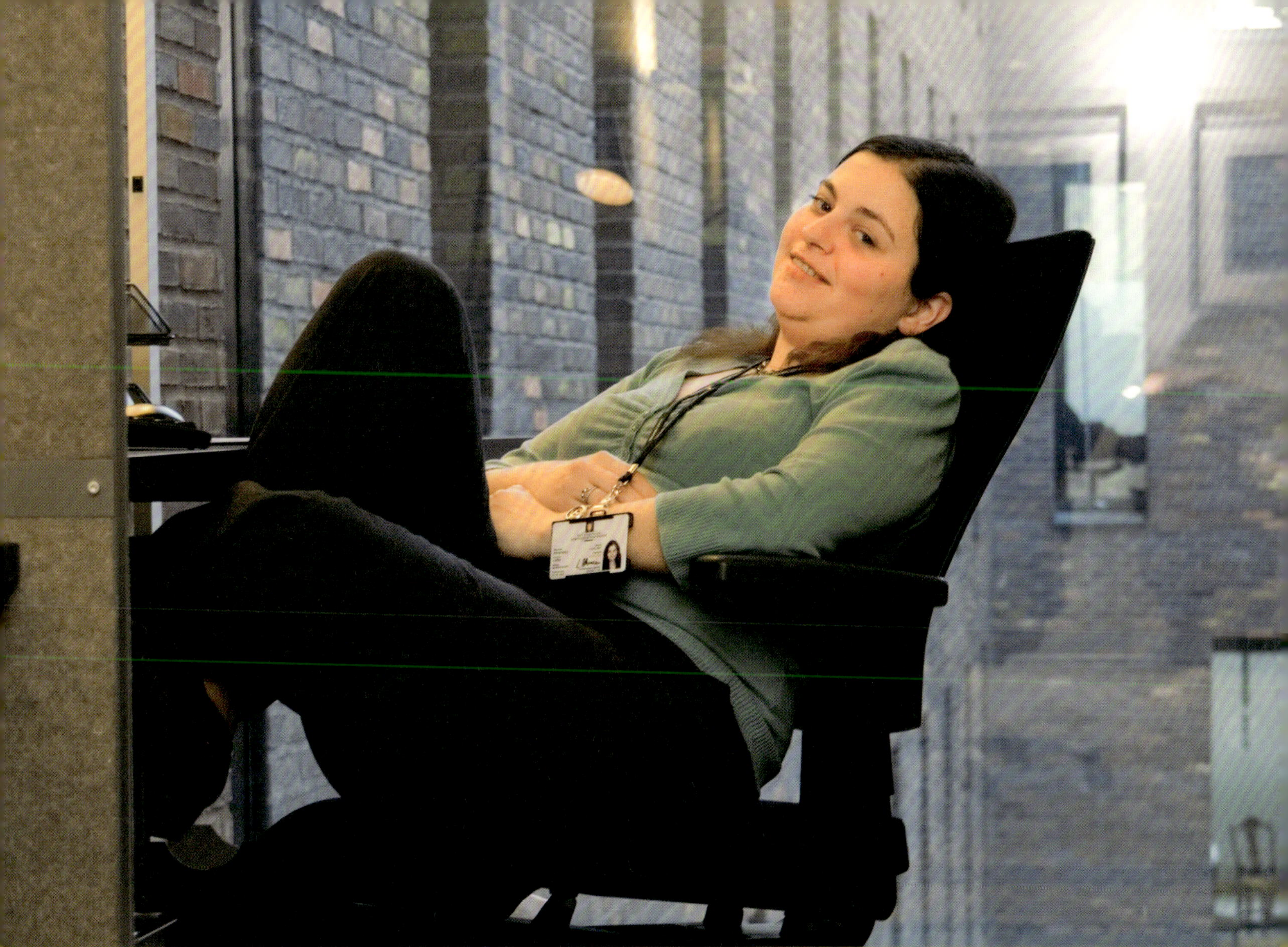

Ministry of Defense, Oslo Swivel chairs: Solis range

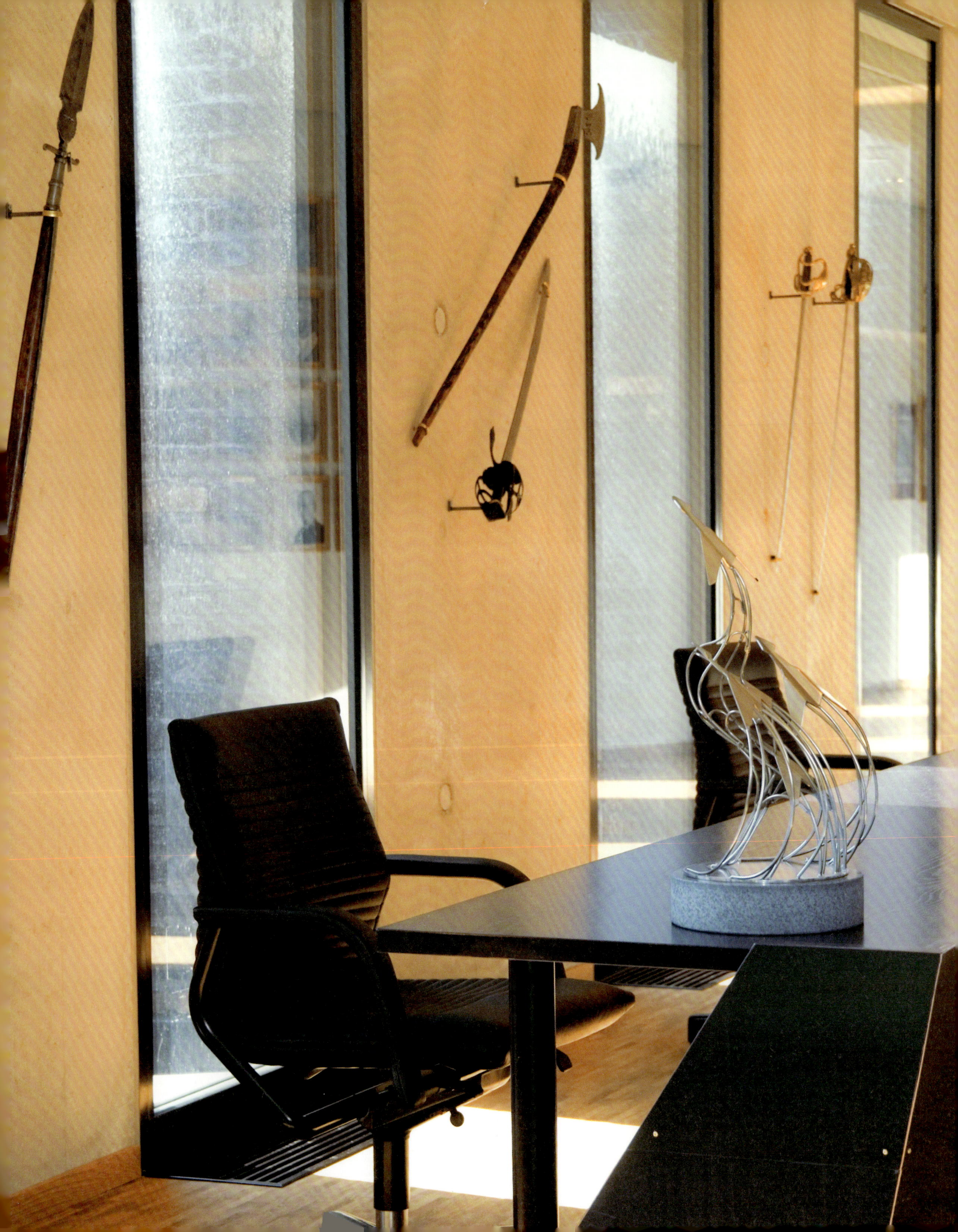

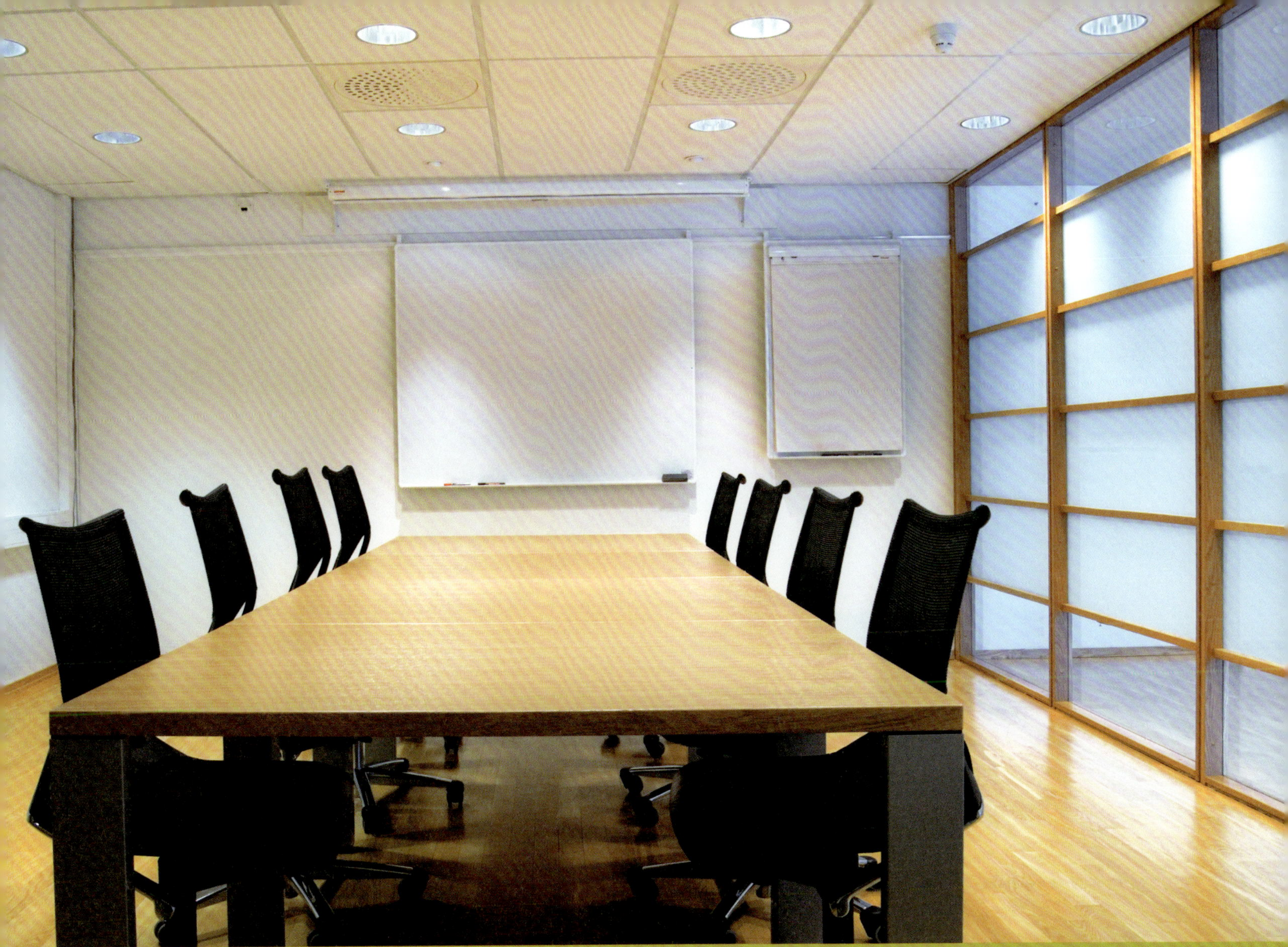

Birgitte Appelong, Interior Architect, Studio AS
"Som interiørarkitekt liker jeg stoler med slanke linjer og ren design, uten mange store detaljer. Komfort og ergonomi er selvfølgelig viktig og må være en integrert del av designet, slik Wilkhahns stoler er."

"As an interior architect, I prefer chairs with slim lines and a clean design. Comfort and ergonomics are important and must be implemented in the design, as Wilkhahn does."

Ministry of Defense, Oslo Swivel chairs: FS range, Modus range; table configuration: Palette range

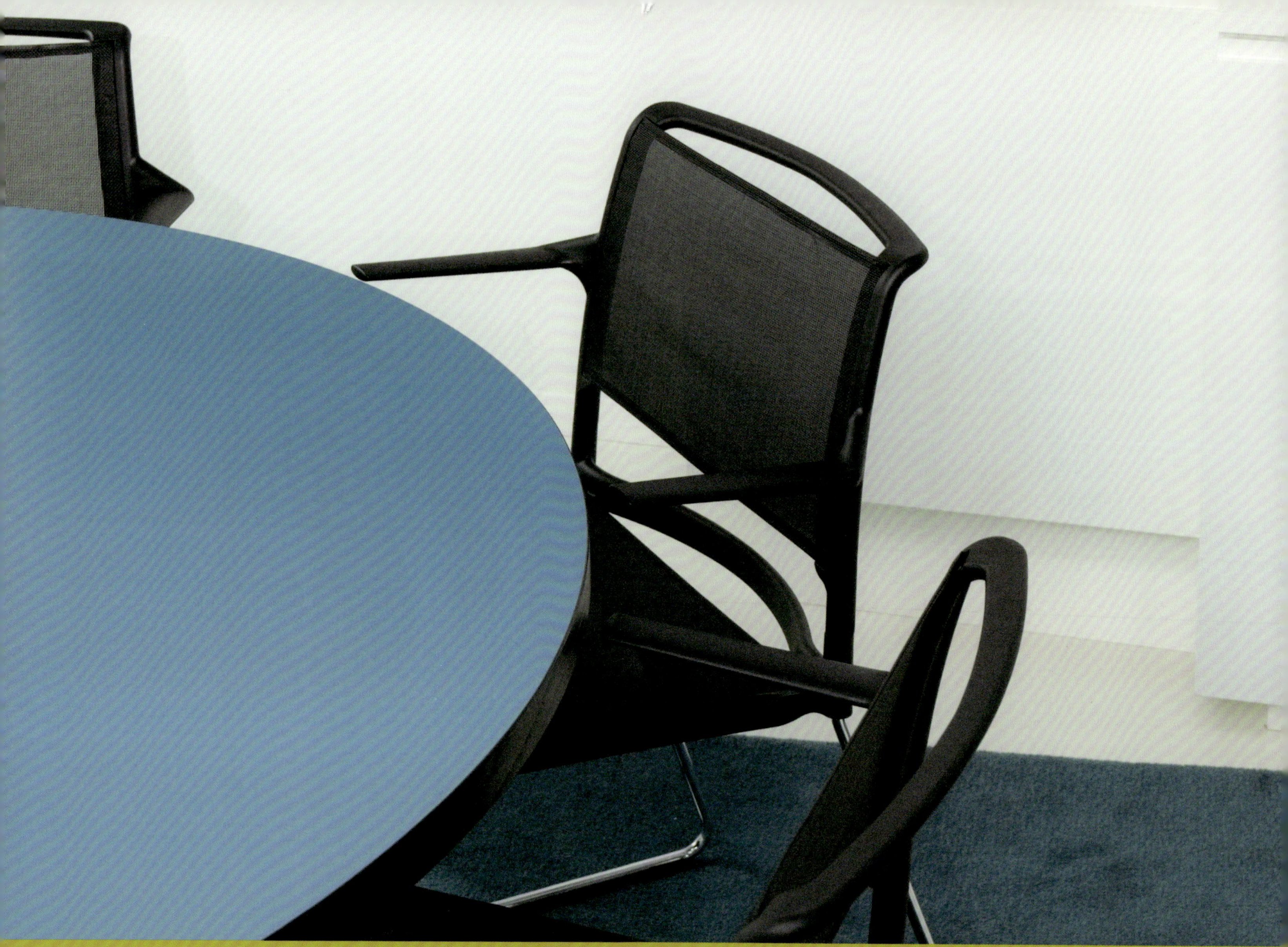

Markus Adler, President
"När vi förverkligande vårt nya kontors- och eventkoncept för Tysk-Svenska Handelskammaren var det viktigt att arbeta med flexibla, moderna och estetiska möbler. Wilkhahns lösningar gav ett avgörande bidrag till det."

"In the process of realizing the new office and event concept for the German-Swedish Chamber of Commerce, it was important for us to work with flexible, modern and aesthetic furniture. Wilkhahn played a crucial role in this process."

German-Swedish Chamber of Commerce, Stockholm Multi-purpose chairs: Aline

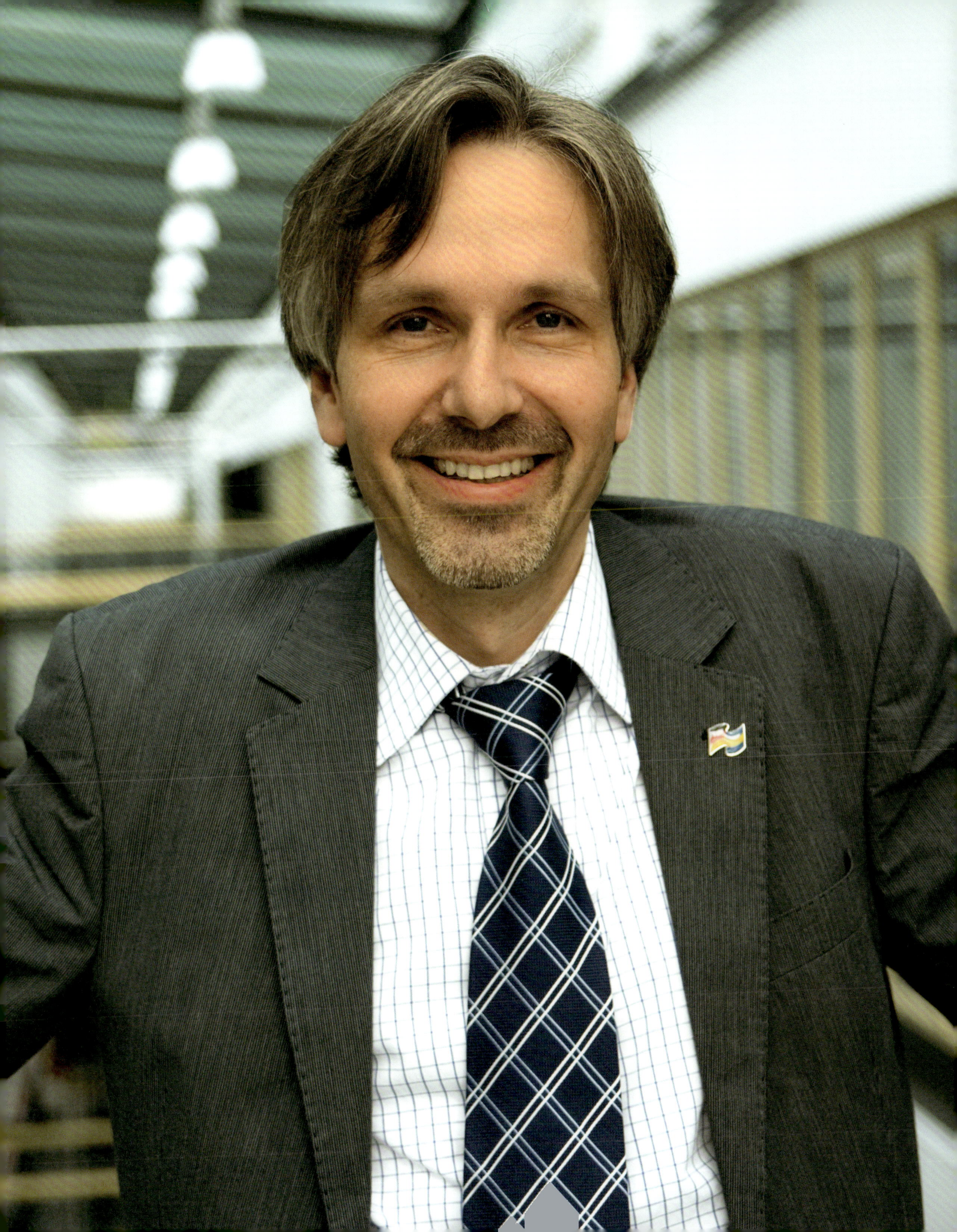

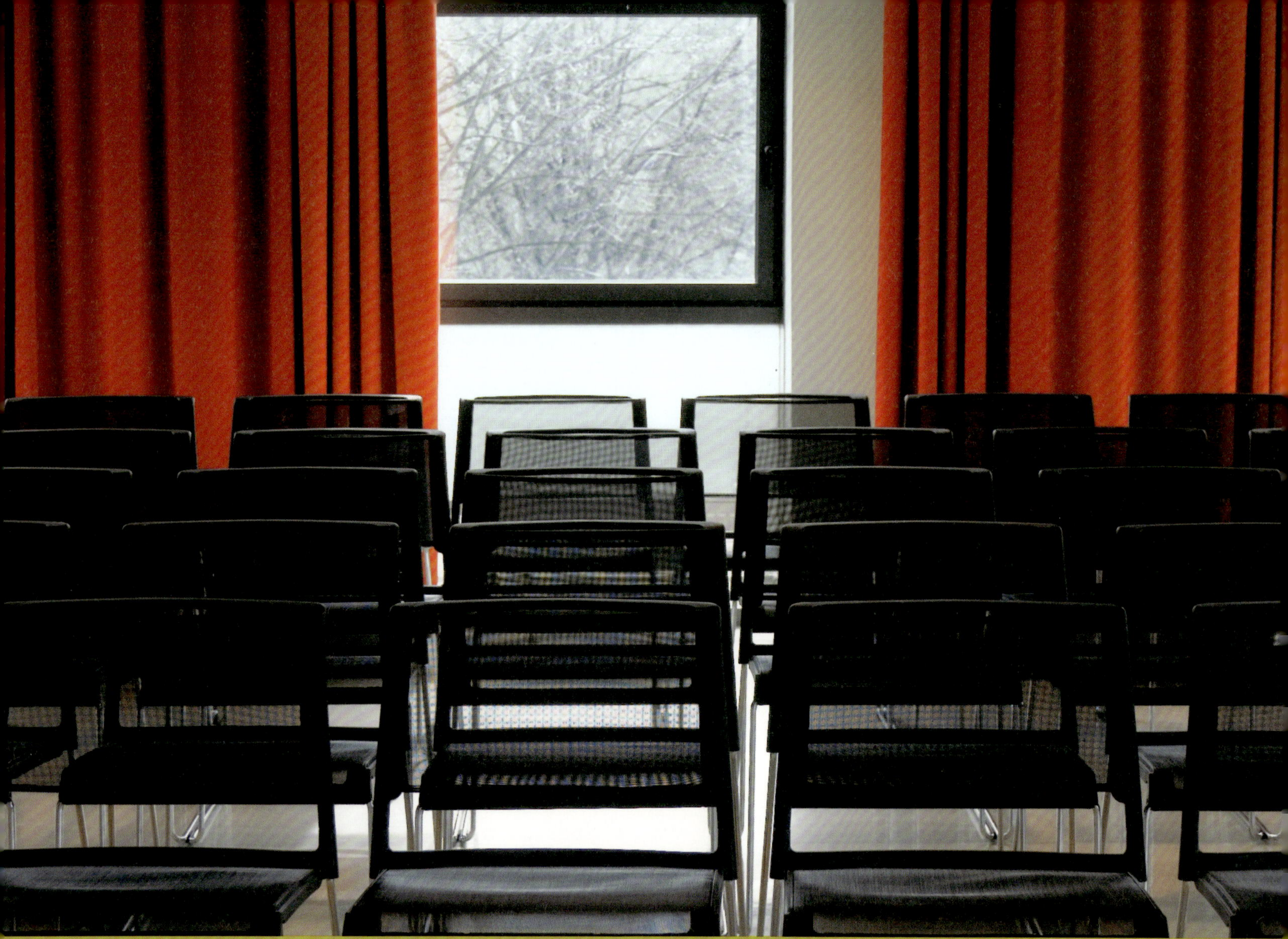

Thomas Eriksson, Architect, Stockholm
"Jag har inte kunnat släppa tanken på denna stol sedan den förevisades för mig av Stefan Hoske på Stockholm Furniture Fair, för ett par år sedan. Sättet den kombinerar en hög teknisk nivå med en häpnadsväckande hög sittkomfort förvånade mig mycket! Som stapelbar stol är den fantastisk och är lätt och enkel att handskas med. Luftig och fräsch.
Passar utmärkt in på vårt nya kontor i Stockholm."

"I have had my mind on this chair since I first saw it presented at the furniture fair in Stockholm some years ago, shown to me by Stefan Hoske. The way it combines a highly technical and competent approach with a stunningly comfortable and soft sitting experience surprised me a lot! It has fantastic stacking performance and is light and easy to handle. Airy and very fresh. Looks great in our new office space in Stockholm."

German-Swedish Chamber of Commerce, Stockholm
Multi-purpose chairs: Aline

Multi-purpose chair: Aline

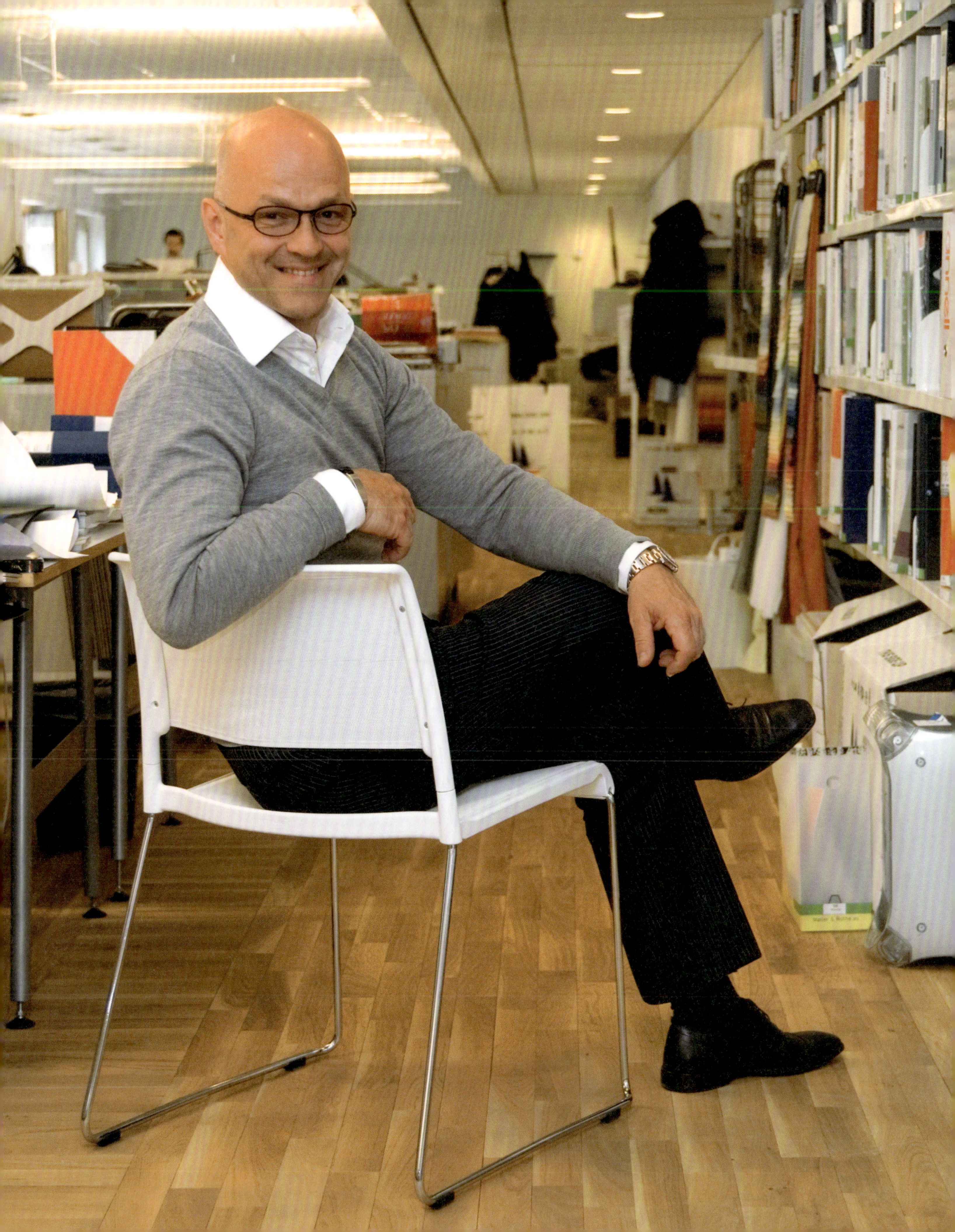

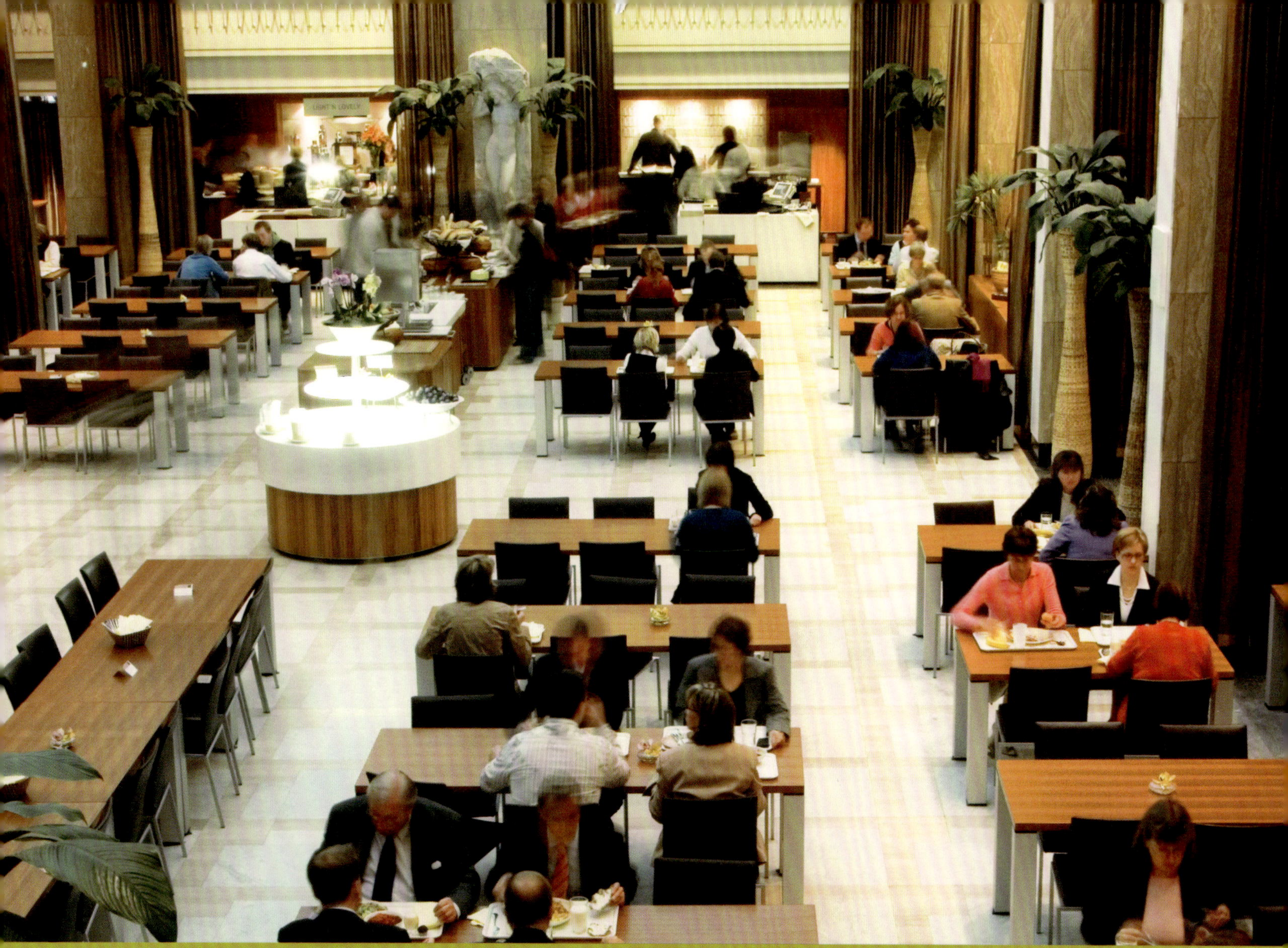

Restaurant Bank, Helsinki Tables: Palette range, Logon range; swivel chairs: Solis range

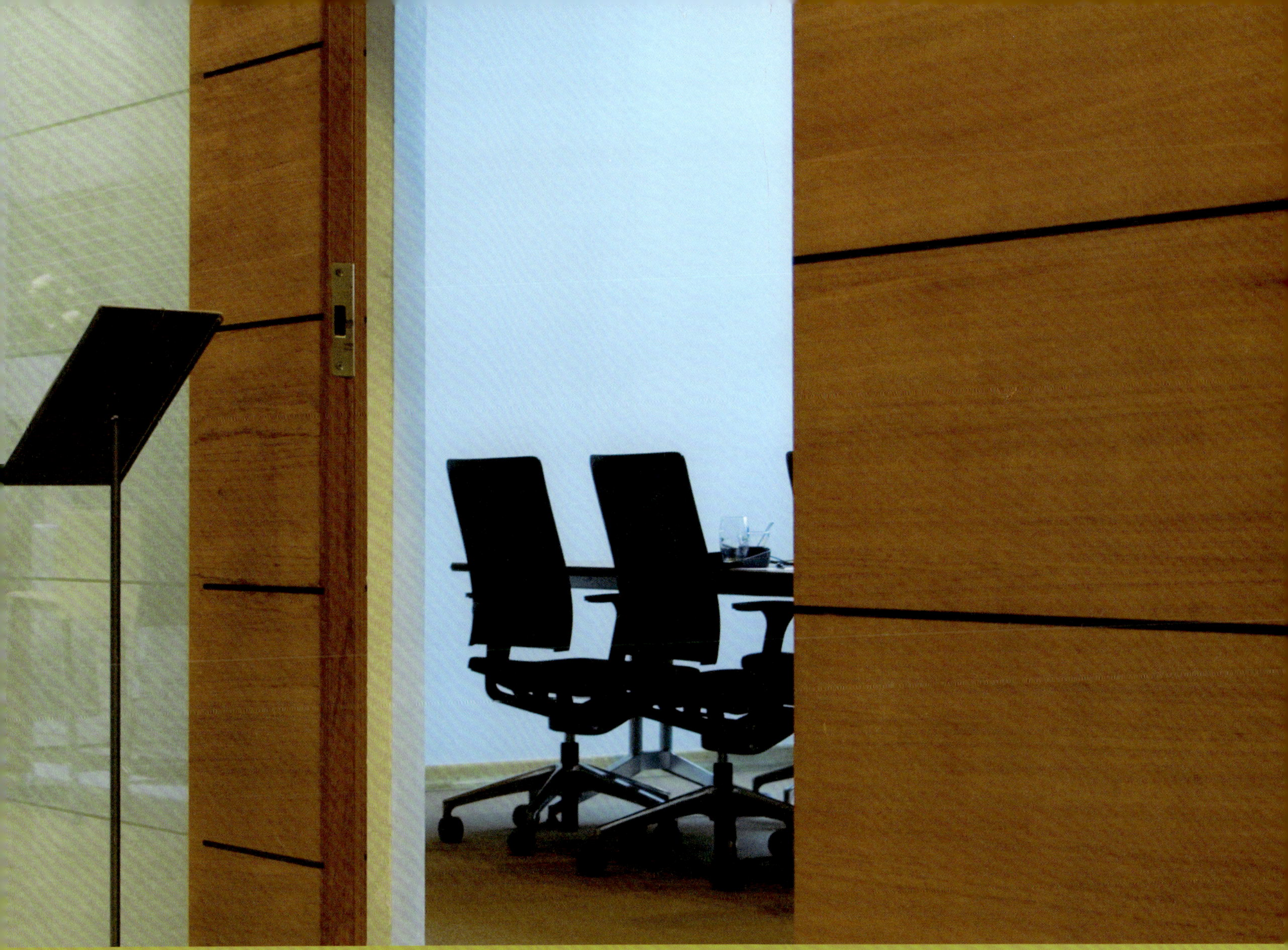

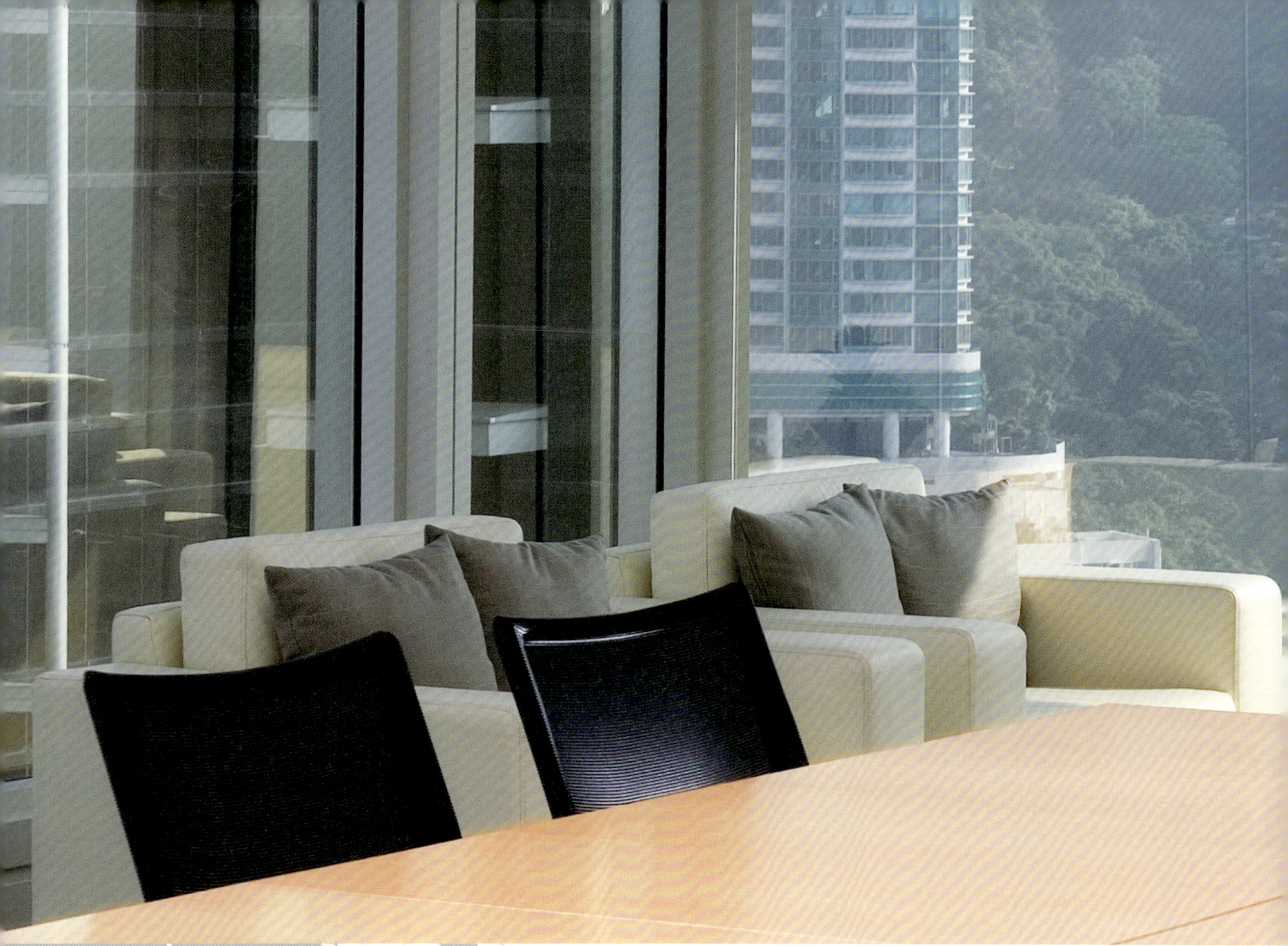
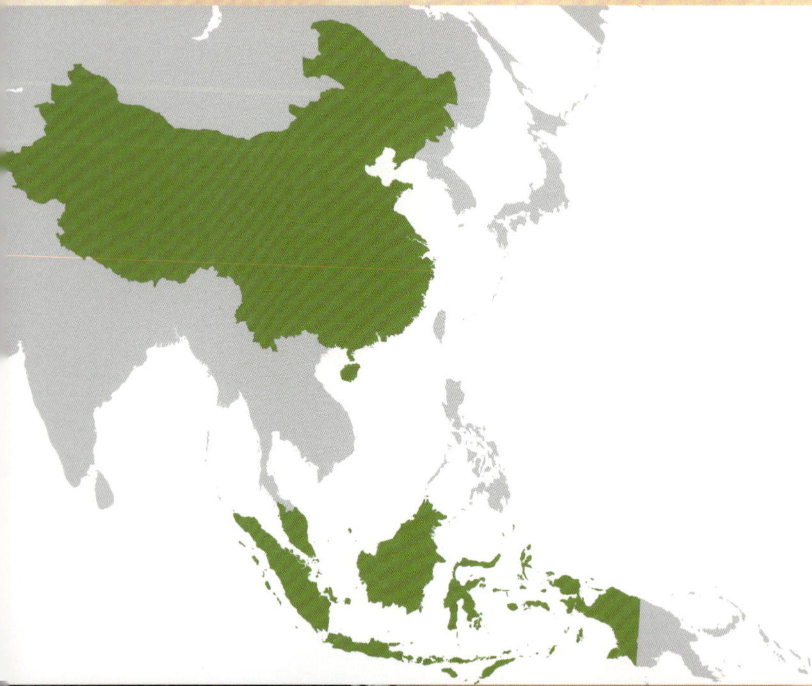
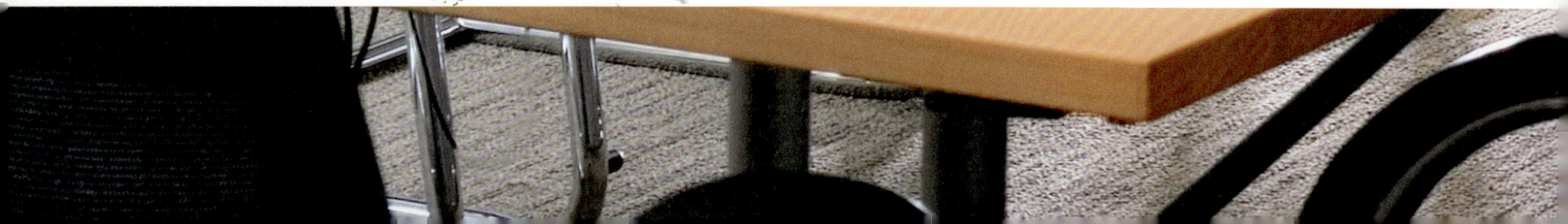

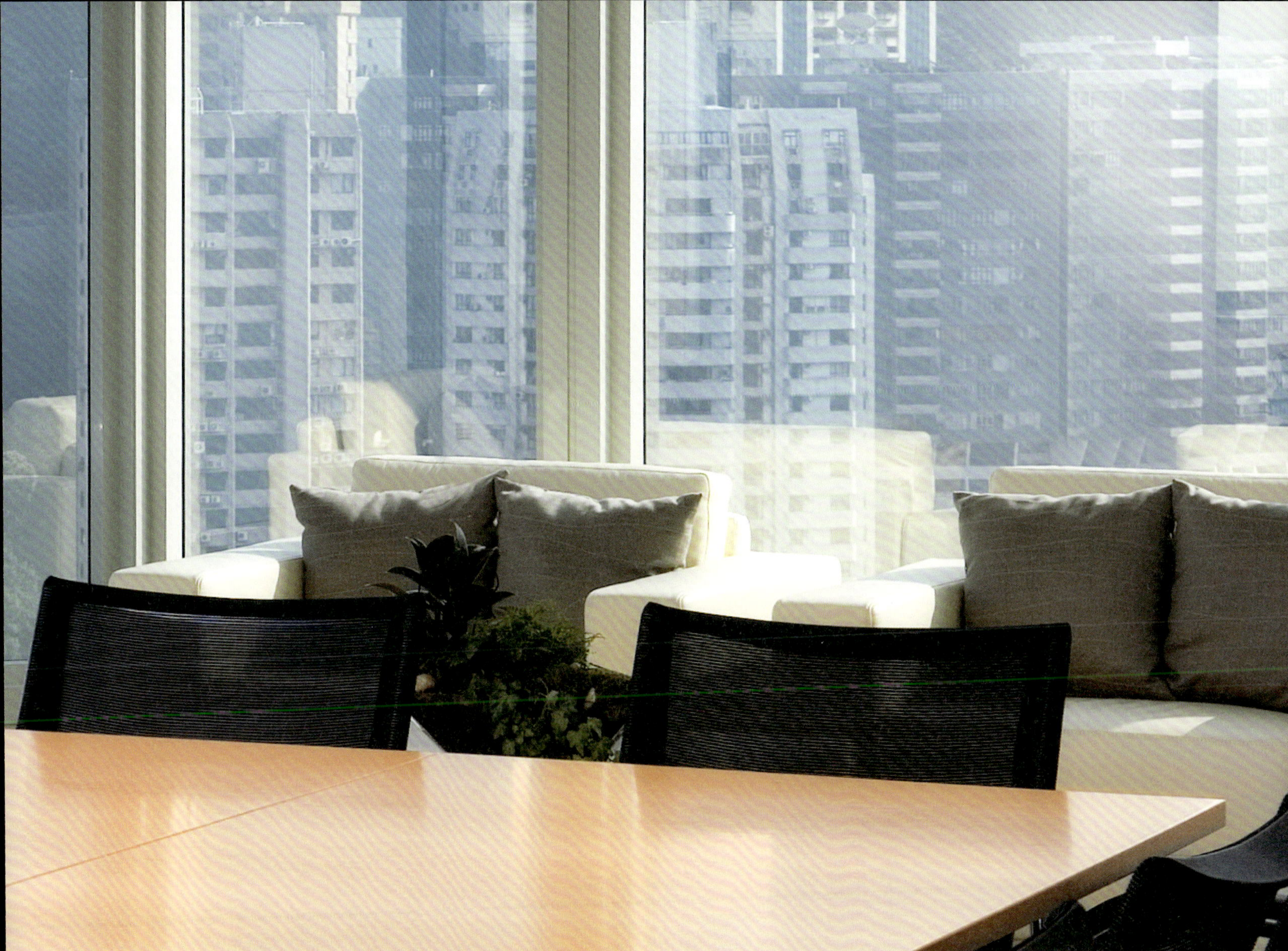

In Asien war Wilkhahn bereits seit den 80er Jahren durch Exportpartner vertreten. Die Einrichtung des Großflughafens Chek Lap Kok in Hongkong mit dem Banksystem Tubis wurde 1997 zum größten Einzelauftrag in der Unternehmensgeschichte. Heute unterhält Wilkhahn eigene Showrooms in Hongkong, Shanghai, Kuala Lumpur und Singapur, die das Unternehmen auch in Asien zu einer begehrten Top-Marke gemacht haben.

Wilkhahn was already represented in Asia by its export partners in the 1980s. The furnishing of the Chek Lap Kok International Airport in Hong Kong in 1997 with Tubis seating units was the largest single commission in the history of the company. Today, Wilkhahn maintains its own showrooms in Hong Kong, Shanghai, Kuala Lumpur and Singapore: these have also made the company a much sought-after exclusive brand in Asia.

Deutsche Bank, Hong Kong Table: Confair folding table; cantilever chairs: Modus range, design firm: Adam Mundy, M Moser Associates

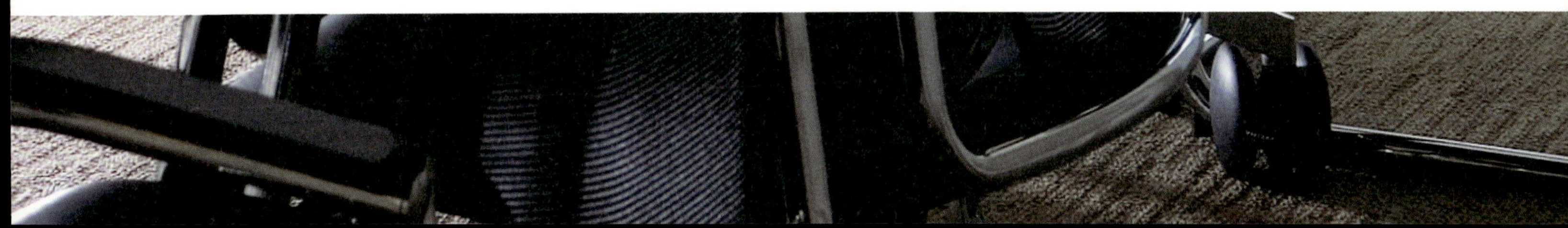

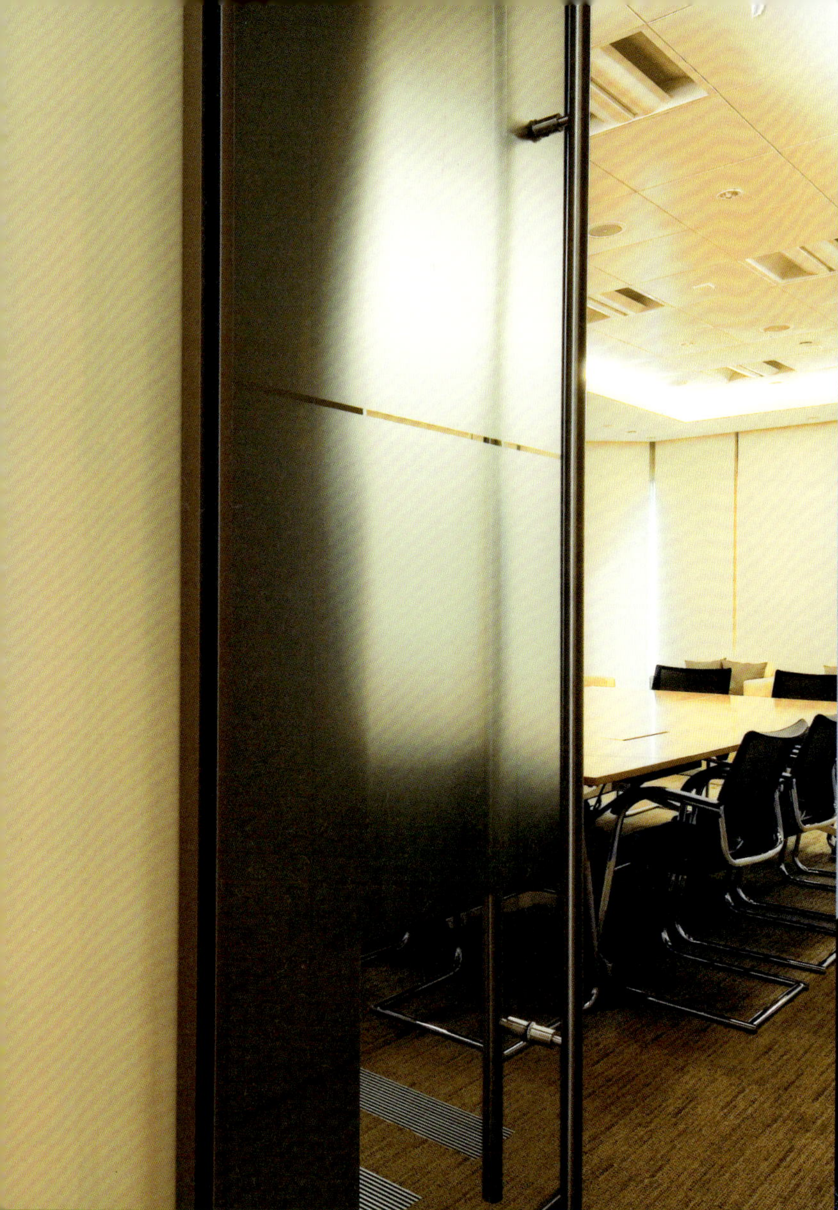

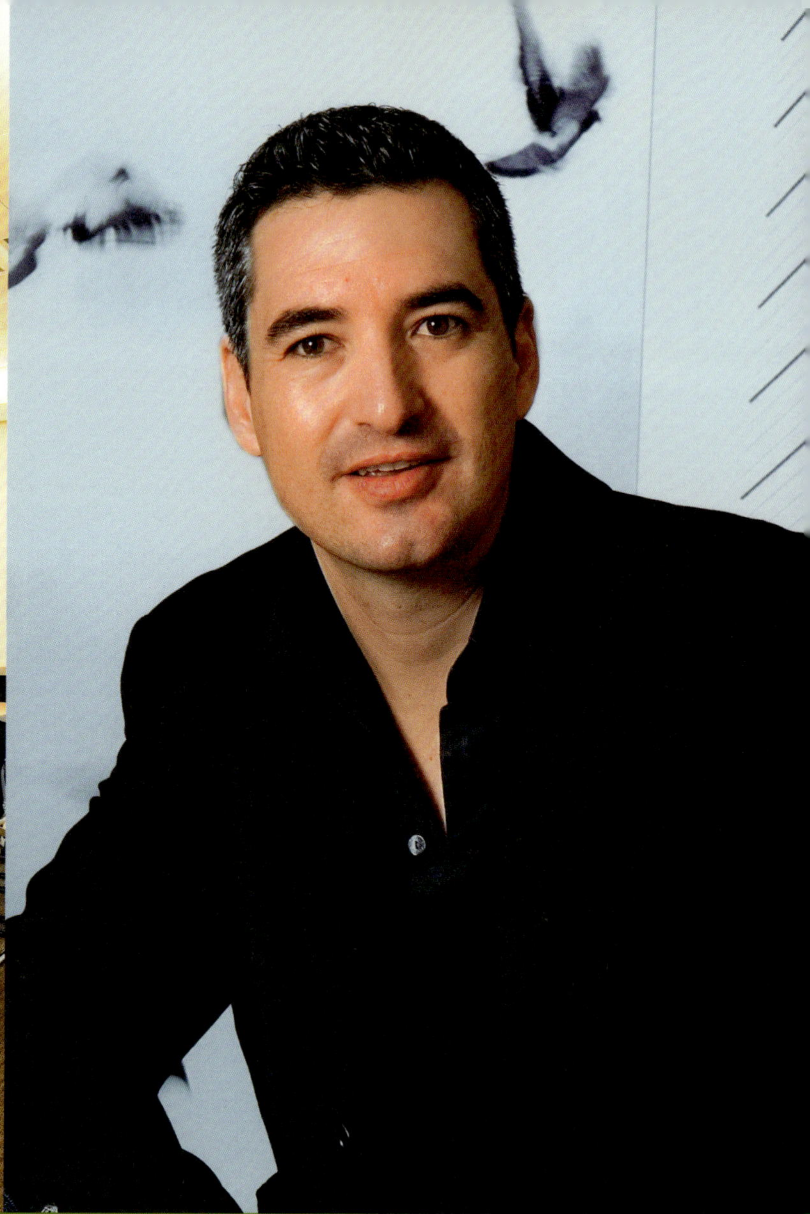

Adam Mundy, Director
" …as we create workplaces for individuals, the useability, sustainability and simplicity of tools we place at the users' resources need to be well considered; Wilkhahn's philosophy equals this …"

Deutsche Bank, Hong Kong
Table: Confair folding table; cantilever chairs: Modus range, design firm: Adam Mundy, M Moser Associates

M Moser Associates, Hong Kong

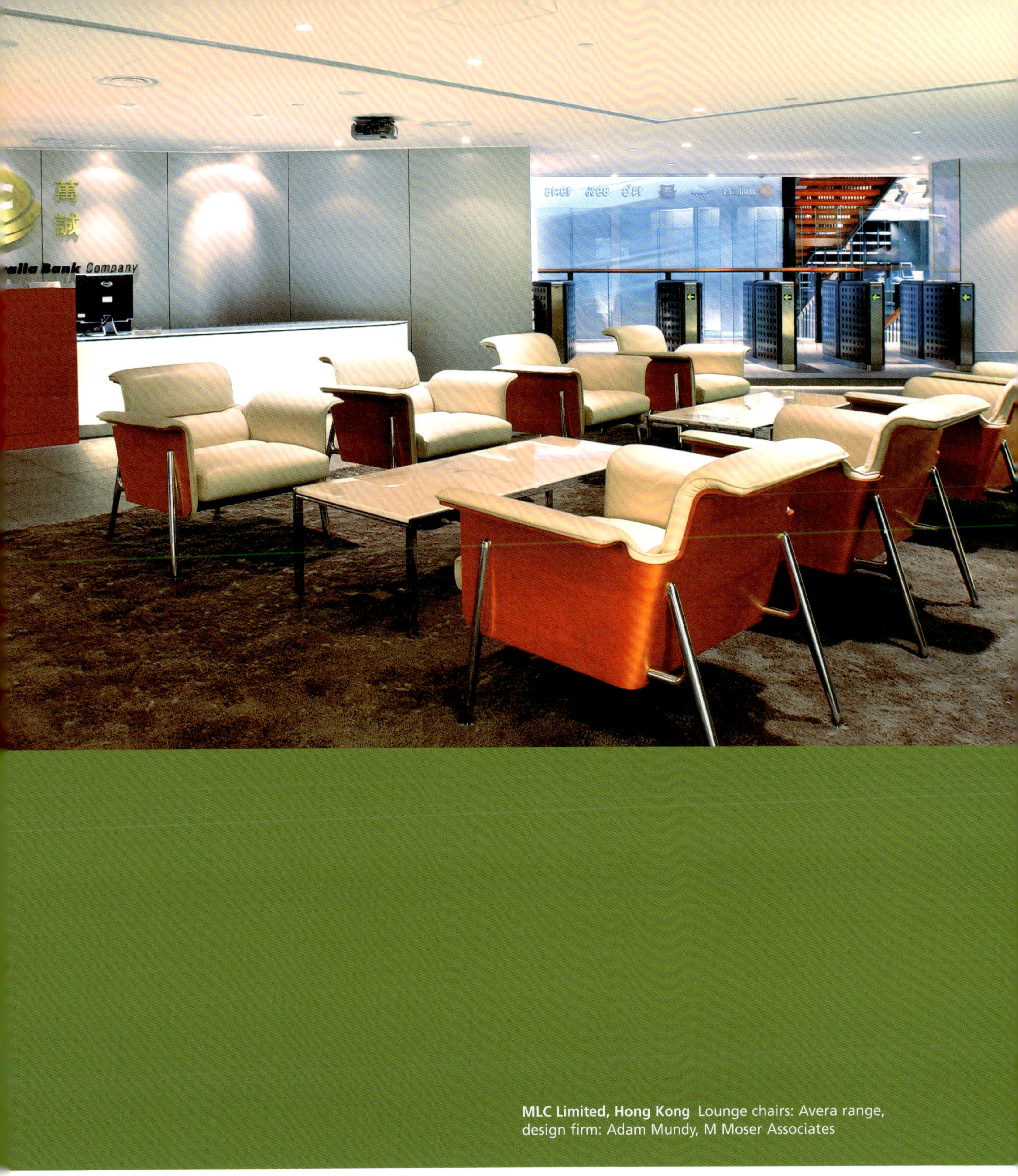

MLC Limited, Hong Kong Lounge chairs: Avera range, design firm: Adam Mundy, M Moser Associates

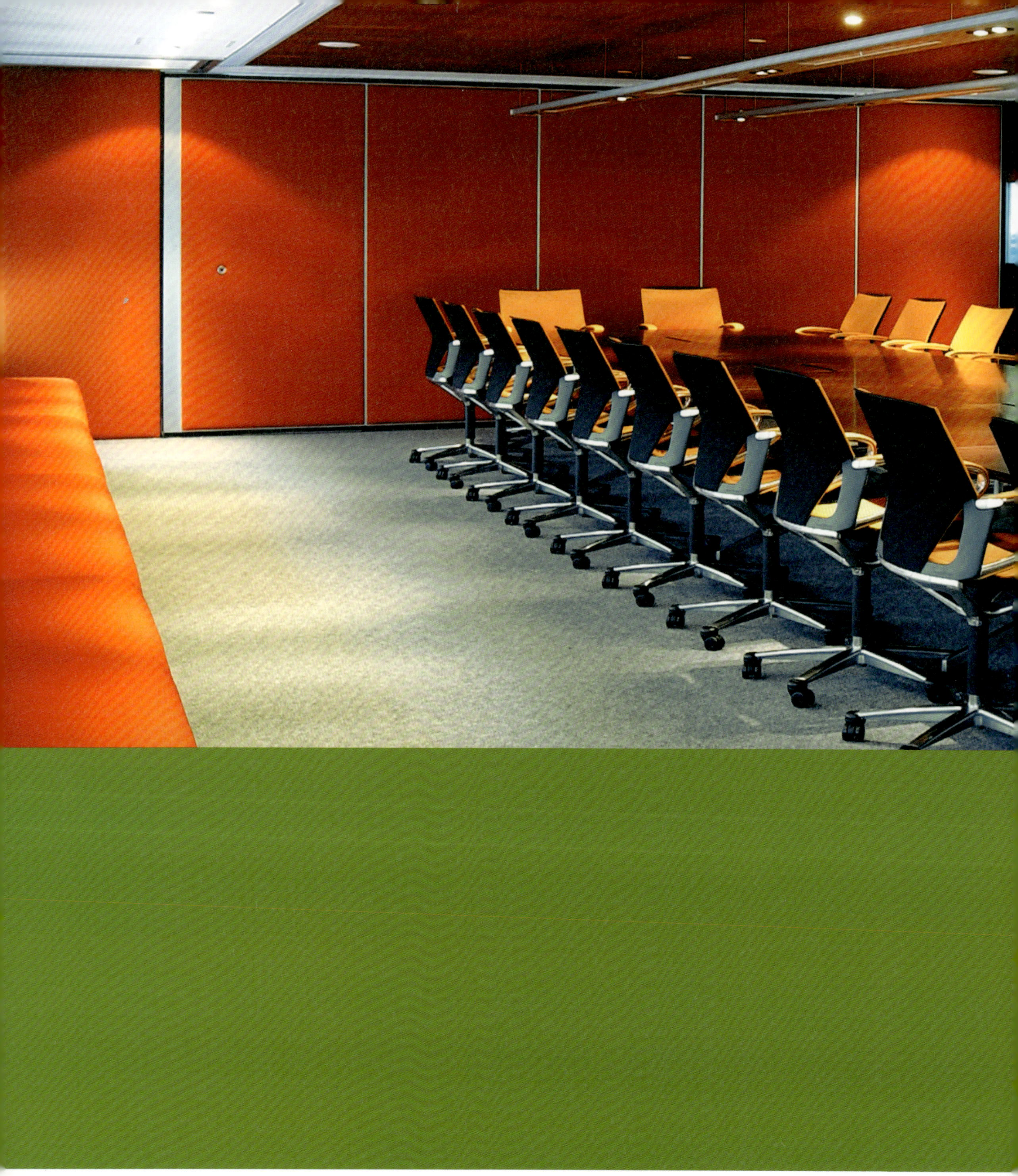

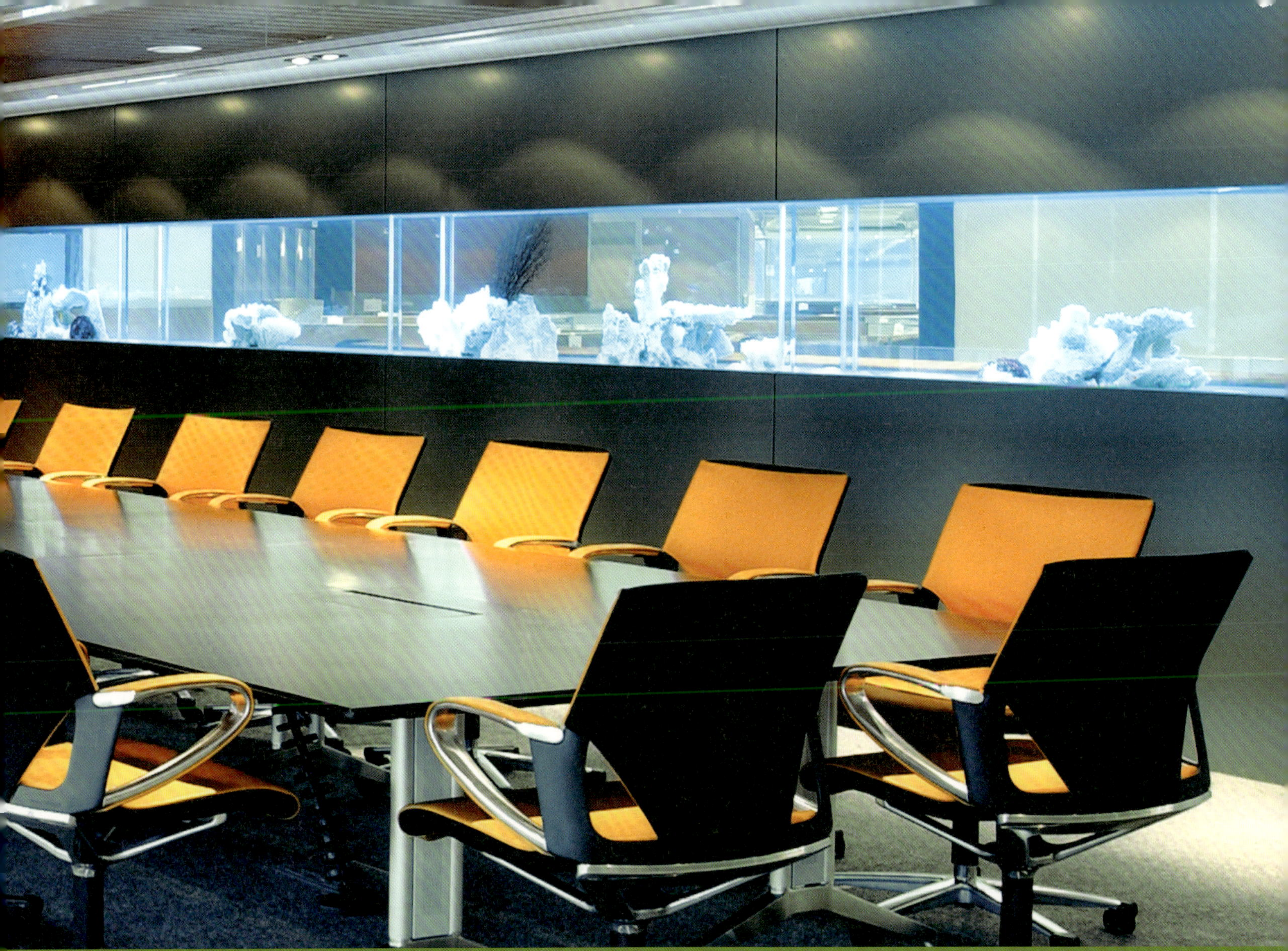

MLC Limited, Hong Kong Table configuration: Logon range; swivel chairs: Modus Medium range, design firm: Adam Mundy, M Moser Associates

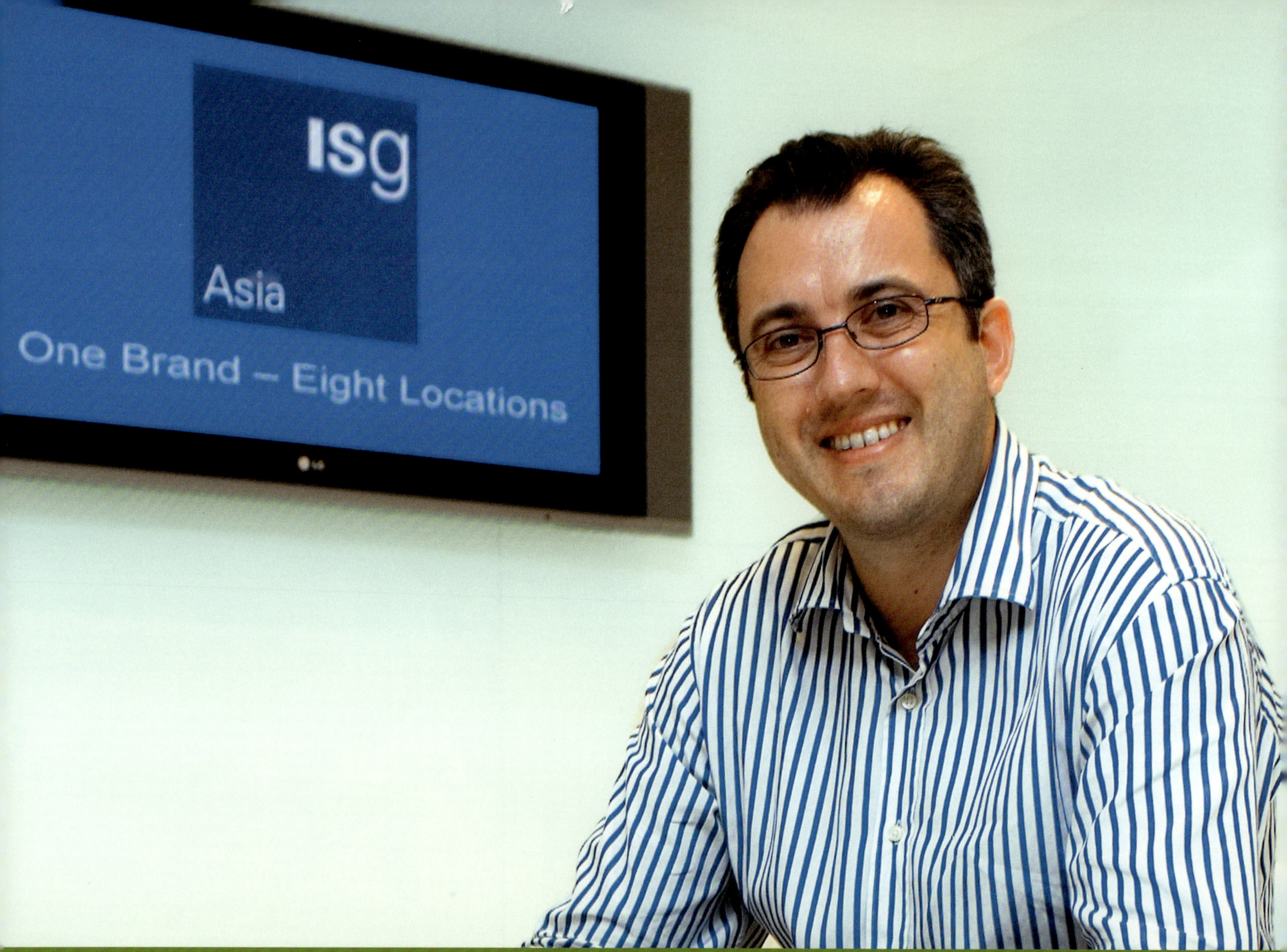

Richard Kimber, Country Director
"When the pressure is on for ISG to deliver and when final installations need to be re-sequenced, brought forward or pushed back, Wilkhahn is flexible, cooperative and works closely with the ISG project team to ensure we hand over as planned. It is always a pleasure to find out that Wilkhahn will be involved and working alongside ISG. Wilkhahn is cooperative, will accept change and will re-programme its installation with minimal fuss to ensure that ISG completes on time. Wilkhahn is attentive when it comes to matters relating to the closing of the project at completion time. Wilkhahn's attitude is one of collaboration and cooperation. ISG strives for quality and Wilkhahn never fails to meet our expectations."

ISG Asia, Singapore

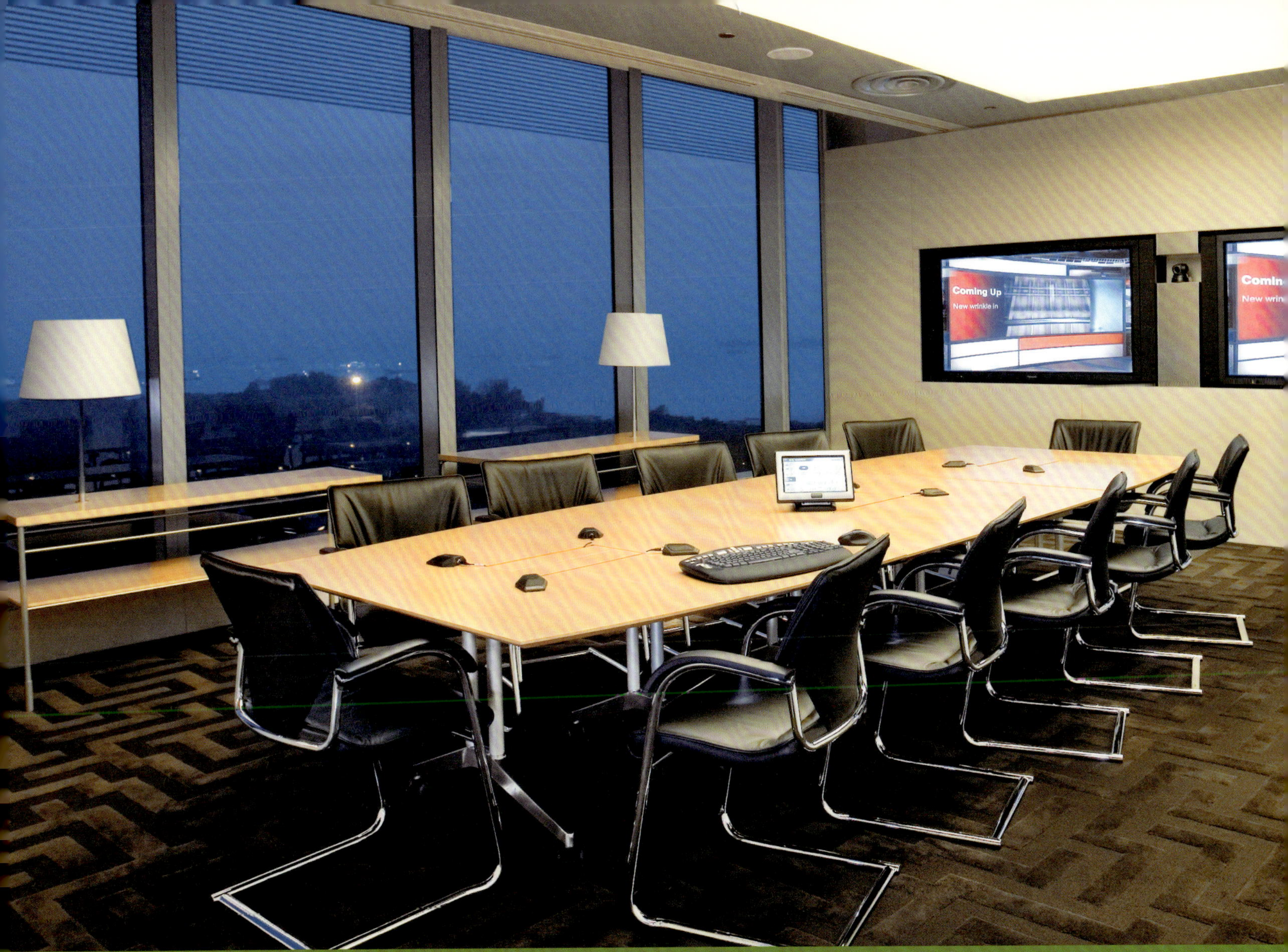

Deutsche Bank, Singapore Table configuration: Logon range; cantilever chairs: Modus Executive range, design firm: Steven Leach + Associates, Singapore

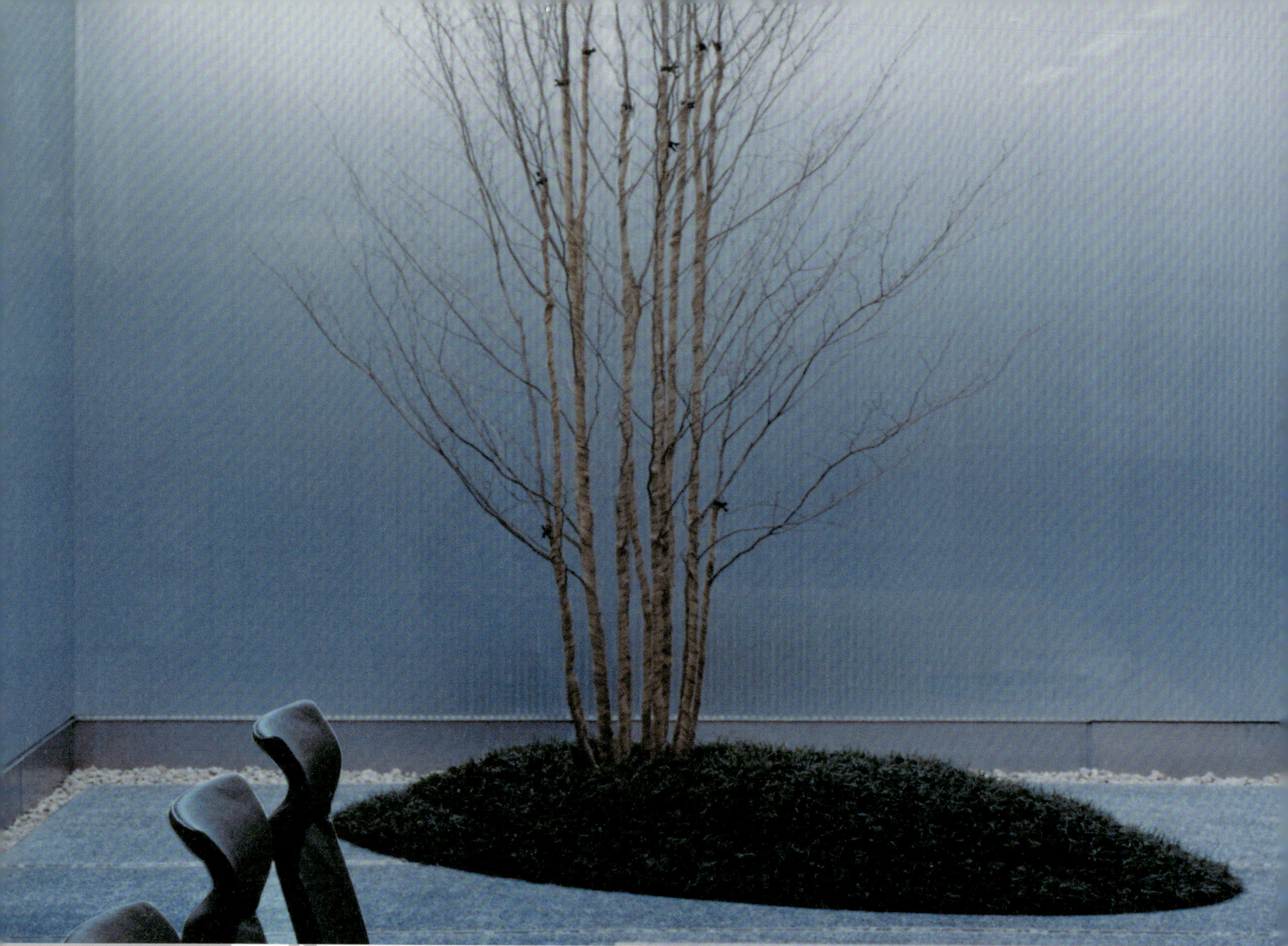

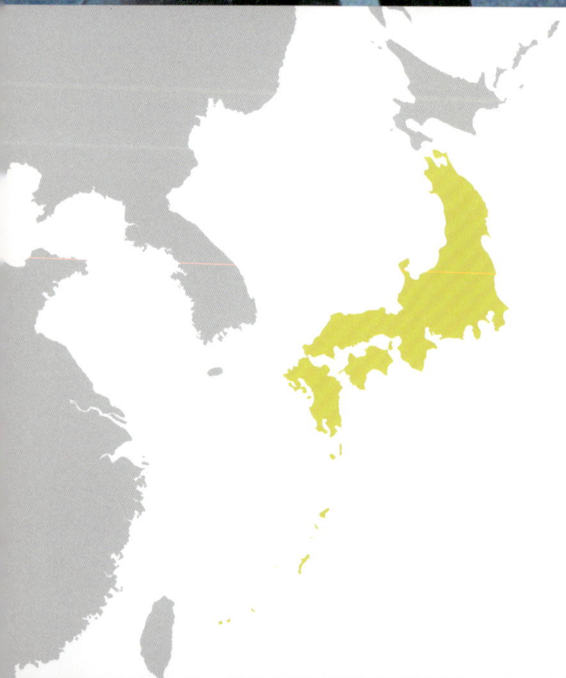

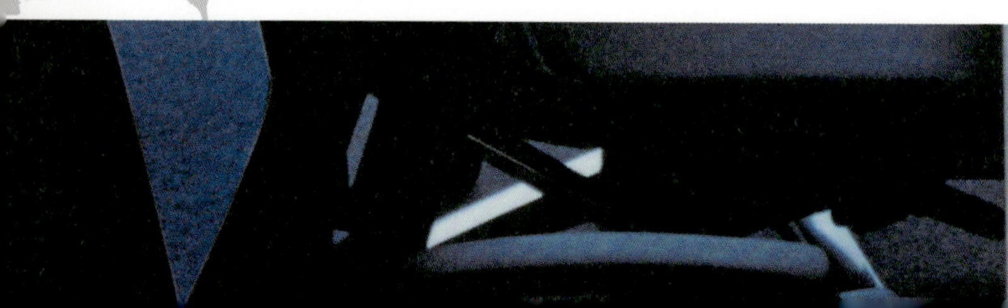

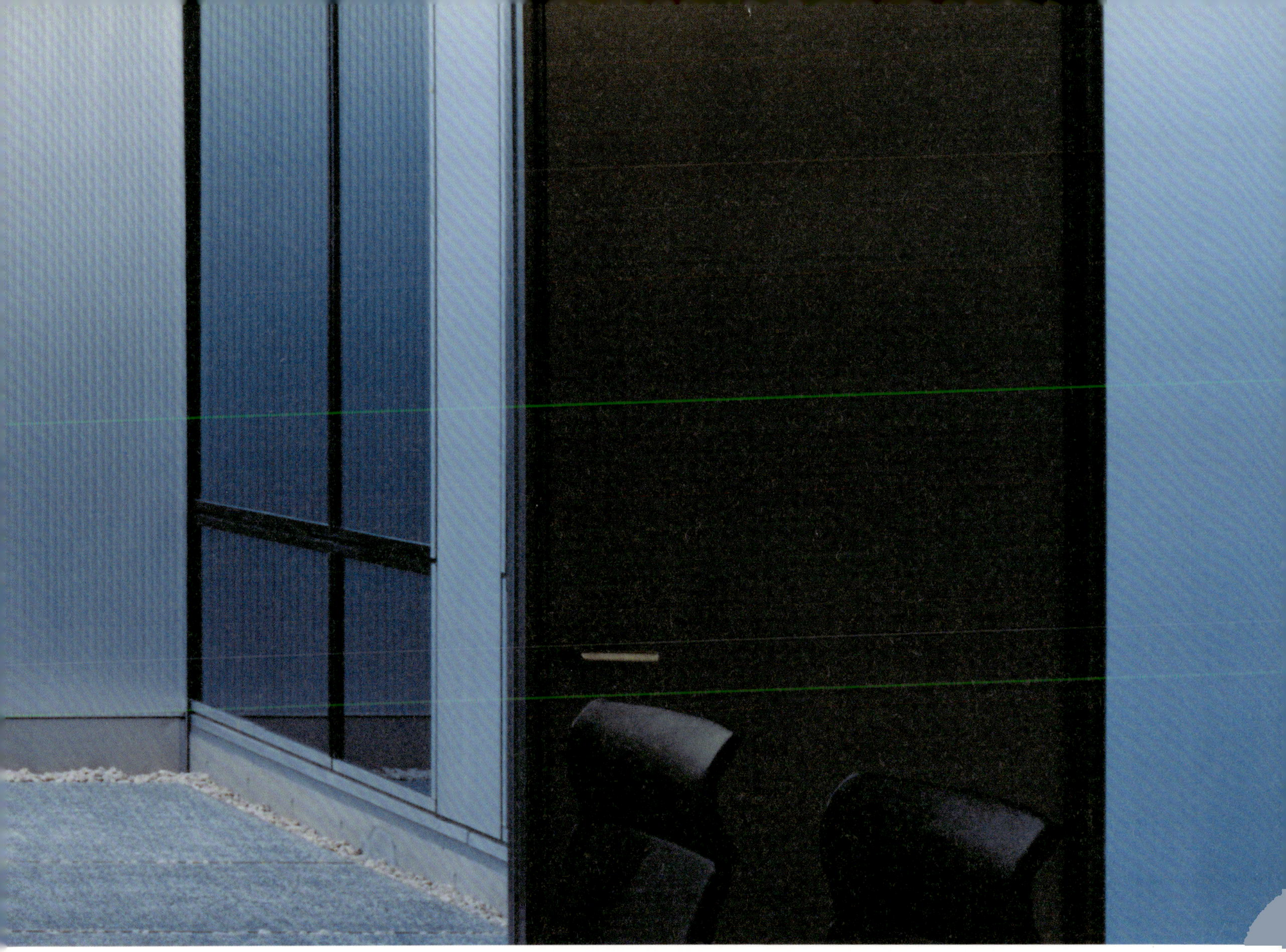

Die Gestaltungsmaximen des Bauhauses und der klassischen Moderne genießen in Japan hohes Ansehen und finden viele Entsprechungen in der japanischen Kulturtradition. Kein Wunder, dass Wilkhahn hier fast zu Hause ist. 1995 wurde mit dem japanischen Kokuyo-Konzern ein Lizenzvertrag geschlossen und Wilkhahn Japan Co., Ltd. wurde gegründet. Mit hohen jährlichen Zuwachsraten gehört Japan inzwischen zu den erfolgreichsten Auslandsmärkten, in denen wunderbare Projekte realisiert werden.

The design principles of Bauhaus and Classic Modernism enjoy a good reputation in Japan and there are many parallels in the Japanese cultural tradition. No wonder Wilkhahn feels almost at home here. A licence agreement was concluded with Kokuyo Group in 1995 and Wilkhahn Japan Co, Ltd. was founded. With its strong annual growth rates, Japan, in the meantime, is one of the most successful foreign markets in which wonderful projects are being realized.

TOYOTA AUTO BODY, Aichi, SHIMIZU CORPORATION and FIELD FOUR DESIGN OFFICE, Tokyo

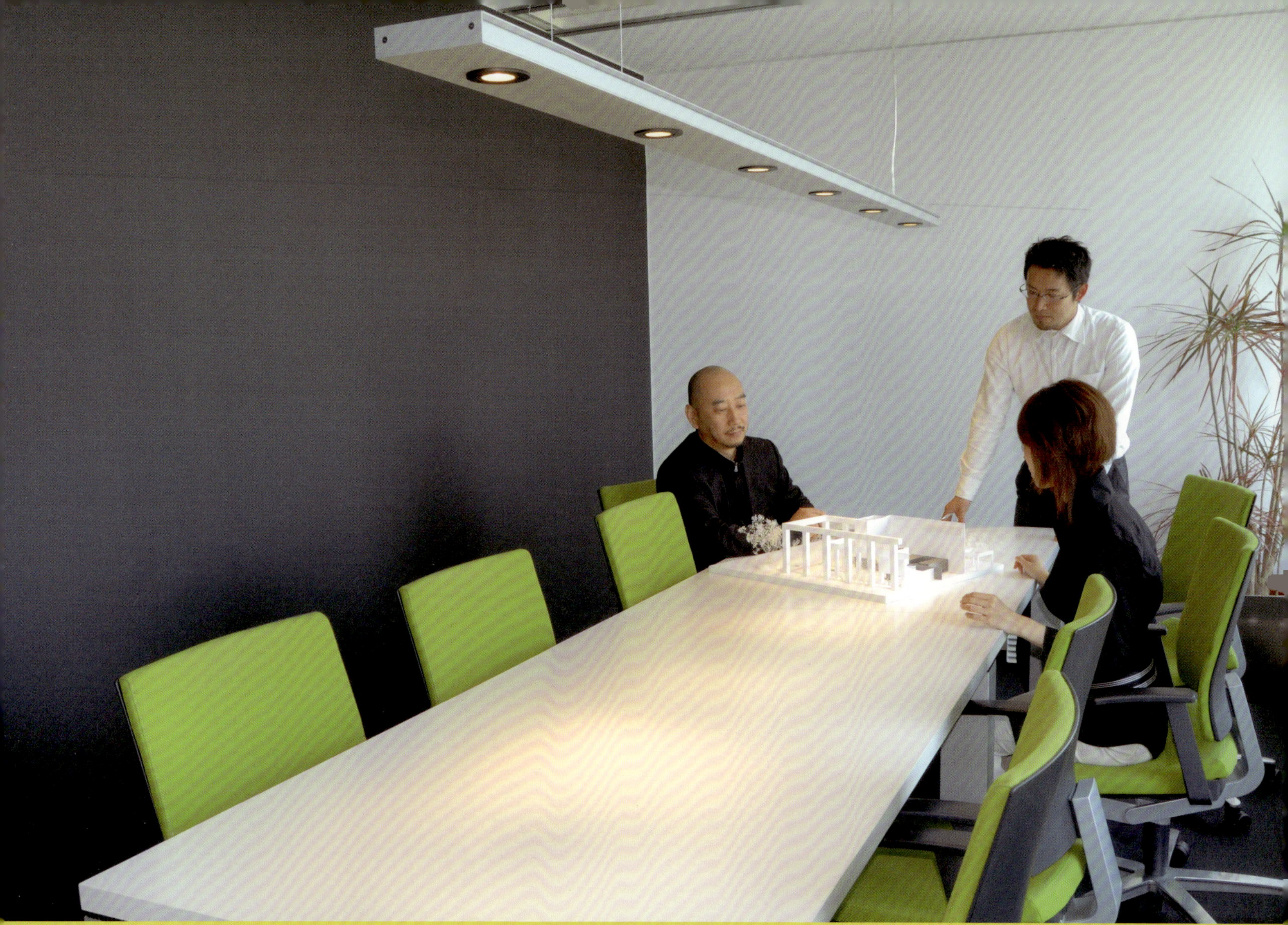

Field Four Design Office, Tokyo
Table configuration: Palette range;
swivel chairs: Solis range,
Modus range

Yoshiharu Shimura

険しい山を登り、突然景色が開けうまい水を飲むとき
自然にころあいのいい切り株を探す。
荒波の磯で魚と戯れたあと、太陽に温められたたおやかな
岩肌に冷えた体を横たえる。
椅子に座った人には、椅子は見えないが体中の触覚を脳と
つなぎそれを感じ取っている。
Wilkhahnのプロダクトは、形や色からではないそういうところから
始まっているように思える。
志村美治

"Climbing a steep mountain,
when suddenly a great landscape opens up. For a drink of tasty water
I am looking for a tree stump, just right for this nature.
After swimming with the fish in the raging waves of the sea,
I lay down my chilled body
on a sunbathed warm and graceful surface of the rock.
When a person sits down, he can not see the chair,
but the sense of touch throughout the body
connects to the brain and he 'feels' it.
The beginning of Wilkhahn products lies there, rather than in form or colour
I believe."

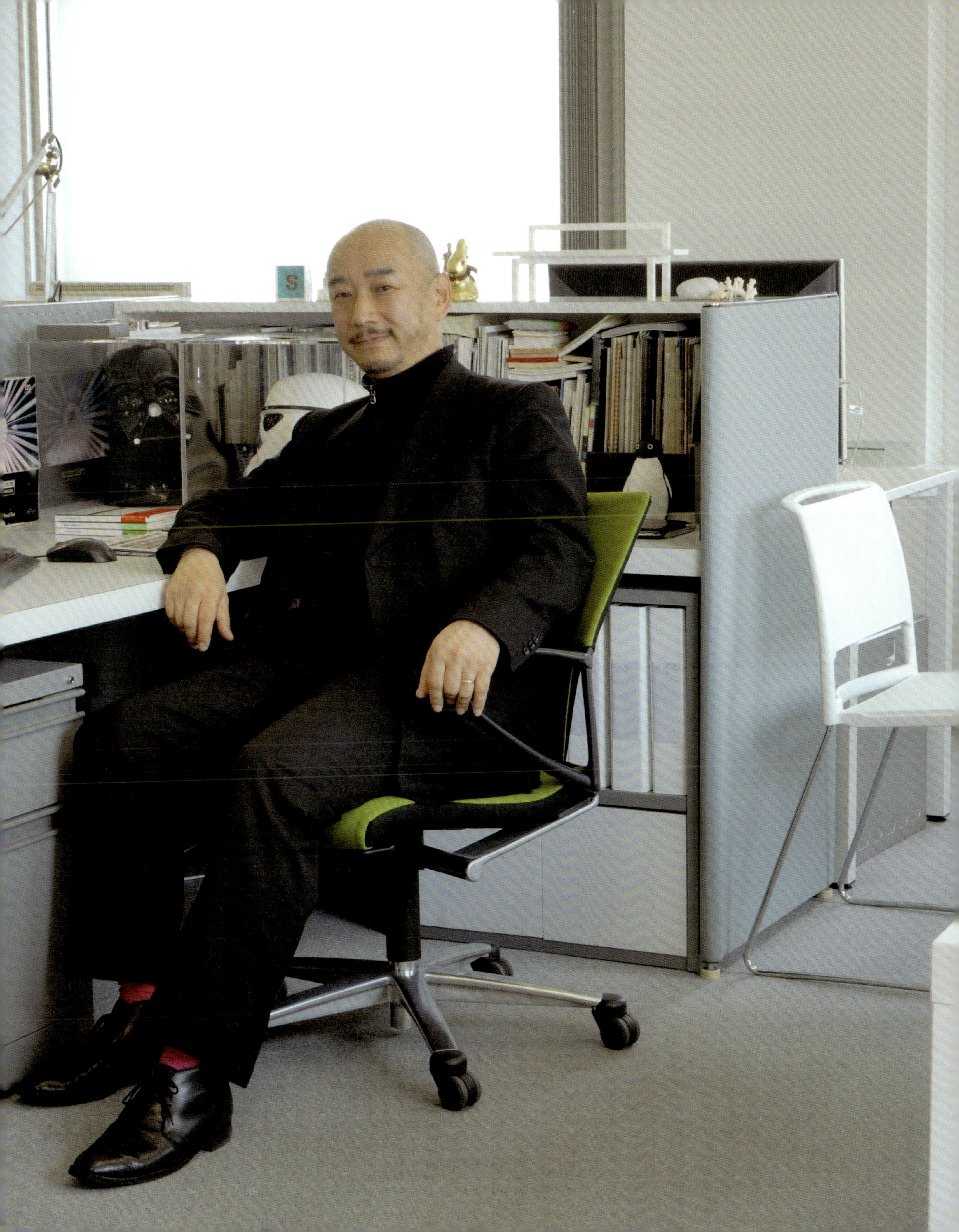

Toyo Ito, Architect, Tokyo

この大学図書館の一階は床が1/20のスロープでつくられています。カフェや雑誌の閲覧スペース、AVバーと呼ばれるAV資料閲覧のカウンター用の椅子としては、高さが自由に調整でき、かつ傾けられるStitzしかないと考えました。
実際に入れてみて、その自由さが学生たちのリラックスした雰囲気を生み出し、思い思いの姿勢を可能にしています。人が立っている姿と座っている姿の中間的な姿勢を初めて可能にしたのが Stitzだと思います。

伊東豊雄

"Most of the first floor of this university library is designed with a constant slope of 1/20. For the café, the magazine reading space, and the so-called AV bar for video documents, I could only think of the Stitz. The height is infinitely adjustable and it allows tilting in any direction desired.
Once all the Stitz leaning aids were installed, this flexible adaptability of the chairs brought forth an atmosphere of relaxation for the students, therefore allowing good posture for contemplation. For me the Stitz is the first type of stool to realize this 'in between' position of standing and sitting so perfectly."

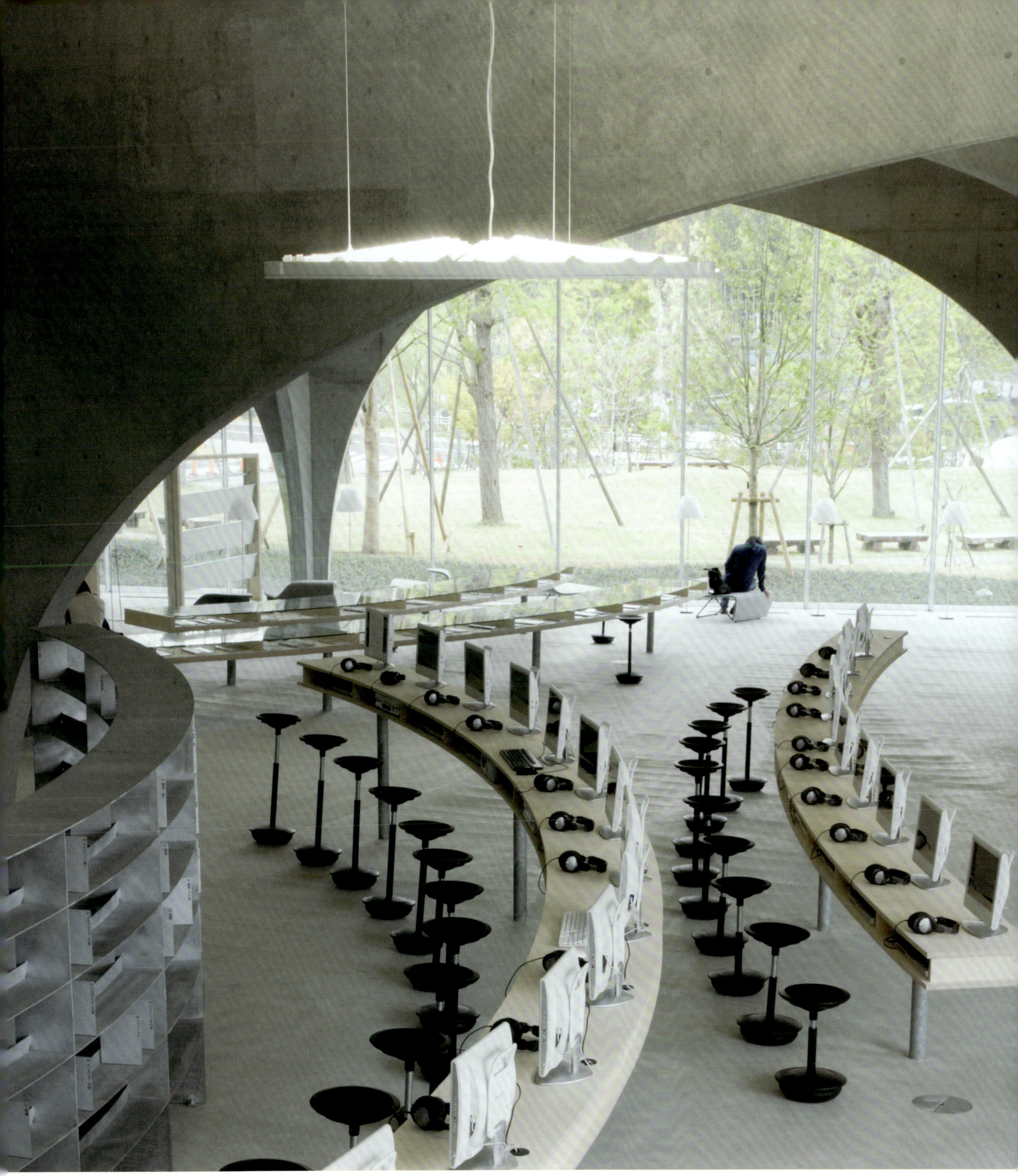

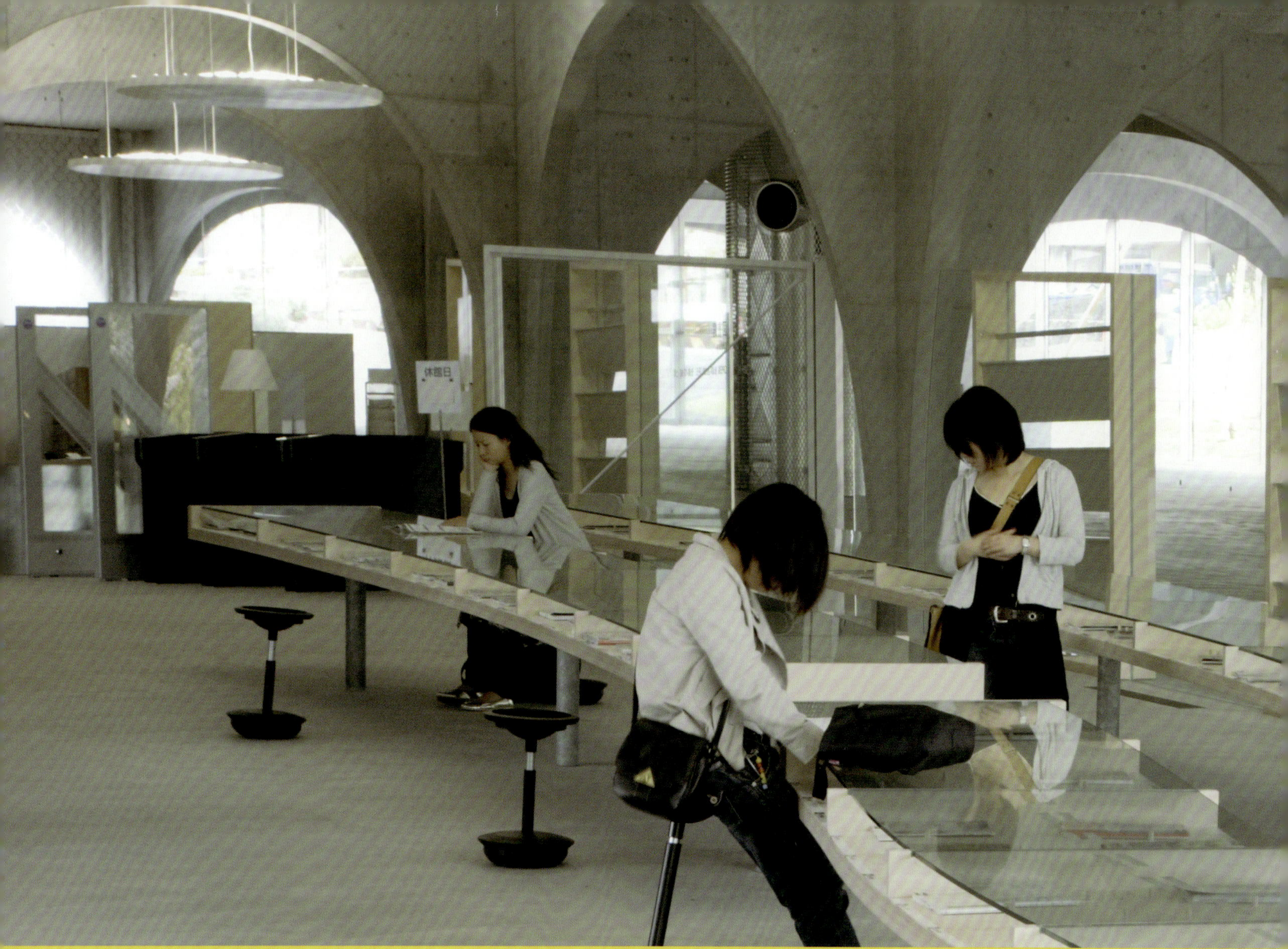

Kazuko Fujie, Interior Planner, Tokyo

傾いた床と開放的で流動的な特徴のあるこの図書館では、何時でもどこでも学生が自由な姿勢で居られることを最も重要視しています。
学生達は、それぞれに高さを変えたり傾きを変えたり、上体の微妙な動きとともに肉体の一部として身体を支えるように使い
こなしています。
このように身体の一部となったStitzは,モバイルな現代人の生活にぴったりだと思います。
藤江和子

"In this library with a gentle slope, with an open and fluid character, the existence of the students' independent attitude, whenever and wherever possible, was of utmost importance. The students are using the body-supporting Stitz as if it was becoming a part of them with every little movement of the upper body, while liberally changing each height and leaning in different directions. I believe that this Stitz becoming one with the human body perfectly fits the current lifestyle of mobility."

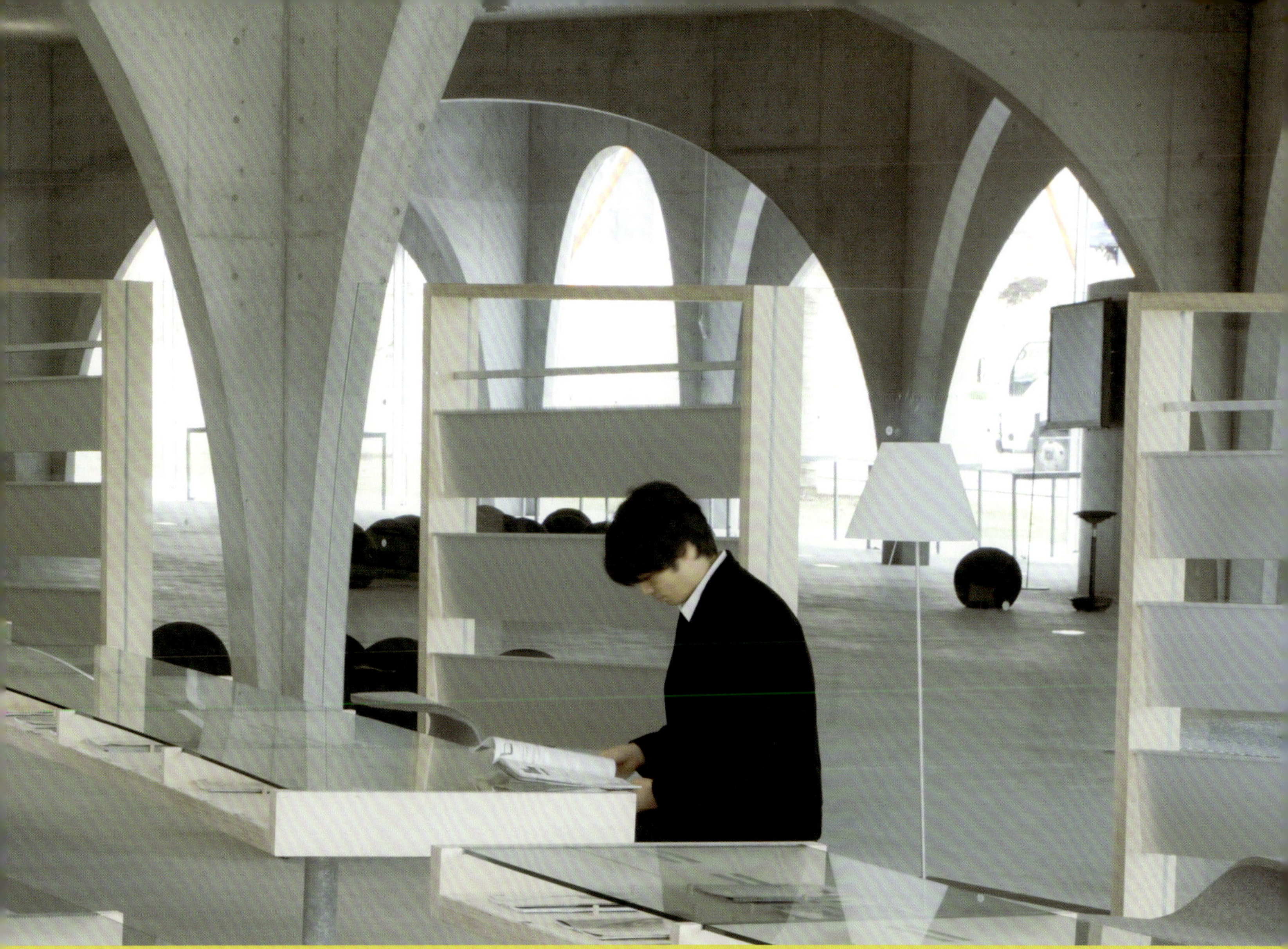

Tama Art University Dynamic stools: Stitz range

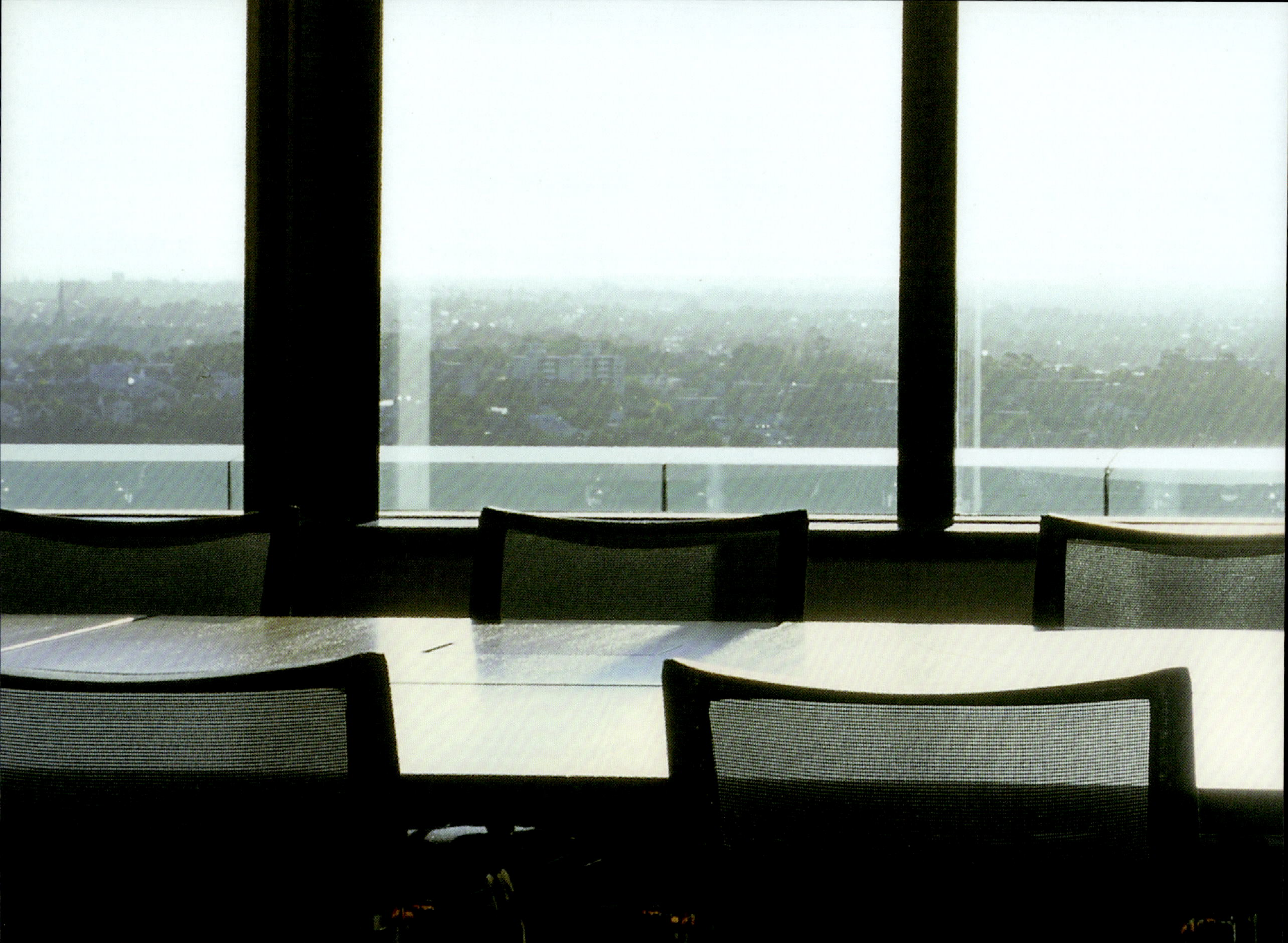

Australien und insbesondere Sydney sind spätestens seit der Olympiade 2000 aus einem Randbereich der globalen Wahrnehmung in ein breites öffentliches Bewusstsein gerückt. Bei Wilkhahn war dies schon viel früher der Fall: Bereits in den 80er Jahren wurde mit großem Erfolg die Bürostuhlreihe FS-Linie durch einen Lizenzpartner gefertigt und zu einem der bekanntesten Bürostuhlprogramme in Australien. 1998 folgte die Gründung von Wilkhahn Asia Pacific mit Standort in Sydney und Showrooms in Brisbane und Melbourne. Der Hauptsitz in Sydney ist auch die Basis für die Wilkhahn-Erfolge in Asien. Das neue Fertigungs-, Verwaltungs- und Ausstellungsgebäude, gestaltet von Bligh Voller Nield Architecture, ist 2007 mit dem renommierten „RAIA NSW Interior Architecture Award" ausgezeichnet worden.

When the Olympic Games took place in Australia in 2000, the country and in particular Sydney finally moved from a peripheral area of global perception into the public eye. In the case of Wilkhahn, this happened much earlier: already in the 1980s, the office chair range FS-Line was manufactured with great success by a licence partner and became one of the most well-known office chair ranges in Australia. The founding of Wilkhahn Asia Pacific in Sydney followed in 1998 with showrooms in Brisbane and Melbourne. The head office in Sydney is also the base of Wilkhahn's success in Asia. The new production, administration and exhibition building, designed by Bligh Voller Nield Architecture, was awarded the renowned "RAIA NSW Interior Architecture Award" in 2007.

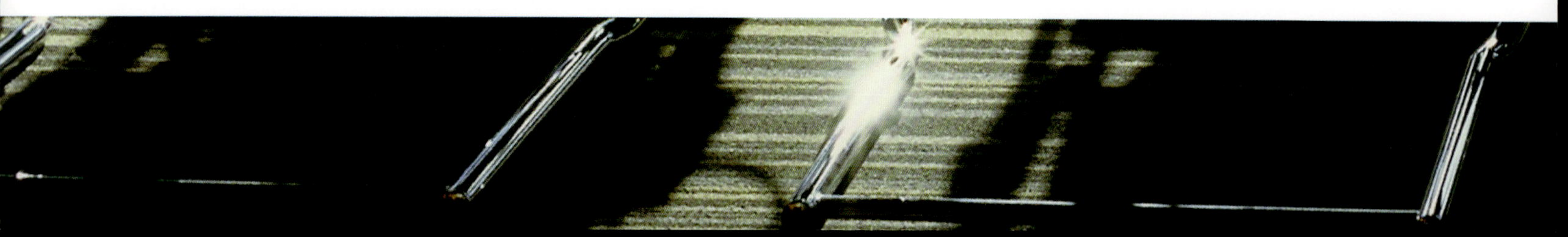

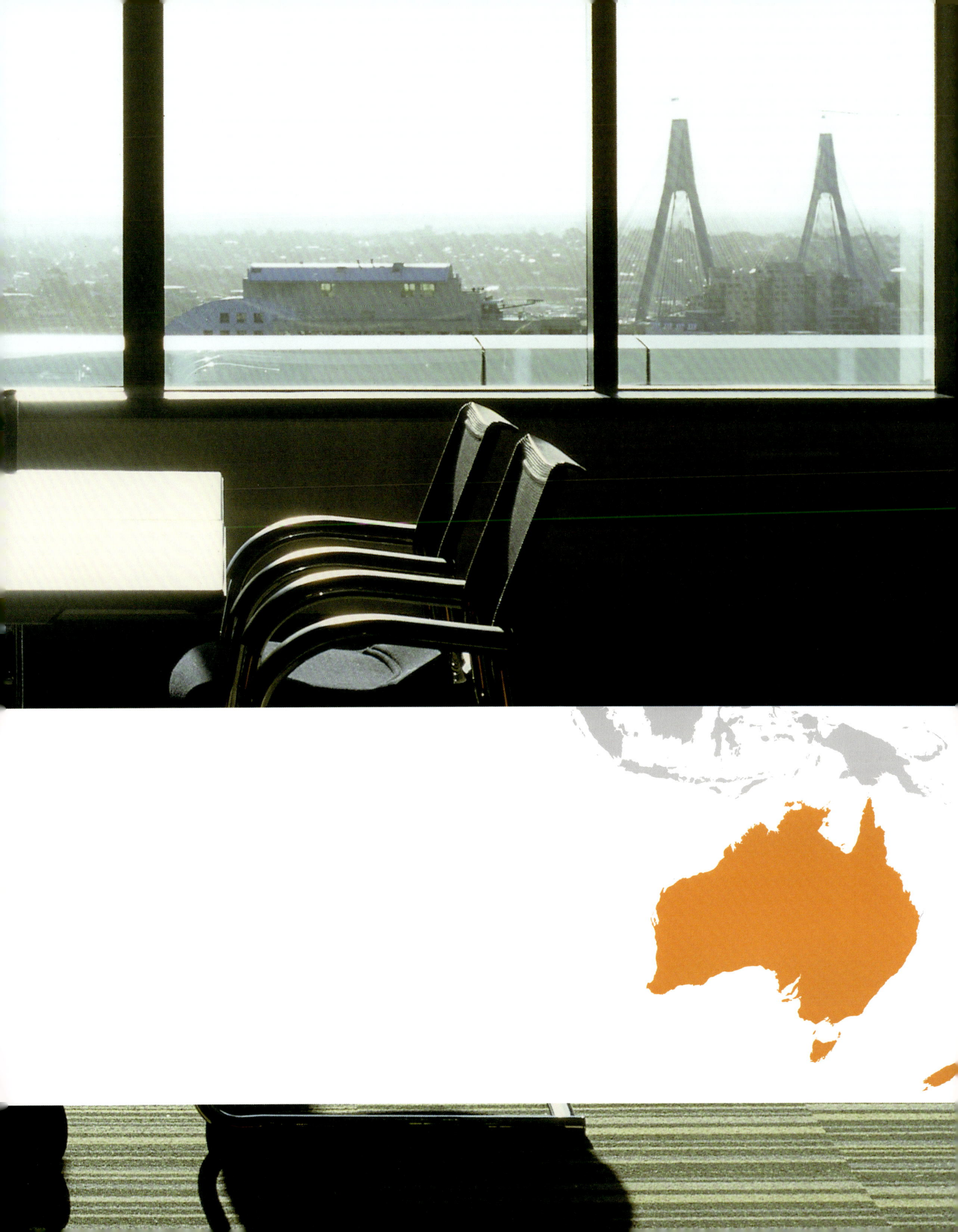

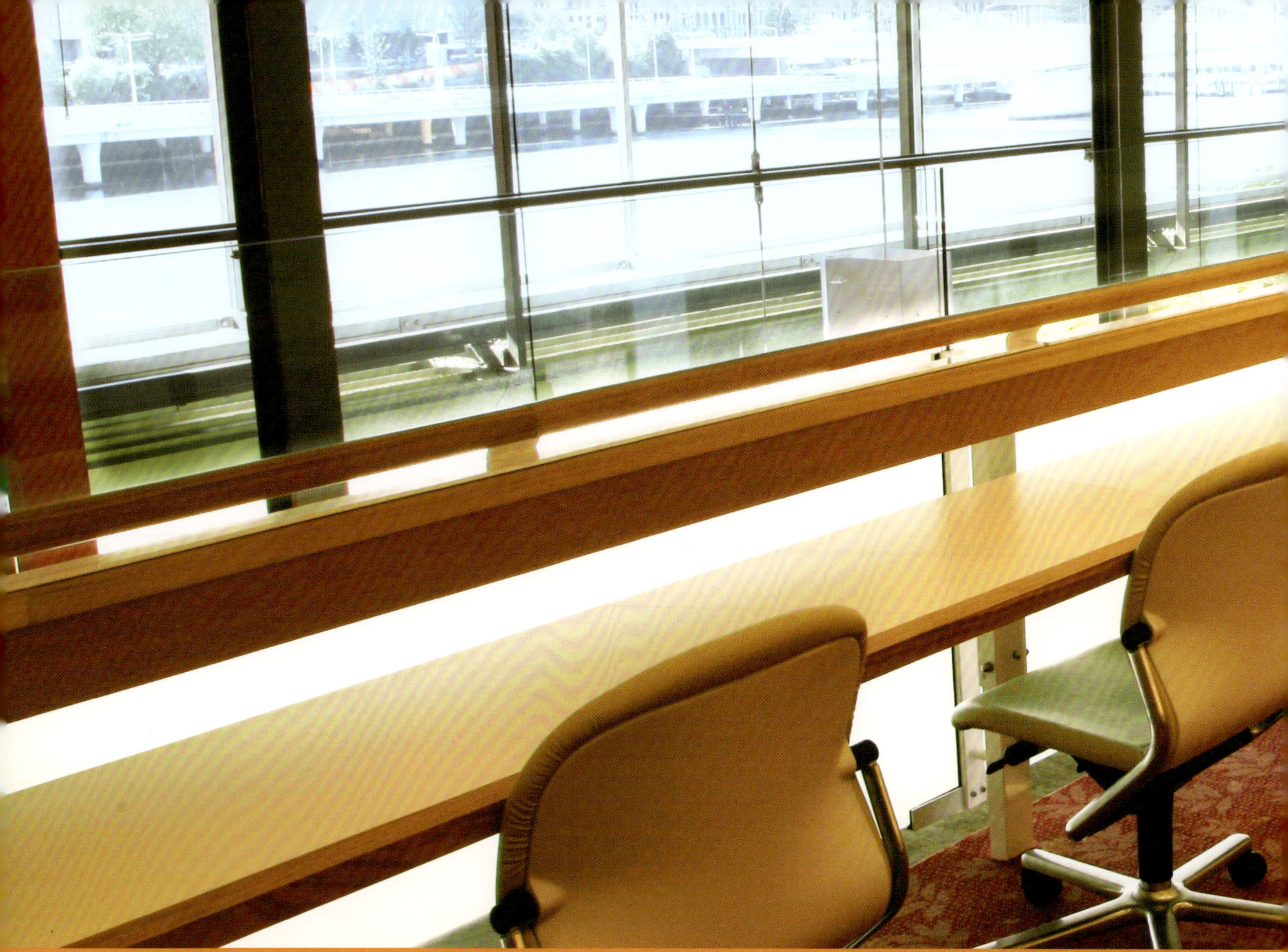

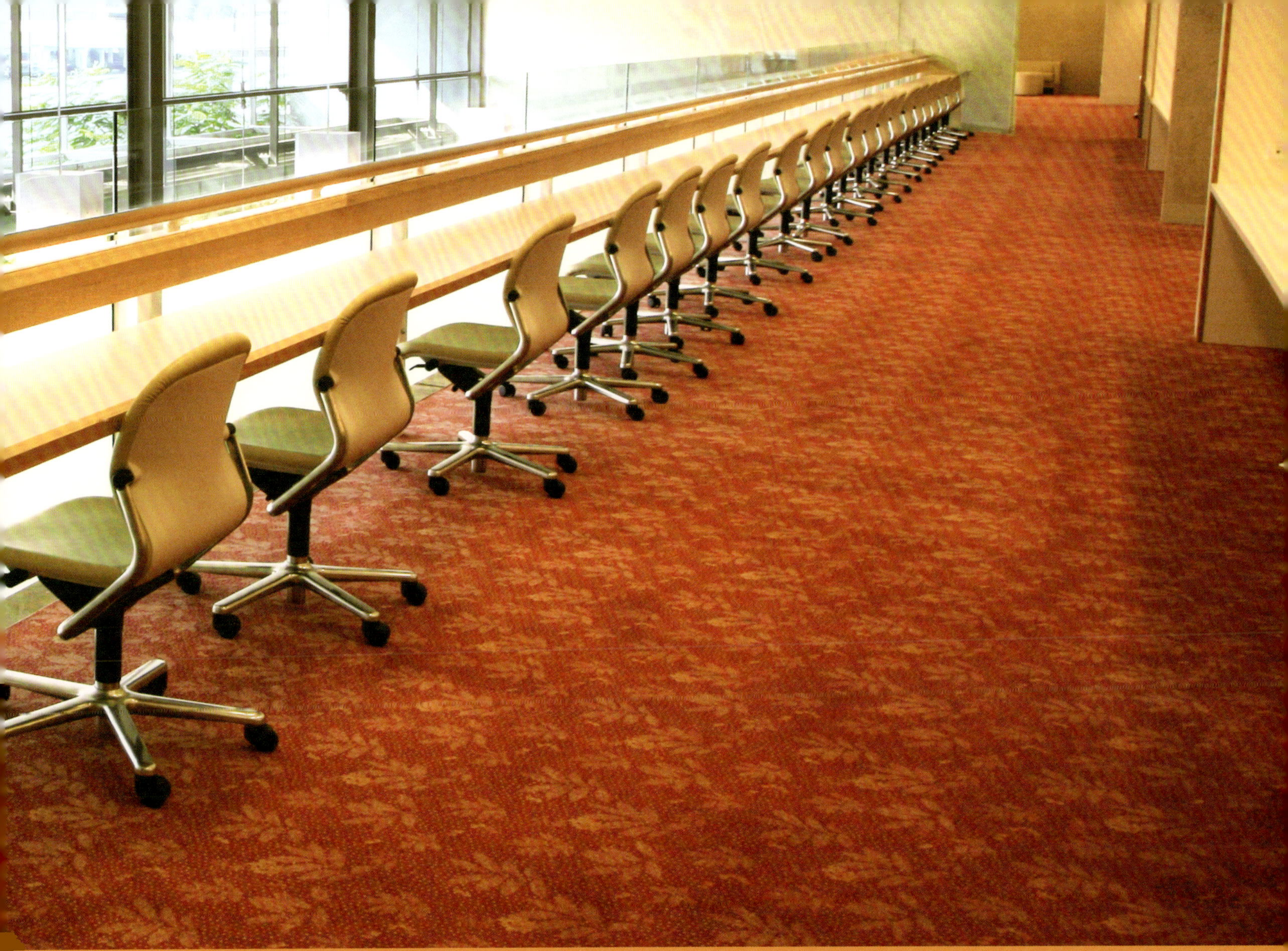

Millennium Arts Project, State Library of Queensland, Brisbane Swivel chairs: FS range, design firm: Greg Lamb, Donovan Hill

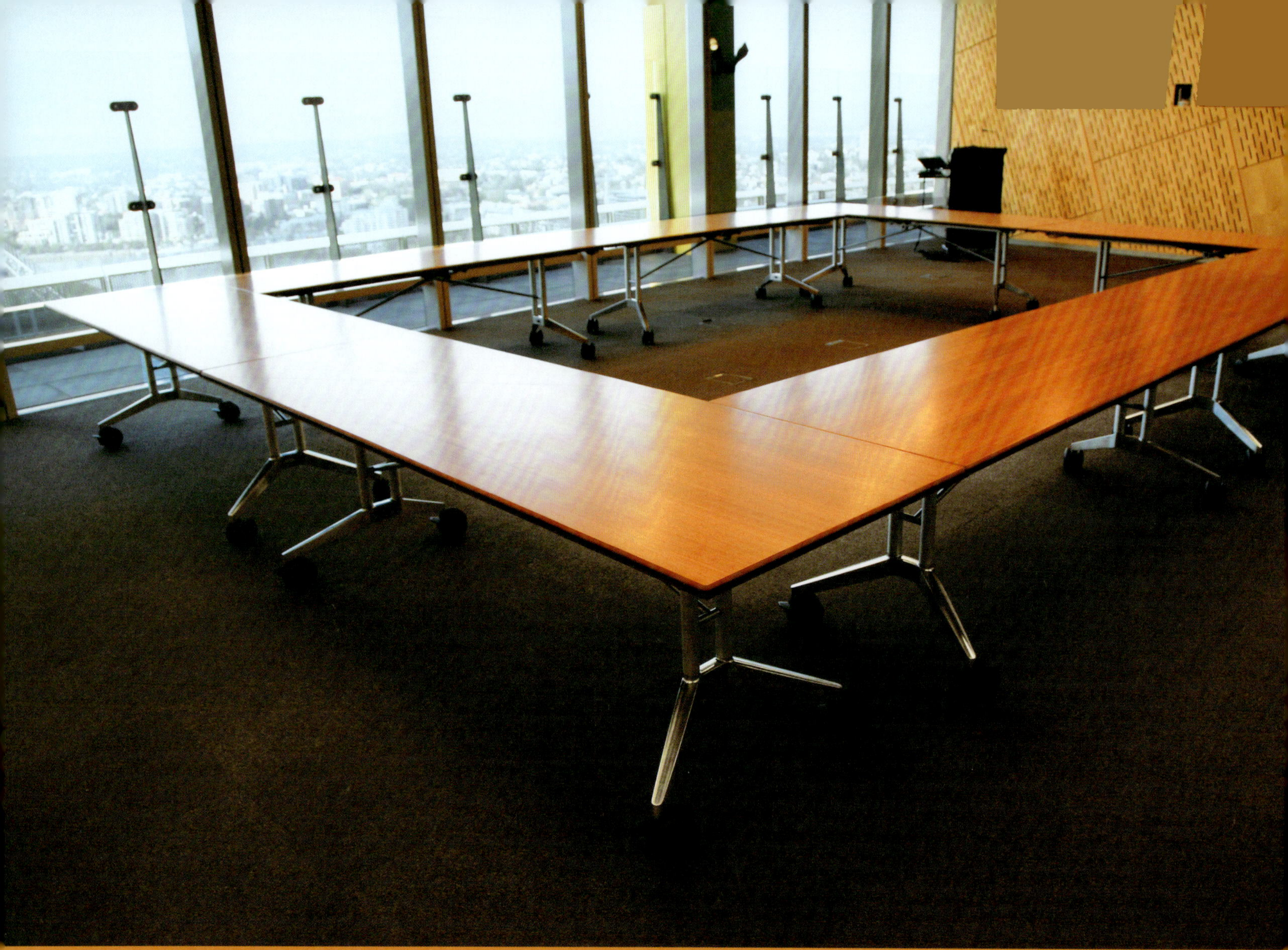

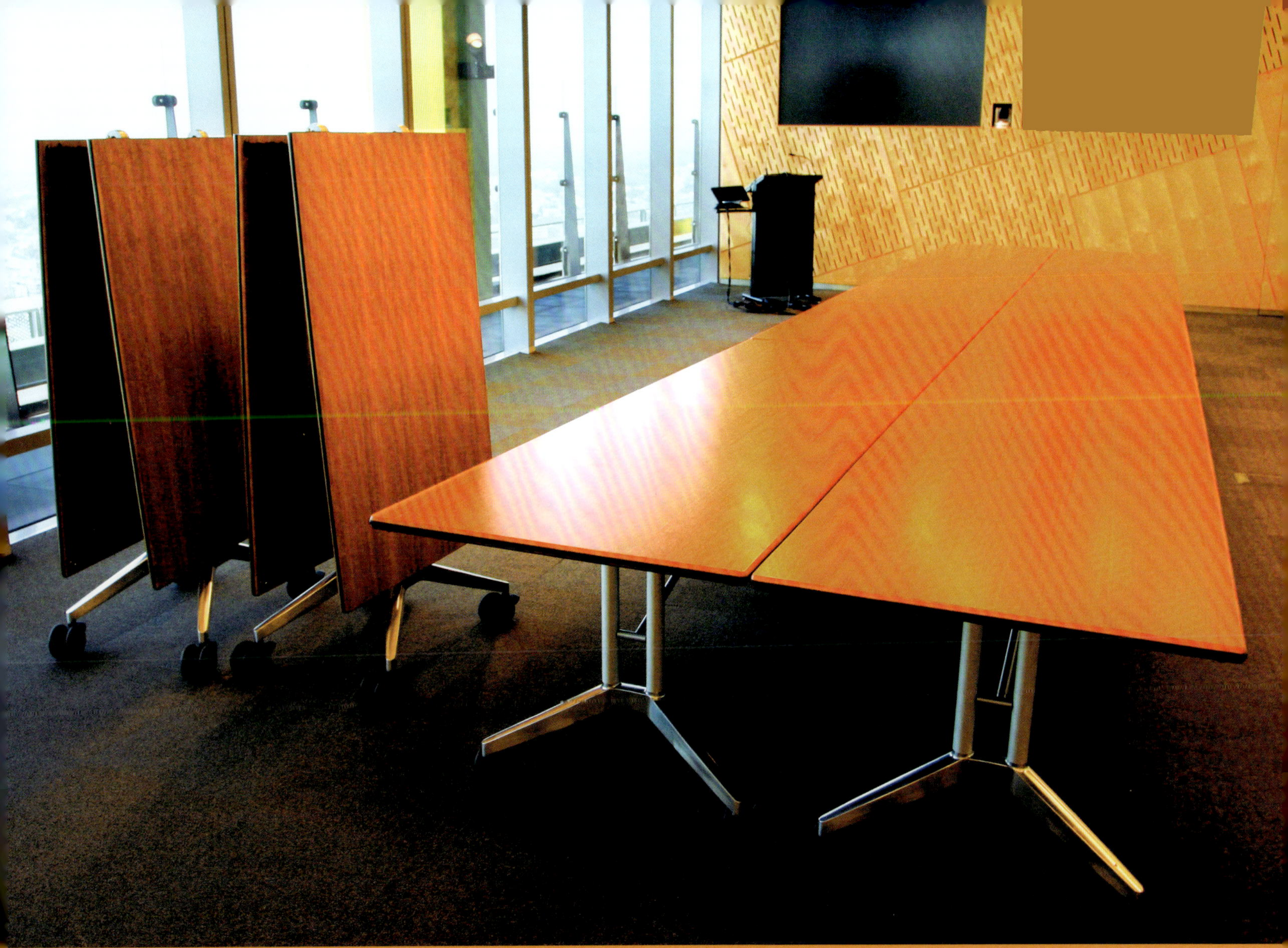

Suncorp-Metway Ltd., Brisbane Table configuration: Confair folding tables, design firm: Wei Shun Lee, Cox Rayner

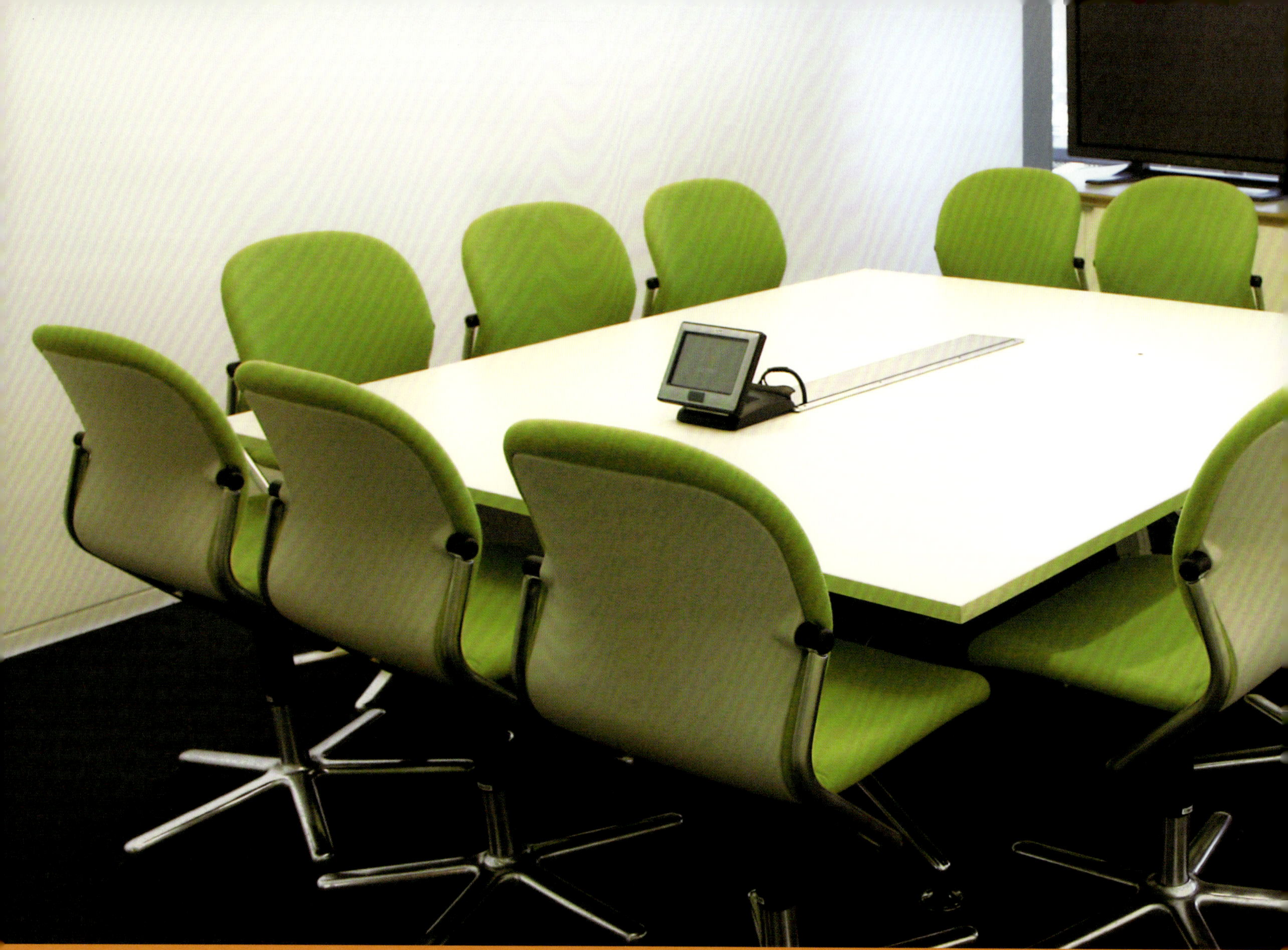

Suncorp-Metway Ltd., Brisbane
Table configuration: Logon range; swivel chairs: FS range, design firm: Wei Shun Lee, Cox Rayner

Suncorp-Metway Ltd., Brisbane City Council, Brisbane

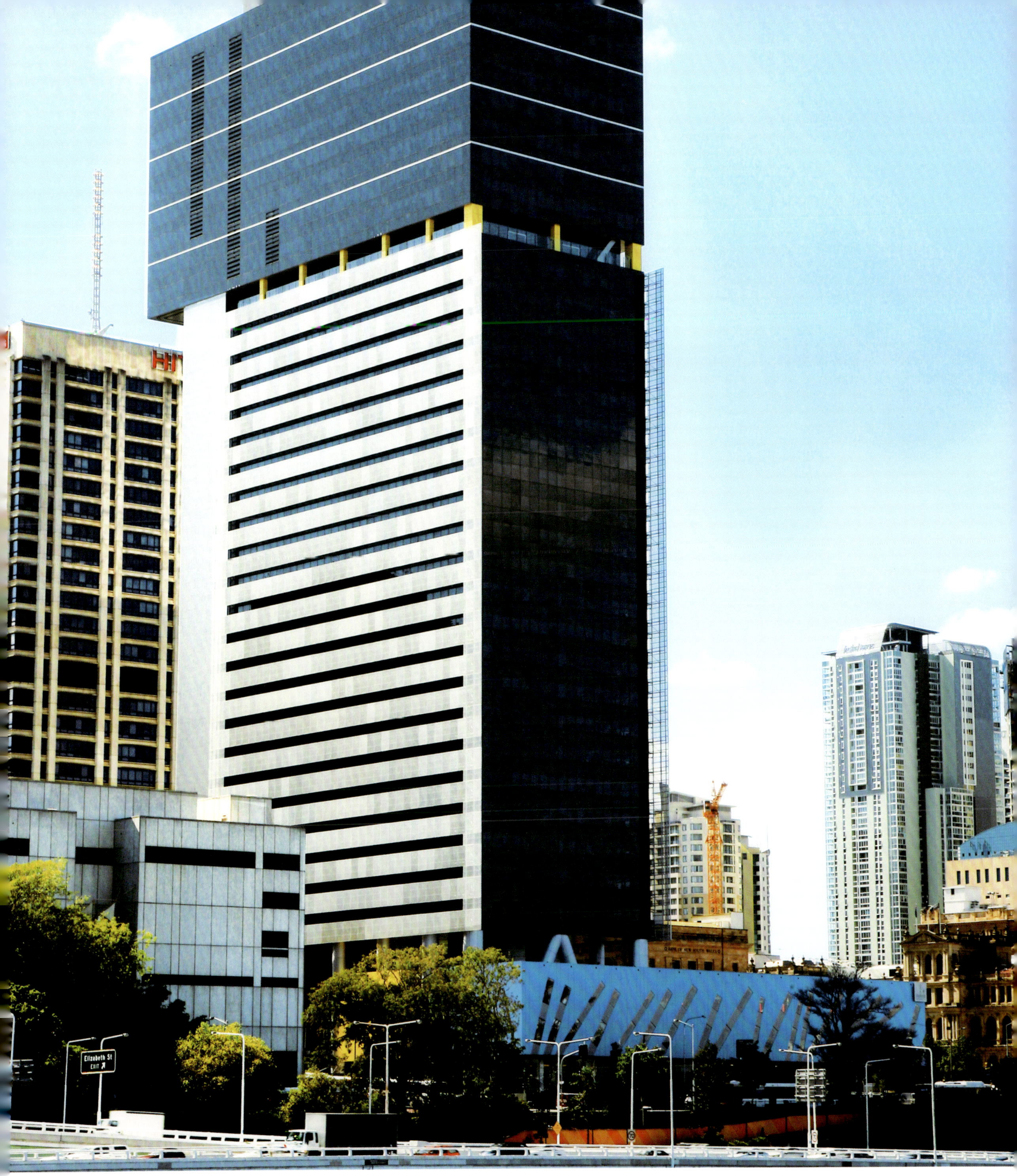
Elizabeth St
EXIT

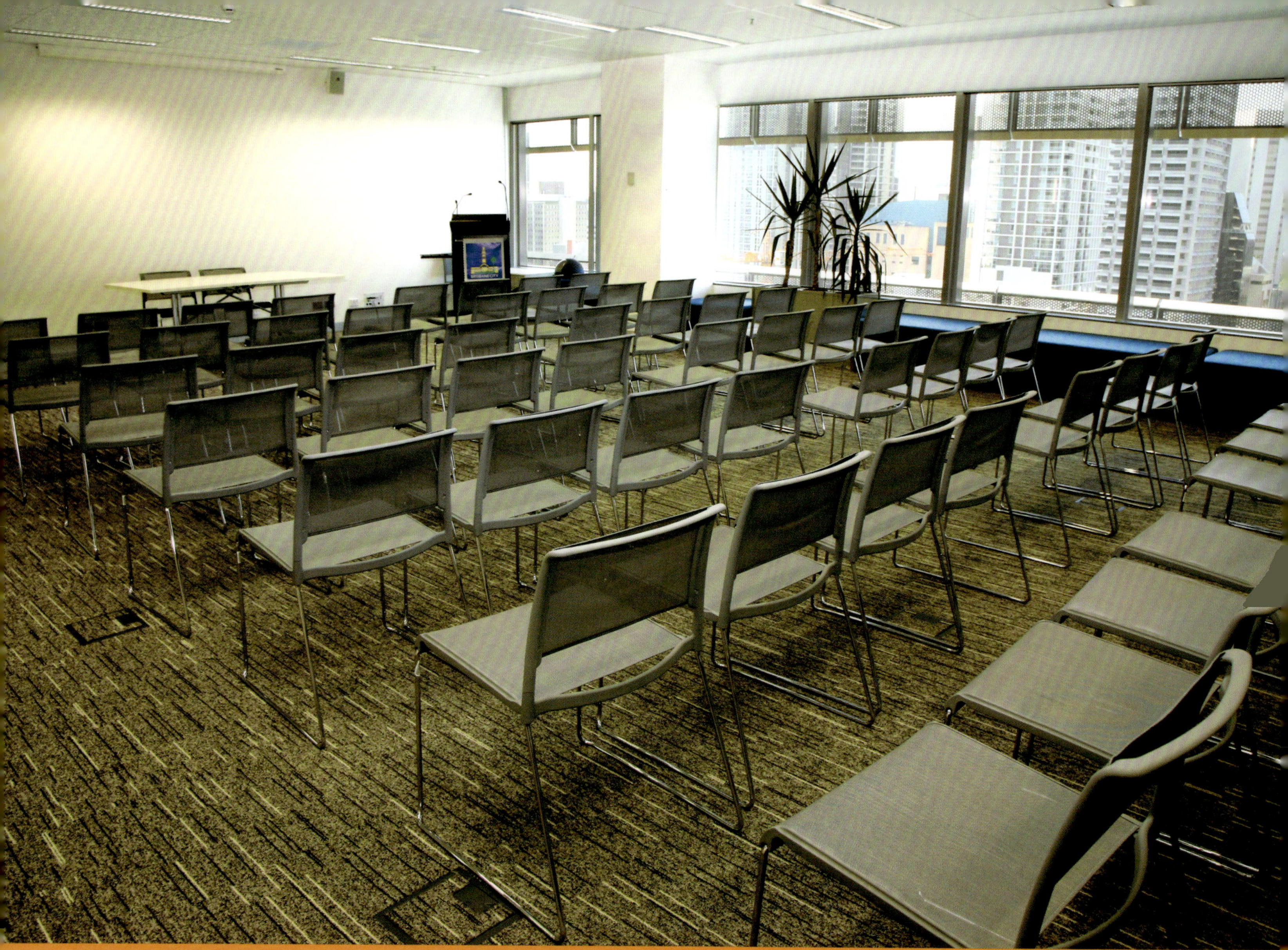

Brisbane City Council, Brisbane Table: Confair folding table; multi-purpose chair: Aline range, designer: Philip Harper, formerly Partner, Woods Bagot

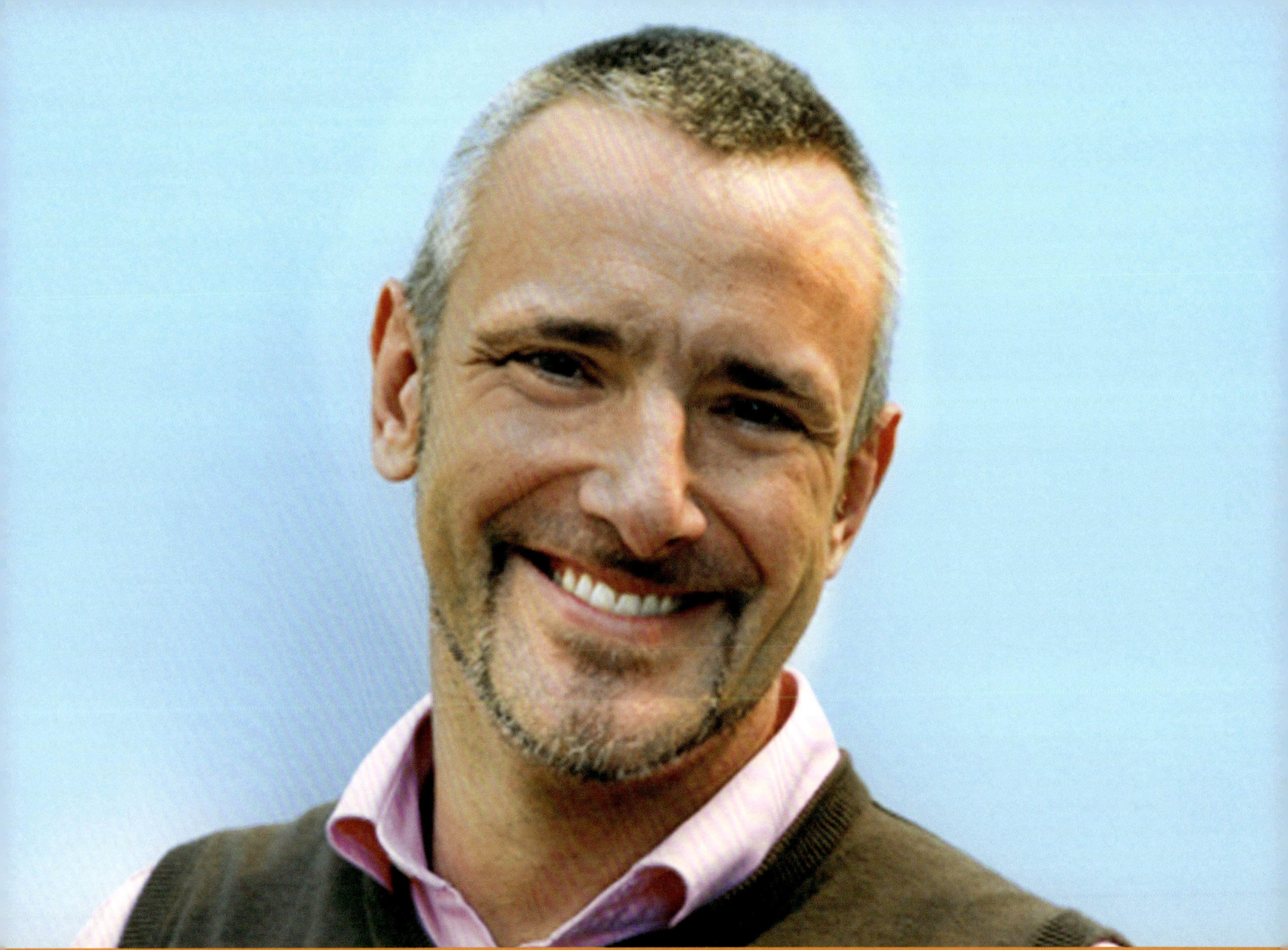

Philip Harper, Associate Director
"Wilkhahn affords designers the ability to create flexible office and conferencing spaces through its ranges of modular systems and seating products. These ranges are created through rigorous research and development and manufactured to the highest quality standards. This can be seen through the relevance and popularity today of the FS range of seating products, created over twenty years ago. A true reflection of Wilkhahn's approach to timeless quality and innovation and testimony to why, this year, we are celebrating Wilkhahn's 100th anniversary of making our work life better through great design."

Cottee Parker Architects Pty Ltd., Brisbane

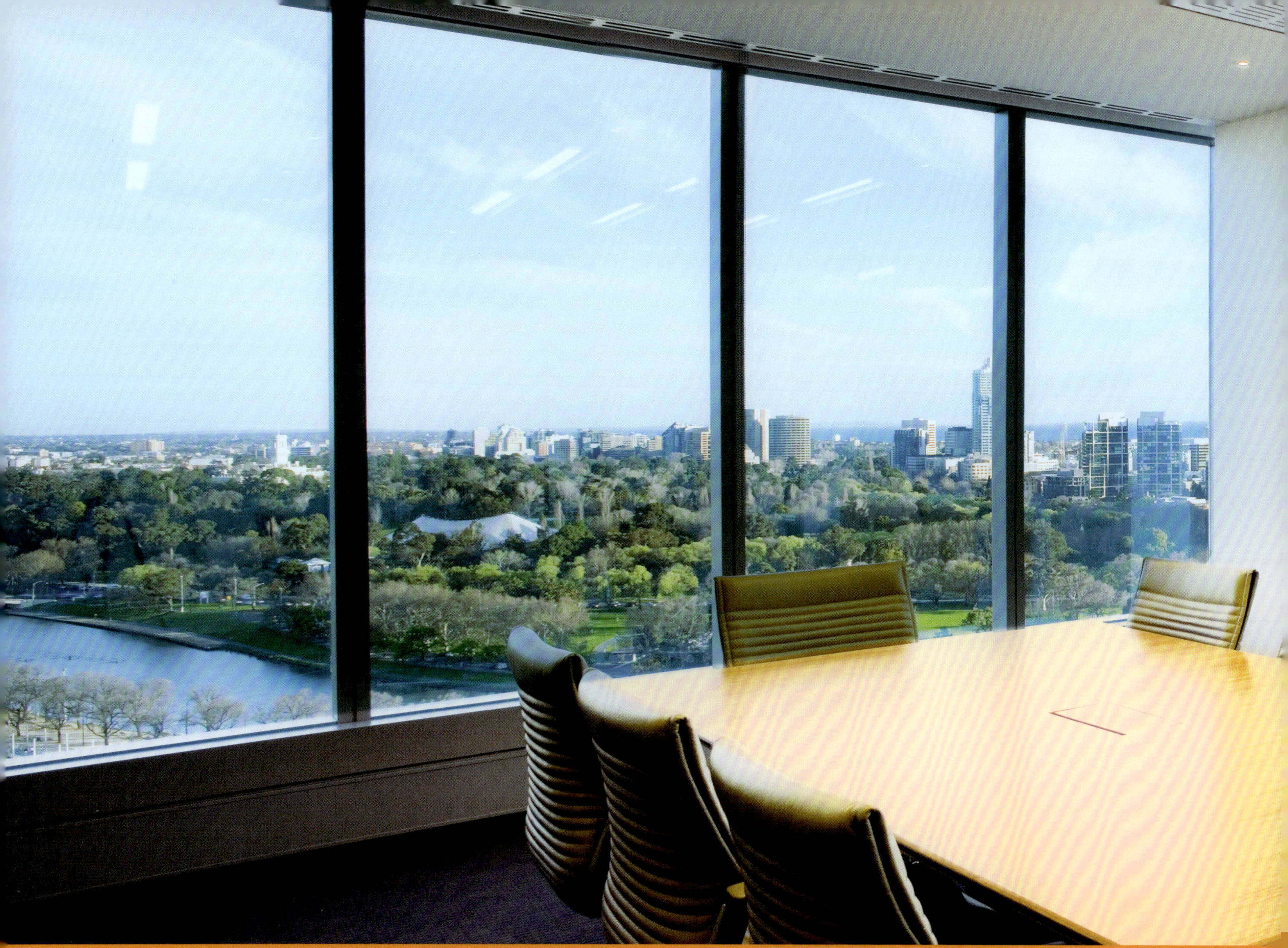

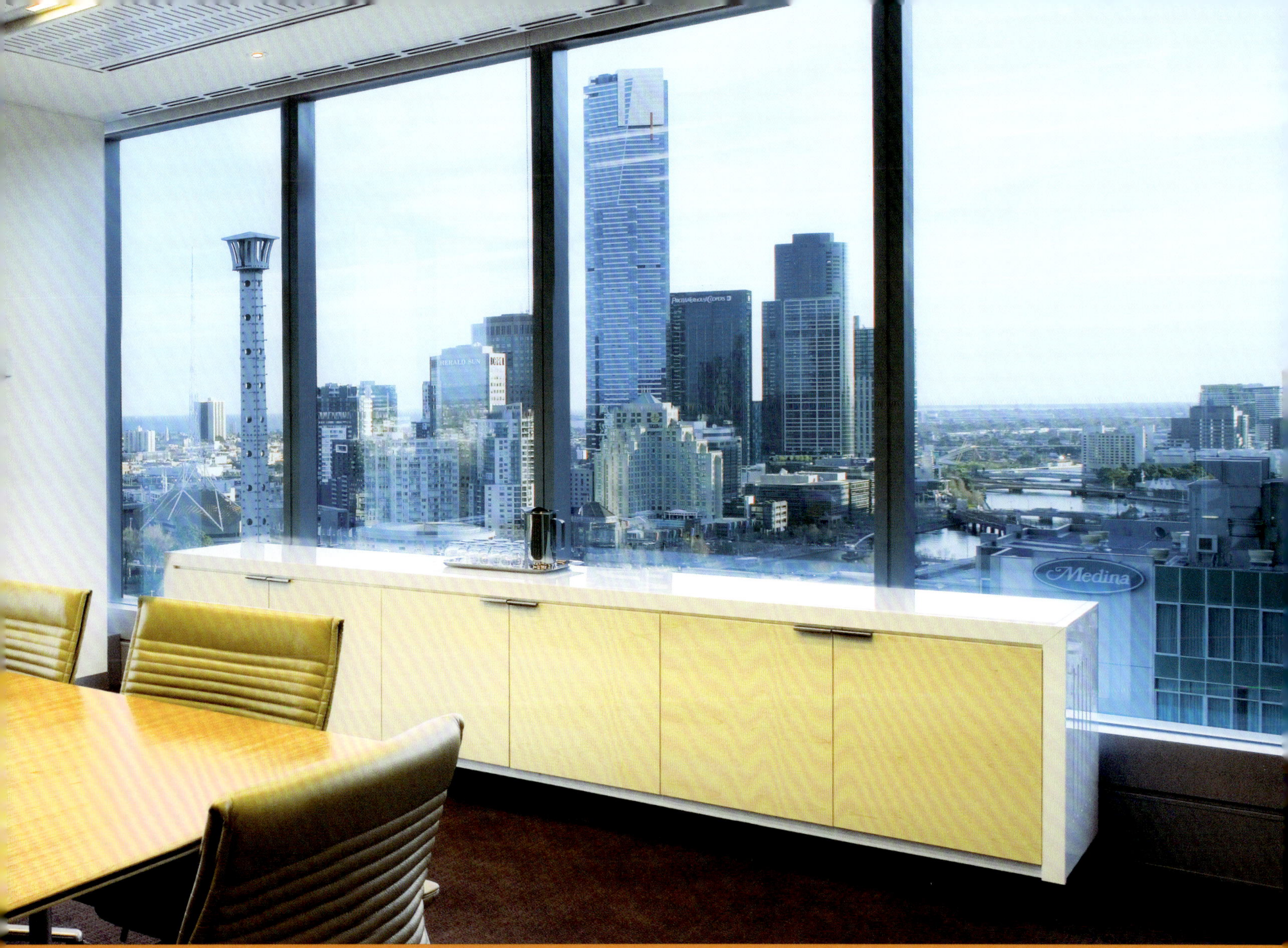

UBS, Melbourne Table: Logon range; swivel chairs: FS range, design firm: Sue Solly, Geyer

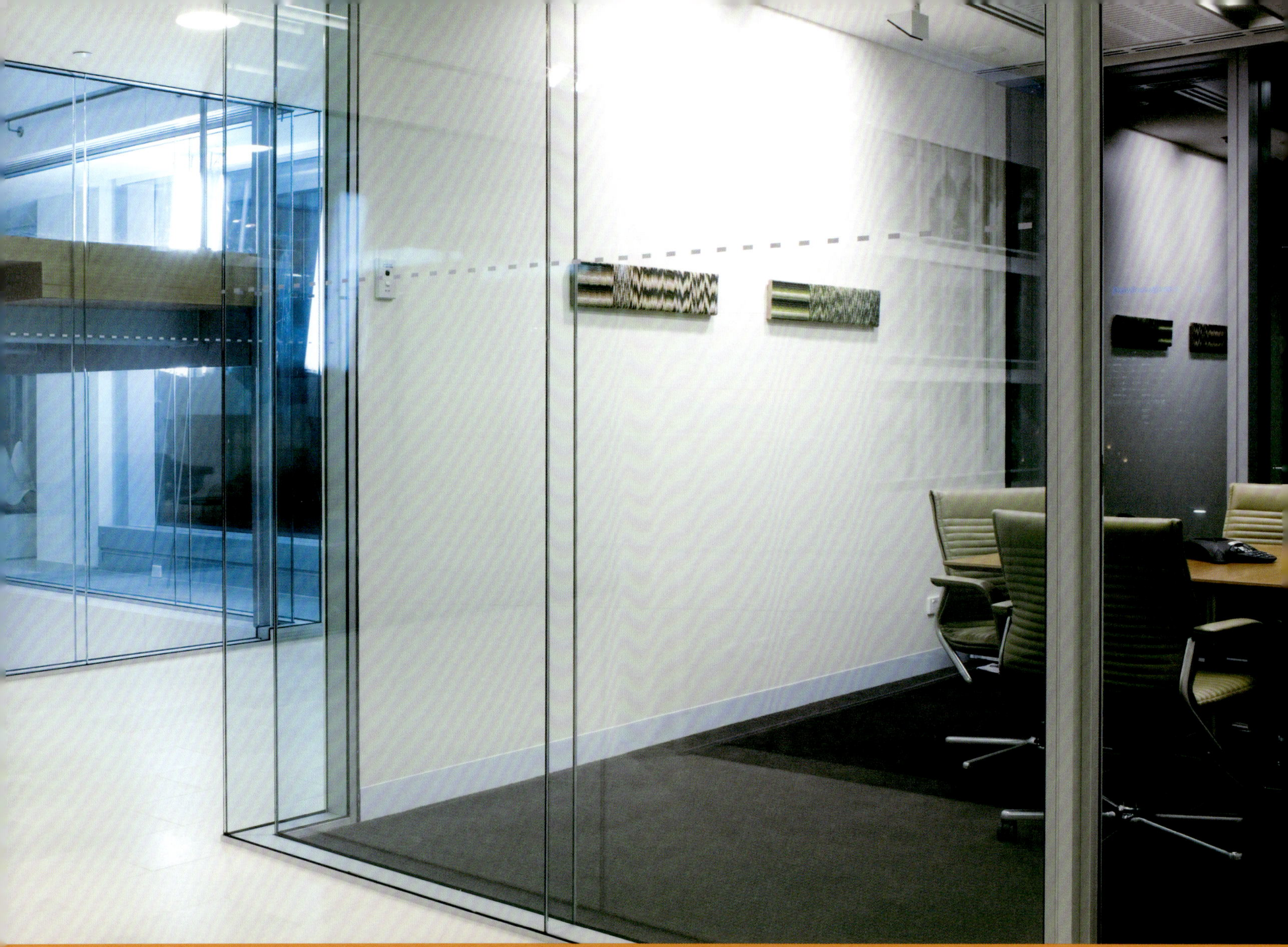

UBS, Melbourne Table: Logon range; swivel chairs: FS range, design firm: Sue Solly, Geyer

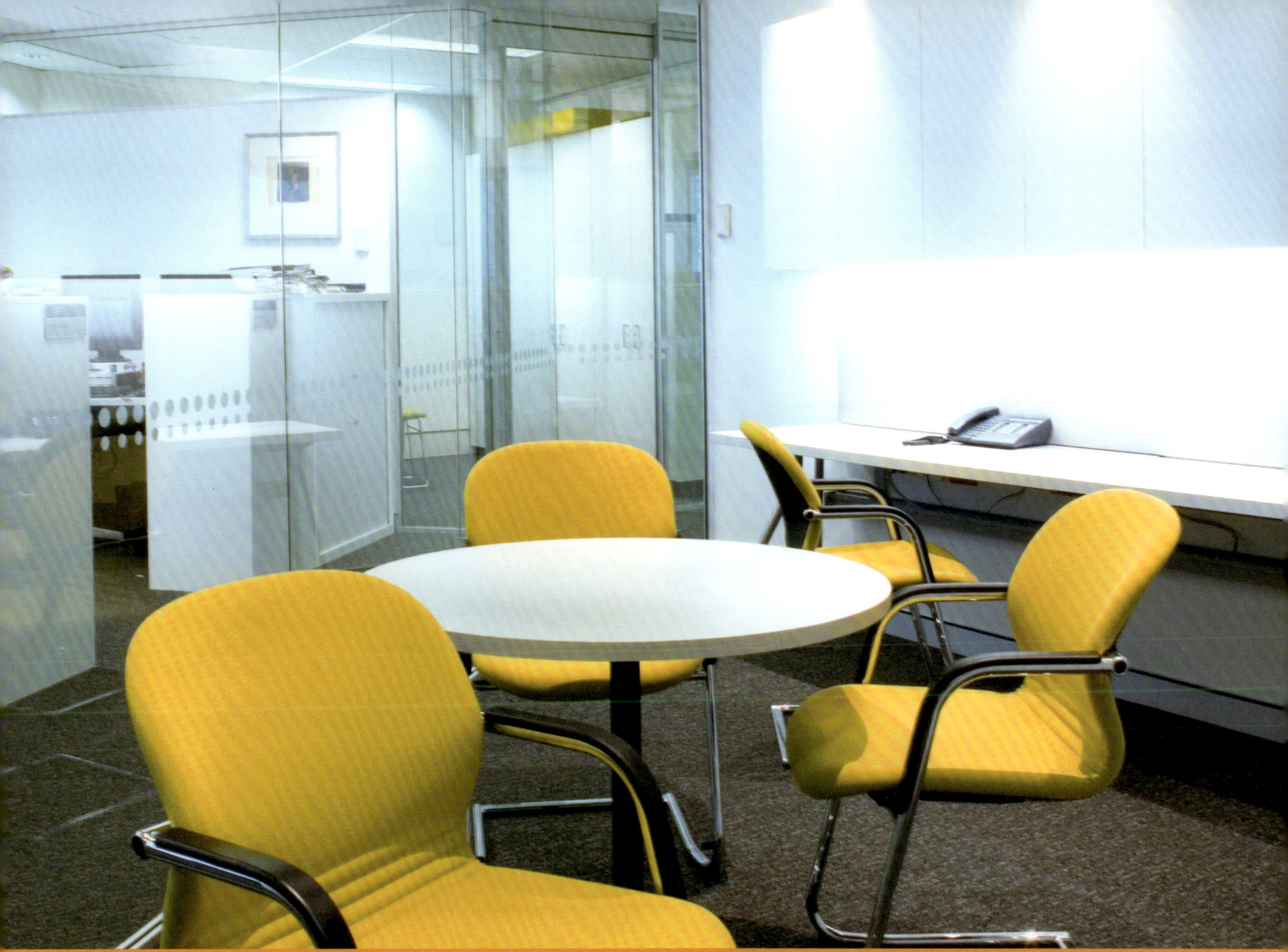

Australian Government Solicitor, Sydney Cantilever chairs: FS range, design firm: Tonia Mudie, MPH

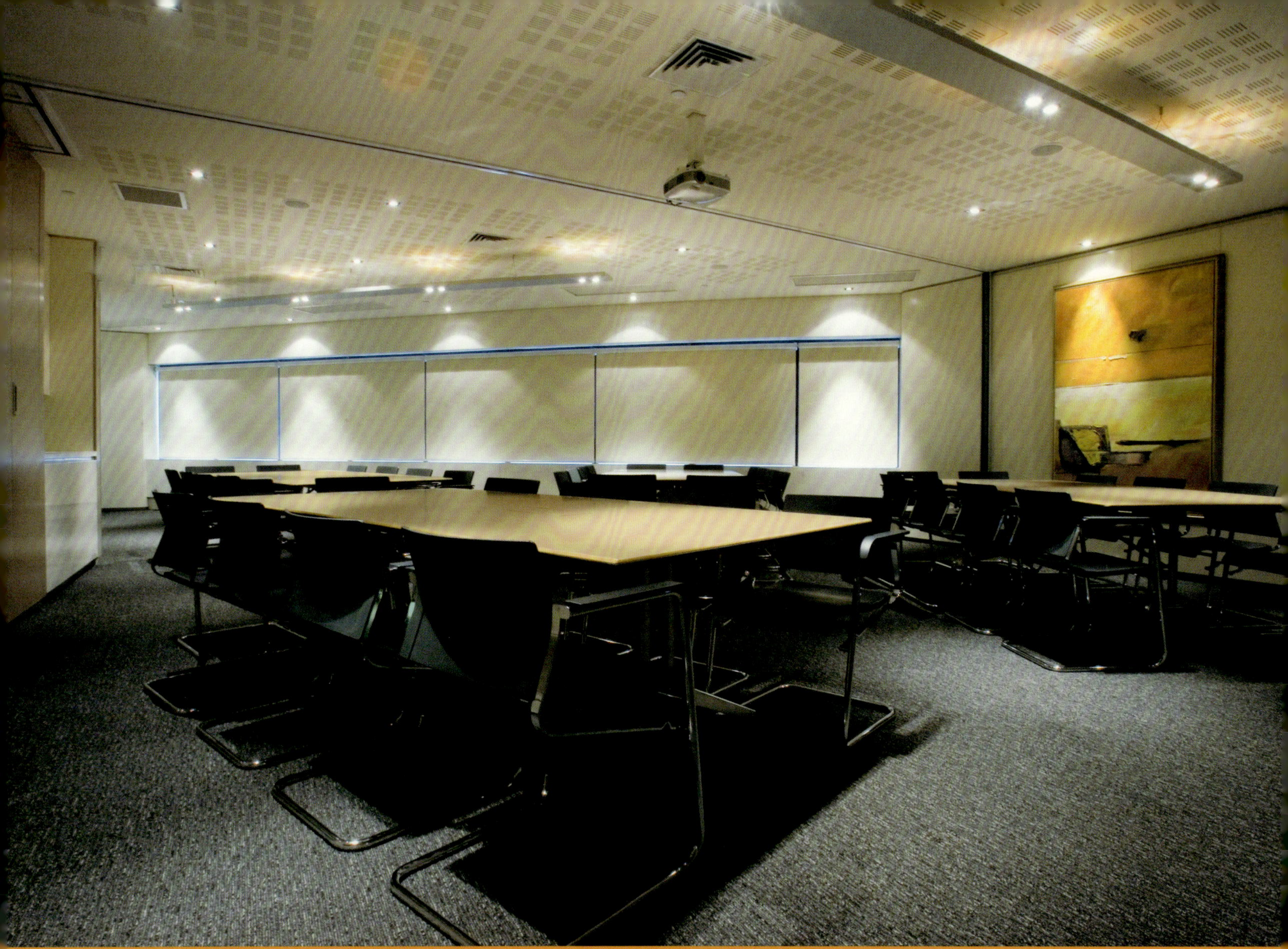

Guy Medland, Director
"The Wilkhahn experience continues to reward my clients with incomparable product functionality and reliability supported by the benchmark in industry professionalism."

Australian Government Solicitor, Sydney
Tables: Confair folding table; cantilever chairs: Sito range, design firm: Tonia Mudie, MPH

Medland Corporate Property, Melbourne
Furniture system: Conrack range; cantilever chair: Cura range

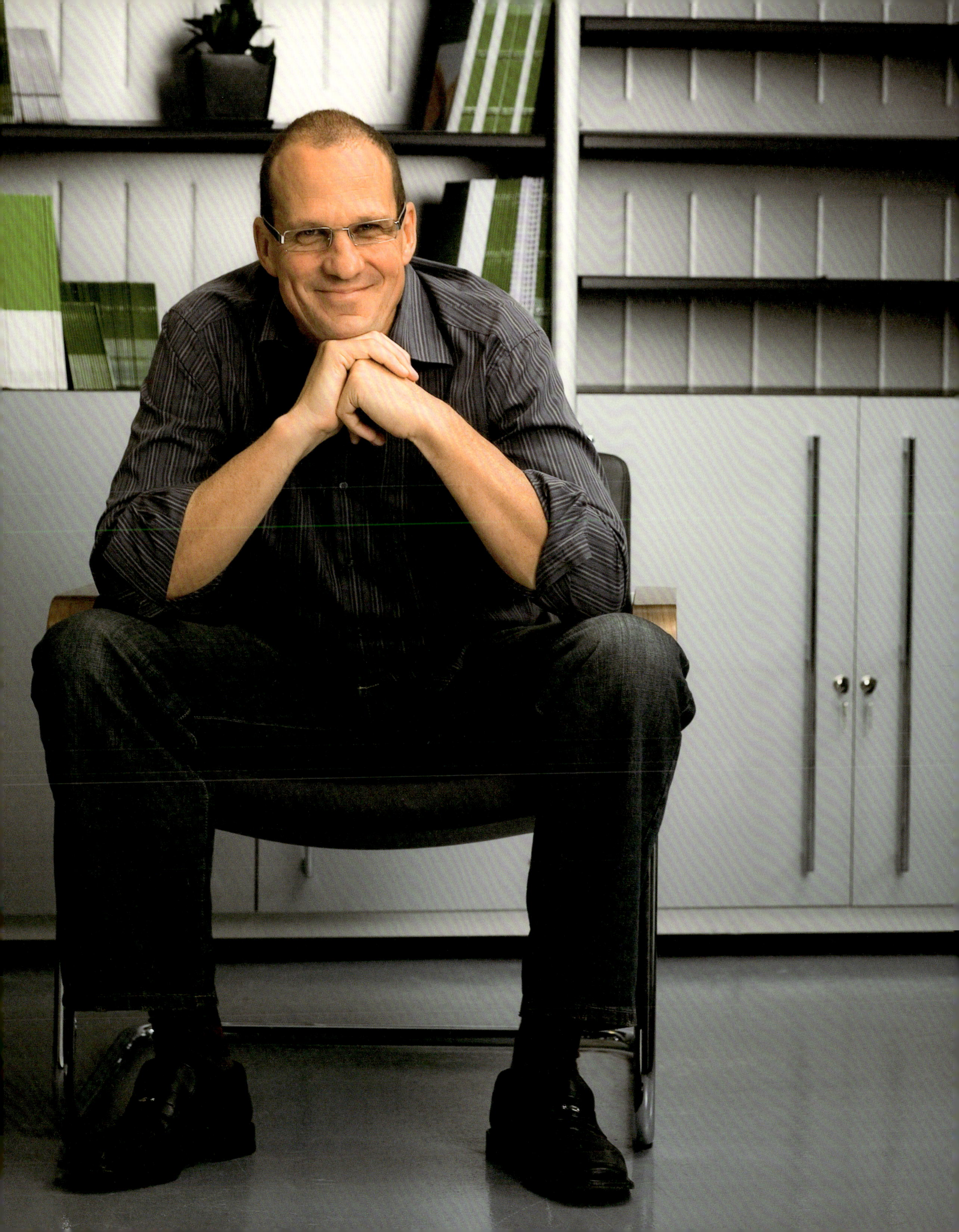

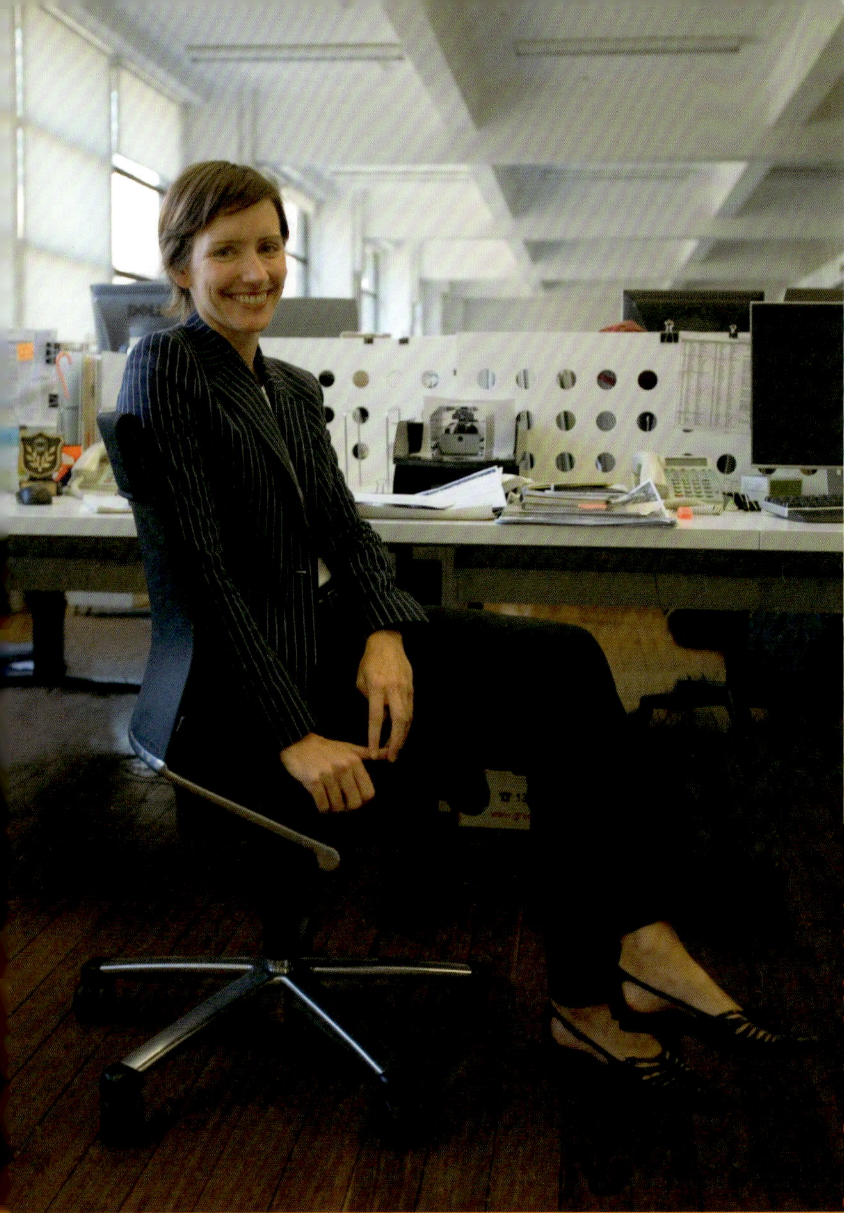

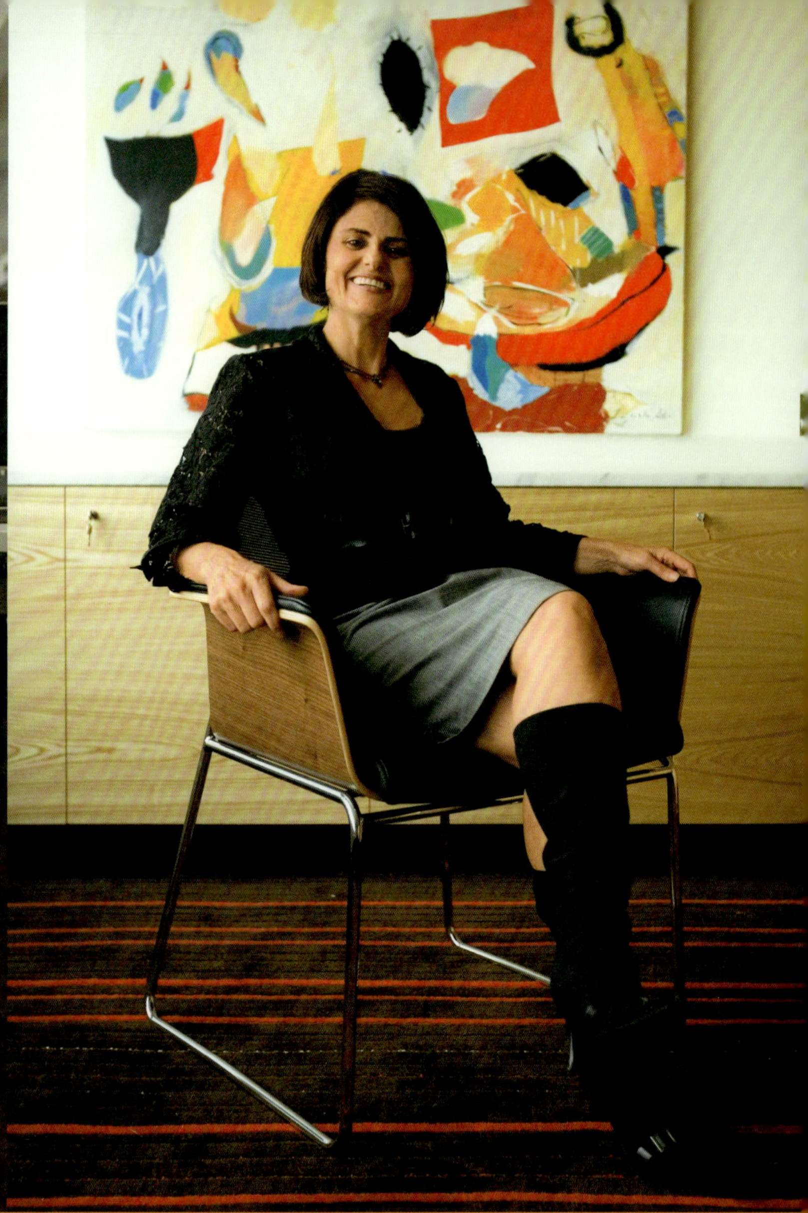

Abbie Galvin, Principal
"I hope this won't seem like cultural stereotyping but Wilkhahn products have a cool, dry German quality about them. Their pieces are sharp, beautifully engineered and perhaps, most importantly from my point of view 'do' whatever it is that they have been designed to do. Oh, and of course, there's a certain something about Wilkhahn's service that just can't be surpassed!"

Georgette Stanton,
Senior Facilities Manager of Project and Design
"What makes a difference for me in selecting a company to work with is the people and the service, but with Wilkhahn it also comes with style."

Bligh Voller Nield Architecture
Swivel chair: Modus Basic range

ING, Sydney
Conference chair: Signo range

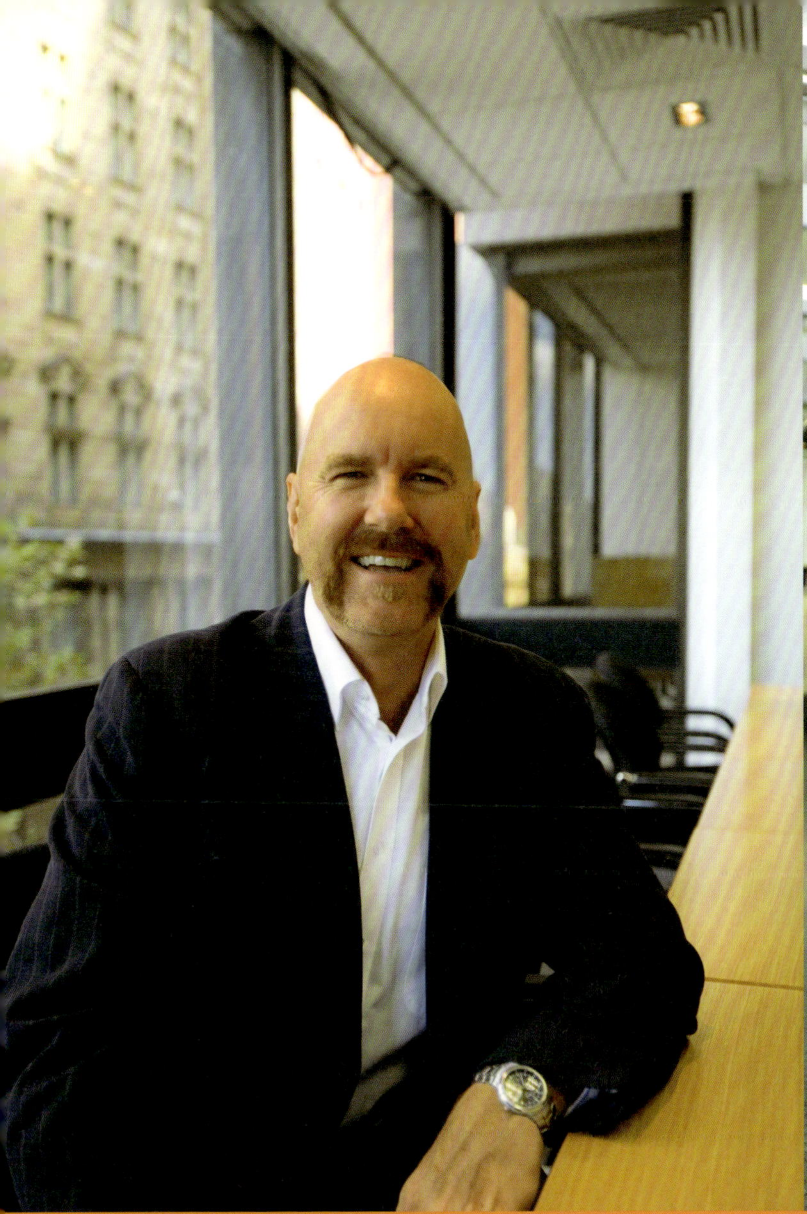

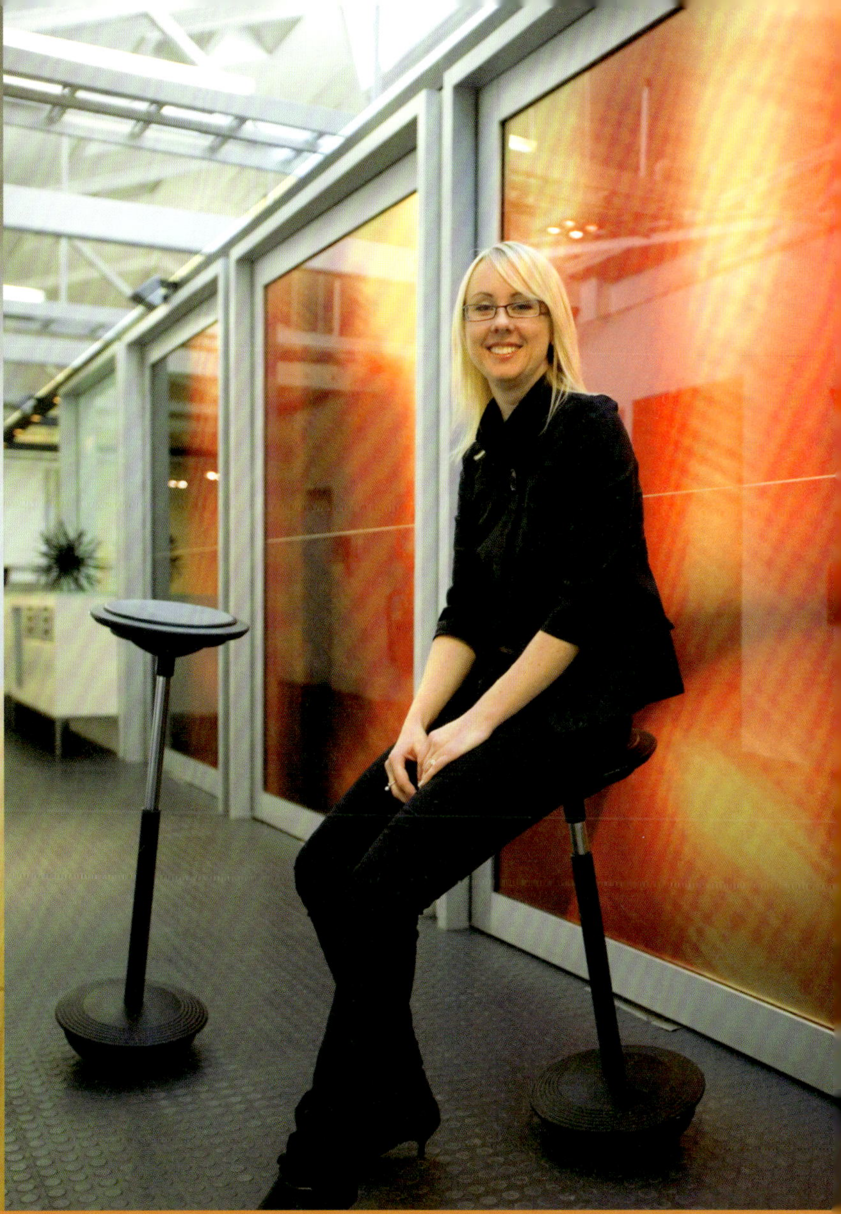

Andrew Isaacson, Design Principal
"The first ever furniture I specified in my professional career was FS sled base conference chairs. That was 25 years ago. The Wilkhahn portfolio has continued to grow, innovate and develop into a really solid range of commercial products that I can confidently recommend to my clients. Their workmanship and delivery is second to none …and 25 years later I am still specifying FS chairs!"

HBO + EMTB, Sydney
Swivel chairs: FS range

Cathy Jamieson, Partner
"Wilkhahn's attention to detail in all their pieces means that all items remain true to the brand. Nothing is unresolved; junctions meet with precision and their focus on quality and sustainability has positioned them as a world leader. Most of all, they are just plain beautiful pieces of furniture!"

Geyer, Sydney
Dynamic stool: Stitz range

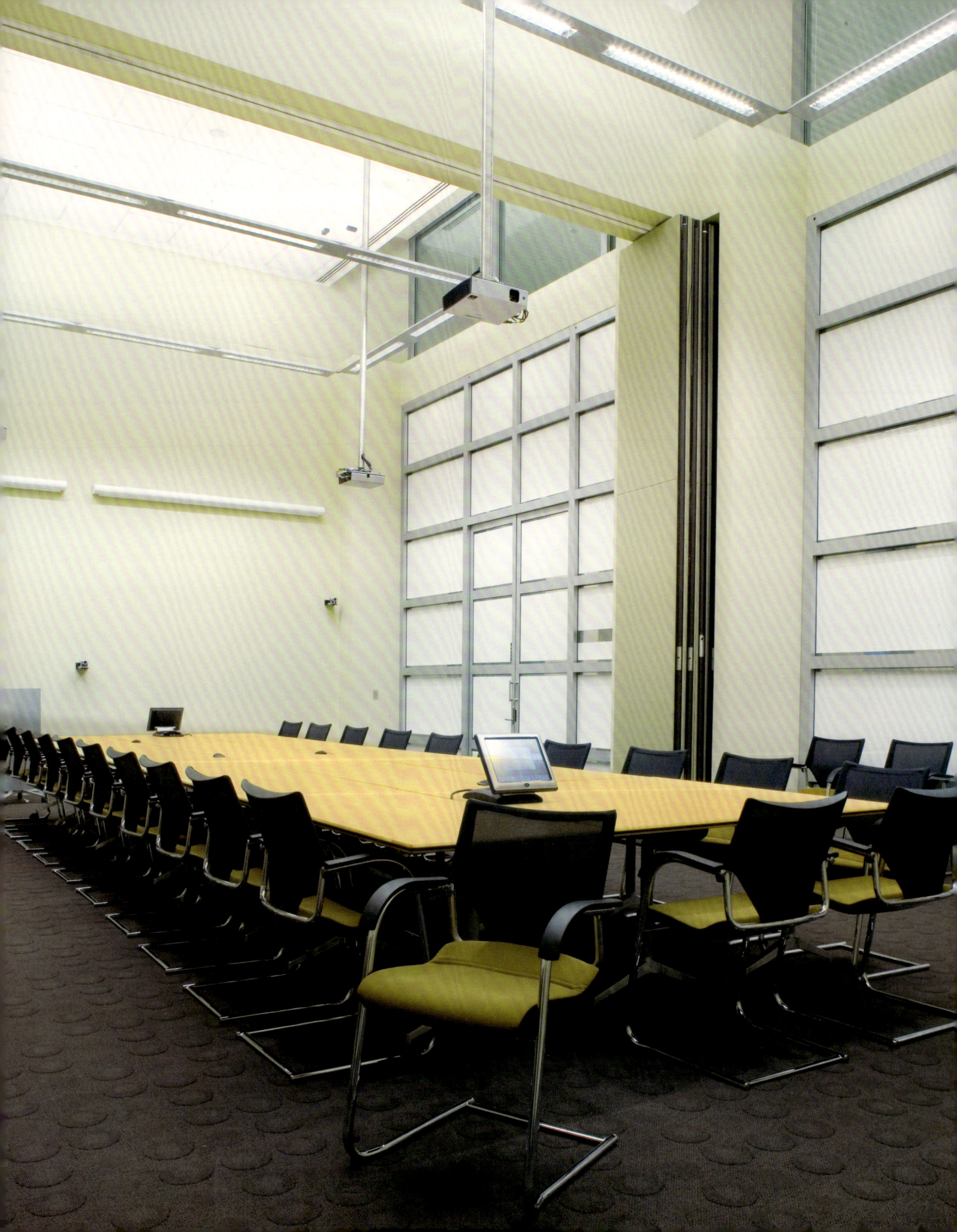

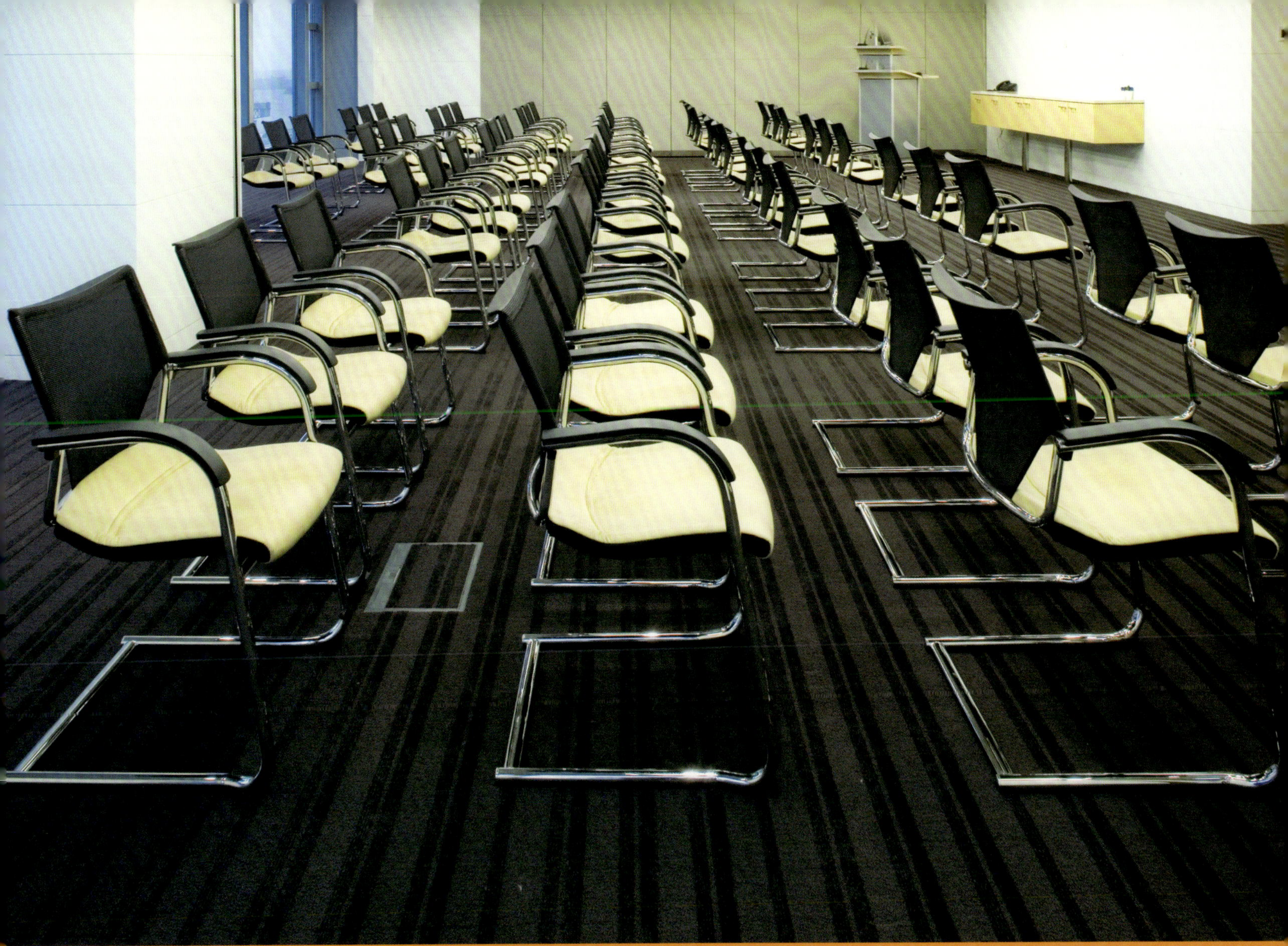

UBS, Sydney Tables: Timetable range; stackable cantilever chairs: Modus range, design firm: Interior Design – EGO Group

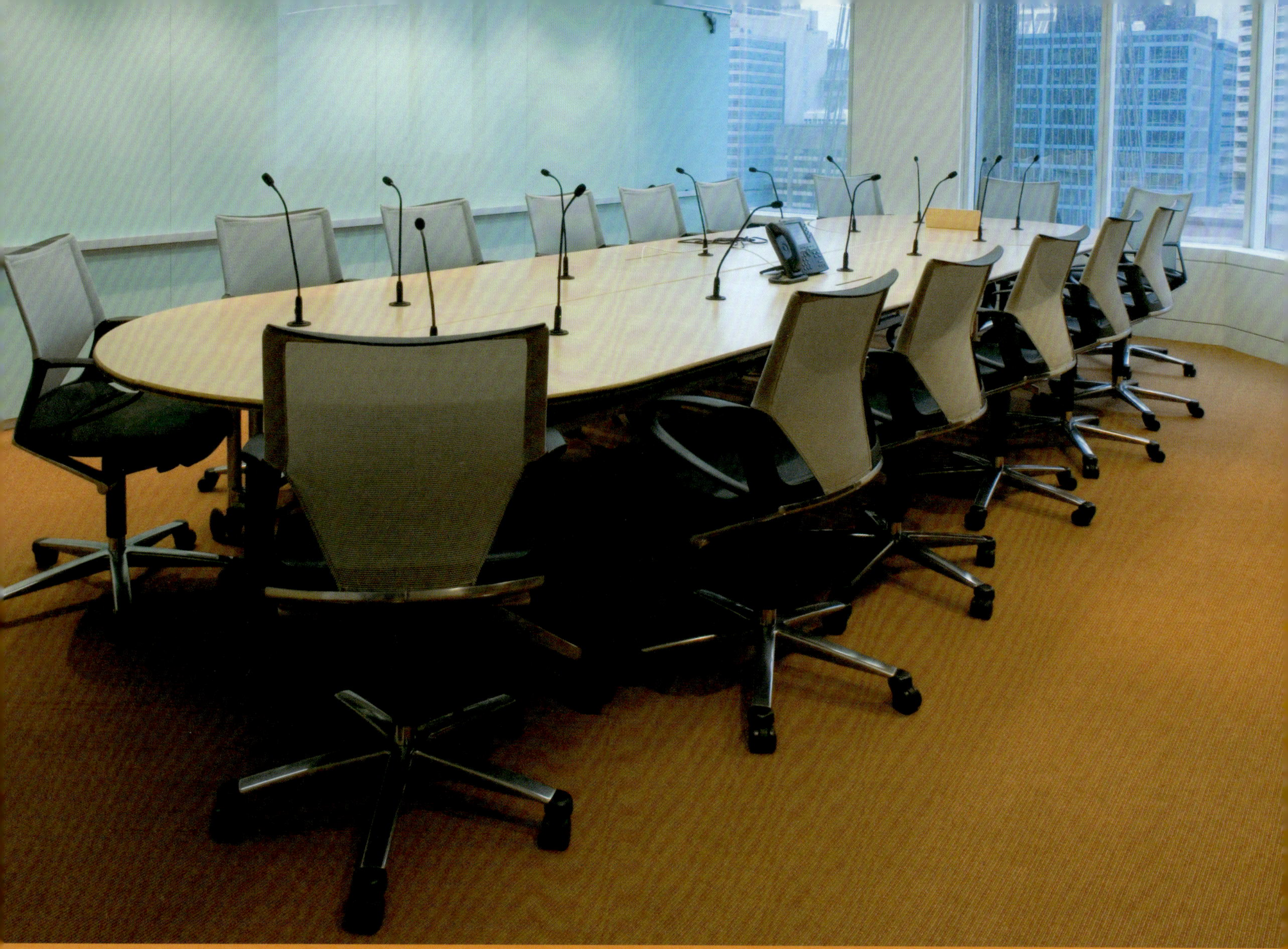

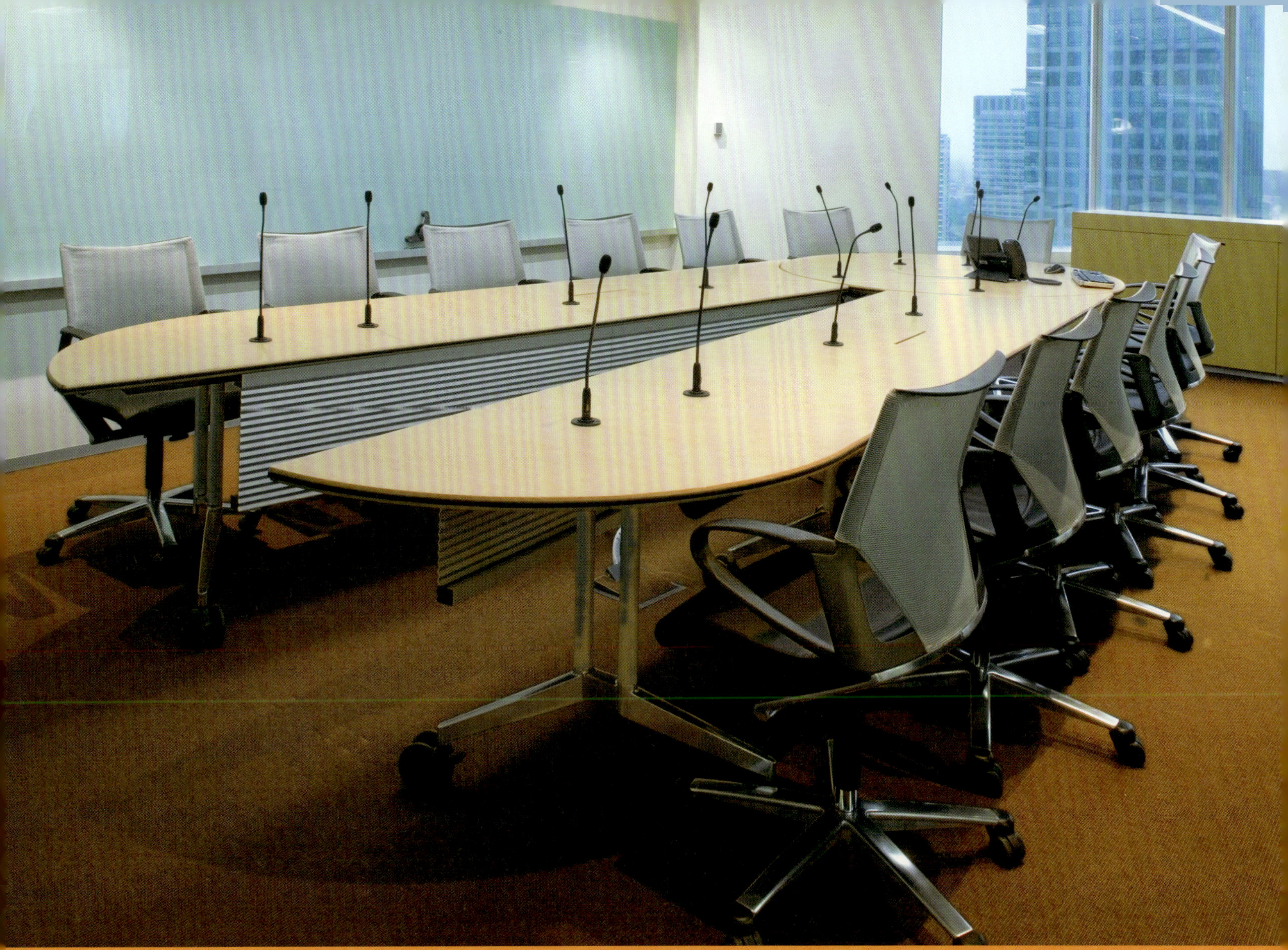

Stockland Corporation Ltd., Sydney Variable table for video conferencing: Logon range; swivel chairs: Modus range, design firm: Abbie Galvin and Laura Robinson, Bligh Voller Nield Architecture

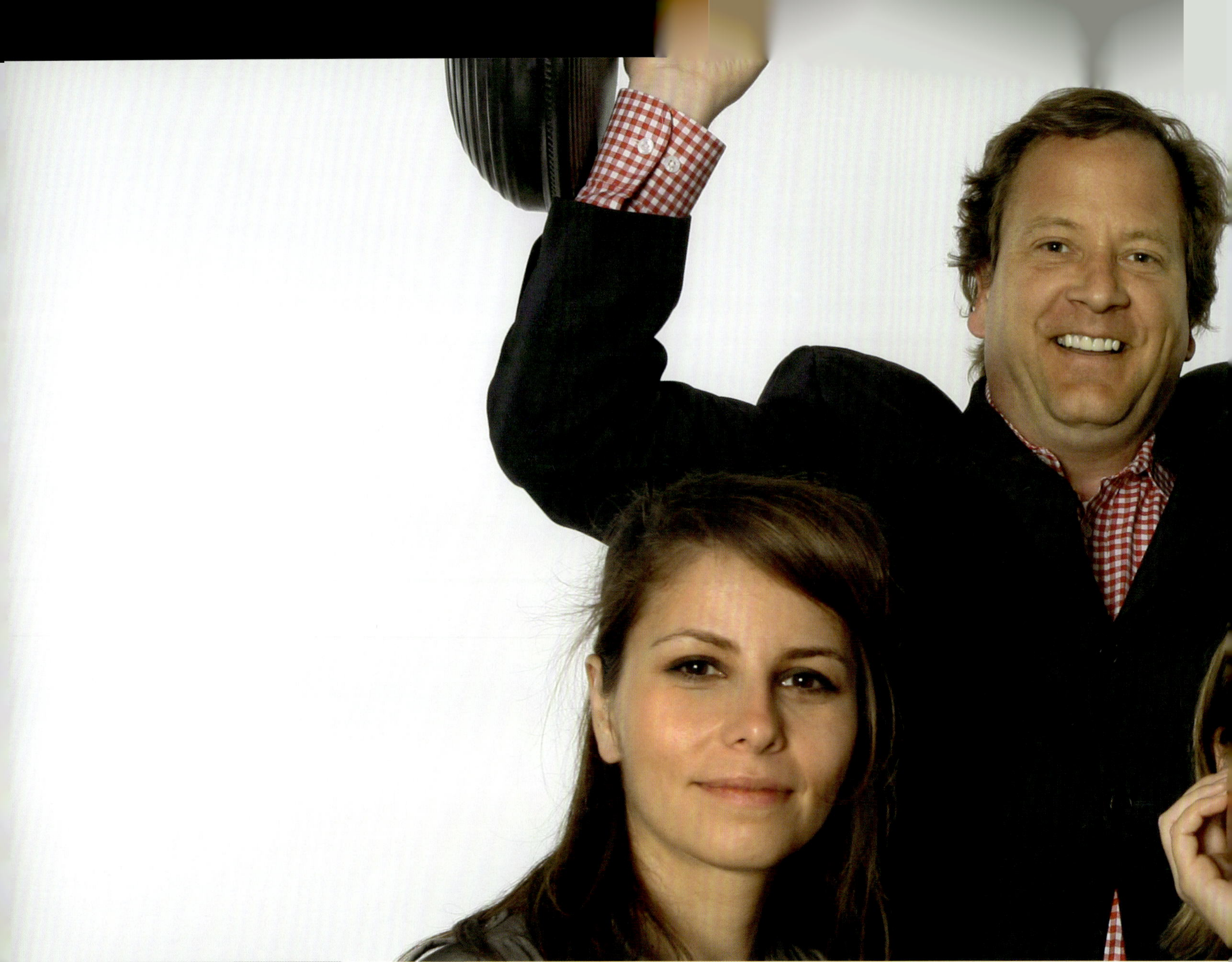

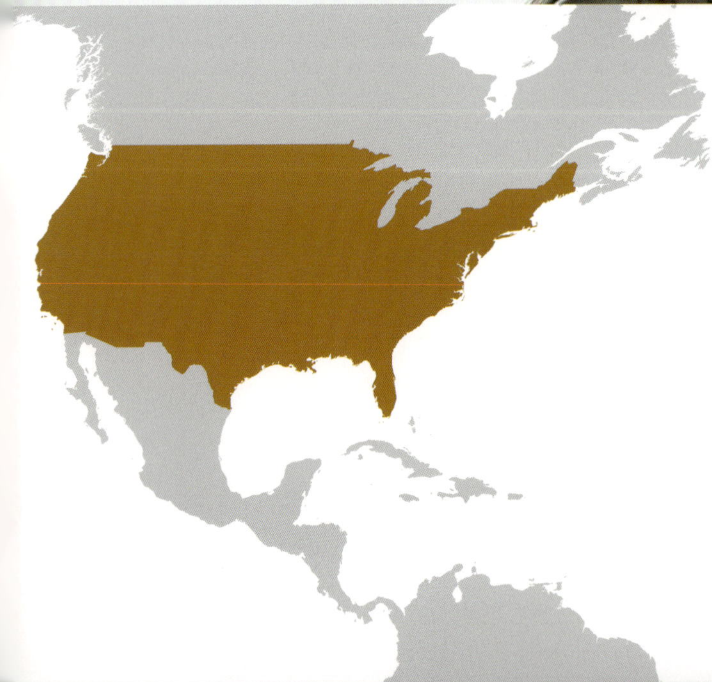

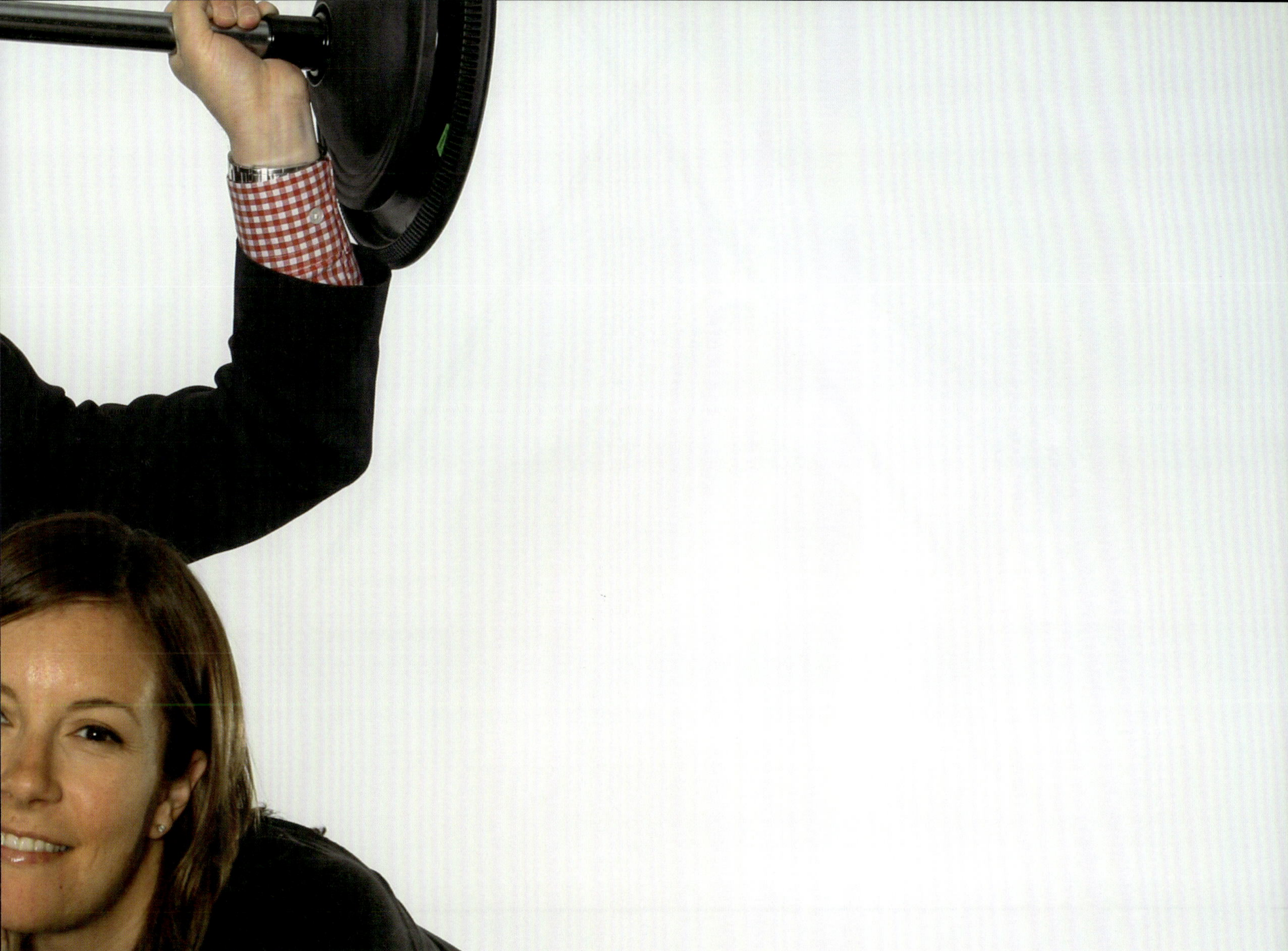

Die Geschichte von Wilkhahn in Amerika ist lang: Bereits Ende der 50er Jahre kam es zu einer Kooperation mit einem nordamerikanischen Hersteller, der sich für „Design made in Germany“ begeisterte. Doch mit den wechselhaften Schicksalen der Kooperationspartner unterlag auch Wilkhahn einem steten Auf und Ab. Seit den 90er Jahren sind wir über den Lizenzpartner Vecta® in diesem riesigen Markt vertreten. Inzwischen sind die internationale Projekterfahrung und die Wilkhahn-Reputation unter Architekten und Innenarchitekten so groß geworden, dass wir Wilkhahn in Amerika mit eigenen Ressourcen unterstützen können. Ein erster Anfang ist gemacht.

Wilkhahn has a long history in America: as early as the end of the 1950s, Wilkhahn collaborated with a North American manufacturer who was inspired by "design made in Germany". The destinies of the cooperation partners were unpredictable, however, and Wilkhahn was also subject to constant ups and downs. Since the 1990s, we have been represented on this huge market by our licence partner Vecta®. In the meantime, Wilkhahn's international project experience and its reputation among architects and interior designers are so considerable that we are able to support Wilkhahn in America with our own resources. A beginning has been made.

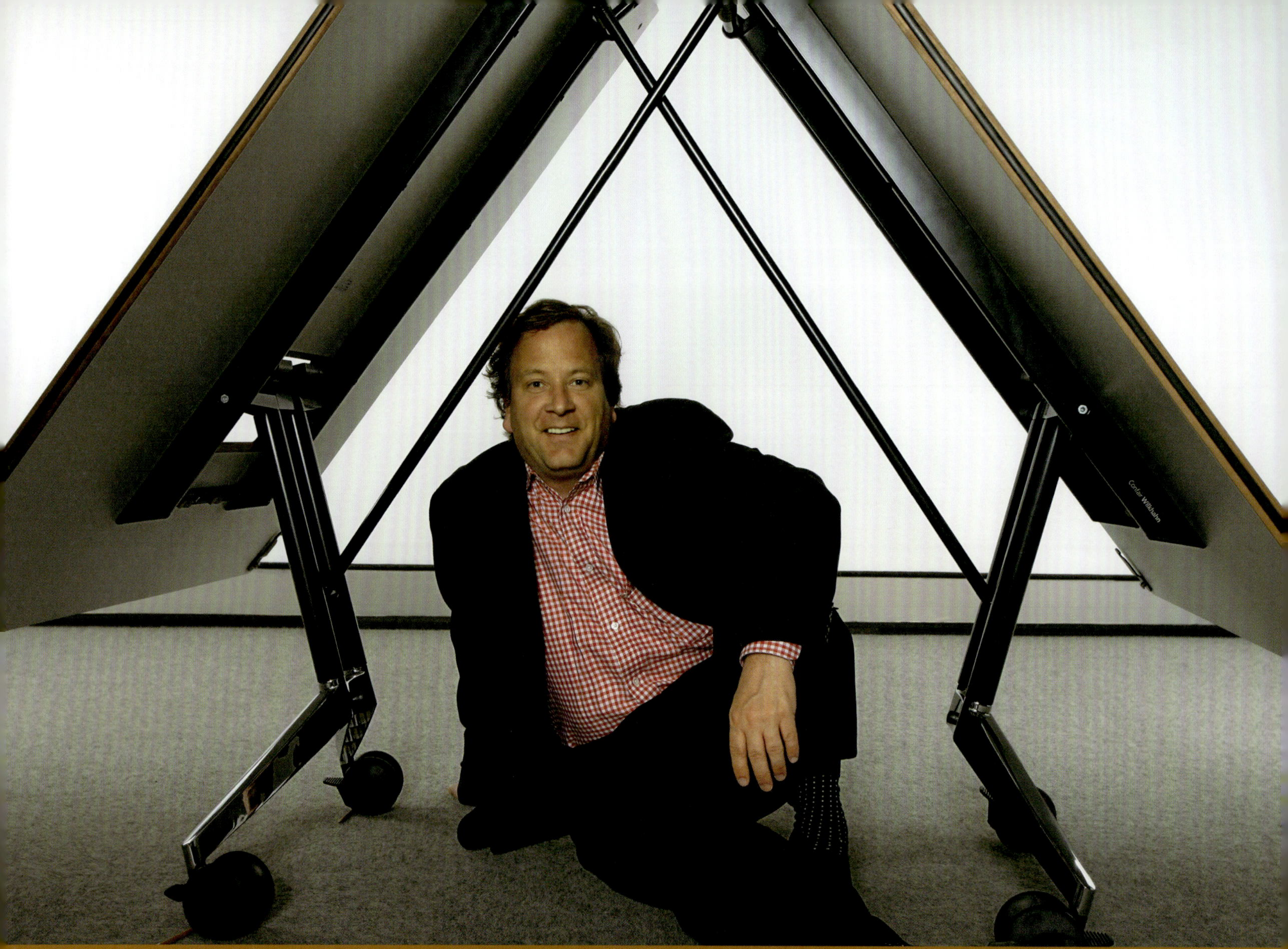

Arthur Fiesel

"From the onset of the development of a new corporate headquarters for Wilson Sporting Goods Co., the Gensler project team emphasized all possible sbd design efforts to create clean, sophisticated, timeless aesthetics that would prompt a new corporate culture and overall business image for Wilson Sporting Goods. In doing so, furniture selections must couple with and complement the mission. Wilkhahn's product portfolio complements Gensler's design perfectly. The beautifully simple profiles for both table and seating options offered the Gensler team ample opportunity to ensure a sacred marriage between the workplace architecture and furniture forms. I truly believe that Wilkhahn represents a company that strives for creating products which offer absolute design while ensuring the best engineering in all of their available product portfolios. On behalf of all my colleagues at Gensler, I thank you for supporting the successful development of a prized project for Wilson Sporting Goods Co."

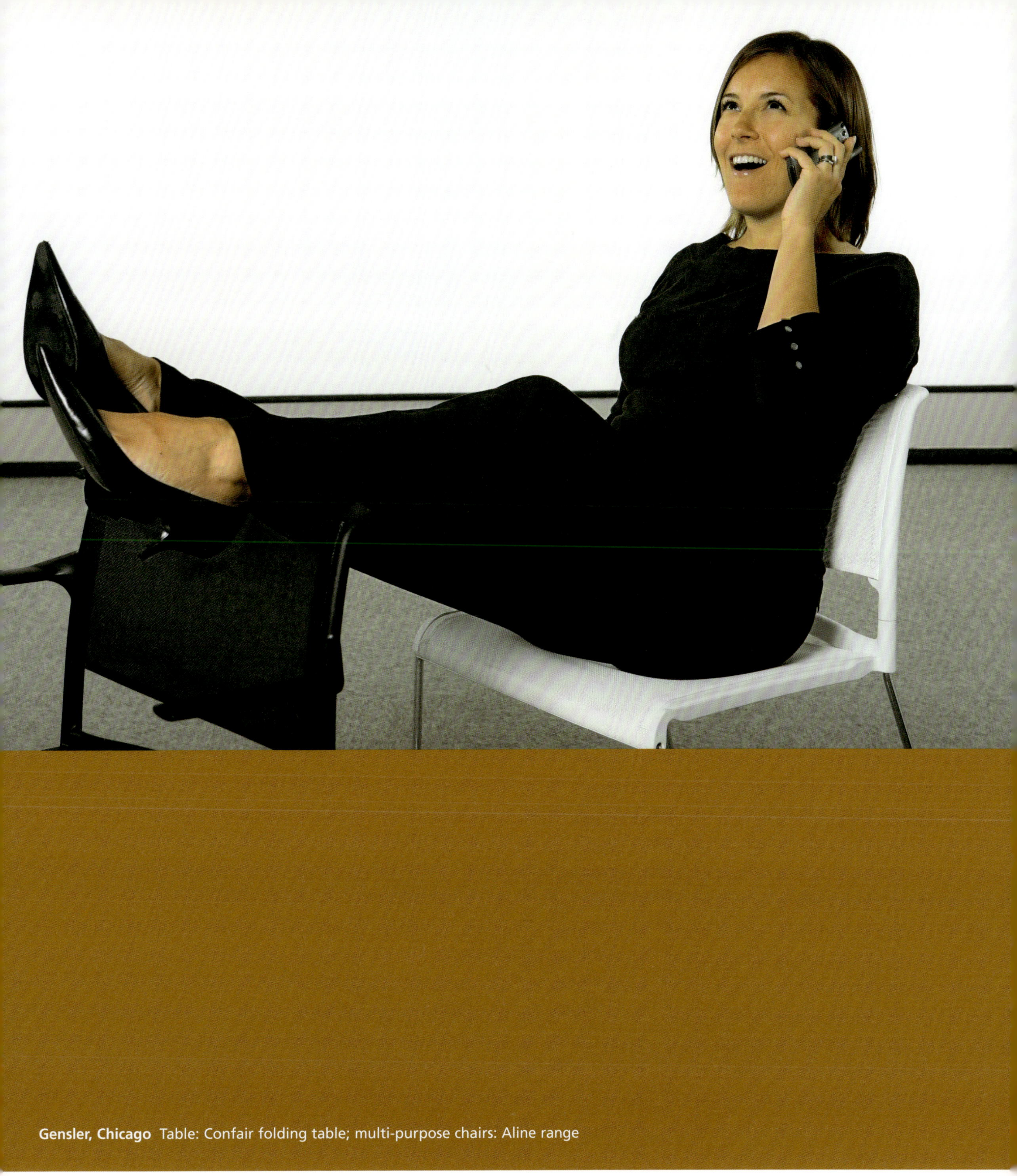

Gensler, Chicago Table: Confair folding table; multi-purpose chairs: Aline range

Ähnlich wie ein Mensch im Laufe seines Lebens seine eigene Identität und Persönlichkeit entwickelt und festigt, hat auch Wilkhahn zahlreiche Entwicklungsstufen durchlaufen: Am Anfang standen die Gründung und das Lernen durch Nachahmung, dann folgte die Phase des Suchens und der Neuorientierung mit zahlreichen Experimenten nach dem Prinzip „Versuch und Irrtum“, bis sich die Identität festigte, das Unternehmen erwachsen wurde und dem eingeschlagenen Weg folgte. Bei aller Konsequenz und Kontinuität der Entwicklung in den letzten Jahrzehnten aber sind es vor allem die Offenheit, die Neugier und der Mut, immer wieder neue Wege zu gehen, die das Unternehmen für Querdenker und kreative Köpfe jenseits des Mainstreams attraktiv machen. Für Menschen, die sich auf der Grundlage eines gemeinsamen Werteverständnisses für die „Idee“ Wilkhahn begeistern, die neue Impulse setzen und die Entwicklung weiter vorantreiben. Auf diese Weise haben zahlreiche Persönlichkeiten die bisherige Wegstrecke begleitet und maßgeblich geprägt – ob als externe Gestalter, Freunde und Partner oder als Manager und Mitarbeiter im Unternehmen.

Impulsgeber und Wegbegleiter: Wilkhahn gestalten
Creating the spirit of Wilkhahn

As people develop and consolidate their own identity and personality in the course of their life, Wilkhahn has also undergone numerous stages of development: the initial phase was concerned with the founding of the company and learning by imitating; this was followed by the phase of searching and new orientation with many an experiment in accordance with the principle of "trial and error" until the identity of the company had become stable, the company had grown up and started to pursue its chosen path. In spite of all the consistency and continuity of developments in the last few decades, it is primarily openness, curiosity and courage to embark on new paths as an ongoing process that makes the company attractive to lateral thinkers and creative minds far beyond mainstream ideas. To people who, on the basis of a mutual understanding of values, are enthusiastic about the "idea" of Wilkhahn, who provide new impetus and spur on development even further. Many personalities have been inspiring companions so far and have had a formative influence on creating the spirit of Wilkhahn – as external designers, friends and partners or as managers and employees in the company.

Handwerkliches Geschick, eine für damalige Verhältnisse fortschrittliche Produktion und ein gewachsenes Qualitätsverständnis waren die Basis, auf der Fritz Hahne und Adolf Wilkening das Unternehmen neu ausrichten konnten. Vor allem aber: eine Mitarbeiterschaft, die begeistert mitzog und die so manches Mal die Risiken, die mit neuen Fertigungstechniken und Materialien verbunden waren, durch größtmöglichen persönlichen Einsatz wettmachte, um die Kunden termin- und qualitätsgerecht zufriedenzustellen.

Craft skills, progressive production methods for that time in history and a solid appreciation of quality formed the basis upon which Fritz Hahne and Adolf Wilkening could reposition the company. But above all: staff who had the enthusiasm to go along with developments and who more than often compensated for the risks that new production methods and materials involved by means of the greatest personal commitment to satisfy customers in terms of adherence to deadlines and quality.

Herbert Hirche, geboren 1910 in Görlitz, studierte nach einer Tischlerlehre von 1930 bis 1933 am Bauhaus in Dessau und Berlin, unter anderem bei Ludwig Mies van der Rohe, in dessen Büro er bis 1938 arbeitete. Der Mitarbeit bei Egon Eiermann und Hans Scharoun folgte 1947 die Berufung als Professor an die Hochschule für angewandte Kunst, Berlin-Weißensee, und 1952 an die Staatliche Akademie für bildende Künste, Stuttgart. Als deren Rektor, als Gründungsmitglied des Deutschen Werkbunds Berlin, als Präsident des Verbandes Deutscher Industriedesigner (VDID) und als Mitglied des Rats für Formgebung gehörte Hirche bis zu seinem Tod 2002 zu den prägenden deutschen Gestaltern der Nachkriegszeit. Neben vielen Möbelprogrammen, etwa dem auf der Weltausstellung 1958 in Brüssel gezeigten Programm 480, hat er für Wilkhahn das Verwaltungsgebäude entworfen (1960).

Herbert Hirche, born in Görlitz in 1910, studied cabinet making at the Bauhaus in Dessau and Berlin from 1930 to 1933, including studying under Ludwig Mies van der Rohe, at whose office he worked until 1938. Collaboration with Egon Eiermann and Hans Schauron was followed by appointment as Professor at the Hochschule für angewandte Kunst, Berlin-Weißensee, and in 1952 at the Staatliche Akademie für Bildende Künste, Stuttgart. As its Rector, as founding member of the Deutscher Werkbund Berlin, as President of the Verband Deutscher Industriedesigner (VDID), and as member of the German Design Council, Hirche was one of the formative German designers of the post-war period until his death in 2002. In addition to many furniture ranges, for example the 480 range which was shown as the World Exposition in Brussels in 1958, he designed an administration building for Wilkhahn (1960).

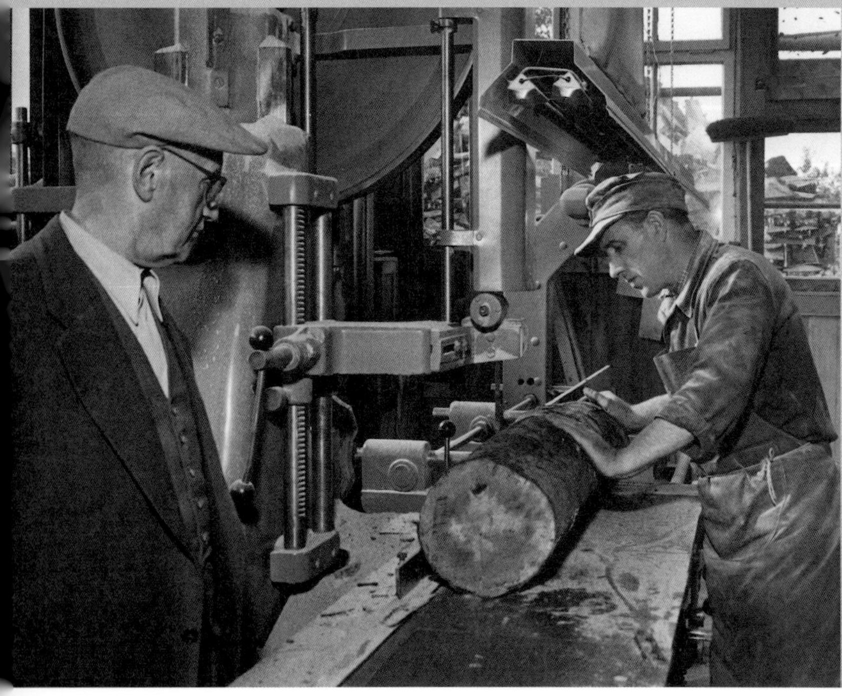

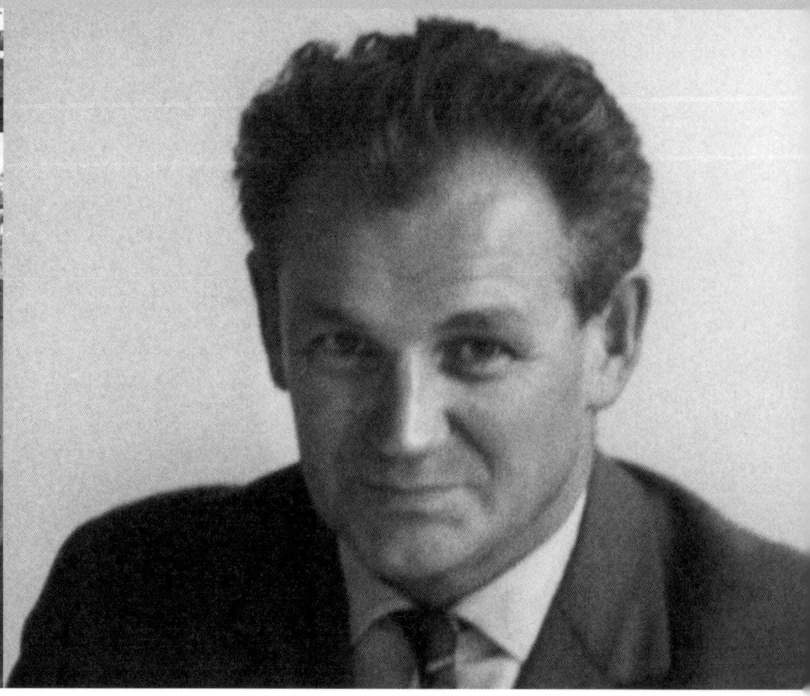

Fritz Hahne, im Gespräch mit dem Gestalter **Hartmut Lohmeyer**. Lohmeyer, Jahrgang 1923 und Sohn eines Theologieprofessors, hatte an der Technischen Hochschule Karlsruhe bei Egon Eiermann Architektur studiert und zwei Jahre an der späteren Alexandria-Universität in Kairo unterrichtet. Als freier Architekt und Möbelgestalter entwickelte er in den 50er und 60er Jahren für Wilkhahn vor allem Möbel für den Wohnbereich, wie zum Beispiel das Programm 400, einen bequemen Sessel mit Holzgestell, der mit abnehmbaren Polstern zu haben war.

Fritz Hahne talking to designer **Hartmut Lohmeyer**. Lohmeyer, born in 1923 and son of a professor of theology, first studied architecture at the Technische Hochschule in Karlsruhe under Egon Eiermann, then two years later he taught at the Alexandria University in Cairo. As a free-lance architect and furniture designer, he developed furniture for the home in particular for Wilkhahn in the 1950s and 1960s, for example the 400 range, a comfortable chair with a wooden frame that was available with detachable cushions.

1955 wurde der 15-jährige **Horst Knigge** eingestellt, der zunächst eine Lehre als Beizer und dann als Polierer absolvierte. Wissbegierig wusste er alle Chancen zu nutzen, die ihm das Unternehmen und die Gewerkschaft boten, um sich weiterzuqualifizieren. Er wurde Industriemeister und von 1982 bis 1990 Produktionsleiter, bevor er den Betriebsratsvorsitz übernahm und schließlich technischer Geschäftsführer wurde. Die Bilderbuchkarriere von Horst Knigge spielte bis zu seinem Ruhestand 2004 bei Wilkhahn eine Schlüsselrolle für die soziale Ausrichtung und für die Entwicklung moderner Arbeitsformen.

15-year old **Horst Knigge** started his employment at Wilkhahn as an apprentice in the wood staining shop in 1955 and then worked as a polisher. Eager to learn, he made use of every opportunity that the company and the trade union offered him to increase his skills. He became an industrial technician, and from 1982–1990 he was Production Manager before becoming Chairman of the Works Council and finally Technical Director. Horst Knigge´s picture book career played a key role in the social focus of the company and in the development of modern forms of work until his retirement in 2004.

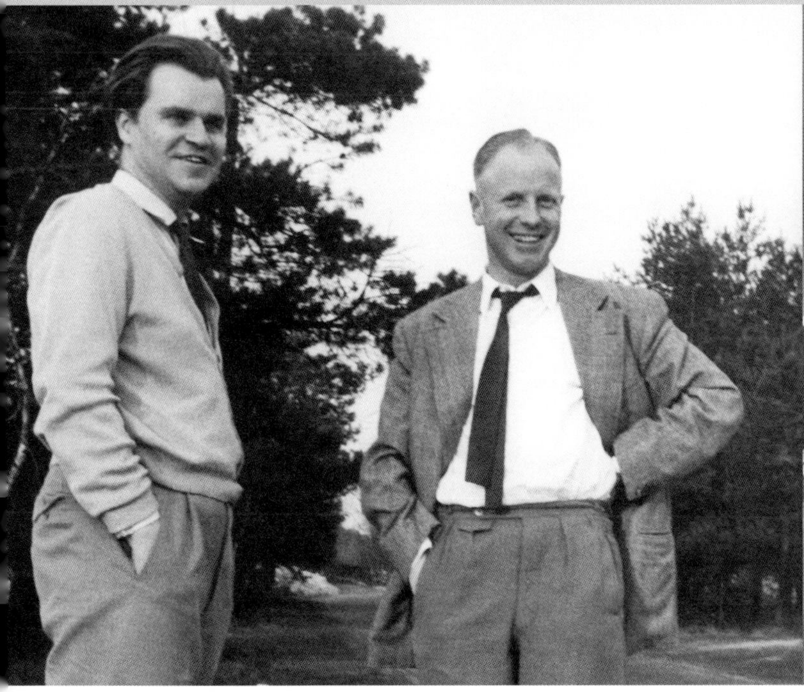

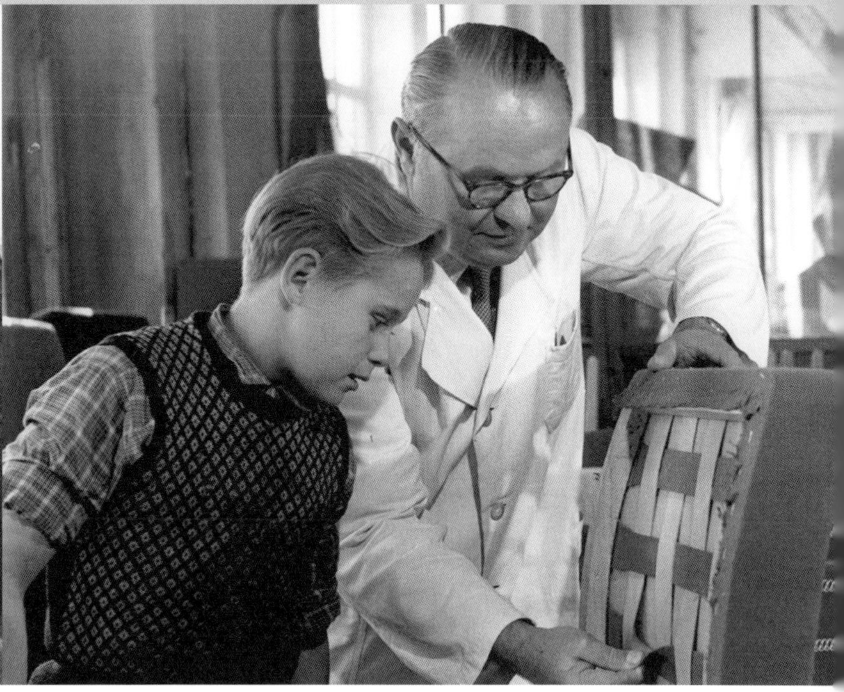

Einer der wichtigsten Gestalter in den 50er und 60er Jahren war **Georg Leowald**. 1908 in Düsseldorf geboren, wurde der studierte Architekt nach dem 2. Weltkrieg als Professor für Architektur an die Hochschule für Bildende Künste, Berlin, berufen und 1949 vom Museum of Modern Art New York mit dem 1. Preis für industriell gefertigte Möbel ausgezeichnet. Ab 1957 lehrte er Produktgestaltung an der Hochschule für Gestaltung Ulm. Er entwarf für Wilkhahn zahlreiche erfolgreiche Produktlinien aus Holz und Sperrholz, Polstermöbel und vor allem die ersten Stühle und Sessel mit Sitzschalen aus glasfaserverstärktem Kunststoff. Als Architekt baute er zudem ein neues Produktionsgebäude, ein Kesselhaus sowie eine Kantelhalle als Freiluftlager für Holzrohteile. Allesamt Bauten, die noch heute überzeugen. In seinen letzten Lebensjahren litt Leowald an einer psychischen Krankheit, der Kontakt brach ab. Er starb 1969.

Georg Leowald was one of the most important designers in the 1950s and 1960s. Born in Düsseldorf in 1908, he studied architecture and, after the Second World War, he was appointed Professor of Architecture at the Hochschule für Bildende Künste (HfBK) in Berlin. In 1949 he was awarded the 1st prize in the competition for industrially manufactured furniture by the Museum of Modern Art, New York. From 1957 he taught product design at the Academy of Design in Ulm (HfG). For Wilkhahn, he designed numerous, successful ranges in wood and plywood, upholstered furniture and in particular chairs with seat shells in glass fibre-reinforced plastic. As an architect, he also designed a new production building, a boiler house as well as a hall made from scantlings as an open-air store for unfinished wooden parts. These were all buildings that are still convincing solutions today. In the latter part of his life, Leowald suffered from a mental illness and contact to Wilkhahn ceased. He died in 1969.

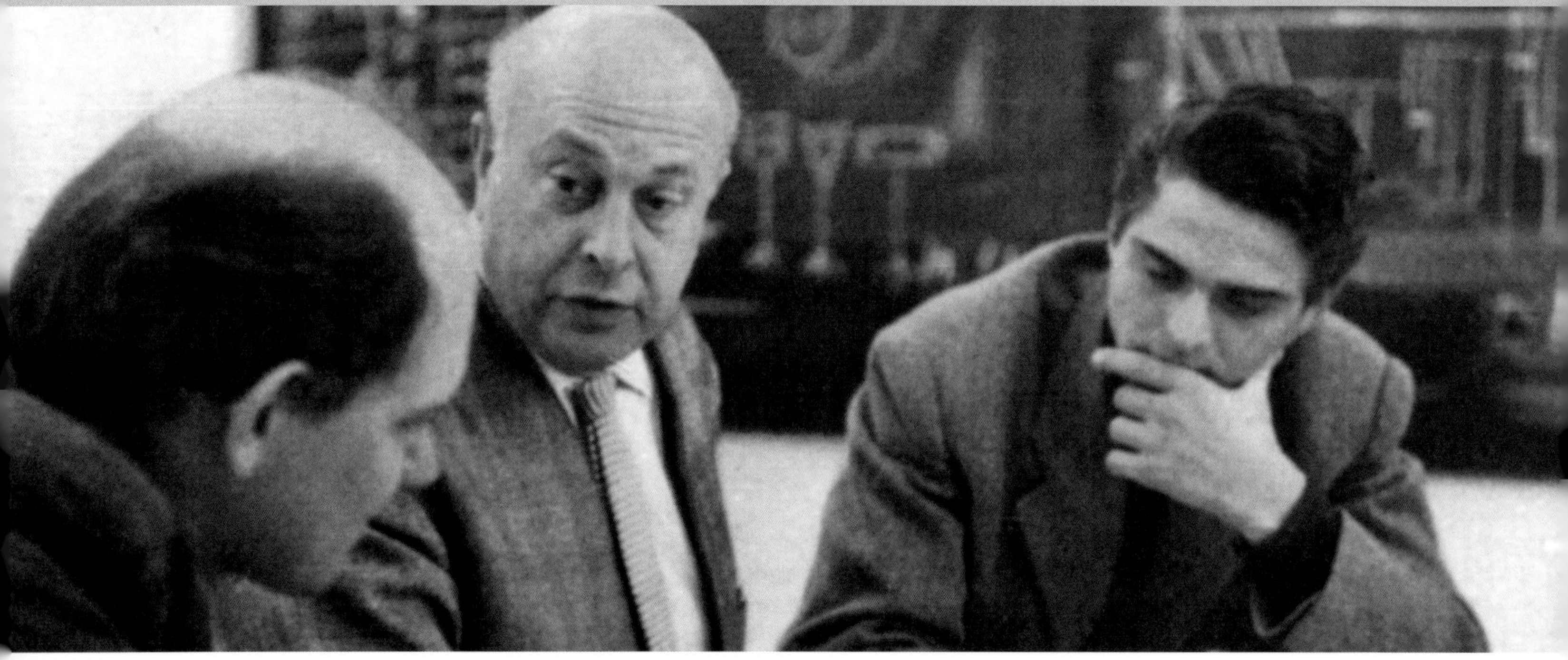

Walter Papst, geboren 1924, studierte an der Muthesius-Werkkunstschule, Kiel. Von 1957 bis 1974 unterhielt er in Köln ein Atelier für Industriedesign und Produktentwicklung. Heute arbeitet Papst als Zukunftsforscher und Schriftsteller vorwiegend in Frankreich. Für Wilkhahn konstruierte er Ende der 50er Jahre einen dreibeinigen Holzstuhl (Mod. 360/2), der für Kinder und Erwachsene produziert wurde. Seine avantgardistisch gestalteten Industrieprodukte aus Polyester erhielten hohe Auszeichnungen (Triennale, Mailand; Musée des Arts Décoratifs, Paris; Gute Industrieform if, Hannover): das Kindermöbelprogramm inklusive der berühmten „Schaukel-Plastik" und die Gartenbank.

Walter Papst, born in 1924, studied at the Muthesius-Werkkunstschule, Kiel. From 1957 to 1974 he had a studio for industrial design and product development in Cologne. Today Papst is concerned with futurology and is an author, primarily in France. For Wilkhahn, he constructed a three-legged wooden chair (model 360/2) at the end of the 1950s that was produced for children and adults. His industrial products in polyester with their avant-garde design won high-calibre awards (Triennale Milan, Museé des Arts Décoratifs, Paris, Gute Industrieform iF, Hanover): his range of children's furniture, including the famous "Rocking Sculpture" and a garden bench.

Roland Rainer, geboren 1910, studierte an der Technischen Hochschule Wien. 1953 wurde er als Professor für Wohnungswesen, Städtebau und Landesplanung an die Technische Hochschule Hannover berufen. Es begann die Zusammenarbeit mit Wilkhahn. Zurück in Wien leitete er bis 1980 die Akademie der bildenden Künste und bis 1999 als Präsident den Österreichischen Kunstsenat. Er verstarb 2004 mit 94 Jahren in Wien. Seine Möbelentwürfe für Wilkhahn waren durch eine natürliche, auf das Notwendige reduzierte Form gekennzeichnet. Der Stapel- und Reihenstuhl aus Buchenholz (21A, 1958–1966) und der Mehrzweckstuhl (702, 1961–1966) mit unverwechselbarem Stahlrohrgestell und nach vorne gerundeter, auch als Armlehnen dienender Rückenlehne setzten neue Maßstäbe.

Roland Rainer, born in 1910, studied at the Technical University in Vienna. In 1953 he became Professor for Housing, Urban Planning and Regional Planning. Collaboration with Wilkhahn began. Back in Vienna, he was Principal of the Akademie der Bildenden Künste until 1980, and until 1999 he was President of the Austrian Art Senate. He died in Vienna in 2004 at the age of 94. His furniture designs for Wilkhahn were characterized by natural form, reduced to essentials. The stackable chair in beechwood for concourse seating (21A, 1958–66) and the multi-purpose chair (702, 1961–1966) with an unmistakable steel tube frame and a backrest with forward curvature that also functioned as armrests set new standards.

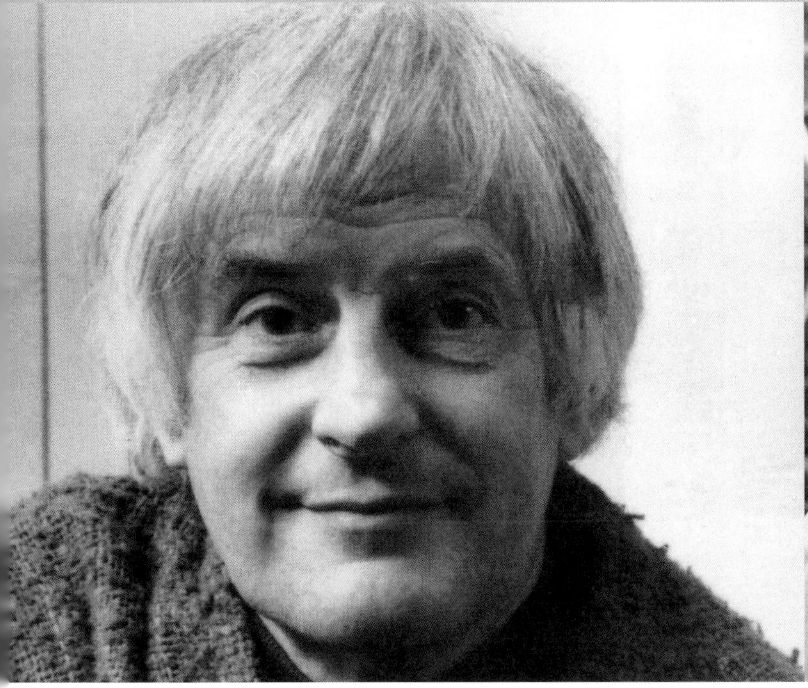

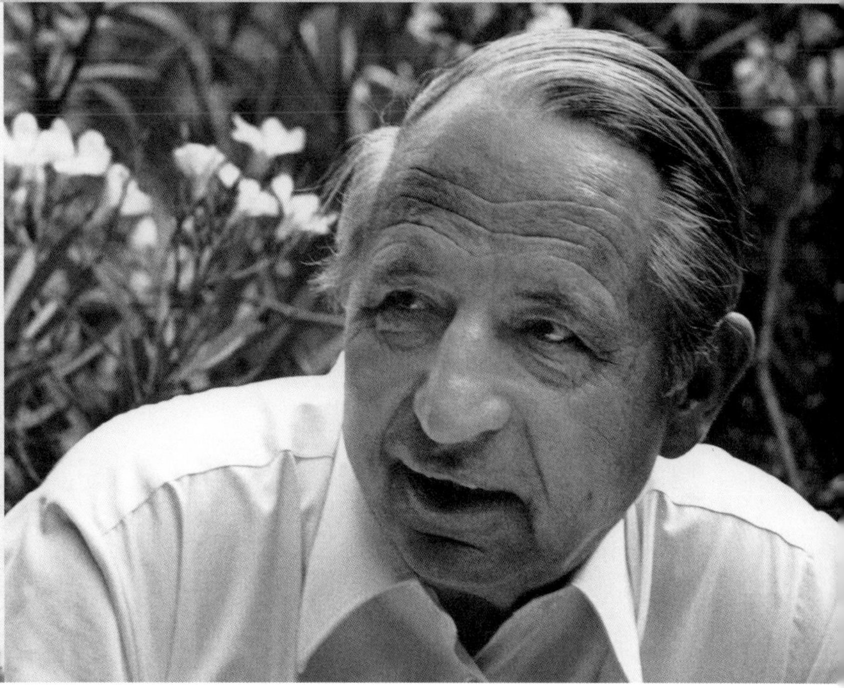

Tomás Gonda, Jahrgang 1926, lehrte seit Anfang der 60er Jahre Visuelle Kommunikation an der Ulmer Hochschule für Gestaltung. 1966 wurde er in die Ford Foundation berufen. Er arbeitete in Mailand als Artdirector bei Rinascente/Upim und von 1977 bis zu seinem Tod 1988 führte er in New York sein Atelier „Gonda Design, Incorporated". 1964 mit der Gestaltung des grafischen Erscheinungsbilds beauftragt, erarbeitete er innerhalb eines Jahres für alle publizistischen Einsatzmittel ein einheitliches, präzise und detailliert definiertes Gesicht. Das streng geometrische Signet stand bis 1980 für das Unternehmen und seine Erzeugnisse, bevor es durch den Schriftzug Wilkhahn als alleiniges Markensymbol abgelöst wurde.

Tomás Gonda, born in 1926, taught visual communications at the Academy of Design in Ulm (HfG). In 1966 he was appointed to the Ford Foundation, he worked as Art Director at Rinascente/Upim in Milan, and from 1977 until he died in 1988 he ran the "Gonda Design, Incorporated" agency in New York. In 1964 he was commissioned with designing the graphic corporate image of Wilkhahn. Within a period of one year, he drew up a finely profiled image for all published material. The strictly linear design mark which he developed was used by Wilkhahn for the company and its products until 1980 when it was replaced by the Wilkhahn lettering as the sole company logo.

Friso Kramer, geboren 1922 in Amsterdam, arbeitete nach Abschluss der Rietveld-Akademie für Angewandte Kunst von 1948 bis 1963 als Designer beim Stahlmöbelhersteller Cirkel. 1963 wurde er Mitbegründer der angesehenen Gruppe Total Design, später lehrte er als Gastdozent an der Königlichen Kunstakademie in Den Haag. Von 1971 bis 1983 war er Artdirector und Vorstandsmitglied der Ahrend-Gruppe sowie Juror in internationalen Designwettbewerben.
Neben dem berühmten Sitzbankprogramm 120, das noch heute in vielen U-Bahn-Stationen zu finden ist, entwarf er für Wilkhahn 1967 einen Stuhl aus glasfaserverstärktem Polyester mit Massivholzgestell und einen gepolsterten GFK-Schalensessel.

Friso Kramer, born in Amsterdam in 1922. After graduating from the Rietveld Academy of Applied Arts, Amsterdam, he worked as a designer for Cirkel, a steel furniture manufacturer, from 1948 to 1963. In 1963 he became a co-founder of the acclaimed Total Design Group. Later on he was a guest lecturer at the Royal Academy of Art in Den Haag. From 1971 to 1983 he was Art Director and Member of the Board of the Ahrend Group and member of the jury for many international design competitions.
In addition to the famous 120 bench system that is still in use in many underground stations today, he designed a chair in glass fibre-reinforced polyester with a solid wood frame and an upholstered glass fibre-reinforced chair for Wilkhahn in 1967.

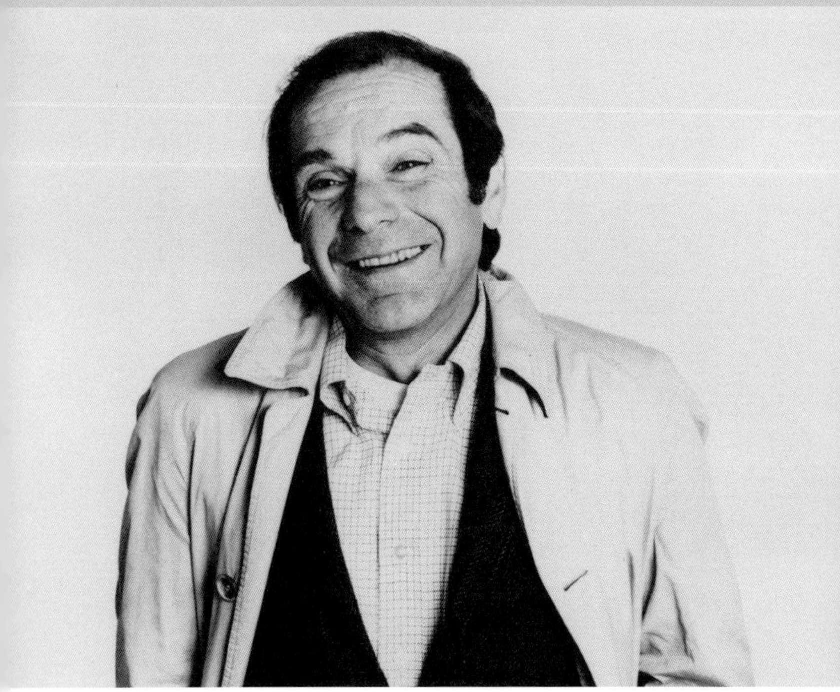

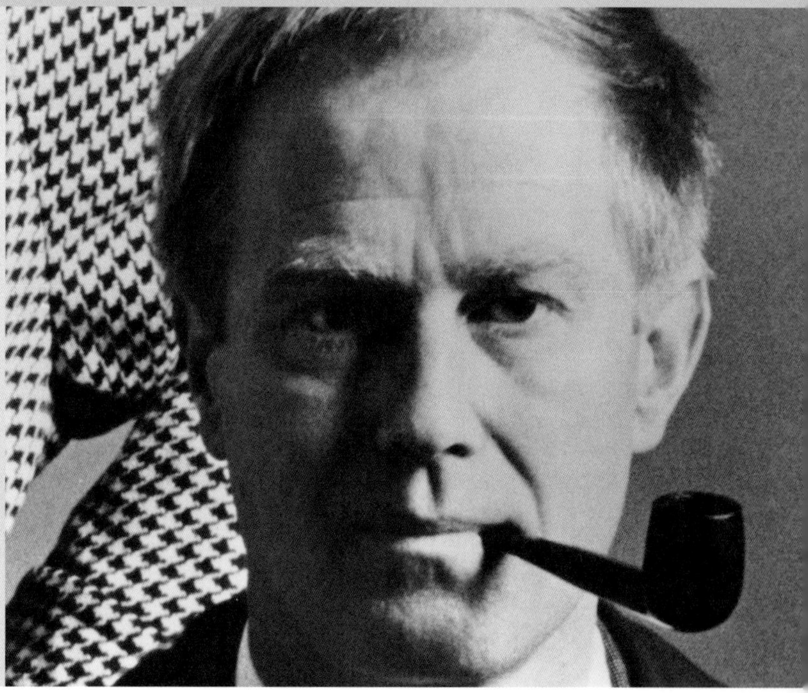

Hans Peter Piehl wurde 1926 geboren. Nach einer Tischlerlehre studierte er an der Werkkunstschule Hannover. Als Innenarchitekt arbeitete er unter anderem mit den Architekten Frei Otto und Egon Eiermann zusammen. Dann wurde er Leiter des Rosenthal-Teams für die Gestaltung der Rosenthal-Studiohäuser und Chefarchitekt der Einrichtungsfirma Behr. Hans Peter Piehl lebt heute in Bad Nenndorf, südwestlich von Hannover.
In seiner Zeit als freiberuflicher Architekt (1964–1989) entwickelte er für Wilkhahn unter anderem die Programme 300, 800 und 7000. Besonders das stapelbare Stuhlmodell aus dem Programm 300 wurde zum Inbegriff eines archetypisch gestalteten, unverwüstlichen Massivholzstuhls und über Jahrzehnte in hohen Stückzahlen produziert. Das Programm erhielt 1966 den Bundespreis „Gute Form". Bis heute sind zahllose dieser Stühle in öffentlichen Gebäuden wie Bürgerhäusern und in sozialen Einrichtungen im Einsatz.

Hans Peter Piehl, born in 1926. After completing an apprenticeship as a cabinet maker, he studied at the Werkkunstschule in Hanover. As an architect, he cooperated with Professor Otto and Professor Eiermann to name but a few. He then became Head of the Rosenthal team for the design of Rosenthal studios and Chief Architect at Behr. Today, Hans Peter Piehl lives in Bad Nenndorf, southwest of Hanover.
During his work as a free-lance architect (1964–1989), he developed the 300, 800, 7000 ranges and others for Wilkhahn. In particular, the stackable chair model from the 300 range became the epitome of an indestructible, solid wood chair with archetypal design and a chair that was produced in large numbers for many years. In 1966, this range was awarded the Federal German prize "Gute Form". Today, countless specimens of this chair are still in use in town houses or social welfare and public facilities.

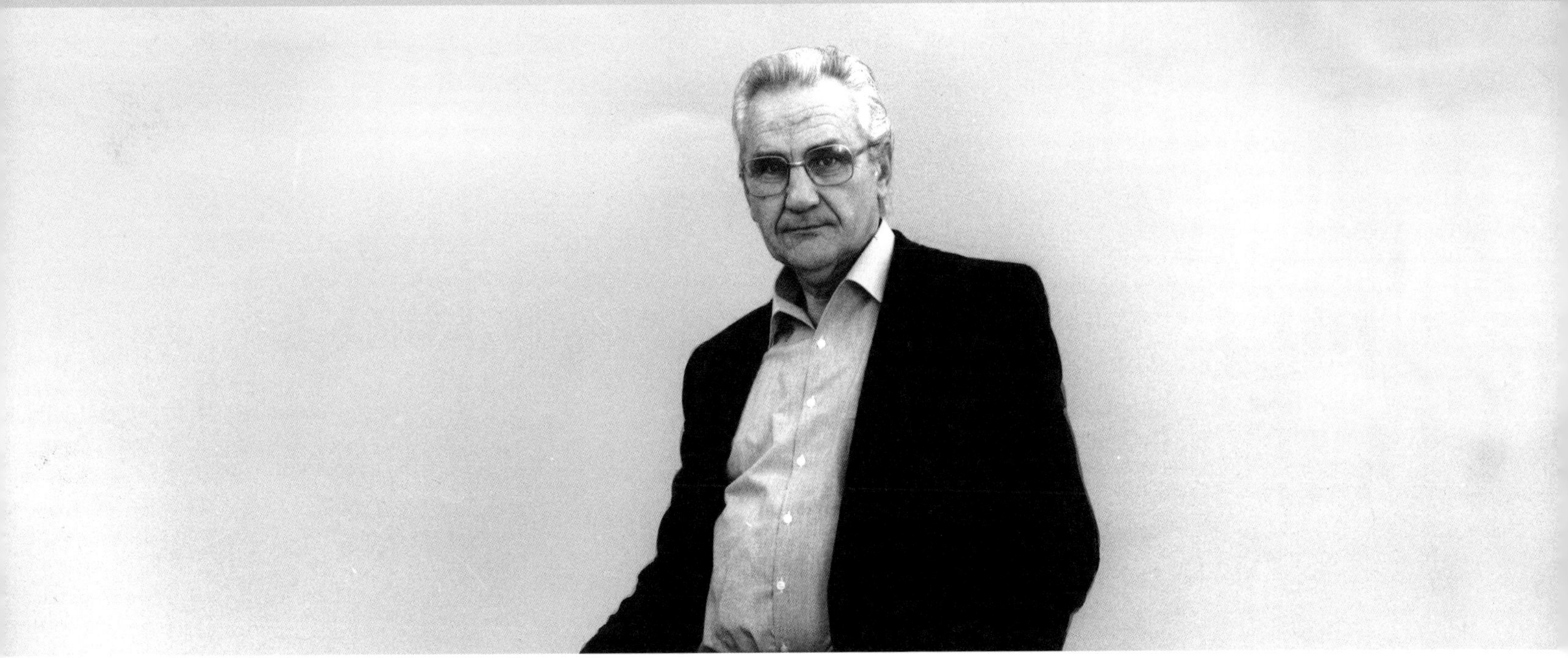

Wilhelm Ritz wurde 1932 als Sohn eines Sägewerksbesitzers in Oberelchingen bei Ulm geboren. Nach seiner Ausbildung in Karosserie- und Fahrzeugbau, Bautechnik, Möbel- und Innenausbau und der Meisterprüfung als Tischler studierte das Multitalent an der Hochschule für Gestaltung Ulm. 1959 machte er sein Diplom über eine neuartige Schichtholzverbindung. Für die Praxisanwendung entwarf er einen hochflexiblen, leichten Schichtholzstuhl: Das Programm 400, von Wilkhahn fast vierzig Jahre lang produziert, war geboren und wurde mit zahlreichen internationalen Designpreisen ausgezeichnet. In seinem später gegründeten Entwicklungsinstitut entstand mit dem Bürodrehstuhl 232 ein weiterer Best- und Longseller. Der Schalenstuhl mit dem markanten Drehgelenk wurde von 1972 bis 1989 produziert.

Wilhelm Ritz, born in 1932, son of a sawmill owner in Oberelchingen near Ulm. After an apprenticeship in body-making and the construction of vehicles, structural engineering, furniture and interior design, and qualifying as a master craftsman in cabinet making, this multi-talent studied at the Academy of Design in Ulm (HfG). In 1959 he wrote his diploma paper on the subject of a new kind of laminated wood. For the purpose of practical application, he designed a highly flexible, lightweight laminated wood chair. This marked the birth of the 400 range, produced by Wilkhahn for almost 40 years. The range won many international design awards. At the development institute which Ritz founded later on, he went on to design yet another best-seller and long-seller in the form of the 232 swivel chair. The shell-structure chair with a distinctive swivel joint was produced from 1972 to 1989.

Bernd Franck, geboren 1938, studierte zunächst Architektur in Berlin und Darmstadt, dann Visuelle Kommunikation an der Ulmer Hochschule für Gestaltung. Seit 1966 arbeitet er als freiberuflicher Grafikdesigner zunächst in Hannover und seit 1973 in Düsseldorf. Sein Arbeitsschwerpunkt ist die Entwicklung von Corporate-Design-Konzepten für Kunden aus Industrie und Dienstleistung. Daneben lehrte er an der Fachhochschule Aachen und als Gastprofessor an der Ohio State University Columbus, Ohio. Von 1968 bis 1984 betreute und pflegte er das grafische Erscheinungsbild von Wilkhahn, ab 1974 Hand in Hand mit seinem Bruder Klaus Franck.

Bernd Franck, born in 1938, first studied architecture in Berlin and Darmstadt then visual communications at the Academy of Design in Ulm (HfG). Since 1966 he has been working as a freelance graphic designer, first in Hanover, since 1973 in Düsseldorf. His work focuses on the development of corporate design concepts for clients from the field of industry and services.
In addition, he lectured at the Fachhochschule Aachen and as guest professor at Ohio State University, Columbus, Ohio. From 1968 to 1984, he supervised and updated the Wilkhahn graphic corporate image; from 1974 onwards he did this work together with his brother, Klaus Franck.

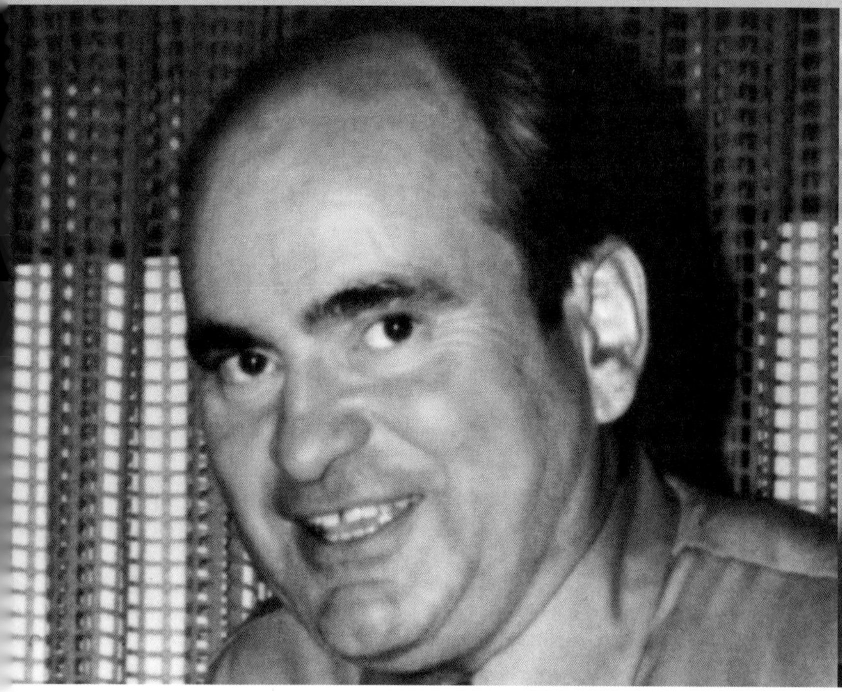

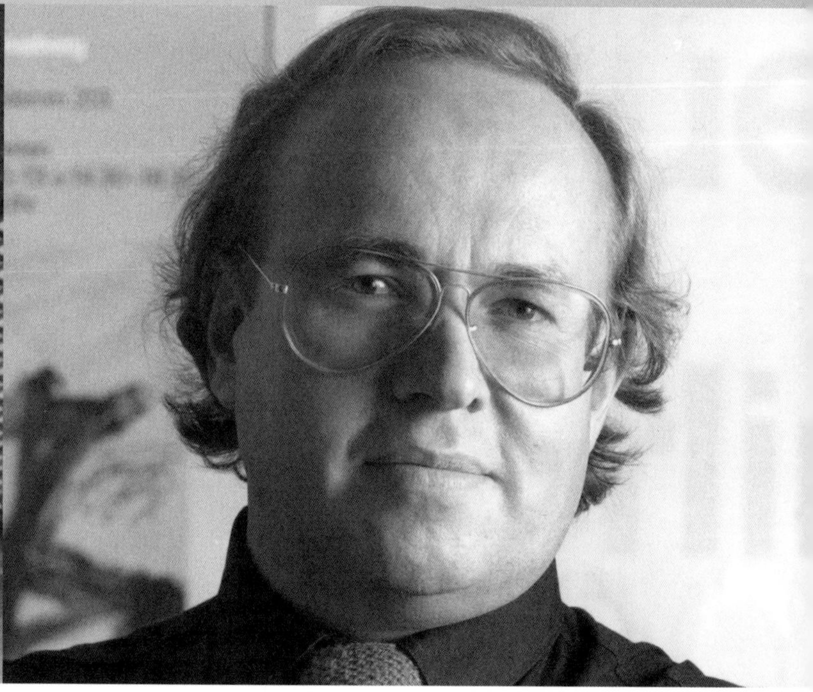

Fritz Hahne gilt als „Menschenfänger" mit der Fähigkeit, besondere Persönlichkeiten um sich zu scharen. Hier ein Gruppenfoto der Geschäftsleitung von 1984 mit einem Teil der Führungsmannschaft: in der ersten Reihe (von links nach rechts) Klaus Franck, Fritz Hahne und Theodor Diener, ganz rechts der Produktionsleiter und spätere Betriebsratsvorsitzende Horst Knigge.

Fritz Hahne with his unique ability to capture people's attention, particularly to have important personalities rallying around him. Group photograph of the Executive Management with some of the top managers, 1984: front row (from left to right) Klaus Franck, Fritz Hahne and Theodor Diener, on the far right, Horst Knigge, Production Manager and later Chairman of the Works Council.

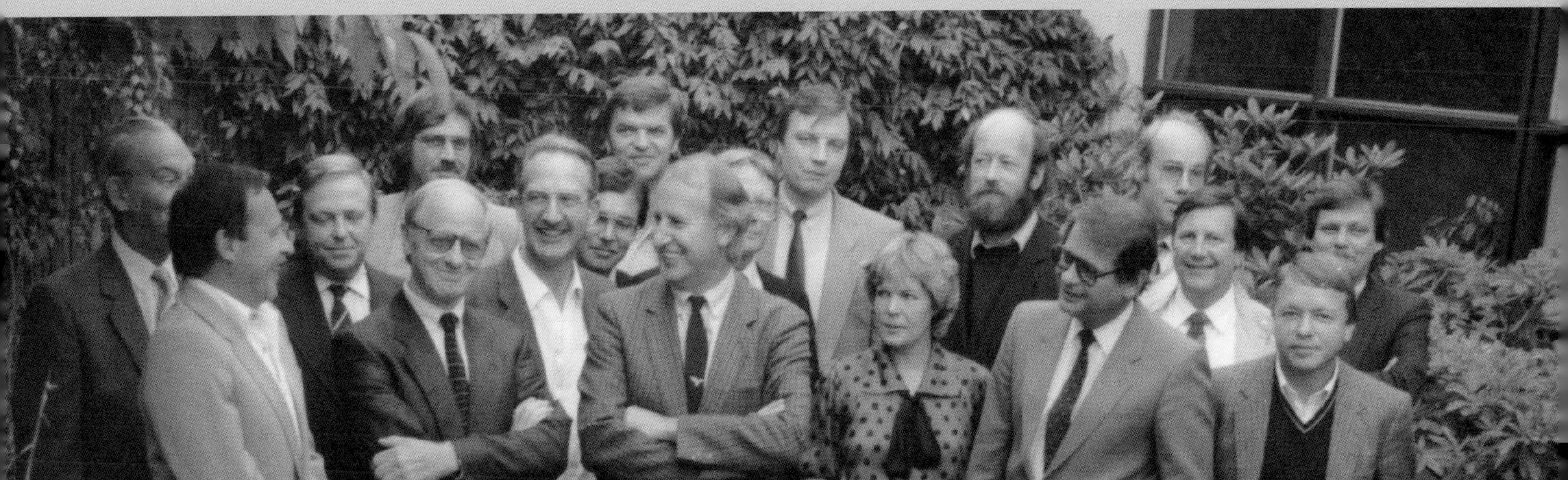

Philip Rosenthal, geboren 1916, promovierte in Oxford zum Master of Arts in Philosophie, Politik und Wirtschaftswissenschaften, trat nach einem Ausflug in die französische Fremdenlegion 1950 in die väterliche Porzellanfirma ein und wurde 1958 Vorstandsvorsitzender der Rosenthal AG. 1969 wurde der Unternehmer als Direktkandidat der SPD in den Bundestag gewählt und unter Karl Schiller für ein Jahr Parlamentarischer Staatssekretär im Wirtschaftsministerium. Sein Streben nach sozialer Gerechtigkeit im Zusammenwirken von Unternehmen und Mitarbeitern war ein dominierendes Thema in seinem Leben. Sein zweites zentrales Anliegen als Mensch und Unternehmer war die „gestaltete Umwelt" mit originaler Kunst und zeitgemäßem Design. 1988 wurde Philip Rosenthal zum Professor für Design an der Bremer Hochschule für Künste berufen. Von 1977 bis 1986 war er Präsident des Rates für Formgebung und Vorsitzender des Bauhaus Archivs. Drei Dinge verbanden Philip Rosenthal und Fritz Hahne: das fortschrittliche Unternehmertum, das starke gesellschaftspolitische Engagement mit den Visionen der Mitarbeiterbeteiligung und die Affinität zu Gestaltungsfragen. Auf dieser Basis entwickelte sich eine Jahrzehnte währende und sich gegenseitig befruchtende Freundschaft, die bis zum Tod Rosenthals 2001 andauerte.

Philip Rosenthal, born in 1916, obtained his Master of Arts in philosophy, politics and economics at Oxford. After a brief period spent in the French Foreign Legion, he joined his father's porcelain factory in 1950, and in 1958 he became CEO of Rosenthal AG. In 1969, entrepreneur Rosenthal won a seat in the German Parliament as a direct candidate for the SPD, and under Karl Schiller he was Undersecretary in the Ministry of Economic Affairs. His striving for social justice in the interrelations between the company and its employees was a dominating theme in his life. His second central concern as a private person and entrepreneur was the "man-made environment" with its original art and contemporary design. In 1988 Philip Rosenthal was appointed Professor of Design at the Bremer Hochschule für Künste. From 1977–1986 he was President of the German Design Council and Chairman of the Bauhaus Archive. Philip Rosenthal and Fritz Hahne had main three things in common: progressive entrepreneurship, strong socio-political commitment with visions of co-determination and great interest in design issues. This was the basis of a mutually inspiring friendship that lasted for decades until Rosenthal died in 2001.

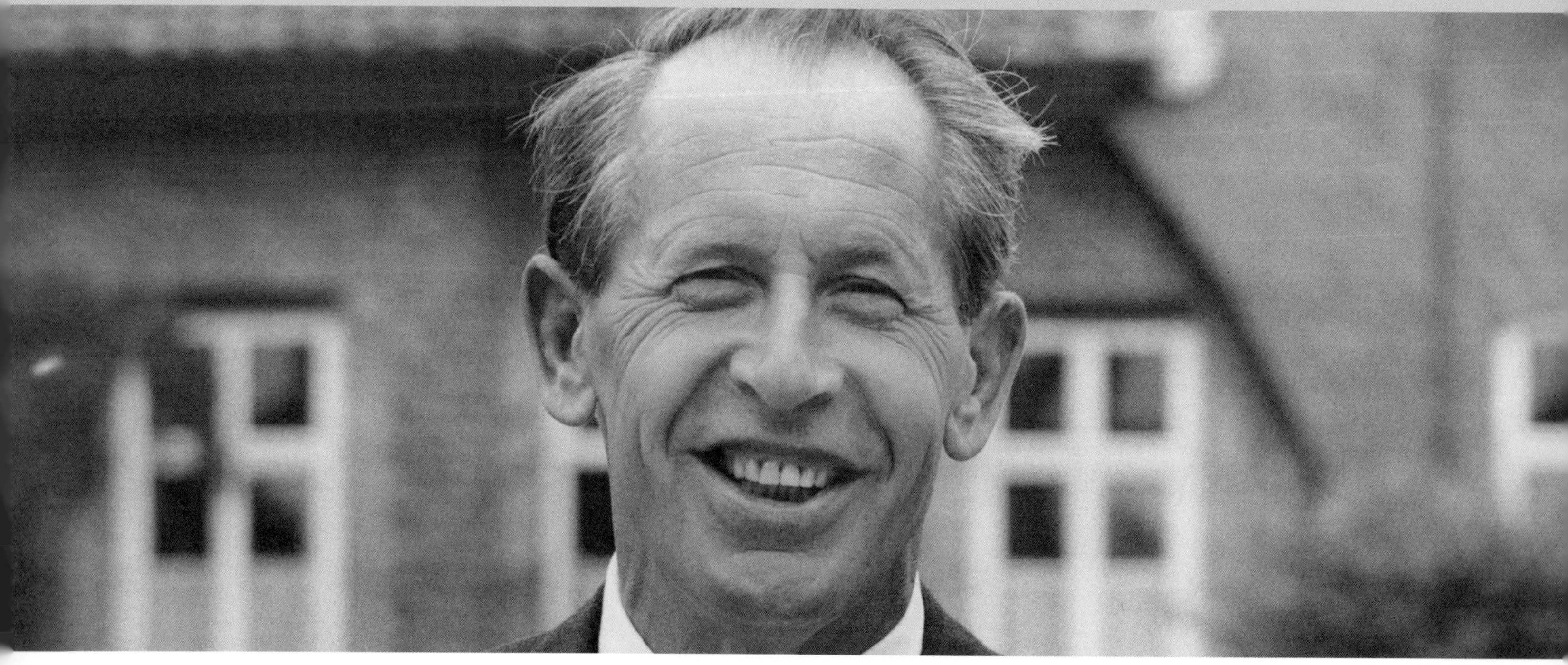

Rudolf Schwarz, Jahrgang 1930 und gelernter Journalist, stieß Mitte der 60er Jahre zu Wilkhahn. Fasziniert von Fritz Hahne und der in jeder Hinsicht innovativen Unternehmensausrichtung arbeitete er sich schnell in die Fachmaterie ein und baute bei Wilkhahn die Informationsabteilung auf. Neben Werbung und Public Relations gehörten dazu die Entwicklung der publizistischen Leitideen, die Konzeption und Durchführung von Verkaufsseminaren, die Organisation externer Vortragsveranstaltungen und die Etablierung von Besuchsprogrammen für interessierte Gruppen aus den Bereichen Architektur, Innenarchitektur und Design. Als Pressesprecher und Leiter der PR-Abteilung von 1984 hat er bis zu seiner Pensionierung 1997 ein tragfähiges Netzwerk mit den Medien in aller Welt etabliert.

Rudolf Schwarz, born in 1930 and a trained journalist, he joined Wilkhahn in the mid-1960s. Fascinated by Fritz Hahne and the corporate approach that was innovative in every respect, he soon became familiar with company-specific matters and built up the "Information Department" at Wilkhahn. In addition to advertising and public relations, this also covered the development of basic publishing ideas, drawing up concepts for sales seminars and running the same, organization of external lecture events and the preparation of visitor programmes for any groups interested from the field of architecture, interior design and product design. From 1984 on, he was Press Spokesman and Head of the PR Department, and by the time he retired in 1997, he had set up a viable network with media throughout the world.

Klaus Franck, Jahrgang 1932, studierte an der Hochschule für Gestaltung Ulm. In der Folgezeit arbeitete er am Institut für Industrialisiertes Bauen, war freiberuflich als Architekt, Grafikdesigner und Buchautor tätig, leitete unter anderem die Gruppe Innenausbau der Lufthansa und lehrte an der Hochschule für Gestaltung Ulm und an der Fachhochschule Hannover. Von 1971 bis 1985 leitete er die Designabteilung bei Wilkhahn, die dann als eigenständiges Unternehmen unter dem Namen „wiege" ausgegründet wurde. Nach weiteren sieben Jahren als Geschäftsführer der wiege folgte der Ausstieg in die erneute Selbstständigkeit. Heute lebt und arbeitet er als freier Designer in Moraira, Spanien. Als Produktdesigner, Innenarchitekt, Artdirector und Autor hat er das „Erbe" der HfG Ulm immer wieder neu übersetzt und die Unternehmensentwicklung entscheidend geprägt.

Klaus Franck, born in1932, studied at the Academy of Design in Ulm (HfG). After that, he worked at the Institut für Industrialisiertes Bauen (Institute for Industrialized Building), he was a free-lance architect, graphic designer and author; amongst other projects, he was Head of Lufthansa's Interior Design Group and lectured at the HfG and the Fachhochschule in Hanover. From 1971 to 1985, he was Head of the Wilkhahn Design Department which was later to become an independent company under the name of wiege. After seven more years as Managing Director of wiege, he left to set up business on his own. Today he lives in Moraira, Spain, and works there as a free-lance designer. As a product designer, interior designer, art director and author, he has constantly retranslated the legacy of the HfG Ulm into a new language and had a formative influence on corporate development at Wilkhahn.

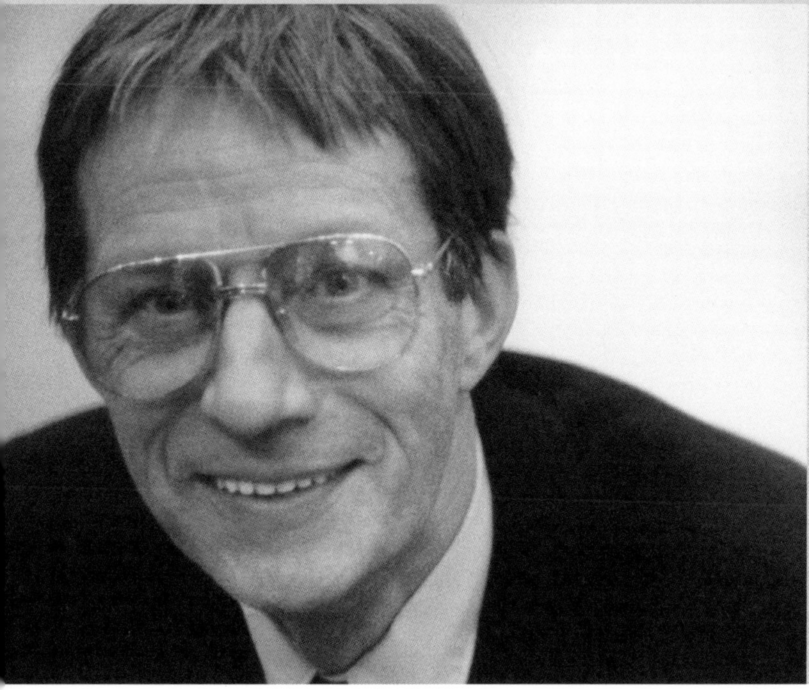

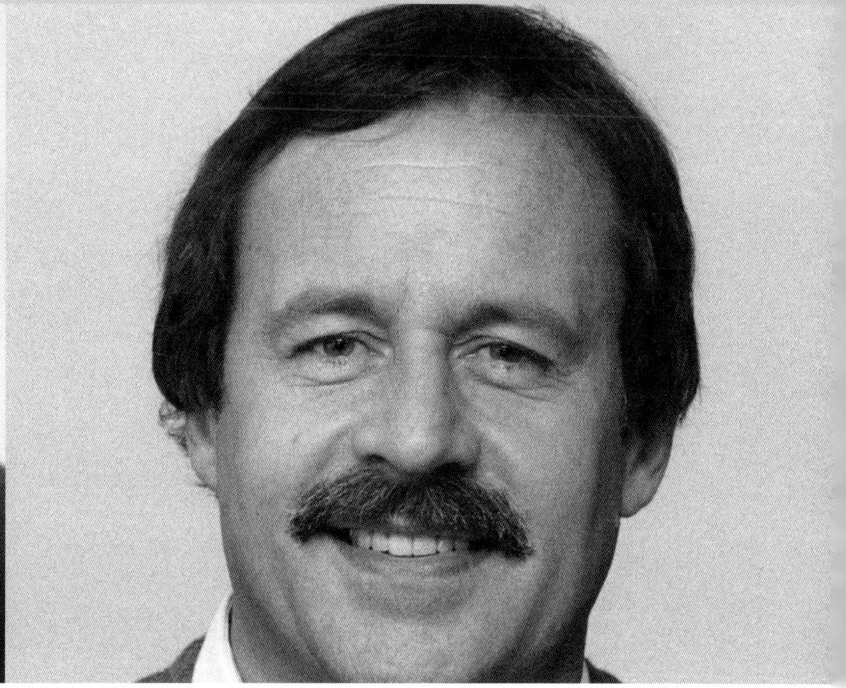

Delta-Design. Michael Conrad, Henner Werner, Detlef Unger und Dieter Raffler hatten gemeinsam an der Hochschule für Gestaltung in Ulm studiert und gründeten 1967 in Stuttgart das Design- und Entwicklungsbüro „Delta-Design". Aus ihrer Arbeit für die Automobilindustrie heraus entwickelten sie diverse Entwürfe für die Möbelindustrie. Delta-Design gestaltete darüber hinaus Haushalts- und Elektrogeräte und Spielzeug – unter anderem auch für die BIG-Spielwarenfabrik (Bobby Car), Fürth. Michael Conrad lehrte an der Hochschule Pforzheim, Henner Werner wirkte an der Fachhochschule Kiel im Fachbereich Gestaltung und entwickelte mit Studenten Anfang der 90er Jahre einen kompakten Linienbus. Dieter Raffler unterrichtete an der Hochschule Anhalt in Dessau im Fachbereich Design.
Für Wilkhahn entwickelte Delta-Design 1967 das Programm 2000, das mit seinen groß dimensionierten Sitzschalen vom Start weg zum begehrten „Chefsesselprogramm" wurde. Die Techniken aus dem Automobilbau hatten es den Designern ermöglicht, auf starke Wölbungen der Schalen zu verzichten und dadurch eine schlanke und geradlinige Kontur zu erzeugen. Das Bürostuhlprogramm 23 (1977–1985) bot mit seiner dreigeteilten Rückenlehne je nach Einsatzbereich unterschiedliche Stuhltypen. Doch die Gestaltung war der Zeit zu weit voraus, um sich in hohen Verkaufszahlen niederzuschlagen.

Delta-Design. Michael Conrad, Henner Werner, Detlef Unger and Dieter Raffler studied together at the Academy of Design in Ulm (HfG), and in 1967 they founded the "Delta-Design" design and development office in Stuttgart. Their work for the car industry resulted in numerous designs for the furniture industry. Delta-Design also designed electrical and household appliances, as well as toys – including products for the BIG-Spielwarenfabrik (bobby car) in Fürth. Michael Conrad taught at the Hochschule In Pforzheim, Henner Werner taught at the Fachhochschule in Kiel in the Design Faculty, and together with his students he developed a compact bus. Dieter Raffler taught at the Hochschule Anhalt in Dessau in the Design Faculty.
For Wilkhahn, Delta-Design developed the 2000 range in 1967 that became a desirable "executive chair range" from the very start due to extremely spacious seat shells. Technology drawn from the car industry had enabled the designers to do without strongly curved seat shells and thus to produce a design with slender, linear contours. The 23 office chair range (1977–1985) provided different types of chairs depending on the application required due to a three-section backrest. But this design was too much ahead of its time to sell well.

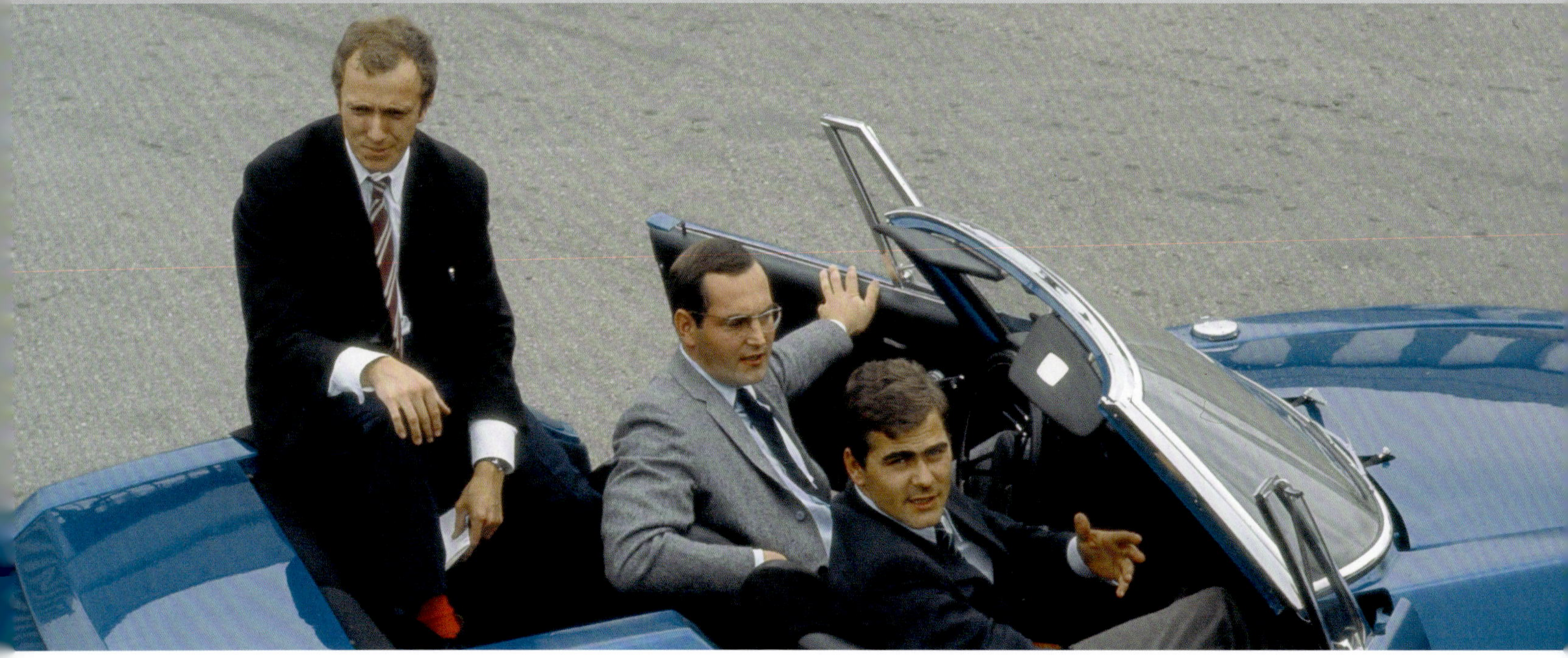

Hans „Nick" Roericht, Jahrgang 1932, studierte an der Hochschule für Gestaltung Ulm, wurde dort selbst Dozent und unterrichtete zeitweise auch an der Universität von Columbus, Ohio. 1976 gründete er mit ProduktEntwicklung Roericht sein eigenes Büro unmittelbar neben der Hochschule für Gestaltung. Von 1973 bis zu seiner Emeritierung 2001 unterrichtete er Industrial Design an der Hochschule der Künste, Berlin. Heute leitet er die „Hans-Roericht-Stiftung", die sich der Talentförderung verschrieben hat. Roehricht gestaltete für Wilkhahn zahlreiche Programme, darunter so wichtige wie das Sesselprogramm 190, das Lounge-Programm 840, das Programm Kendo und das Bürostuhlprogramm Picto. Mit der Stehhilfe „Stitz" entwickelte er einen prägnanten Klassiker, der bis heute als Ikone für die Wilkhahn-Gestaltungsmaximen gilt.

Hans "Nick" Roericht, born in 1932, studied at the Academy of Design in Ulm (HfG) where he later became a lecturer. He also lectured at the University of Columbus, Ohio. In 1976 he founded his own studio, ProduktEntwicklung Roericht, directly adjacent to the HfG. From 1973 until he became an Emeritus Professor in 2001, he taught industrial design at the Hochschule für Künste in Berlin (University of Fine Arts). Today, he manages the "Hans Roericht Foundation" that is dedicated to promoting talented designers. Roericht designed numerous ranges for Wilkhahn, including the 190 chair range, the 840 lounge range, the Kendo range and the Picto office chair range. With the "Stitz" leaning aid, he developed a distinctive design classic that has been regarded as an icon of Wilkhahn design principles until today.

Theodor Diener, Jahrgang 1936, studierter Elektrotechniker, hatte als Vertriebsingenieur bei Braun, Boveri & Cie und als Manager im Krupp-Konzern vielfältige Erfahrungen gesammelt, zuletzt als Direktor der Friedrich Krupp GmbH, bevor er 1982 Vorsitzender der Geschäftsführung und Mitgesellschafter bei Wilkhahn wurde. Nach seinem Ausscheiden 1993 gründete er Diener Consult und war unter anderem Geschäftsführer der EXPO 2000. Theodor Diener löste Fritz Hahne ab, der 62-jährig aus dem operativen Tagesgeschäft ausschied und den Vorsitz im Wilkhahn-Verwaltungsrat übernahm. Unter Dieners Leitung wurden das Management professionalisiert, das Marketing ausgebaut und die Internationalisierung systematisch vorangetrieben.

Theodor Diener, born in 1936, studied electrical engineering and as a sales engineer at Brown, Boveri & Cie from 1960 to 1966 and a manager in the Krupp Group he gained wide experience; after that he was Director of Friedrich Krupp GmbH before he became Managing director and Managing Partner at Wilkhahn in 1982. After leaving the company in 1993, he set up Diener Consult and, amongst others, held the position of Managing Director of EXPO 2000. Theodor Diener succeeded Fritz Hahne when the latter retired from the operative side of daily business at the age of 62 to take over the Chairmanship of the Wilkhahn Administrative Board. It was under Diener's leadership that Management at Wilkhahn became more professional, marketing was expanded and the internationalization of the company was promoted systematically.

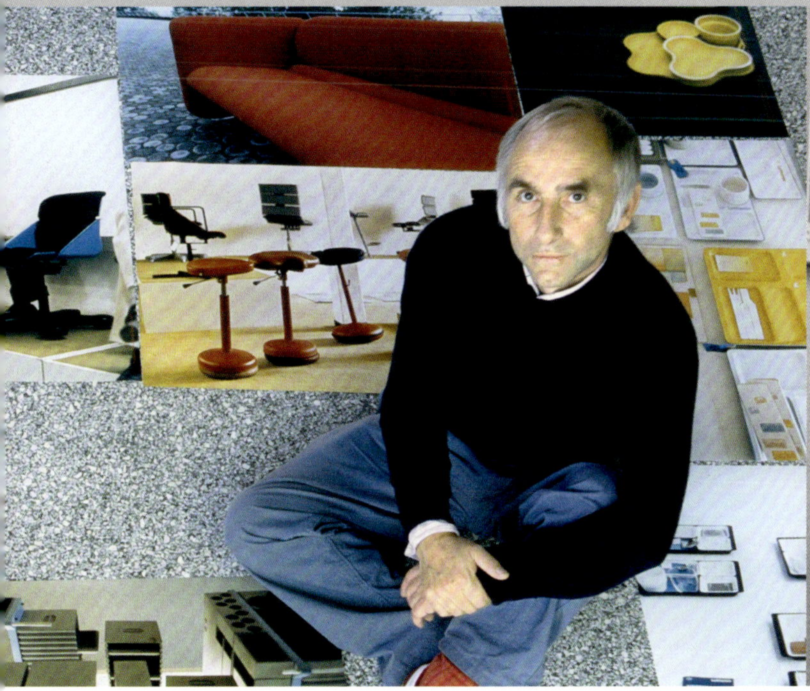

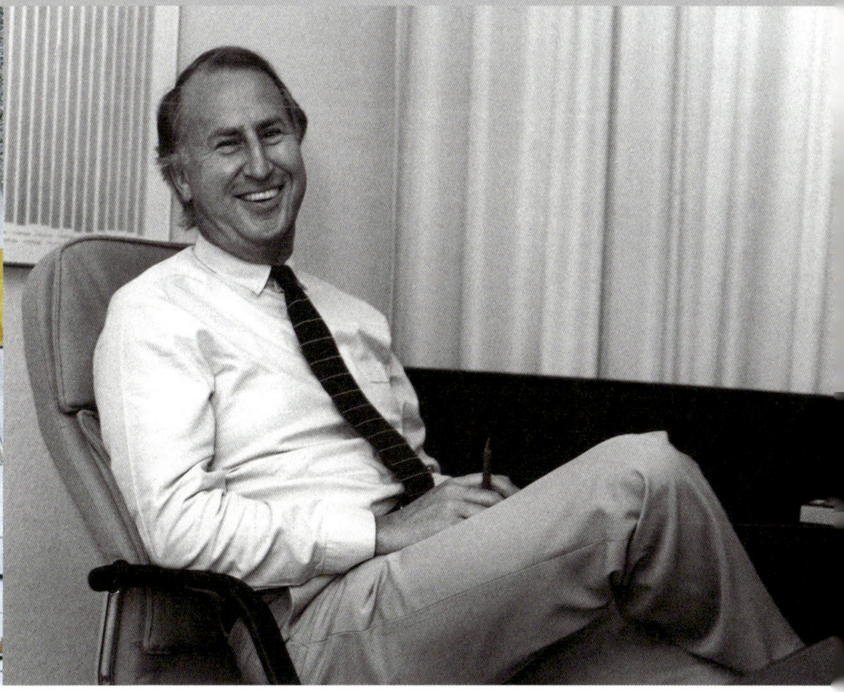

Im Duo zu Spitzenleistungen: Klaus Franck und Werner Sauer haben über 20 Jahre hinweg Wilkhahn-Designgeschichte geschrieben. Neben zahlreichen Erfolgsprogrammen, die das kongeniale Zweiergespann zunächst in der Designabteilung und dann in der seit 1985 selbstständigen Entwicklungsgesellschaft wiege entworfen haben, haben insbesondere die FS-Linie (1980) und das Programm Modus (1994) den Ruf von Wilkhahn als führendem Hersteller für hochwertige Bürostühle geprägt. Beide Programme setzten weltweit neue Standards für das Sitzen am Schreibtisch und in der Konferenz. **Werner Sauer**, Jahrgang 1950, studierte Industrial Design an der Folkwang Hochschule, Essen, und Experimentelle Umweltgestaltung an der Staatlichen Hochschule für Bildende Künste, Braunschweig. Erst freiberuflich, ab 1978 fest angestellt, arbeitete er zunächst in der Designabteilung und dann bis 1993 in der wiege, bevor er sich 1994 mit einem eigenen Designstudio in Springe bei Hannover selbstständig machte. Einem Lehrauftrag für Produktdesign an der Fachhochschule Hildesheim / Holzminden / Göttingen folgte die Berufung als Professor, im Jahr 2000 als Dekan und seit 2003 als geschäftsführender Dekan. Werner Sauer ist Vorstandsmitglied im Deutschen Werkbund Nord.

A duo with top achievements: Klaus Franck and Werner Sauer have written Wilkhahn design history for more than 20 years now. In addition to numerous successful ranges, which this congenial team designed, starting in the Design Department and then in the wiege Design Studio, an independent design and development company since 1985, the FS-Line (1980) in particular and the Modus range (1994) have played a major role in shaping Wilkhahn's reputation as a leading producer of high-quality office chairs. Both ranges have set new international standards for sitting at the desk and in conferences. **Werner Sauer**, born in 1950, studied industrial design at the Folkwang Hochschule in Essen, and experimental environmental design at the Staatliche Hochschule für Bildende Künste in Brunswick. First on a freelance basis, then employed from 1978, he started off working in the Design Department and then continued at wiege until 1993 before setting up his own design studio in Springe near Hanover. He began as a lecturer at the Fachhochschule Hildesheim / Holzminden / Göttingen, then advanced to Professor, then Dean in 2000, and since 2003 he has been Executive Dean. Werner Sauer is a Member of the Board of the Deutscher Werkbund Nord.

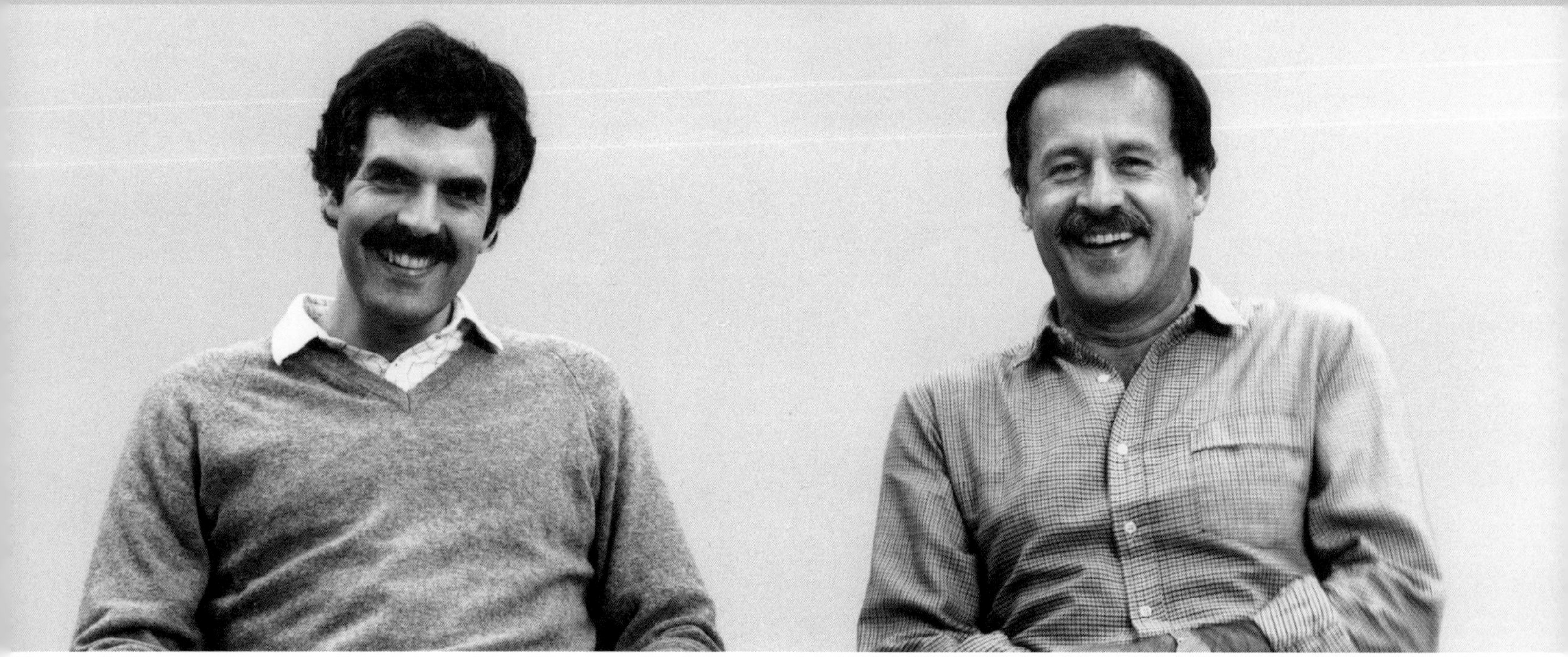

Seit der Arbeit von Tomás Gonda für Wilkhahn gehörte ein durchgängig gestaltetes Corporate Design zum festen Bestandteil des Wilkhahn-Erscheinungsbilds. Ganz im Sinne der Ulmer Hochschule für Gestaltung bildeten seitdem Produktgestaltung, Grafikdesign und Unternehmenskommunikation eine konsistente Einheit. Das von Bernd Franck gemeinsam mit seinem Bruder Klaus Franck weiterentwickelte Konzept wurde 1986 mit dem **Corporate Design Award** der if Hannover ausgezeichnet. Im Bild: Übergabe des Preises an Klaus Franck (links) und Fritz Hahne (Mitte).

Since Tomás Gonda's work for Wilkhahn, a coherent corporate design has been an integral part of the Wilkhahn image. Very much in line with the thinking of the Academy of Design in Ulm (HfG), product design, graphic design and corporate communication have formed a consistent entity since that time. The concept, further developed by Bernd Franck and his brother Klaus Franck, won the **Corporate Design Award** of the iF Hanover in 1986. The photograph shows the award being received by Klaus Franck (left) and Fritz Hahne (centre).

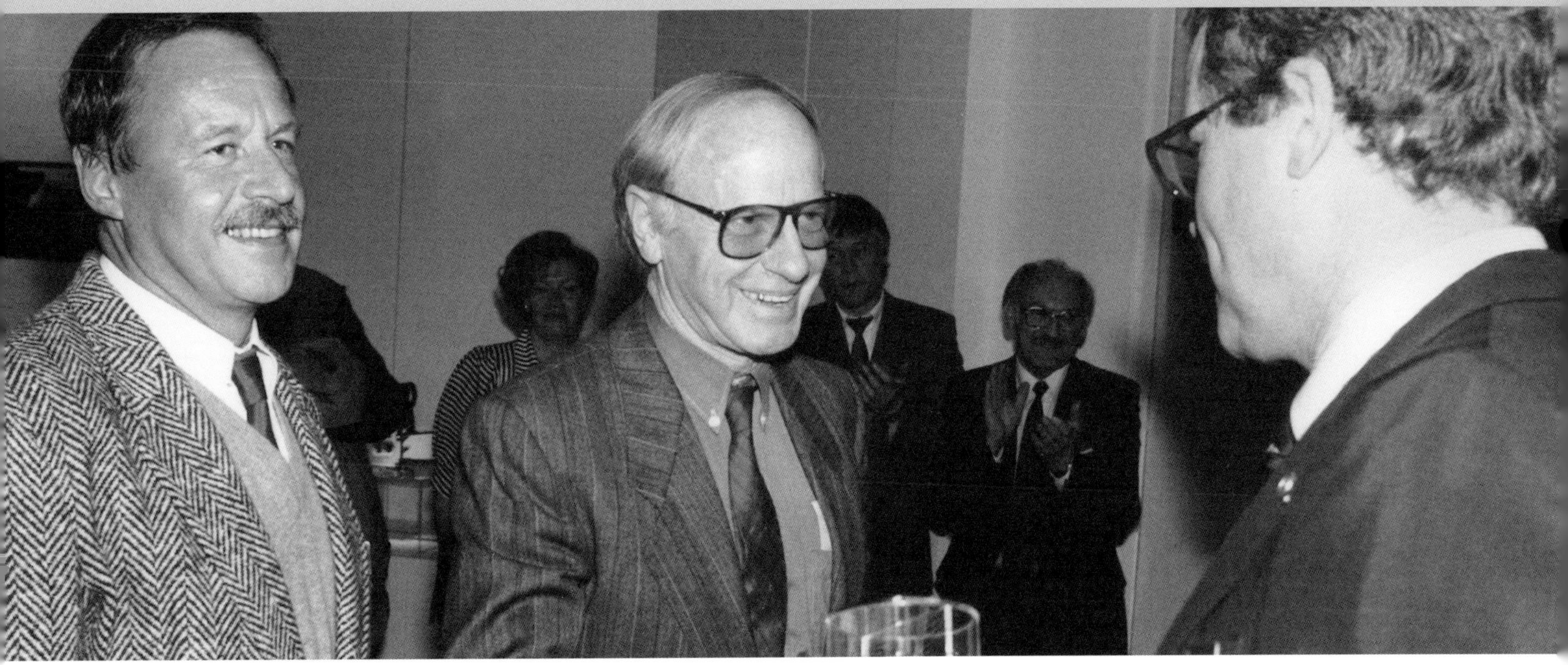

Reiner Moll, geboren 1947, studierte Industriedesign bei Professor Karl Dittert und Professor Nuss an der Fachhochschule Schwäbisch Gmünd. Hier gründete er 1971 das Büro „moll design", das beispielsweise für Dornbracht, Philips, AEG Licht, Villeroy & Boch und Wolf Heiztechnik arbeitet. moll design wurde mit vielen Preisen im In- und Ausland ausgezeichnet, unter anderem 1991 vom Design Zentrum Nordrhein Westfalen als „Design-Team des Jahres". Für Wilkhahn entwickelte Moll das technologisch innovative Mehrzweckstuhlprogramm Binar (1979–1990), bei dem eine komfortable Sitzschale aus hohl geblasenem Kunststoff durch einfaches Auf- und Einklipsen mit einem raffiniert ausgebildeten Stahlrohrgestell verbunden wurde.

Reiner Moll, born in 1947, studied industrial design under Prof. Karl Dittert and Prof. Nuss at the Hochschule Schwäbisch Gmünd. This is where he set up the "moll design" office in 1971 which worked for Dornbracht, Philips AEG Licht, Villeroy & Boch and Wolf Heiztechnik, to name but a few. moll design won many awards at home and abroad, including the award from the Design Centre North-Rhine/Westphalia in 1991 as "Design Team of the Year". For Wilkhahn, Moll developed Binar (1979–1990), a technologically innovative multi-purpose chair range with a comfortable seat shell in blow-moulded plastic that simply clipped into an ingeniously designed steel tube frame.

Herbert Ohl, Jahrgang 1926, studierte Malerei und Grafik an der Akademie der bildenden Künste, Karlsruhe, anschließend bei Egon Eiermann Architektur an der Technischen Hochschule. Bis Ende der 60er Jahre arbeitete er als freier Architekt und Designer in Ulm, wo er von 1964 bis 1968 als Rektor der Hochschule für Gestaltung wirkte. Arbeiten im Sachverständigenrat für Bauforschung, am Institut für Industrialisiertes Bauen, am Institut für Umweltgestaltung und in der Gute-Form-Preis-Kommission begleiteten seine Entwürfe als Industriedesigner. Ohl erhielt zahlreiche Preise wie den New Yorker Low-Cost-Housing-Prize 1969 und den Gute-Form-Preis 1983. 1984 wurde er als Professor an die Fachhochschule Pforzheim berufen. Seine Leidenschaft für „Sitzen auf Netzen" führte bei Wilkhahn zu den Programmen O-Linie (1982) und Nuvola (1998).

Herbert Ohl, born in 1926, studied painting and graphic design at the Akademie der bildenden Künste in Karlsruhe, and went on to study architecture under Egon Eiermann at the Technische Hochschule there. Until the end of the 1960s, he worked as a free-lance architect and designer in Ulm where he was Rector of the Academy of Design (HfG) from 1964 to 1968. His projects as an industrial designer were accompanied by work on the Council of Experts on Building Research (Sachverständigenrat für Bauforschung) at the Institute for Industrialized Building, the Institute for Environmental Design and the "Gute-Form" Commission. Ohl won many awards such as the New York Low-Cost-Housing-Prize in 1969 and the "Gute-Form" Prize in 1983. In 1984 he was appointed Professor at the Fachhochschule in Pforzheim. His passion for "sitting on netting" resulted in the Wilkhahn O-Line (1982) and Nuvola (1998) ranges.

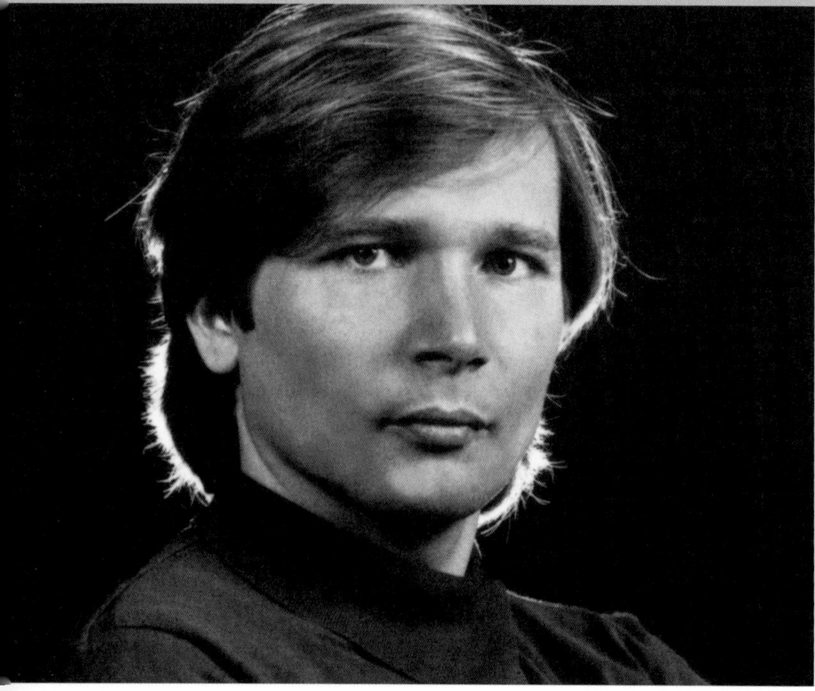

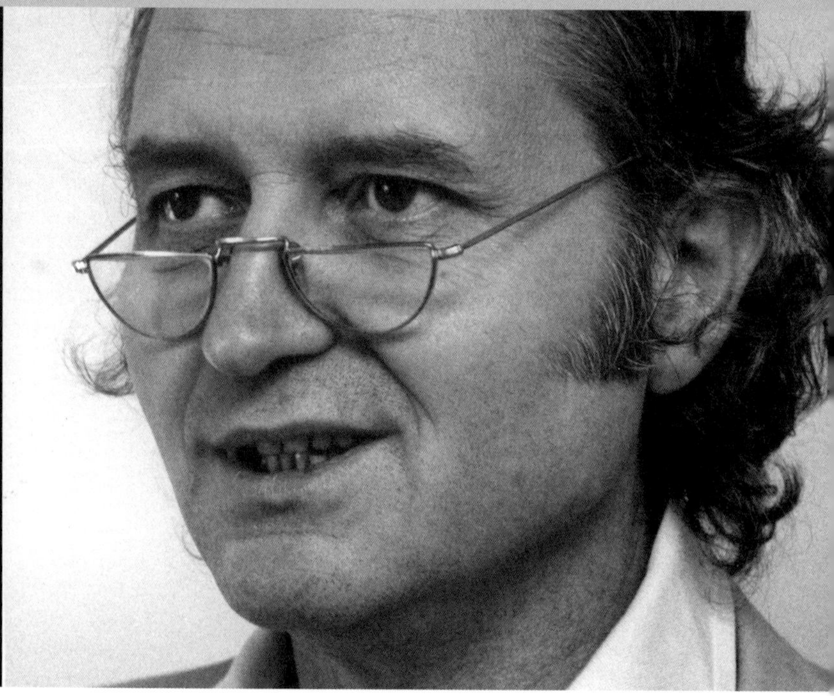

Burkhard Schmitz, Jahrgang 1957, studierte Design an der HdK Berlin. Als „Roericht-Schüler" arbeitete er nach dem Diplom als Designer bei ProduktEntwicklung Roericht (PER), Ulm. 1992 gründete er mit Franz Biggel das Büro „7.5 Projektgruppe". Einer Lehrtätigkeit im Fachbereich Neue Medien an der Fachhochschule Schwäbisch Gmünd folgte 1993 die Berufung als Professor für das Entwerfen interaktiver Systeme an die Universität der Künste, Berlin, der er seit 2006 als Vizepräsident vorsteht. Schmitz entwickelte für Wilkhahn unter anderem das Programm Kendo mit der Kombination aus Massivholzgestell und perforierter Polypropylen-Rückenschale (1988). Wegweisend wurde das mit Franz Biggel entwickelte Bürostuhlprogramm Picto, das nach den neuen ökologisch orientierten Designrichtlinien gestaltet wurde (1992).

Burkhard Schmitz, born in 1957, studied design at the Hochschule der Künste in Berlin He was a pupil of Nick Roericht's, and after obtaining his diploma he worked at ProduktEntwicklung Roericht (PER), Ulm. In 1992 he founded the "7.5 Project Group" together with Franz Biggel. A lectureship at the Fachhochschule Schwäbisch Gmünd in the Design Faculty was followed by appointment as Professor for designing interactive systems at the Universität der Künste, Berlin (Berlin University of the Arts) in 1993 where he has been Vice President since 2006. For Wilkhahn, Schmitz developed several ranges, including the Kendo range, that is marked by a combination of a solid wood frame and a perforated polypropylene back shell (1988). One groundbreaking development was the Picto office chair range which he developed in collaboration with Franz Biggel, a chair that was designed in accordance with new, ecologically oriented design principles (1992).

Franz Biggel, Jahrgang 1957, studierte Maschinenbau an der Fachhochschule Technik, Ulm. Von 1984 bis 1993 arbeitete er freiberuflich im Designbüro ProduktEntwicklung Roericht, bevor er gemeinsam mit Burkhard Schmitz das Büro „7.5 Projektgruppe" gründete, das im Möbelbereich unter anderem auch Produkte für Pohlschröder und Herman Miller entwickelte. Dem Lehrauftrag an der Hochschule der Künste, Berlin, (1989–1992) folgte die Berufung als Professor für Gestaltung von Produktsystemen, mit dem Schwerpunkt der technologieorientierten, konstruktiven Produktentwicklung an die Fachhochschule Schwäbisch Gmünd.

Franz Biggel, born 1957, studied mechanical engineering at the Fachhochschule Technik in Ulm. From 1984 to 1993 he did free-lance work for the ProduktEntwicklung Roericht Design Studio before he founded the "7.5 Project Group" together with Burkhard Schmitz. This Group developed furniture products for Pohlschröder and Herman Miller, to name but a few. A lectureship at the Hochschule der Künste Berlin (1989–1992) was followed by an appointment as Professor for the design of product systems at the Fachhochschule Schwäbisch Gmünd with a focus on technology-oriented, constructional product development.

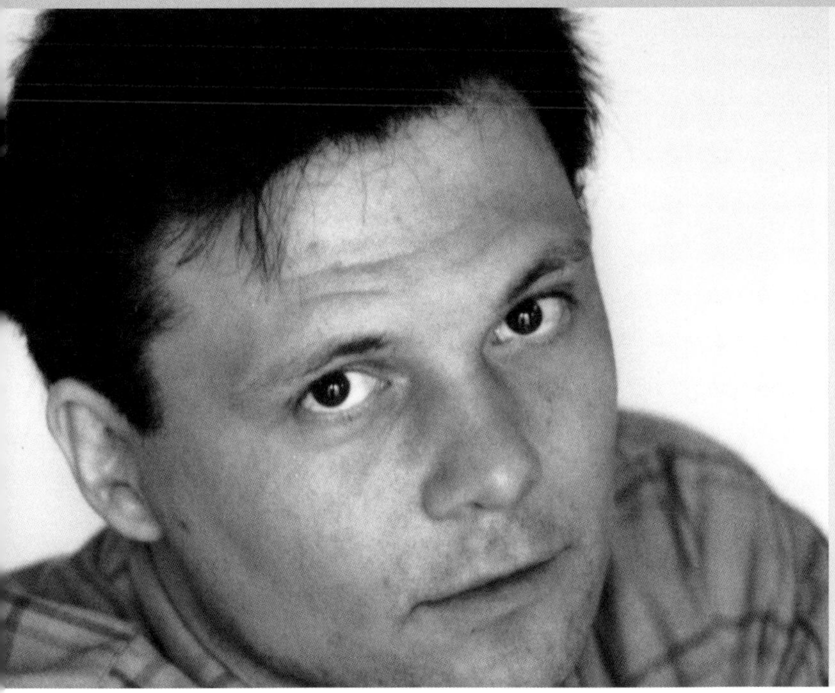

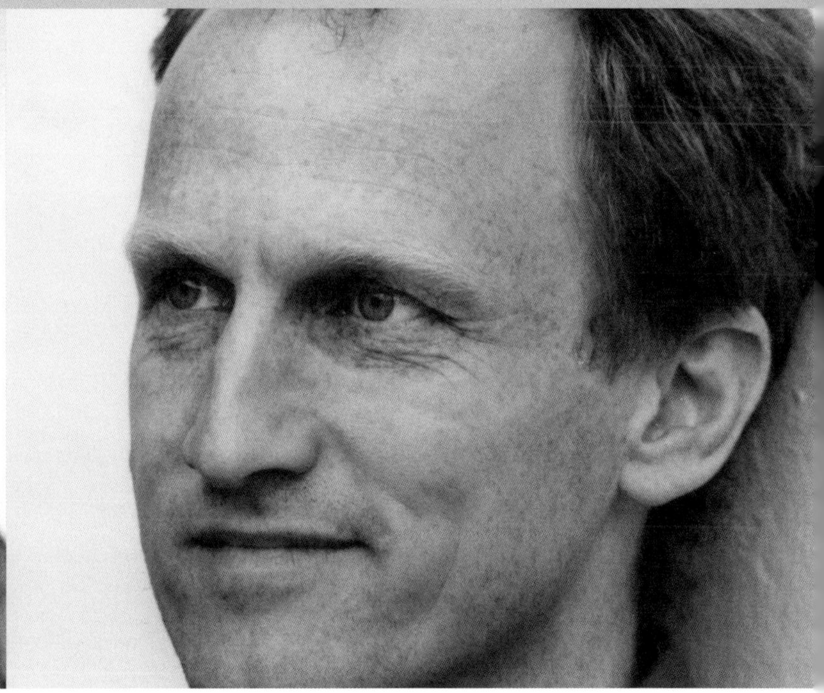

Auf weltweiter Mission in Sachen Wilkhahn: Karl Schneider (Mitte), von 1973 bis zu seiner Pensionierung 1995 Exportleiter bei Wilkhahn, und Theodor Diener (rechts) bei einer Ausstellungseröffnung. Als Sohn eines schwäbischen Missionars hatte **Karl Schneider** seine frühe Kindheit in Afrika verbracht, war in den USA aufgewachsen und hatte bis 1982 vor allem Geschäftskontakte in die europäischen Nachbarländer aufgebaut. Als „Missionar" in Sachen Wilkhahn in aller Welt unterwegs hatte er maßgeblichen Anteil an der Verdreifachung des Auslandsumsatzes innerhalb von gut zehn Jahren. Vor allem: Er etablierte Wilkhahn in vielen Auslandsmärkten so gut, dass hier eigene Tochtergesellschaften gegründet werden konnten. Nachdem in Spanien und Frankreich bereits Tochtergesellschaften bestanden, folgten innerhalb kurzer Zeit Österreich, die Schweiz, Großbritannien und die Niederlande. In den USA, in Südafrika, Japan und Australien wurden Lizenzpartner gefunden und in über dreißig Ländern Vertriebspartnerschaften vereinbart.

On a global mission for Wilkhahn: Karl Schneider (centre), Head of Export at Wilkhahn from 1973 until his retirement in 1995, and Theodor Diener (right) at the opening of an exhibition. The son of a Swabian missionary, **Karl Schneider** had spent his early childhood in Africa, had grown up in the USA, and by 1982 had built up business contacts to neighbouring European countries in particular. As a "missionary" for the Wilkhahn cause throughout the world, he had made a substantial contribution to foreign sales being tripled within just ten years. But above all: he established Wilkhahn in many foreign markets so efficiently that independent subsidiary companies could be founded there. After subsidiary companies had already been set up in Spain and France, Austria, Switzerland, Great Britain and the Netherlands followed soon afterwards. Licence partners were appointed in the USA, South Africa, Japan and Australia, and sales partnerships were agreed upon in more than thirty countries.

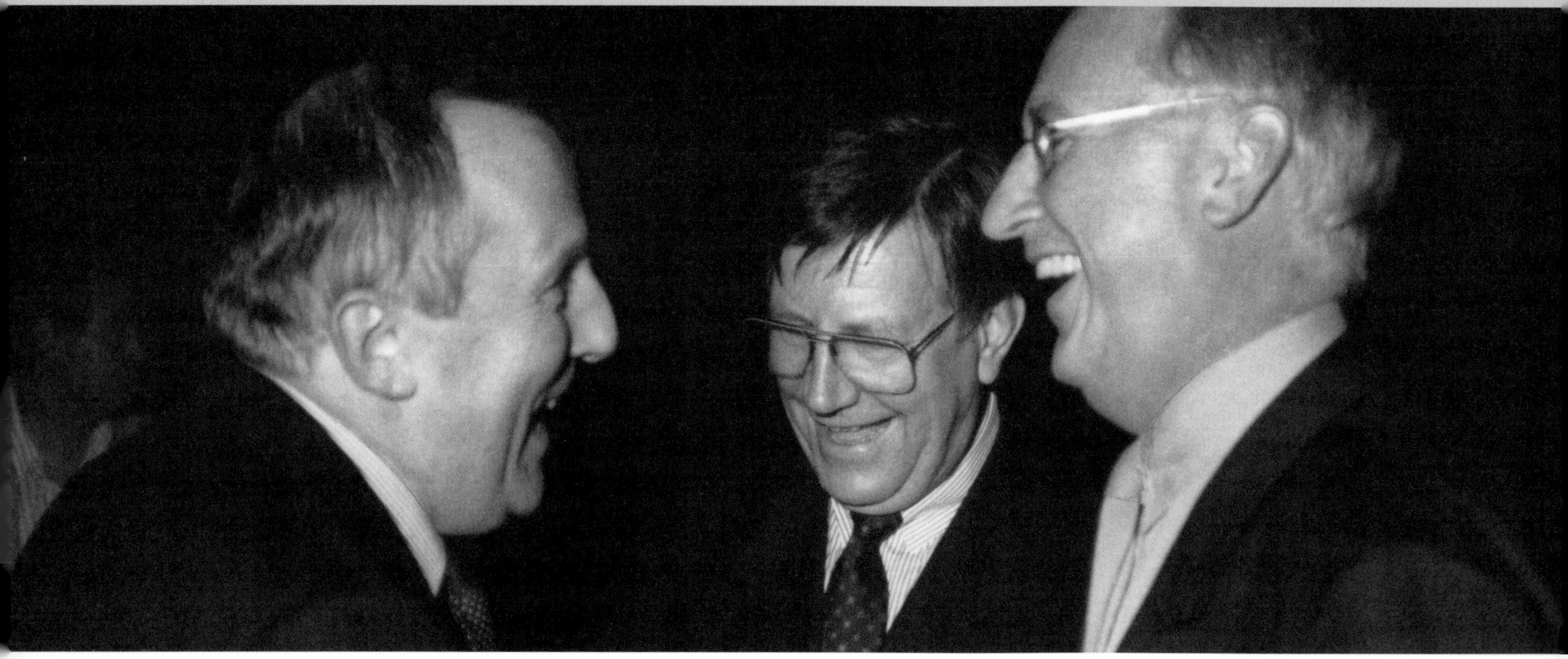

Die beiden Leitfiguren der wiege von 1985 bis 1992: Klaus Franck und Werner Sauer, die den umfassenden Gestaltungsanspruch von Wilkhahn sowohl in den Produkten wie auch auf Messen, Ausstellungen und im Grafikdesign umsetzten. Bis heute vollzieht sich der **Design- und Entwicklungsprozess nach der „Ulmer Lehre"**: Genaue Erforschung und präzise Analyse des Nutzungskontexts, die Formulierung der Produktidee und die Entwicklung von ersten Funktions- und Designmodellen. Dann erst beginnt der eigentliche Entwicklungsprozess, der von Konstruktions- und Designmodellen im Maßstab 1 : 1, von Testvorrichtungen und Materialproben begleitet wird.

The two key figures at wiege from 1985 to 1992: Klaus Franck and Werner Sauer who implemented Wilkhahn's all-embracing design principles into products as well as at trade fairs, exhibitions and in graphic design. Until today, the design and **development process at Wilkhahn is carried out in line with "Ulm teachings"**: thorough research and precise analysis of the application profile required, formulation of the product idea and development of initial functional and design models. Only then does the real development process begin that is accompanied by constructional and design models on a 1 : 1 scale, testing equipment and material sampling.

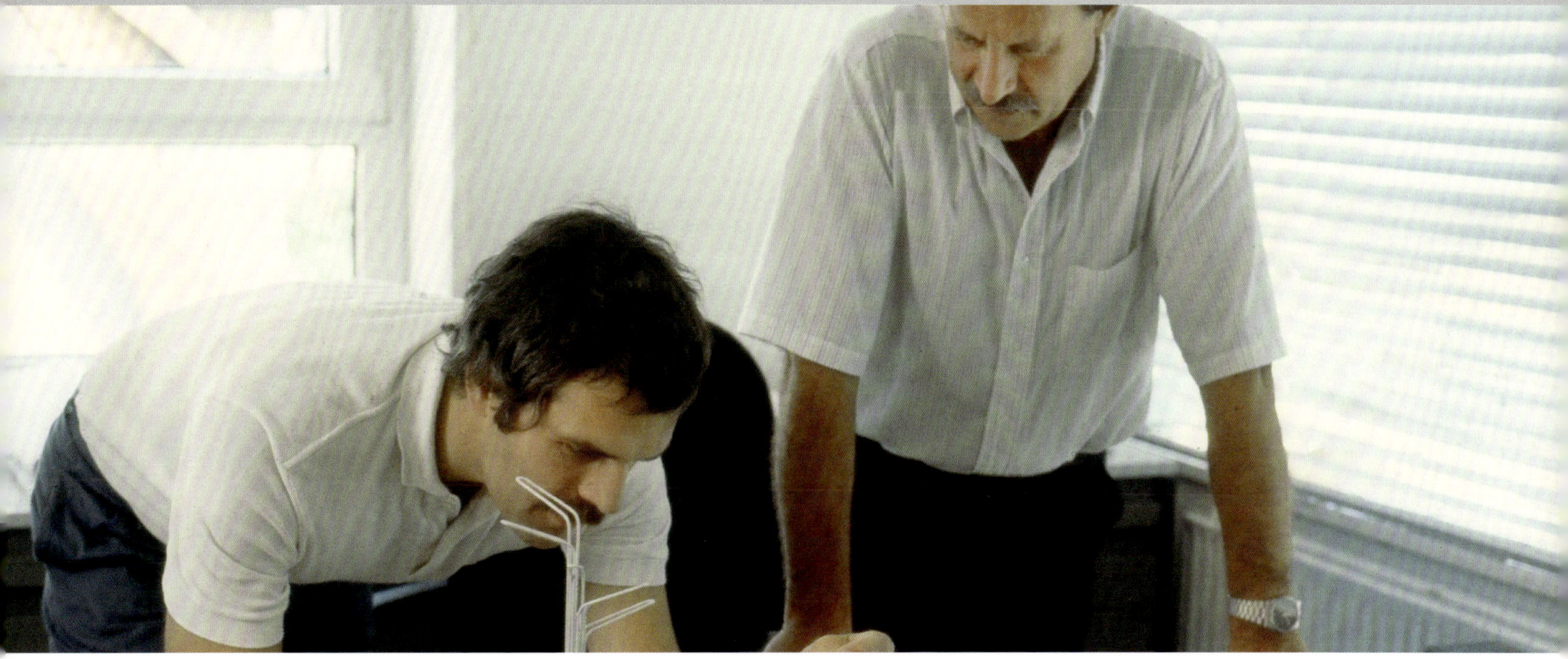

Die über Jahrzehnte gewachsene und profilierte Ausrichtung von Wilkhahn hat das Unternehmen in der internationalen Architektur- und Designszene zu einer festen, klar umrissenen Größe gemacht. Auf der Basis gemeinsamer Wertvorstellungen entwickelte sich ein spannender Diskurs, aus dem manche Freundschaft entstand. Das Bild zeigt Fritz Hahne im Gespräch mit **Herbert Schultes**. Schultes, Jahrgang 1938, hatte in München ein Ingenieur- und Designstudium abgeschlossen und in der Folgezeit vor allem dem Design von Siemens zu Weltgeltung verholfen. Von 1961 bis 2000 war er bei Siemens als Designer, Chefdesigner und schließlich Geschäftsführer der dann ausgegründeten Designabteilung tätig. Wie Fritz Hahne wurde auch Herbert Schultes für sein Lebenswerk mit dem Bundespreis für Förderer des Designs (2002) ausgezeichnet.

The positioning of Wilkhahn, which has grown and taken on a profile over the course of decades, has led to the company's becoming an established, clearly defined leading figure on the design scene. Based on mutual ideals, an exciting dialogue emerged that resulted in many a friendship. The photograph shows Fritz Hahne talking to **Herbert Schultes**. Schultes, born in 1938, studied mechanical engineering and design in Munich and then had a formative influence on helping Siemens' design to achieve world fame. From 1961 to 2000 he worked as a designer, Head Designer and finally as Managing Director in the Siemens Design Department, which had by then become an independent company. Like Fritz Hahne, Herbert Schultes was also presented with the "Federal German Award for Design Leadership" (2002).

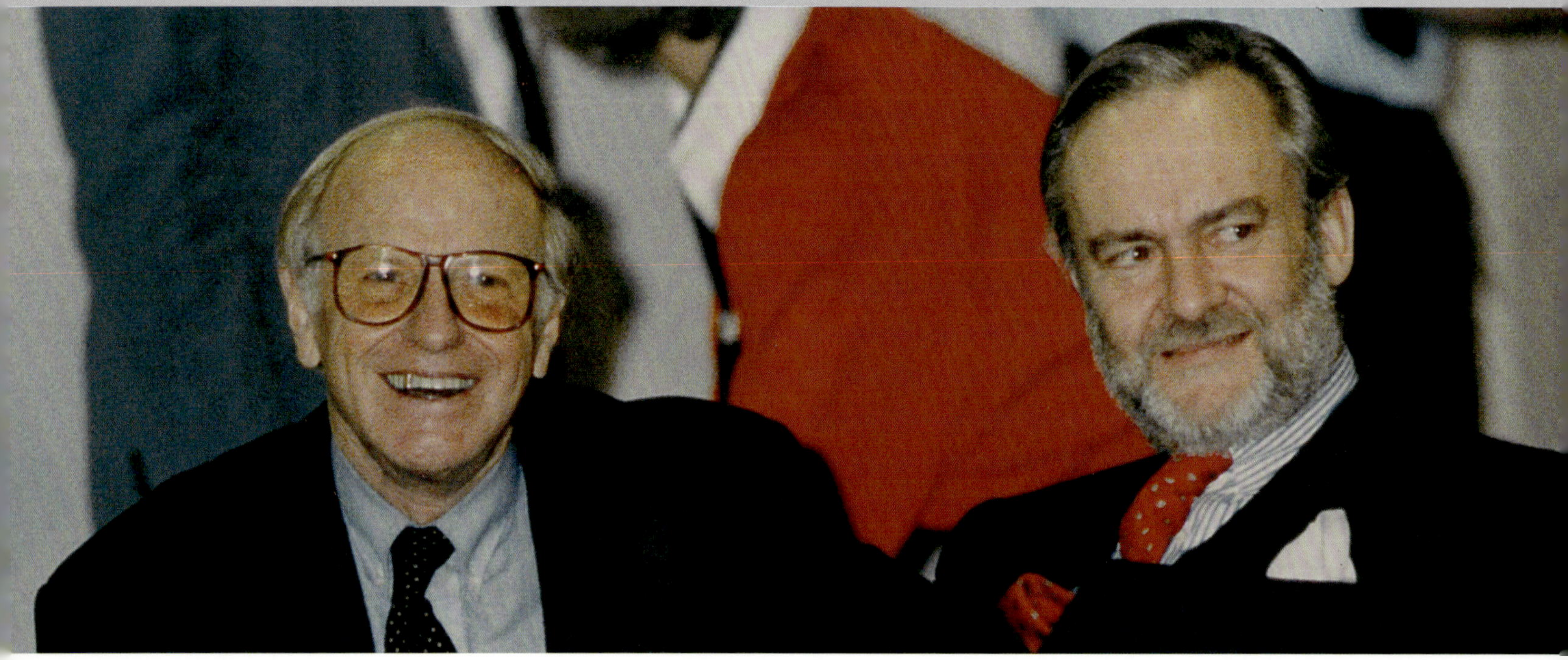

Die Wurzeln der Moderne liegen in Deutschland. Der Deutsche Werkbund (gegründet 1907) und das Bauhaus (1919–1933) haben die Grundlagen der klassischen Moderne entwickelt. Um dem deutschen Design nach dem 2. Weltkrieg wieder ein Forum zu bieten, wurden Institutionen wie der Rat für Formgebung und verschiedene Design Zentren, wie das **iF International Forum Design Hannover** und das **Design Zentrum Nordrhein Westfalen** gegründet, die heute auch international sehr renommierte Designpreise vergeben. Die Abbildung zeigt eine Veranstaltung anlässlich des 40. Geburtstages des iF Hannover. Mit dabei (von rechts nach links) Gerhard Schröder, damals Ministerpräsident von Niedersachsen, der renommierte Industriedesigner Prof. Herbert Lindinger, Fritz Hahne und Rudolf Stilcken, renommierter PR-Berater deutscher Top-Unternehmen.

The roots of the Modern Age can be found in Germany. The Deutscher Werkbund (founded in 1907) and Bauhaus (1919–1933) developed the foundation of Classic Modernism. With a view to providing a forum for German design after the Second World War, the Rat für Formgebung (German Design Council) and various design centres were founded such as the **iF International Forum Design Hanover** and the **Design Centre North-Rhine / Westphalia**, which today still awards, internationally highly acclaimed design prizes. The photograph shows an event to celebrate the 40th anniversary of the iF Hanover, attended by Gerhard Schröder (then Premier of Lower Saxony), Prof. Herbert Lindinger, a renowned industrial designer, Fritz Hahne and Rudolf Stilcken, an acclaimed PR adviser to top German companies.

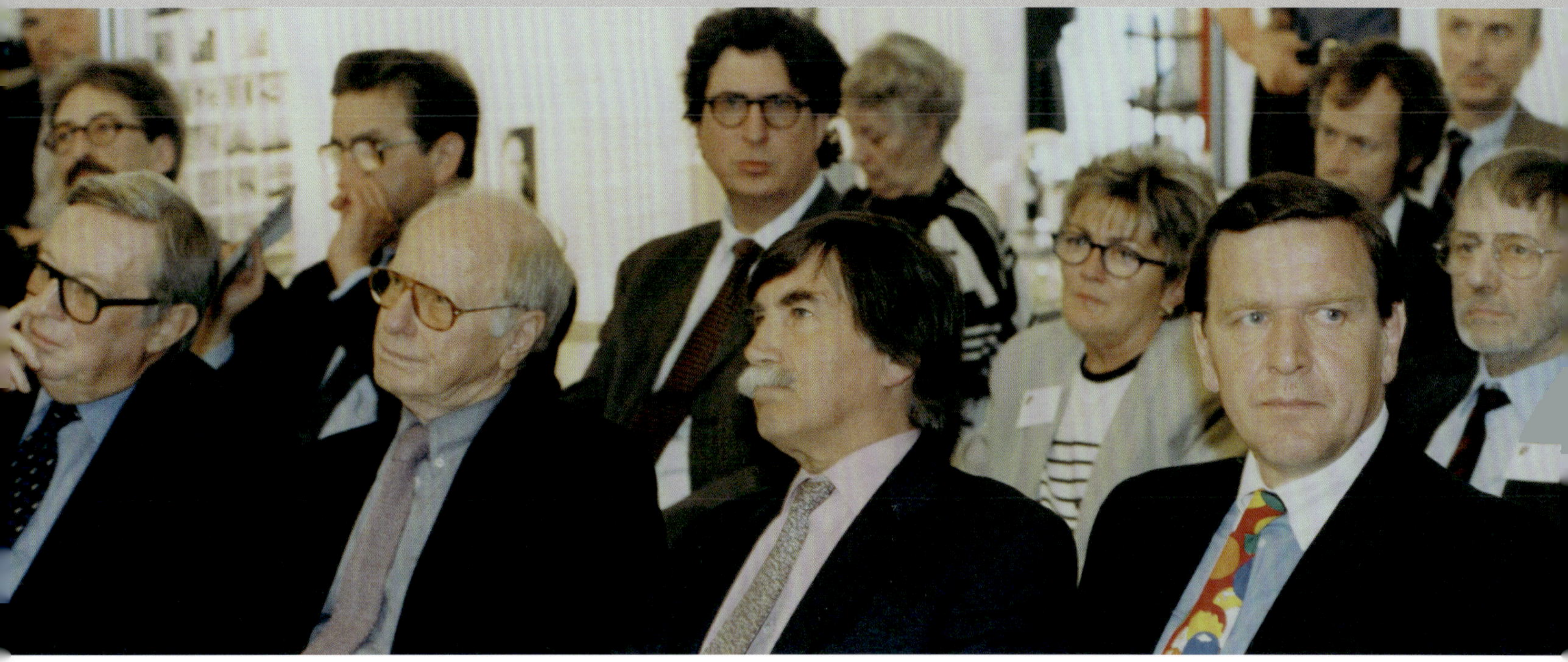

Markus Ferstera, Jahrgang 1954, hatte sich bei Wilkhahn seit 1982 vom Leiter des Innendienstes zum Vertriebsleiter entwickelt. Nach dem Ausscheiden von Theodor Diener übernahm er gemeinsam mit dem ehemaligen Bertelsmann-Manager Dr. Jürgen Krämer die Geschäftsführung. Als Geschäftsführer Gesamtvertrieb und Fertigung forcierte er gemeinsam mit Horst Knigge Entwicklungsprojekte, wie etwa eine Produktoffensive, die auf der Orgatec 1994 zur Präsentation von vier neuen Programmen führte. Wie für viele andere wurde Wilkhahn auch für Markus Ferstera zum Karriere-Sprungbrett: 1998 folgte er der Berufung als Geschäftsführer eines weitaus größeren Unternehmens der Küchenindustrie.

Markus Ferstera, born in 1954, had moved up at Wilkhahn from Head of Indoor Sales in 1982 to Sales Director. When Theodor Diener left the company, he took over Executive Management together with former Bertelsmann manager Dr. Jürgen Krämer. As Director Sales and Production he promoted development projects together with Horst Knigge, such as the product campaign that led to the launch of four new ranges at Orgatec 1994. As for many other people, Wilkhahn became the stepping stone to Markus Festera's subsequent career: in 1998 he went on to become the Managing Director of a much larger company in the kitchen appliances industry.

Karlheinz Rubner, Jahrgang 1948, studierte zunächst Maschinenbau an der Technikerschule für Maschinenbau in Göppingen, dann Design an der Fachhochschule Schwäbisch Gmünd.
Nach dem Diplom folgten die Mitarbeit in den Designbüros Schlagheck & Schultes Design, München, und bei ProduktEntwicklung Roericht, Ulm, zwischen 1980 und 1986 Lehraufträge an der Hochschule der Künste, Berlin, und an der Fachhochschule Schwäbisch Gmünd. Seit 1983 hat Rubner sein eigenes Designbüro. Für seine Arbeiten erhielt er zahlreiche Auszeichnungen, für das Programm Palette unter anderem den if Award 1990 und den Preis „Best of Catagory" des Design Zentrums Nordrhein Westfalen 1991. Das Programm markierte den Weg zum führenden Anbieter hochwertiger und intelligenter Konferenztischsysteme.

Karlheinz Rubner, born in 1948, first studied design and mechanical engineering at the Technikerschule für Maschinenbau in Göppingen, then design at the Fachhochschule Schwäbisch Gmünd. After completing his studies, he started to work for Schlagheck & Schultes Design in Munich, and for ProduktEntwicklung Roericht in Ulm. From 1980 to 1986 he lectured at the Hochschule für Künste in Berlin and at the Fachhochschule Schwäbisch Gmünd. Since 1993, Rubner has had his own design studio. He has won many awards for his work, for example, the iF Award 1990 for the Palette range and the "Best of Category" from the Design Centre North-Rhine/Westphalia in 1991. This range paved the way for Wilkhahn's becoming one of the leading providers of high-quality and intelligent conference table systems.

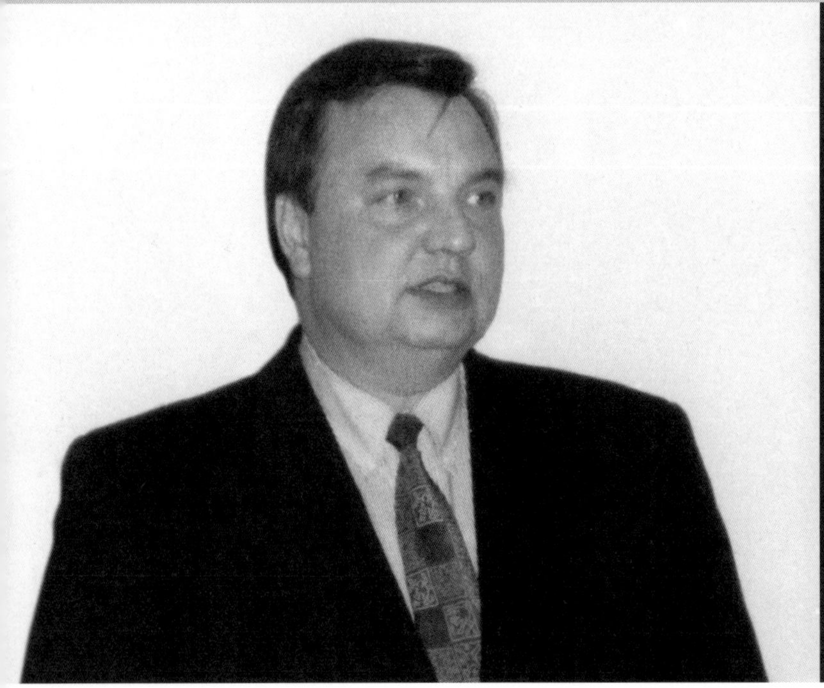

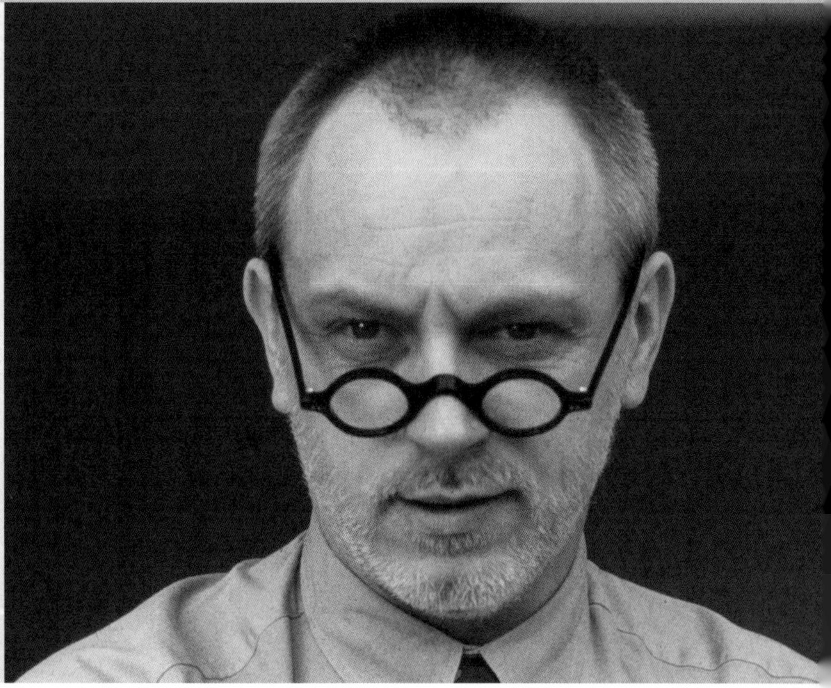

Zu einem der Höhepunkte und zum großen persönlichen Erfolg von Theodor Diener wurde 1992 **die Verleihung des Deutschen Marketingpreises** an das Unternehmen. Der seit 1972 jährlich ausgelobte Preis, über den eine hochkarätige Jury aus Wirtschaft und Politik entscheidet, wurde an Wilkhahn vergeben, „weil das Unternehmen ein überzeugendes Beispiel dafür ist, wie mit einer Unternehmenskultur, die umfassendes Marketing und erstklassige Produkte bündelt, eine Sonderstellung erreicht wird, die Märkte in der ganzen Welt gewinnen lässt." Gerhard Schröder, Ministerpräsident von Niedersachsen und späterer Bundeskanzler, reiste zur Preisverleihung an, um Theodor Diener persönlich zu gratulieren.

The award of the German Marketing Prize to the company in 1992 was one of the highlights in the corporate history and meant great personal success for Theodor Diener. This prize, awarded yearly since1972 by a jury from the world of business and politics, was awarded to Wilkhahn because "the company is a convincing example of how a corporate culture can use the joint forces of an extensive marketing programme and first-class products to create a special exhibition that enables markets to be won throughout the world." Gerhard Schröder attended the award ceremony to congratulate Theodor Diener in person.

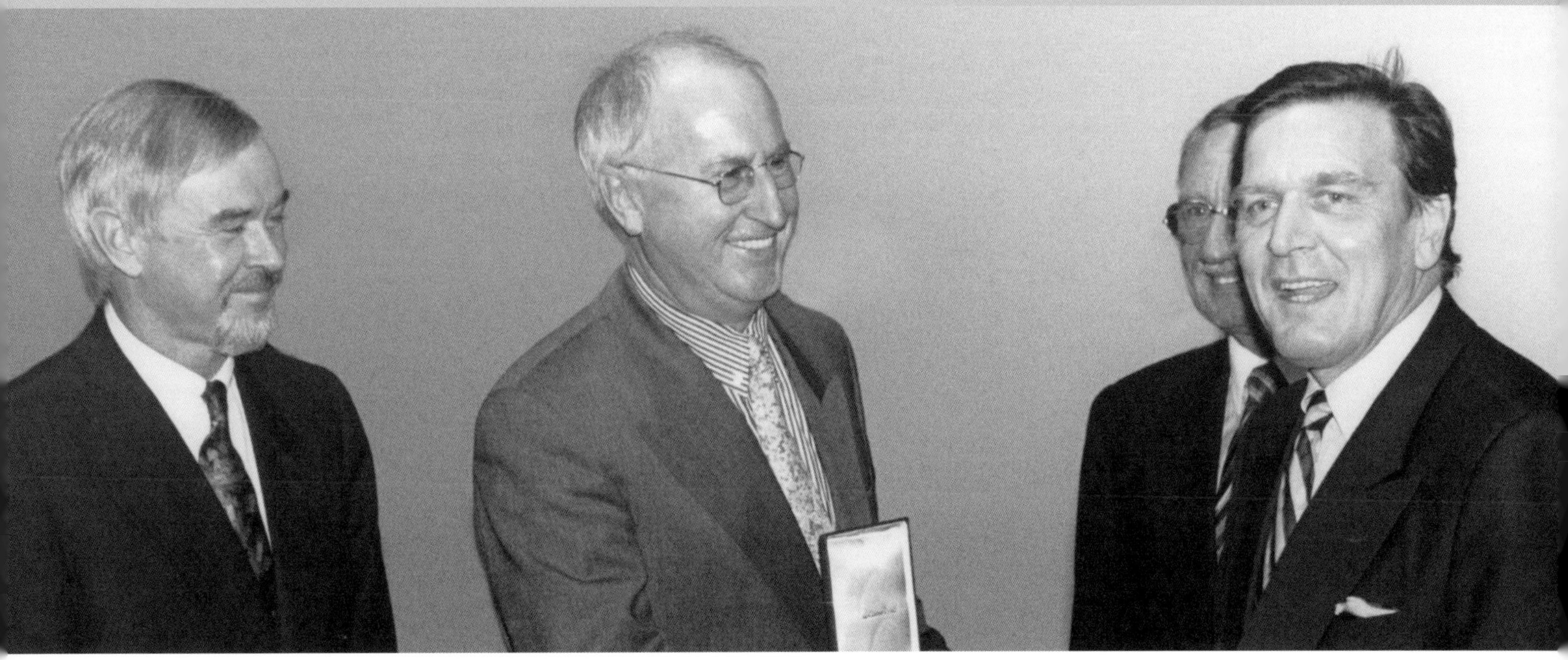

Horst Knigge, Jahrgang 1931, wurde als Betriebsratsvorsitzender in den 90er Jahren zur treibenden Kraft bei der Konzeption und Einführung moderner Arbeitsformen bei Wilkhahn. Sein Selbstverständnis als Co-Manager der Geschäftsführung und seine Mitgliedschaft im Verwaltungsrat machten ihn zu einer der wichtigsten Persönlichkeiten bei Wilkhahn. Neben der materiellen und immateriellen Beteiligung der Mitarbeiter wurden durch ihn die ökologische Ausrichtung und ein moderner Gesundheitsschutz fest in der Mitarbeiterschaft verankert. Seine konstruktive Mitarbeit an Veränderungsprozessen, die intime Kenntnis der Produktion und der Rückhalt im Unternehmen führten 1998 zur Berufung als Geschäftsführer und Technischer Leiter. 2004 schied Knigge nach 49 Jahren Mitarbeit in den Ruhestand aus.

Horst Knigge, born in 1931. As Chairman of the Works Council he was the driving force in the concept formulation and introduction of modern forms of work at Wilkhahn in the 1990s. His understanding of his role as Co-manager to Executive Management and his membership of the Administrative Board made him one of the key figures at Wilkhahn. In addition to material and non-material employees' participation, it is thanks to him that ecological focus and a modern system of health care and protection have been firmly established at the company. His constructive cooperation on change processes, his first-hand knowledge of production and the support and respect he gained within the company resulted in his being appointed a member of Executive Management and Technical Director in 1998. In 2004, Horst Knigge retired after 49 years of service at Wilkhahn.

Dr. Jochen Hahne, Jahrgang 1958, studierte in Hannover Wirtschaftswissenschaften. Dem Abschluss als Diplomökonom folgten mehrere berufliche Stationen bei Beratungsunternehmen, als kaufmännischer Leiter und späterer Geschäftsführer eines Unternehmens in der Mikroelektronik-Branche. Parallel promovierte er am Institut für angewandte Innovationsforschung, Bochum. Mit 35 Jahren trat er in das Unternehmen ein, übernahm die Vertriebsverantwortung für Fernost und wurde kaufmännischer Geschäftsführer. Mit der Gründung von Wilkhahn Asia Pacific schuf er die strategische Basis für den Markterfolg in Australien und Asien. Seit Januar 2000 verantwortet Jochen Hahne als Vorsitzender der Wilkhahn-Geschäftsführung auch die Produktentwicklung.

Dr. Jochen Hahne, born in 1956, studied economics in Hanover. After graduating, he gained experience at various consulting companies, as Commercial Director and later as Managing Director of a company in the micro-electronics industry. Parallel to this, he completed his PhD at the Institut für angewandte Innovationsforschung in Bochum. At the age of 35 he joined Wilkhahn, took over sales responsibility for the Middle East and became Commercial Director. With the founding of Wilkhahn Asia Pacific, he created the strategic basis for marketing success in Australia and Asia. Since January 2002, Jochen Hahne has been Wilkhahn Chairman and CEO, responsible for product development too.

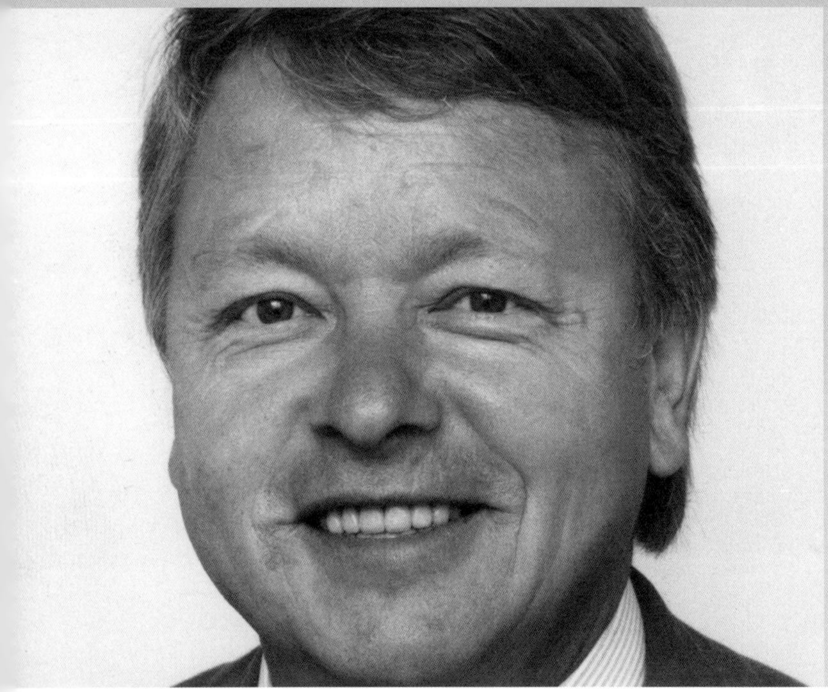

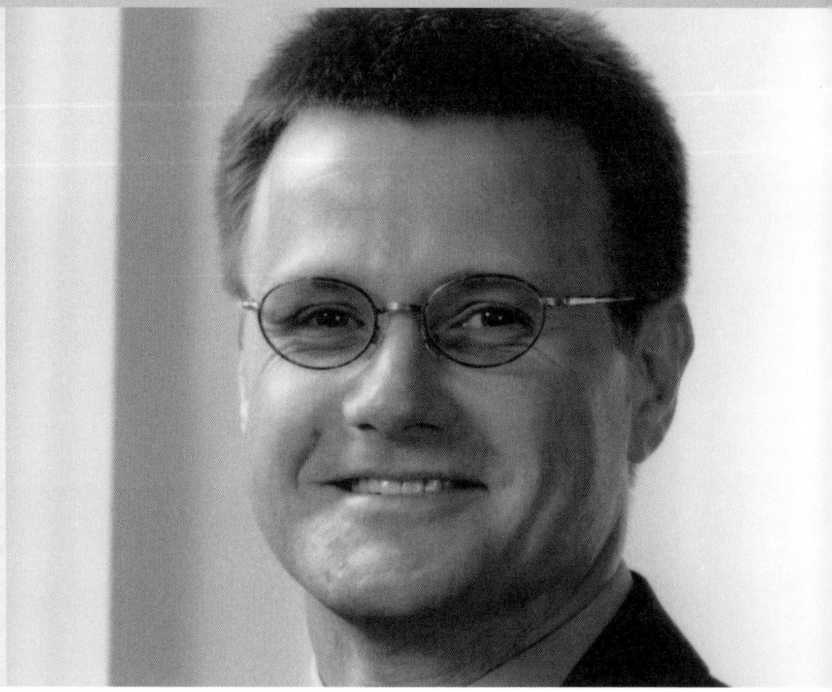

Carl-August Sautier, geboren 1947, hatte als gelernter Kaufmann bei Wilkhahn eine erstaunliche Karriere vom Verkäufer zum Marketingleiter durchlaufen. 1988 wurde er Leiter der Produktentwicklung, die 1990 in den neu geschaffenen Geschäftsführungsbereich „Innovation und Ökologie" integriert wurde. Zu den herausragenden Leistungen des engagierten Quer- und Vordenkers zählen die Entwicklungsleitung für das Bürostuhlprogramm Picto, der als erster Bürostuhl der Welt konsequent ökologische Belange berücksichtigte, und für das Programm Confair, das einen neuen Standard für die dynamische und selbstorganisierte Konferenzeinrichtung setzte. Nach knapp 30 Jahren bei Wilkhahn entschied sich Sautier 1997, eine neue berufliche Herausforderung als Geschäftsführer eines anderen Unternehmens anzunehmen.

Carl-August Sautier, born in 1947. After completing commercial training, he went on to pursue an amazing career at Wilkhahn, moving up from a salesman to Head of Marketing. In 1988 he became Head of Product Development which was integrated into the newly founded division of "Innovation and Ecology" in 1990. Some of the outstanding achievements of this committed lateral and proactive thinker include managing the development of the Picto office chair, the first office chair in the world to consider ecological needs in a consistent manner, and the Confair range that set a new standard for dynamic and self-organized conference furniture. After almost 30 years at Wilkhahn, Sautier decided to leave the company in 1997 to accept a new challenge as director of another company.

Guido Englich, geboren 1961, studierte in Berlin zunächst Germanistik und Theaterwissenschaften, dann Industrial Design an der Hochschule der Künste, Berlin. Von 1987 bis 1990 folgten die Mitarbeit bei ProduktEntwicklung Roericht, Ulm, und bis 1992 eine Kooperation mit Baumm + Partner Produktentwicklung, München. Er gründete 1988 die Design-Werkstatt MacLab, seit 1990 arbeitet er freiberuflich im eigenen Designbüro. Nach Lehraufträgen an der Hochschule der Künste, Berlin, erhielt er zunächst eine Gastprofessur und 2004 die Berufung als Professor für Strategische Produkt- und Konzeptentwicklung an die Burg Giebichenstein Hochschule für Kunst und Design Halle. Guido Englich begleitet Wilkhahn seit vielen Jahren. Unter seiner Federführung entstanden unter anderem das „Einrichtungs-Handbuch für Kommunikationsräume" und die Ausstellungsgrafik des EXPO-2000-Projekts.

Guido Englich, born in 1961, first studied German and drama in Berlin, then industrial design at the Hochschule für Künste in Berlin. From 1987 to 1990 he worked together with ProduktEntwicklung Roericht in Ulm, and until 1992 he collaborated with Baumm + Partner Produktentwickung in Munich. In 1988 he founded Design-Werkstatt MacLab and since 1990 he has been free-lancing in his own design studio. Following lectureships at the Hochschule der Künste, Berlin, he was a guest professor there and in 2004 was appointed Professor for Strategic Product and Concept Development at the Burg Giebichenstein Hochschule für Kunst und Design in Halle. Guido Englich has been working for Wilkhahn for many years. Amongst other projects, he was the creative mind behind the "Interior and specifiers' handbook for communication areas" and the graphics side of the EXPO 2000 project.

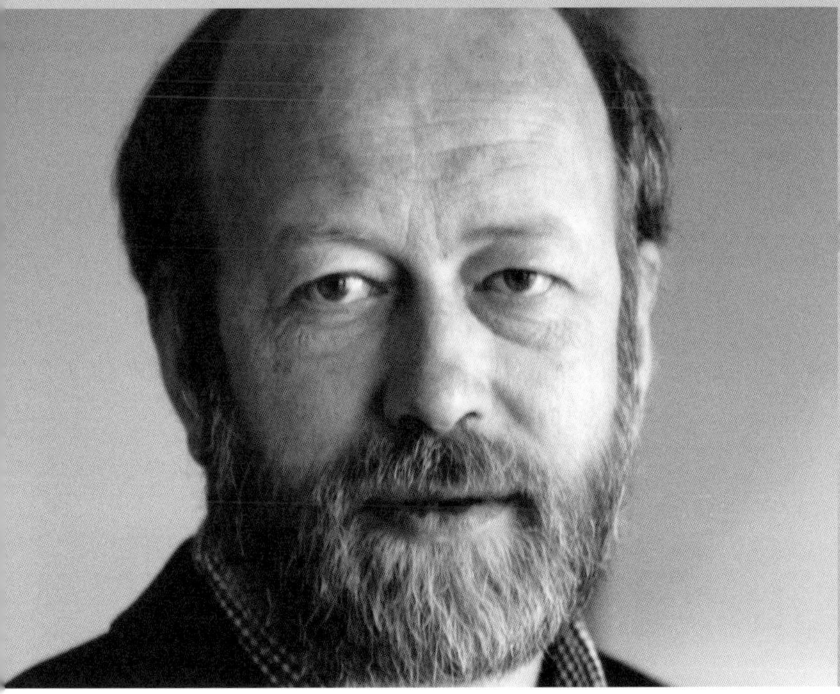

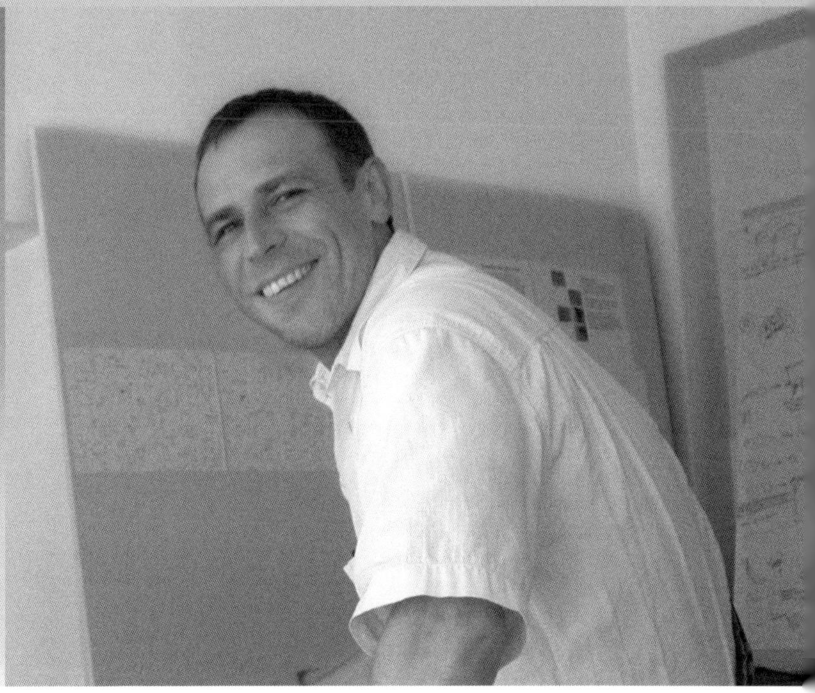

Fritz Frenkler, Jahrgang 1954, studierte in Braunschweig Industriedesign. 1983 begann er als Designer bei Frog Design, übernahm die Leitung des deutschen Büros und baute die Niederlassung in Tokio auf. 1992 wurde er in der Nachfolge von Klaus Franck Geschäftsführer der wiege, 1999 Chefdesigner der Deutschen Bahn AG. 2001 gründete er in Frankfurt das Büro „f/p design". Zu seinen Kunden zählen unter anderem AEG, Alessi, Apple, Hans Grohe, Yamaha, Wilkhahn und Volkswagen. Fritz Frenkler gehört der Kommission des International Council of Societies of Industrial Design und weiteren Designinstitutionen an. Seit 2006 ist er Professor am Lehrstuhl Industrial Design an der Technischen Universität München. Frenkler hat die wiege als eigenständiges und profiliertes Designunternehmen etabliert, international positioniert und zahlreiche Designpreise für die wiege-Kunden gewonnen.

Fritz Frenkler, born in 1954, studied industrial design in Brunswick. In 1983 he started as a designer at Frog Design, took over the management of the German office and built up the branch in Tokyo. In 1992 he succeeded Klaus Franck as Managing Director of wiege, and in 1999 he became Head Designer of Deutsche Bahn AG. In 2001 he founded the "f/p design" office in Frankfurt. His clients included AEG, Alessi, Apple, Hans Grohe, Yamaha, Wilkhahn and Volkswagen. Fritz Frenkler was a member of the Commission of the International Council of Societies of Industrial Design at other design institutions. Since 2006 he has been Professor for Industrial Design at the Technical University of Munich. Frenkler established wiege as an independent and high-profile design company, positioned it internationally and won numerous design awards for wiege clients.

Manfred Schmitz, Jahrgang 1940, machte eine kaufmännische Lehre und absolvierte im Anschluss die Kölner Möbelfachschule. Nach einer Blitzkarriere beim amerikanischen Büromöbelhersteller Herman Miller baute er als Geschäftsführer das deutsche Tochterunternehmen des Schweizer Möbelsystemherstellers USM zu einer der bekanntesten Marken im Möbelmarkt auf. 1995 übernahm Schmitz den Vorsitz der Wilkhahn-Geschäftsführung. Öffentlicher Höhepunkt war die Entgegennahme des **Deutschen Umweltpreises**, mit dem die ganzheitliche Umweltorientierung des Unternehmens Wilkhahn ausgezeichnet wurde. Wichtiger für ihn war, dass es in kurzer Zeit gelang, die Prozesse zu restrukturieren, das Profil des Unternehmens zu schärfen und die Übernahme durch Jochen Hahne vorzubereiten.

Manfred Schmitz, born in 1940, completed a commercial apprenticeship and then attended the Möbelfachschule in Cologne. After a lightning career at American office furniture manufacturers Hermann Miller, in his capacity as Director he expanded the German subsidiary company of the Swiss furniture system manufacturer USM to make it one of the best known brands in the furniture market. In 1995 Schmitz became Wilkhahn Managing Director. The public highlight of his career was receiving the **German Ecology Prize** on behalf of the company, awarded for its holistic environmental focus. It was however more important for him to have succeeded in restructuring processes within a very short period, in honing corporate profile and paving the way for Jochen Hahne to take over.

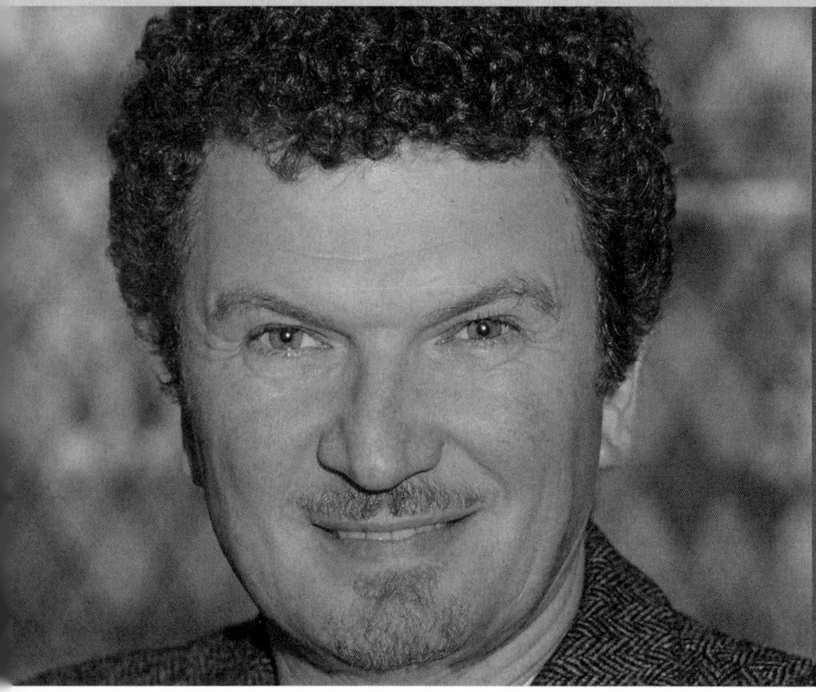

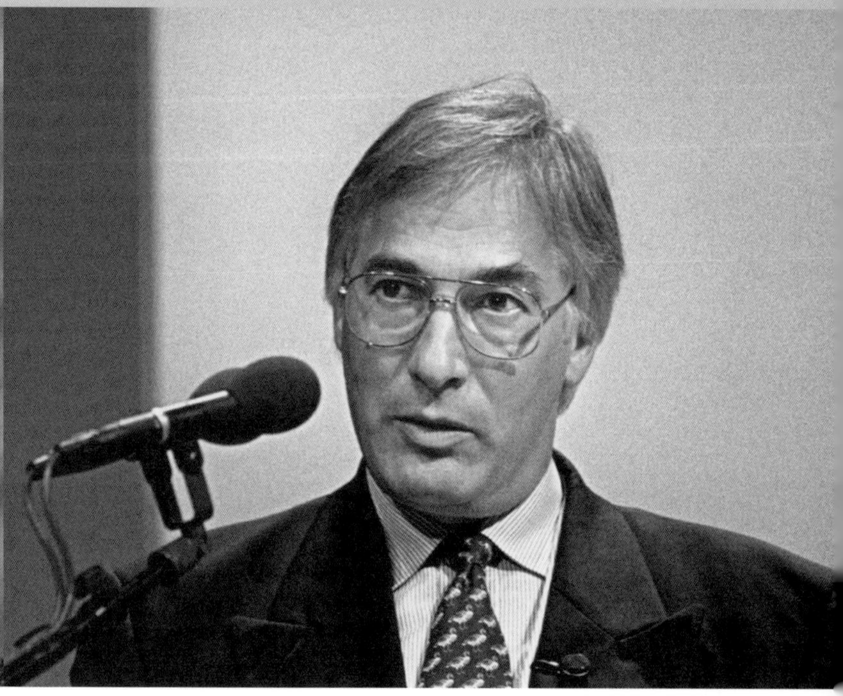

Burkhard Remmers, geboren 1960, studierte Deutsch und Geschichte an der Universität Augsburg. Nach dem Staatsexamen übernahm er beim Möbelsystemspezialisten USM Deutschland den Bereich Marketing und Presse. 1995 wechselte er zu Wilkhahn. Er war wesentlicher Initiator des Forschungs- und Entwicklungsprojekts „Future Office Dynamics" und hat neben dem Tagesgeschäft mit zahlreichen Fachpublikationen, mit Buchveröffentlichungen und mit Vorträgen die Position von Wilkhahn als führendem Unternehmen weltweit ausgebaut, beispielsweise mit der Konzeption und Umsetzung eines der ersten Nachhaltigkeitsberichte. Seit 2000 verantwortet er die Bereiche Kommunikation und Unternehmensentwicklung, seit 2006 auch das Internationale Marketing.

Burkhard Remmers, born in 1960, studied German and history at the University of Augsburg. After graduating he took over the Marketing and Press Department at furniture system specialists USM Germany. In 1995 he changed to Wilkhahn. He was the key initiator of the R & D project "Future Office Dynamics", and in addition to daily business, he has helped to expand Wilkhahn's position as a leading international company with the aid of numerous special publications, for example with the concept and implementation of one of the first sustainability reports, and specialized publications of books and a series of lectures. Since 2000 he has been responsible for Communications and Corporate Development, and since 2006, for International Marketing too.

Andreas Störiko, geboren 1965 in Köln, studierte Architektur und Produktdesign an der Staatlichen Akademie der Bildenden Künste, Stuttgart. Noch während seines Auslandsstipendiums an der Cranbrook Academy of Art, Bloomfields Hills, Michigan, entwarf er für Wilkhahn den berühmten Falttisch im Programm Confair. Es folgte eine kurze Assistenzzeit bei Richard Sapper, Mailand, bevor er 1994 in Mailand sein eigenes Designstudio gründete. Seine Entwürfe gehören zu den bestverkauften und mit vielen internationalen Designpreisen ausgezeichneten Programmen von Wilkhahn. Nach dem Falttisch entwickelte er das Tischsystem Logon (1996), den mobilen, schwenkbaren Seminartisch Timetable (2000) und mit dem Programm Aline (2004, 2006) einen weiteren Bestseller. Störiko lebt und arbeitet in Mailand, München und Stockholm.

Andreas Störiko, born in Cologne in 1965, studied architecture and product design at the Staatliche Akademie der Bildenden Künste in Stuttgart. While still on a scholarship abroad at the Cranbrook Academy of Art Bloomfields Hills, Michigan, he designed the famous folding table in the Confair range for Wilkhahn. A short period as assistant to Richard Sapper, Milan, followed before he set up his own design studio in Milan in 1994. His designs are some of the best-selling Wilkhahn ranges and have won many international design awards. After the folding table, he developed the Logon table system (1996), Timetable, a mobile, pivotable seminar table (2000) and the Aline range (2004, 2006), yet another best-seller. Störiko lives and works in Milan, Munich and Stockholm.

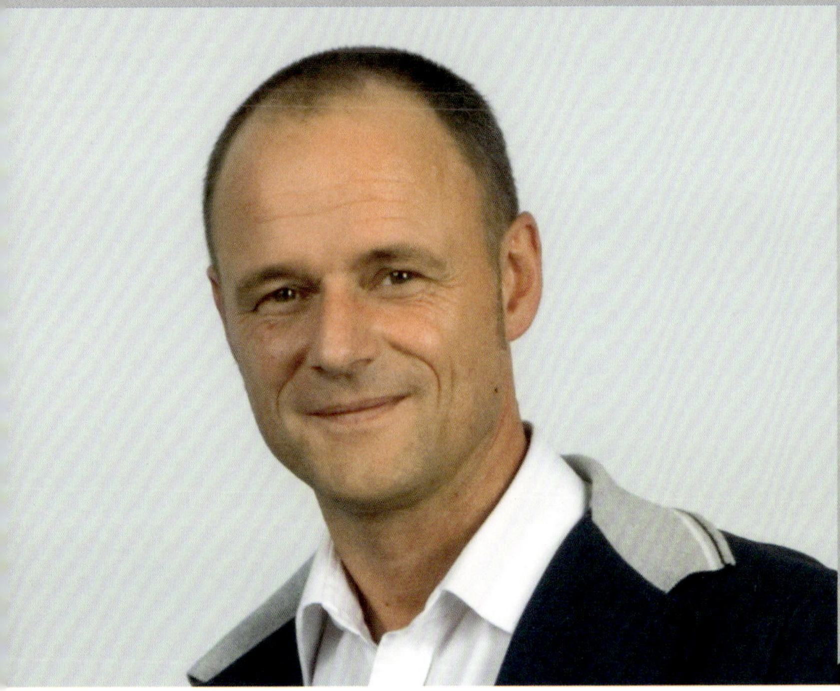

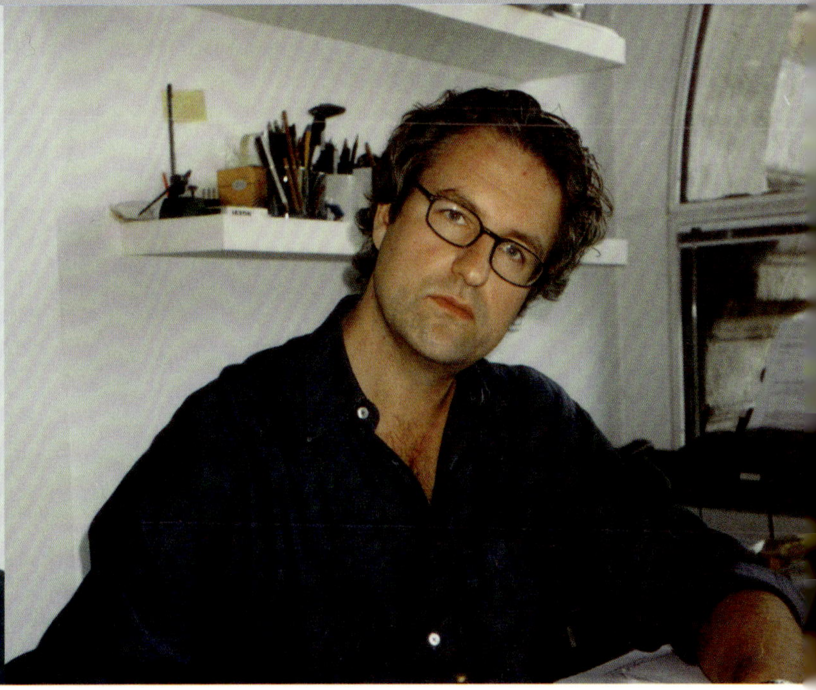

Mit der Ausgründung der **wiege Entwicklungsgesellschaft mbH** verfolgte Fritz Hahne eine ebenso einfache wie wichtige Idee: Zum einen sollte für externe Designer die Hemmschwelle gesenkt werden, ihre Entwürfe bei Wilkhahn vorzustellen. Zum anderen sollten durch ein breites Betätigungsfeld auch außerhalb der Möbelgestaltung der Horizont erweitert und neue Impulse gegeben werden. Beides ist Wirklichkeit geworden. Die wiege steht im fruchtbaren Wettbewerb mit anderen Designern, mit denen Wilkhahn arbeitet, und ihr Geschäftsfeld umfasst Kunden aus ganz unterschiedlichen Branchen und Bereichen: von Messegesellschaften über Unterhaltungselektronik- und Automobilhersteller bis zu öffentlichen Auftraggebern, von der Produktgestaltung über Messeauftritte bis zum Interfacedesign.

By founding the **wiege Entwicklungsgesellschaft mbH**, Fritz Hahne pursued an idea that was both simple and significant: on the one hand, the inhibition threshold for external designers to present their ideas to Wilkhahn was to be lowered. On the other hand, a wide field of activity outside furniture design should serve to widen horizons and give new impetus. Both became true. wiege competes productively with other designers with whom Wilkhahn works, and its field of activity includes clients from quite different branches and areas: ranging from exhibition management, home entertainment products, car manufacturers to public commissions, product design, exhibition platforms and interface design.

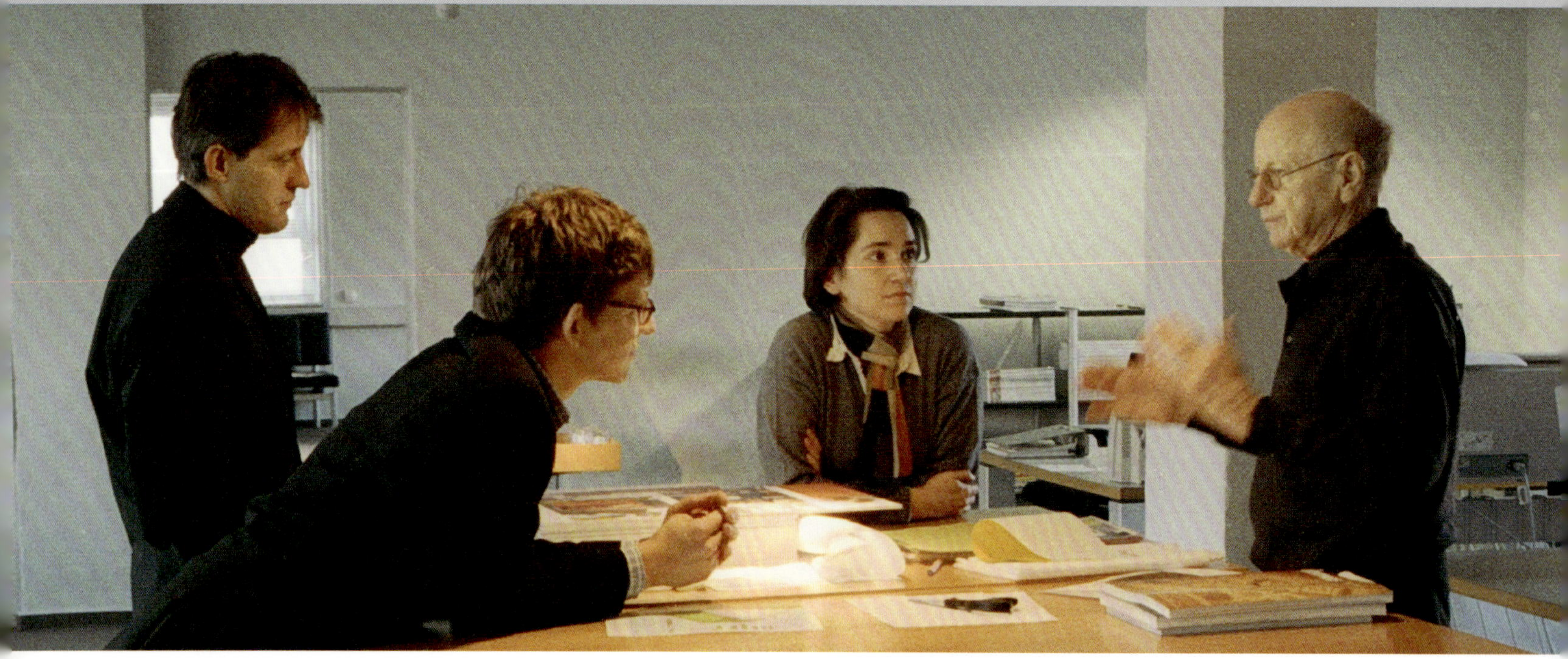

Fritz Hahne, Jahrgang 1920, gilt als Initiator der Orientierung am Design-Anspruch. Er hat die Entwicklung des Unternehmens entscheidend geprägt. Die 1971 eingeführte Beteiligung der Mitarbeiter am Betriebsergebnis geht ebenso auf ihn zurück wie der partnerschaftlich orientierte Führungsstil und die ökologische Ausrichtung. 1982 zog sich Fritz Hahne aus dem Tagesgeschäft zurück, übernahm bis Mitte 1994 den Vorsitz des Verwaltungsrates und ist heute dessen Ehrenvorsitzender. Für sein sozialpolitisches Engagement wurde Fritz Hahne mit dem Bundesverdienstkreuz ausgezeichnet. Viele Jahre war er Vorstandsmitglied des Rates für Formgebung, des Internationalen Design Zentrums Berlin (IDZ) und Kuratoriumsmitglied des Bauhaus-Archivs. 1992 wurde er vom European Design Council mit der Nominierung für den Sonderpreis des Europäischen Design-Preises geehrt, das Land Niedersachsen verlieh ihm das Große Niedersächsische Verdienstkreuz. Seit 1998 ist er außerordentliches Mitglied des Bundes deutscher Architekten, seit 2005 Ehrenmitglied des Deutschen Werkbunds Nord. Höhepunkt der Würdigung seiner Verdienste war 1999 die Verleihung des **Bundespreises für Förderer des Designs** durch Siegmar Mosdorf, Parlamentarischer Staatssekretär im Bundesministerium für Wirtschaft und Technologie.
Fritz Hahne lebt mit seiner Frau, Prof. Carola Stockmann, in Hannover.

Fritz Hahne, born in 1920, is regarded as the initiator of concentration on a special design approach, and he had a formative influence on corporate development. He was responsible for the introduction of a profit-sharing scheme for employees in 1971, a style of management that focused on fair cooperation and the ecological positioning of the company. In 1982 Fritz Hahne withdrew from daily operations, became Chairman of the Administrative Board until 1994 and is today its Honorary Chairman.
He was awarded the Bundesverdienstkreuz (Order of Merit of the Federal Republic of Germany) for his socio-political commitment. For many years he was a member of the Board of the German Design Council, the International Design Centre in Berlin (IDZ) and a member of the Board of Trustees of the Bauhaus Archive. In 1992 he was nominated for the special prize of the European Design Award by the European Design Council, and the state of Lower Saxony awarded him the Great Order of Merit of Lower Saxony. Since 1998 he has been a special member of the Association of German Architects, and since 2005 Honorary Member of the Deutscher Werkbund Nord. The highlight of his merit awards was in 1999 when he was presented with **the Federal German Award for Design Leaderships** by Siegmar Mosdorf, Parliamentary Undersecretary at the German Ministry of Economics and Technology.
Fritz Hahne lives in Hanover together with his wife, Prof. Carola Stockmann.

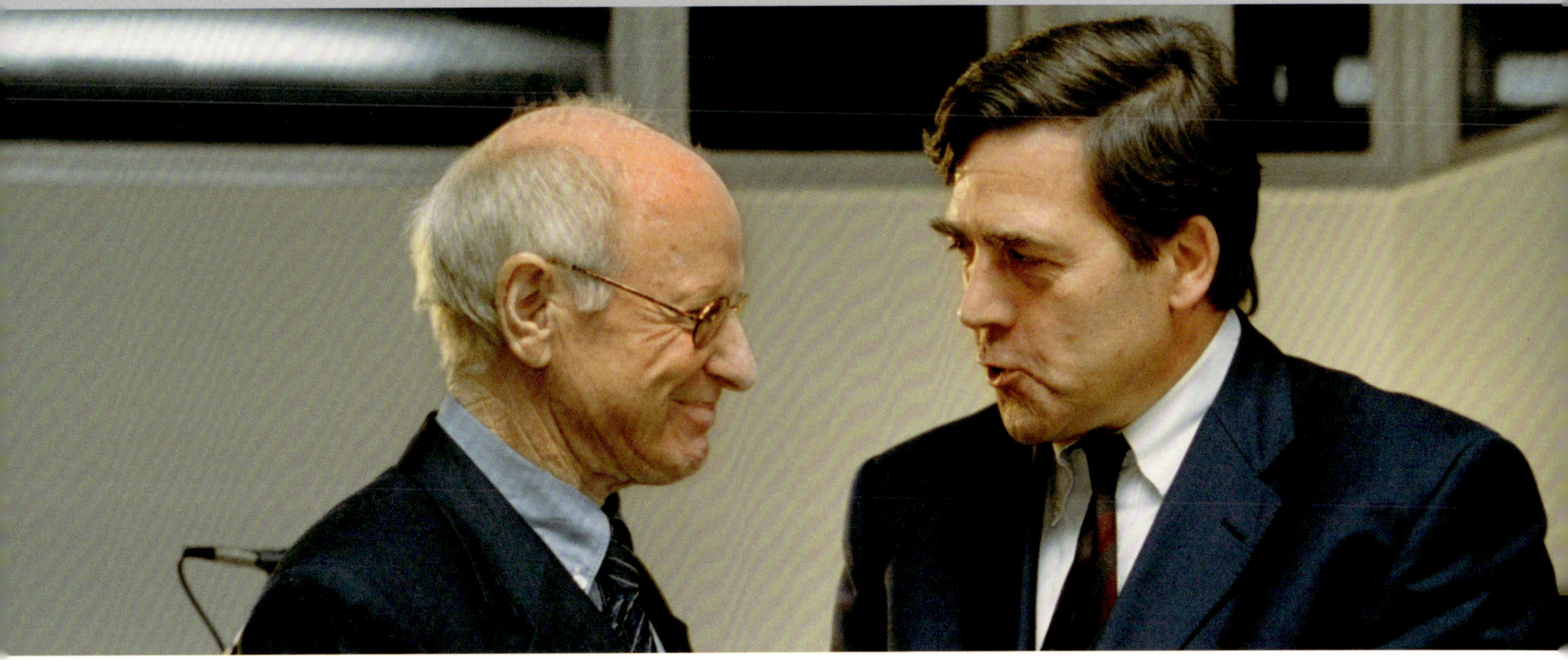

Michael Englisch (Mitte), geboren 1961, studierte Industrial Design an der Fachhochschule für Gestaltung, Schwäbisch Gmünd. Er diplomierte mit einer Laseranlage zur Materialbearbeitung und war Dozent für CAD, bevor er 1988 als Designer beim Büromöbelhersteller VOKO angestellt wurde. 1991 wurde er Leiter Design + Entwicklung beim Tochterunternehmen Elka, 1992 Leiter von VOKO Design und 1995 VOKO Design- und Marketingchef. Nach der Gründung des Designbüros industrial design Michael Englisch 1998 wurde er 1999 Nachfolger von Fritz Frenkler und übernahm die Geschäftsführung der wiege Entwicklungsgesellschaft mbH. Mehr als jeder seiner Vorgänger versteht Michael Englisch den Designprozess vor allem als Teamaufgabe. Unter seiner Leitung entwickelte die wiege das erfolgreiche Freischwingerprogramm Sito, die vielfach ausgezeichneten Bürostuhllinien Solis und Solis F, die Bürostuhlfamilie Neos, den Freischwingersessel Cura und das neue Tischsystem Travis.

Michael Englisch (centre), born in 1961, studied industrial design at the Fachhochschule für Gestaltung Schwäbisch Gmünd. His diploma paper was on a laser device for material processing: He was a lecturer for CAD before joining office furniture manufacturer VOKO as a designer in 1998. In 1999 he became Head of Design and Development at Elka, a subsidiary company. In 1992 he became Head of VOKO Design and in 1995 Head of Design and Marketing at the same company. After founding the "industrial design" office in 1998, Michael Englisch succeeded Fritz Frenkler in 1999 and took over the management of the wiege Entwicklungsgesellschaft mbH. More than anyone else, Michael Englisch understands the design process primarily as a team task. Under his leadership, wiege developed the successful Sito cantilever range, the prize-winning Solis and Solis F office chair ranges, the Neos office chair family, the Cura cantilever chair and the new Travis table system.

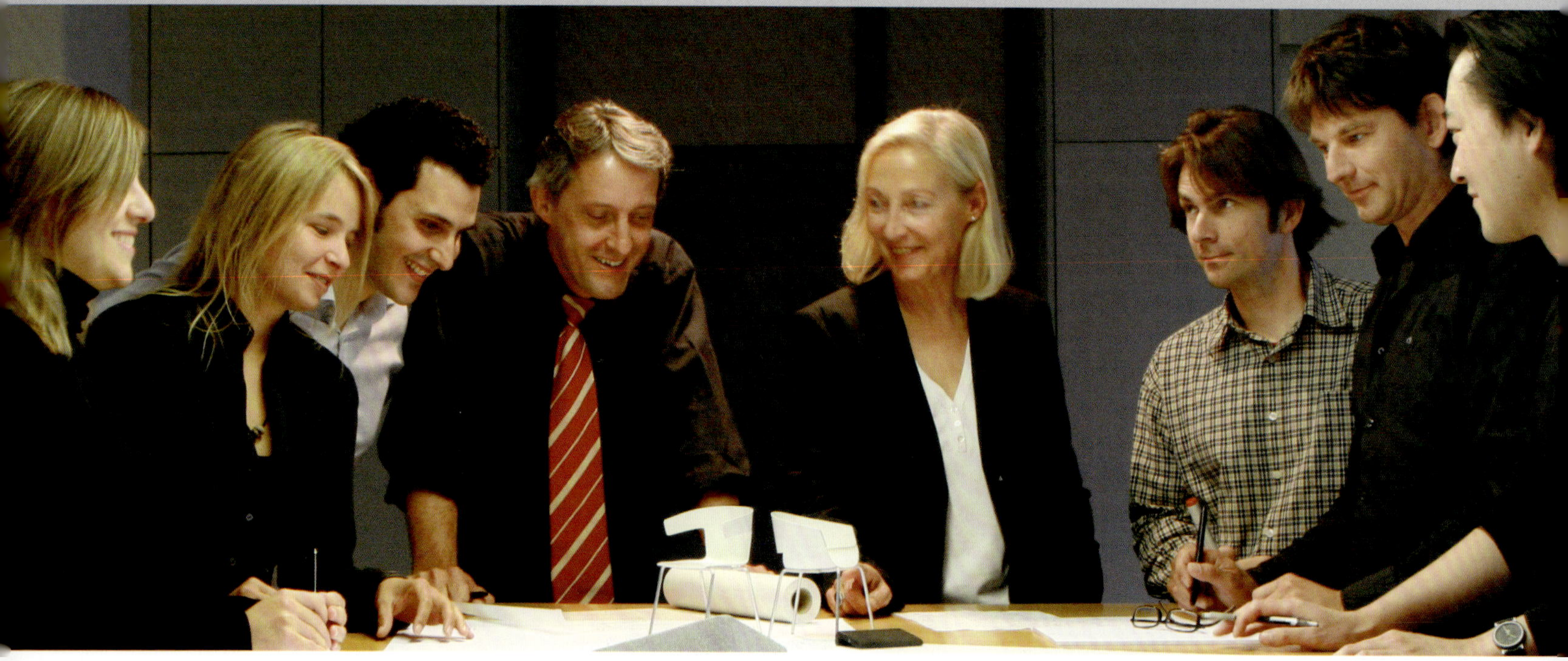

Norbert Ruf, geboren 1968, studierte Industrial Design an den Fachhochschulen Dortmund und Hannover. 1997 begann er bei Wilkhahn als Vertriebsmitarbeiter im Export und übernahm 1999 die Geschäftsführung von Wilkhahn Österreich. Im Jahr 2001 wurde er mit dem Designmanagement betraut, um zukünftige Entwicklungsrichtungen auszuloten und passende Design- und Entwicklungsprojekte für Wilkhahn zu erschließen. Die Entwicklung des Aline-Programms ist im Wesentlichen sein Verdienst. Um seine besonderen Fähigkeiten ohne die Zwänge einer Unternehmensorganisation einbringen zu können, arbeitet Norbert Ruf seit 2007 freiberuflich als Trend- und Designscout für Wilkhahn.

Norbert Ruf, born in 1968, studied industrial design at the Fachhochschulen in Dortmund and Hanover. In 1997 he started in sales at Wilkhahn in the Export Department and took over the management of Wilkhahn Austria in 1999. In 2001 he became responsible for design management to sound out future development trends and secure suitable design and development projects for Wilkhahn. He was largely responsible for the development of the Aline range. Since 2007 Norbert Ruf has been free-lancing as a trend and design scout for Wilkhahn as this allows him to use his special abilities without the restriction of a corporate organization.

Rainer Schilling, Jahrgang 1961, studierte Kommunikationsdesign in Würzburg. Von 1986 bis 1995 arbeitete er als Art- und Creative-Director in internationalen Agenturen. 1995 wurde er Design-Chef der Weltausstellung EXPO 2000 und war verantwortlich für das gesamte Erscheinungsbild. Als Executive Director für Branding, Designstrategie und Corporate Design der Agentur design-net und der Peter Schmidt Group betreute er ab 2000 Kunden wie die Lufthansa, Nomos Uhren, Deutsche Messe AG und Wilkhahn. 2004 gründete er sein eigenes Büro Design-Brand. Im Jahr 2000 wurde das Wilkhahn-Markenbild gründlich analysiert und behutsam modernisiert. Seitdem arbeitet er als Designer eng mit dem Internationalen Marketing zusammen und berät Wilkhahn in allen Bereichen der Markenführung.

Rainer Schilling, born in 1961, studied communication design in Würzburg. From 1986 to 1995 he worked as art and creative director in international agencies. In 1995 he became Head of Design of the World Exposition EXPO 2000, responsible for the entire corporate image. From 2000 on, as Executive Director for Branding, Design Strategy and Corporate Design at the agency of "design-net" and the Peter-Schmidt Group, he looked after clients such as Lufthansa, Nomos Watches, Deutsche Messe AG and Wilkhahn. In 2004 he founded his own office named Design-Brand. In 2000 he worked on the thorough analysis of the Wilkhahn corporate identity which was modernized with care. Since then he has been working closely with International Marketing as a designer and acts as a consultant to Wilkhahn in all areas concerning market launches.

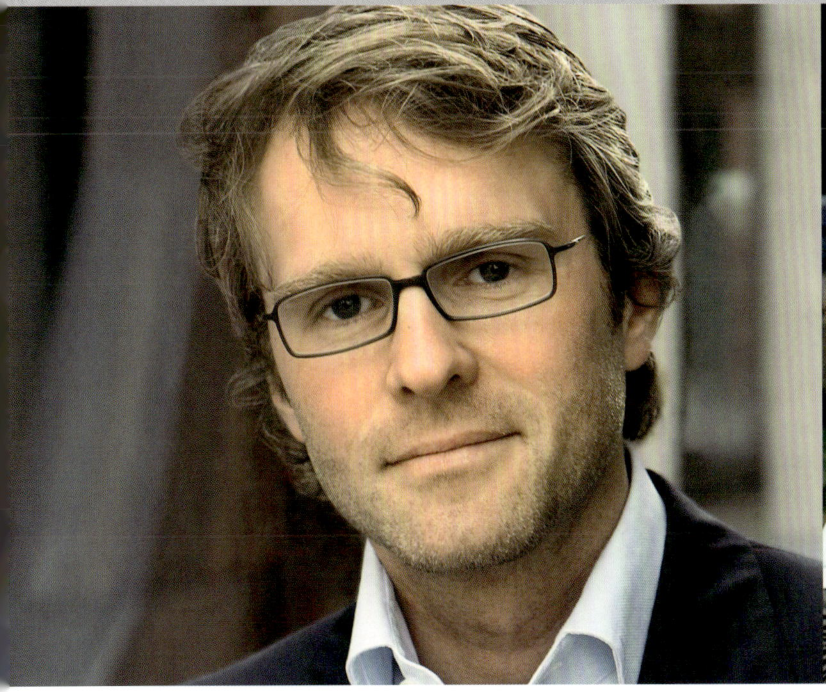

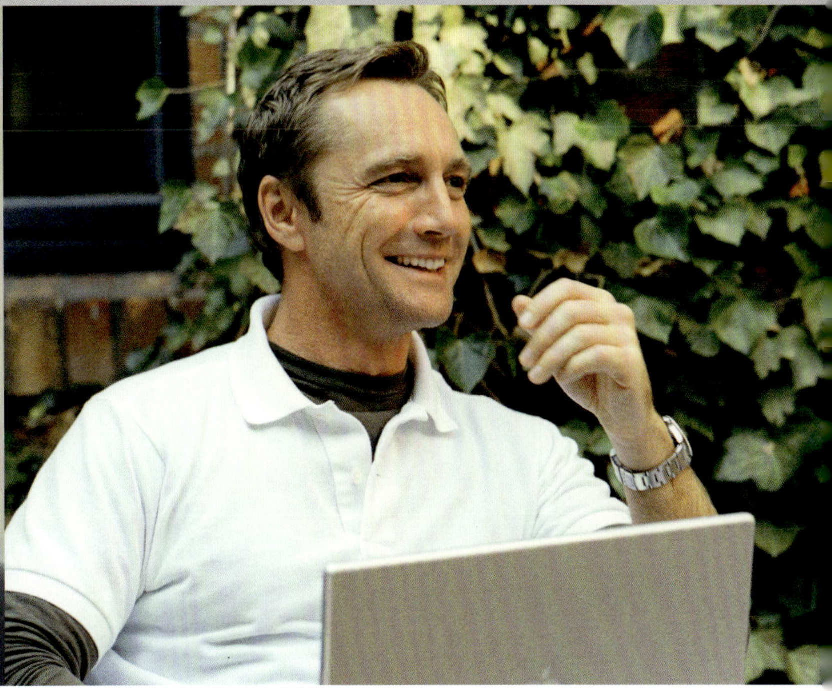

Dr. Dorothea Giesecke-Kuppe, Jahrgang 1958, studierte Wirtschaftswissenschaften mit den Schwerpunkten Arbeitsökonomie und Finanzen an der Universität Hannover. Nach der Promotion zum Thema Arbeitsmarktpolitik übernahm sie die kaufmännische Leitung eines mittelständischen Unternehmens. 1998 wechselte sie als Leiterin Finanzen und Controlling zu Wilkhahn. Seit 2002 verantwortet sie zusätzlich den gesamten Personalbereich. Neben dem Finanzmanagement der internationalen Wilkhahn-Gruppe ist die Weiterentwicklung der Sozialorientierung zentrales Thema ihrer Arbeit. Mit der modernen Ausgestaltung der betrieblichen Altersversorgung ist bereits ein großer Schritt gemacht. Sie ist mit Rosine Hölsch und den beiden Geschäftsführern Mitglied im zentralen Lenkungskreis, der die Unternehmensentwicklung gestaltet.

Dr. Dorothea Giesecke-Kuppe, born in 1958, studied economics specializing in labour economics and finance at the University of Hanover. After obtaining her PhD in labour market policies she took over the commercial management of a medium sized company. In 1998 she changed to Wilkhahn as Head of Finance and Controlling. Since 2002 she has also been responsible for the whole area of human resources. In addition to financial management of the international Wilkhahn Group, the further development of social focus is one of the main areas of her work. A modern company pension scheme already marks a great step in this direction. Together with Rosine Hölsch and the two Managing Directors, she is a member of the Central Steering Committee which determines corporate development.

Rosine Hölsch, Jahrgang 1964, studierte nach einer Lehre als Mechanikerin Wirtschaftsinformatik an der Fachhochschule Konstanz. Von 1991 bis 1996 leitete sie die Abteilung Fertigungssteuerung eines großen deutschen Bürostuhlherstellers, dann die Fertigung eines Büromöbelherstellers in Johannesburg, Südafrika, bevor sie von 1999 bis 2000 die Produktion von Wilkhahn Asia Pacific in Sydney aufbaute. Es folgten die Betriebsleitung bei Wilkhahn Spanien in Castellon und seit 2005 die Leitung des gesamten Bereichs Technik im Wilkhahn-Stammwerk bei Bad Münder. Mit hoher Fachkompetenz in einer bislang ausschließlich von Männern dominierten Position hat sie in kurzer Zeit großes Ansehen erworben, wichtige Veränderungsprozesse eingeleitet und durch gezielte Investitionen die Produktion modernisiert.

Rosine Hölsch, born in 1964, studied business IT at the Fachhochschule in Constance following an apprenticeship as a mechanic. From 1991 – 1996 she managed the Operation Scheduling Department of a large German office chair manufacturer and was then Head of Production at an office chair manufacturers in Johannesburg, South Africa, before building up production at Wilkhahn Asia Pacific in Sydney from 1999 – 2000. This was followed by plant management at Wilkhahn Spain in Castellon, and since 2005 she has been Head of the entire Technical Department at Wilkhahn Headquarters near Bad Münder. With her refreshing manner and a high degree of competence in her field, which has so far been dominated exclusively by men, she has become well regarded within a short time, has initiated important change processes and modernized production by means of focused investments.

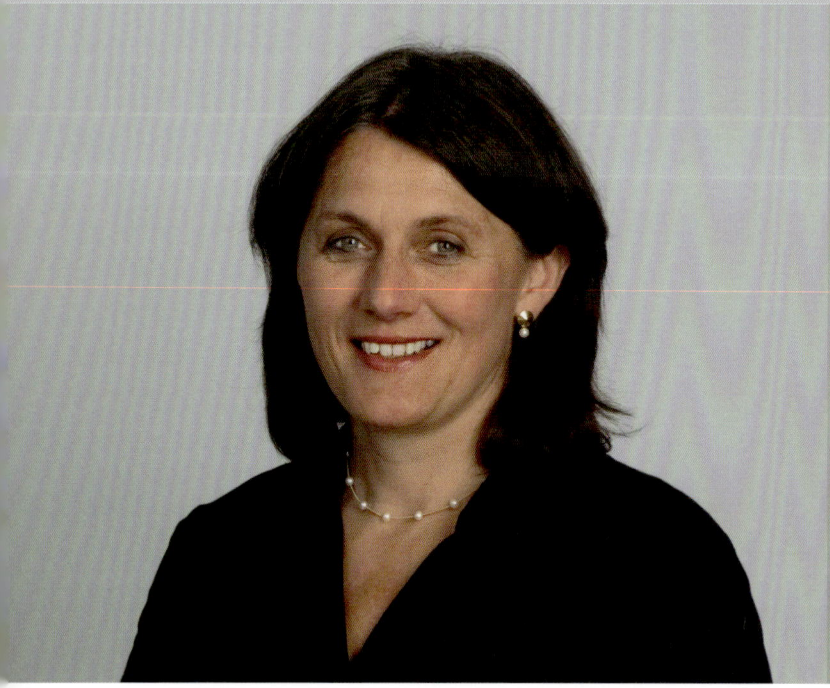

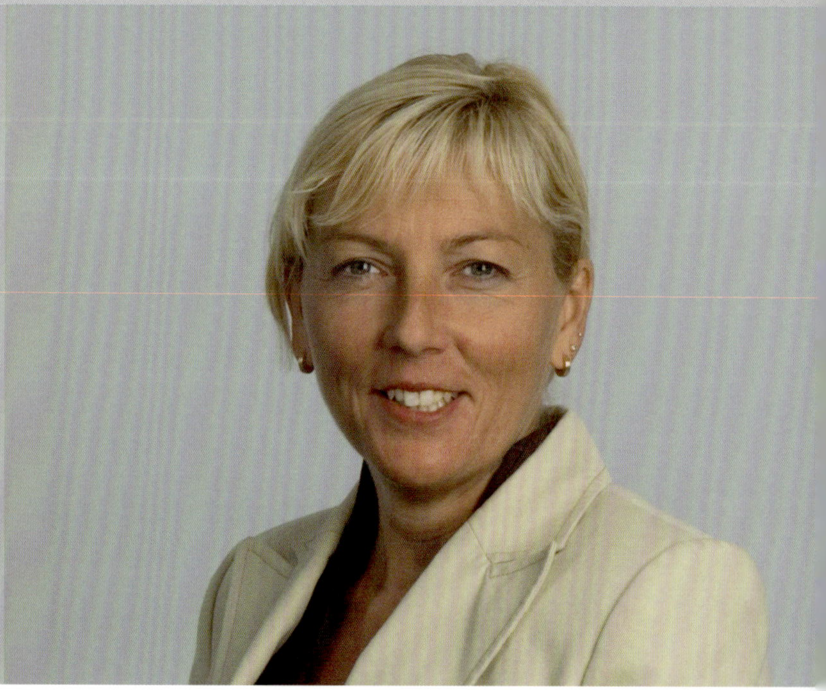

Holger Jahnke, Jahrgang 1964, studierte nach der Ausbildung zum Industriekaufmann an einer Fachschule für Bilanzierung und Steuerrecht Betriebswirtschaft mit Schwerpunkt Marketing und Außenhandel. Seine beruflichen Stationen führten über die Winter Bankeinrichtungen GmbH und den Büromöbelhersteller Planmöbel zum deutschen Bürostuhlhersteller Klöber, bei dem er 1999 zum Geschäftsführer Vertrieb berufen wurde.
Ende 2002 wurde Holger Jahnke bei Wilkhahn Geschäftsführer für Internationales Marketing und Vertrieb, im Jahr 2006 Mitgesellschafter des Unternehmens. Gestaltungsfelder sind die Entwicklung bestehender und neuer internationaler Märkte, die globale Vernetzung der Vertriebsstruktur und innovative Marketing- und Produktstrategien, um zusätzliche Marktsegmente für Wilkhahn zu erschließen.

Holger Jahnke, born in 1964, completed a commercial training at a special college for accounting and axation law and then went on to study business management specializing in marketing and foreign trade. His employment experience started with Winter Bankeinrichtungen GmbH; then he joined office furniture manufacturer Planmöbel, and moved on to German Office chair manufacturer Klöber where he was appointed Sales Director in 1999. At the end of 2002, Holger Jahnke became Managing Director International Marketing and Sales at Wilkhahn, and in 2006 Managing Partner. His fields of operation cover the development of existing and new international markets, the global networking of the sales structure and innovative marketing and product strategies for gaining additional market segments for Wilkhahn.

Dr. Jochen Hahne führt das Unternehmen als Familiengesellschafter in der dritten Generation. Seit der Übernahme des Vorsitzes der Geschäftsführung in 2000 hat er entscheidende Weichen für die Zukunft gestellt: Die Ausgründung der innovativen Roomware®-Konzepte in das eigenständige Tochterunternehmen foresee™ sichert den Innovationsvorsprung bei modernster Konferenz- und Medientechnik. Mit dem Tischprogramm Travis wird die Kompetenz im Bereich hochwertiger Konferenzeinrichtungen untermauert. Und mit dem Programm Aline erreicht Wilkhahn erfolgreich neue Marktsegmente. „Wilkhahn Grün" ist auch für die Zukunft Programm: Im Jubiläumsjahr wird ein Blockheizkraftwerk mit Kraft-Wärme-Kopplung, das mit nachwachsenden Rohstoffen betrieben wird, dafür sorgen, dass bei Wilkhahn nahezu CO_2-neutral gefertigt wird!

Dr. Jochen Hahne runs the company as Managing Partner in the third generation of the family business. Since becoming Chairman in 2000, he has set a clear course for the future: transferring the innovative Roomware® concepts to an independent subsidiary company, foresee™, with a view to securing the company's innovative lead in state-of-the-art conference and media technology. The Travis table range serves to emphasize competence in the conference furniture field. And the Aline range has given Wilkhahn successful access to new market segments. "Wilkhahn Green" is the main focus for the future too: in this anniversary year, the combined heat and power plant with power-heat coupling, which operates with renewable raw material, will ensure that production at Wilkhahn will be virtually CO_2 neutral.

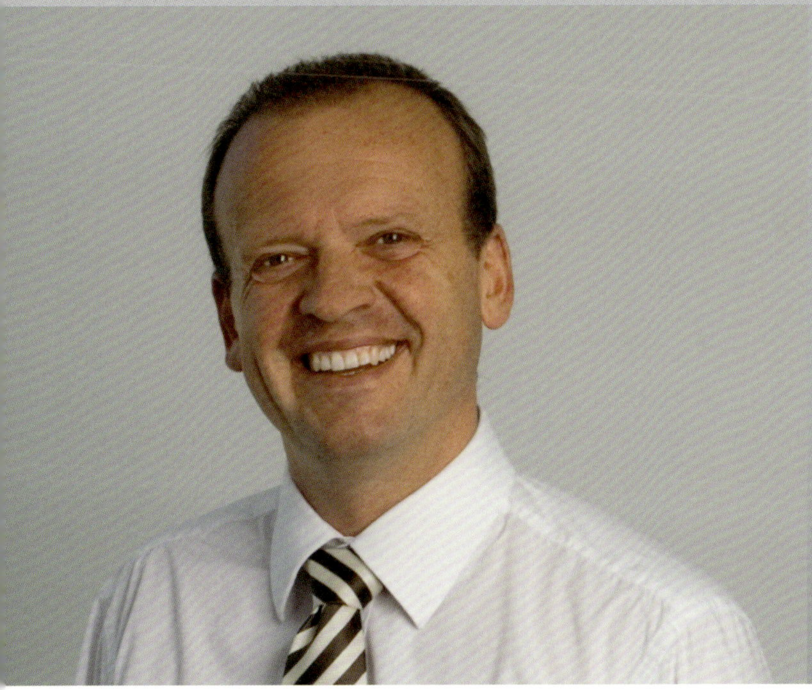

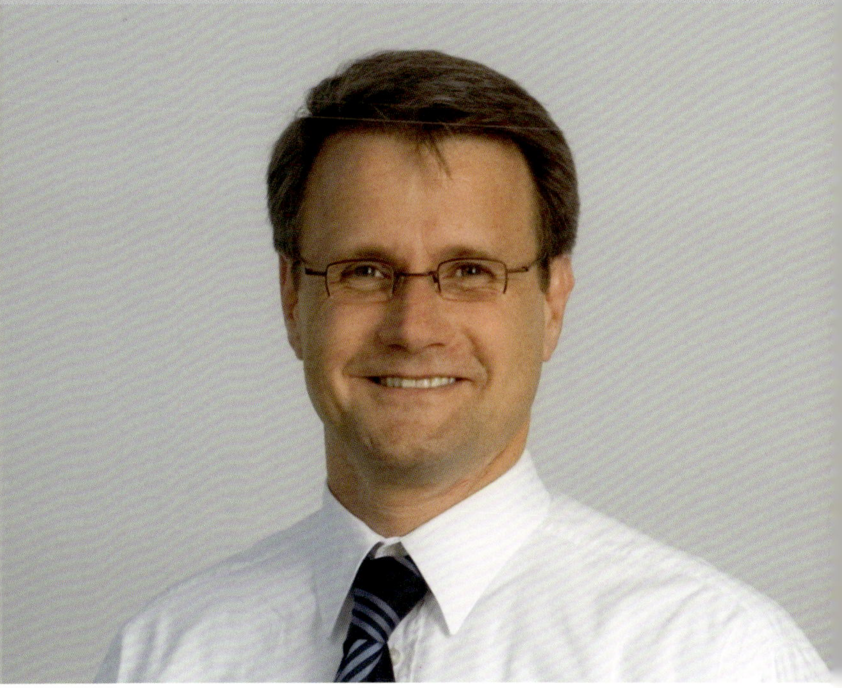

Die Unternehmenswerte bei Wilkhahn haben sich über einen langen Zeitraum entwickelt. Sie umfassen das, was produziert wird, die Frage nach welchen Grundsätzen die Beziehungen innerhalb des Unternehmens, mit Kunden und mit Marktpartnern gestaltet werden, und die ethischen Grundlagen, wie mit Umweltgütern verfahren wird.

Wahrhaftigkeit in der Produktgestaltung
Das Ziel langlebige Produkte zu entwickeln, ihren Gebrauchswert zu erhöhen und die Verschwendung zu reduzieren, fand Wilkhahn bereits Mitte der 50er Jahre im Gründungsmanifest der Ulmer Hochschule für Gestaltung. Die Zusammenarbeit mit den „Ulmern" hat die Grundlagen der Produktentwicklung bis heute geprägt. Dabei geht es in erster Linie um die Sinnhaftigkeit der Produkte, verbunden mit dem Anspruch, einen gültigen Beitrag zur Kultur der Zeit zu leisten. „Less is more" oder „reduce to the max" sind die Leitbegriffe, die Wilkhahn immer wieder neu für die Zukunft übersetzt. Mit den ökologisch orientierten Designleitlinien sind seit Anfang der 90er Jahre Aspekte der Umweltverträglichkeit zum selbstverständlichen Bestandteil der Produktentwicklung geworden.

Fairness in der Zusammenarbeit
In den 50er Jahren begann mit der Einführung einer betrieblichen Altersversorgung und mit Leitsätzen wie „Keine Anweisung ohne Begründung" die Ausrichtung als Unternehmen, das die Menschen in den Mittelpunkt stellt. Die Entwicklung eines kooperativen Führungsstils, der die Mitarbeitervertretung als Co-Management für die Unternehmensgestaltung versteht, erzeugte eine besondere interne Unternehmensatmosphäre, die sich Anfang der 70er Jahre mit der Einführung einer Beteiligung der Mitarbeiter am Unternehmensgewinn auch materiell ausdrückte. Die Entwicklung neuer Arbeitsformen mit teilautonomer Gruppen- und Projektarbeit sorgt seit Anfang der 90er Jahre auch für eine immaterielle Beteiligung aller Beschäftigten an der Unternehmensgestaltung. Zum wesentlichen Bestandteil der Sozialorientierung wurde in der Folgezeit auch das fest etablierte Gesundheitsmanagement.

Ökologische Verantwortung
Schon sehr früh entwickelte sich bei der Unternehmensführung ein Bewusstsein für die Endlichkeit der Ressourcen und die entsprechend notwendige Schonung allgemeiner Umweltgüter. Wilhelm Wagenfeld, einer der Protagonisten des Deutschen Werkbundes hatte bereits in den 50er Jahren beklagt, dass die schönste, reinste Form eines Glases nichts nutze, wenn man kein sauberes Wasser mehr habe, um es zu füllen. Die Warnungen des Club of Rome, das gestiegene öffentliche Bewusstsein und die Zusammenarbeit mit Frei Otto, einem weltbekannten Pionier für natürliches Bauen, verankerten die ökologische Verantwortung bei Wilkhahn Ende der 80er Jahre in einem denkwürdigen Beschluss des Verwaltungsrats: „Im Zweifelsfall hat das ökologische Anliegen Vorrang vor schnellem Gewinn." Unter der Überschrift „Wilkhahn grün" begann ein umfassender ökologischer Wandel, der sich auf alle Unternehmensprozesse erstreckte und für den Wilkhahn mit dem Deutschen Umweltpreis ausgezeichnet wurde.
Angesichts der zunehmenden Komplexität des Unternehmens und der gewandelten Rahmenbedingungen vor dem Hintergrund der Internationalisierung und Globalisierung wurden zur Jahrtausendwende die Markenwerte von Wilkhahn genau analysiert, präzisiert und behutsam an die heutige Zeit angepasst. Was früher eine Besonderheit war, ist heute manchmal selbstverständlich. Umgekehrt können Faktoren, die in der Vergangenheit keine Rolle spielten, heute über die Zukunft des Unternehmens entscheiden.

Wie bei der Produkt- und Marktentwicklung gilt es, auch bei den Unternehmenswerten auf den Grundlagen der Tradition immer wieder zeitgemäße Übersetzungen für die Zukunft zu finden.

Unternehmenswerte
Corporate values

At Wilkhahn, corporate values have developed over a long period of time. They embrace everything that is produced, the question as to the principles upon which the organization of relations within the company, with customers and with market partners is based, and also ethical principles regarding how to deal with environmental goods.

Truthfulness in product design
Wilkhahn found its aim to develop lasting products, increase their utility value and reduce waste in statements in the Founding Manifesto of the Ulmer Hochschule für Gestaltung (Academy of Design, Ulm) from the mid-1950s. Cooperation with Ulm circles has shaped the basic principles of product development until today. The main concern is with the sense and purpose of products, combined with the aspiration to make a valid contribution to contemporary culture. "Less is more" or "reduce to the max" are the guiding principles that Wilkhahn continually transports to future developments. Based on ecologically oriented design principles, aspects of environmental compatibility have been a natural, integral part of product development since the early 1990s.

Fairness in cooperation
In the 1950s, the positioning of the company with the main focus on people started with the introduction of a company pension scheme and with maxims such as "no orders without an explanation". The development of a cooperative style of management that recognizes employees' representatives as being co-managers who share in shaping the company generated a unique internal climate at the company this was expressed in material terms at the beginning of the 1970s with the introduction of profit-sharing for employees. At the beginning of the 1990s, the development of new forms of work with partly autonomous group and project work also ensured the non-material participation of all employees in shaping the company. This was followed by a firmly established health management scheme as an integral part of social orientation.

Ecological responsibility
Company Management developed an awareness of the limited nature of resources and the correspondingly necessary preservation of general environmental goods at a very early stage. In the 1950s, Wilhelm Wagenfeld, one the protagonists of the Deutscher Werkbund, had already drawn attention to the dilemma that the most aesthetic, purist form of a glass was useless if there were no longer any clean water with which to fill it. The warnings issued by the Club of Rome increased public awareness, and collaboration with Frei Otto, a pioneer of natural building techniques of international acclaim, led to a memorable resolution of the Administrative Board in the late 1980s, firmly incorporating ecological responsibility into Wilkhahn corporate culture: "In the case of doubt, the ecological aspect has priority over quick profit." Entitled "Wilkhahn Green", wide-reaching ecological change was initiated that embraced all corporate processes – change for which Wilkhahn was awarded the German Ecology Prize.
In view of the increasing complexity of the company and changing conditions in the wake of globalization, Wilkhahn corporate values were analyzed thoroughly at the turn of the millennium, made more precise and carefully adjusted to present times. What was special or an exception in the past often goes without saying today. Vice versa, factors that were not relevant in the past can decide the company's future today.

As with product and market development, it is always of paramount importance to find contemporary renderings and solutions of corporate values too on the basis of tradition that will be able to meet future challenges.

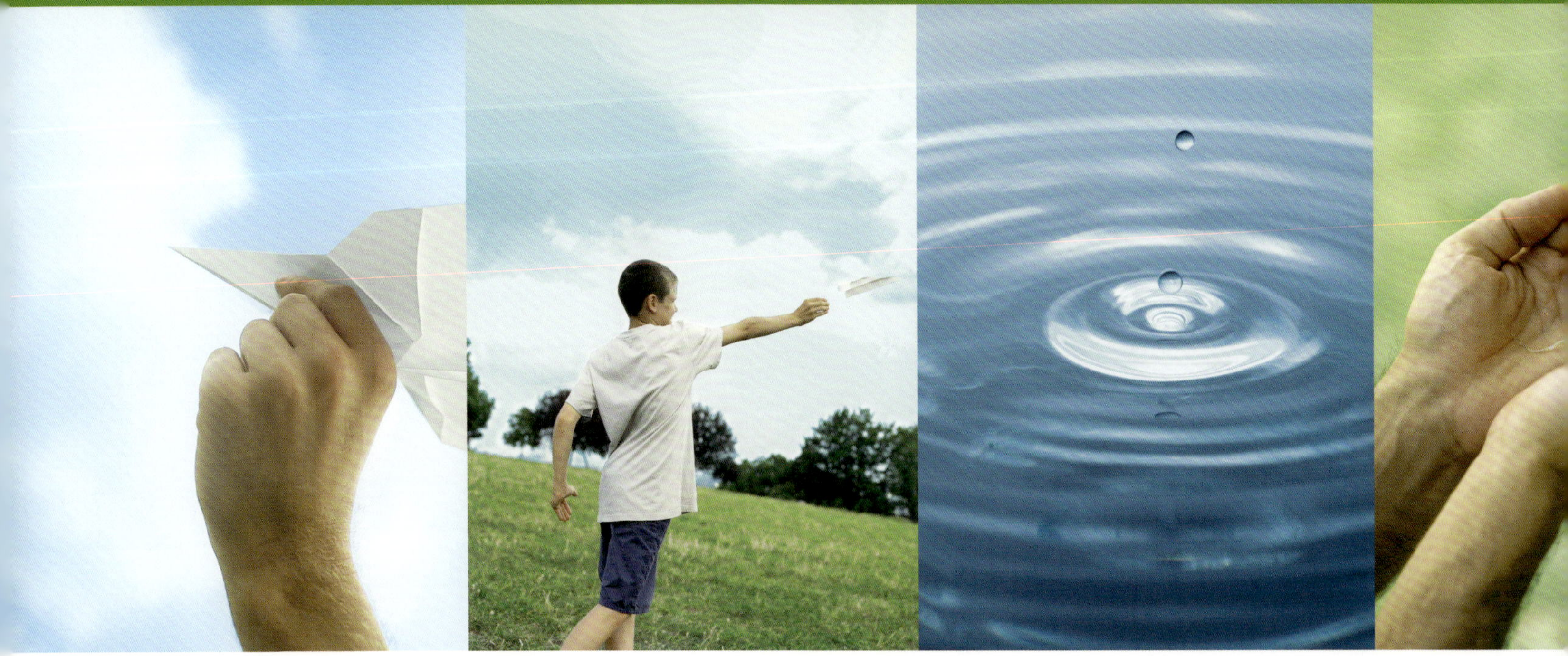

Design design
Innovation innovation
Qualität quality
Verantwortung responsibility
Service service
Tradition tradition
Körpergefühl physical well-being

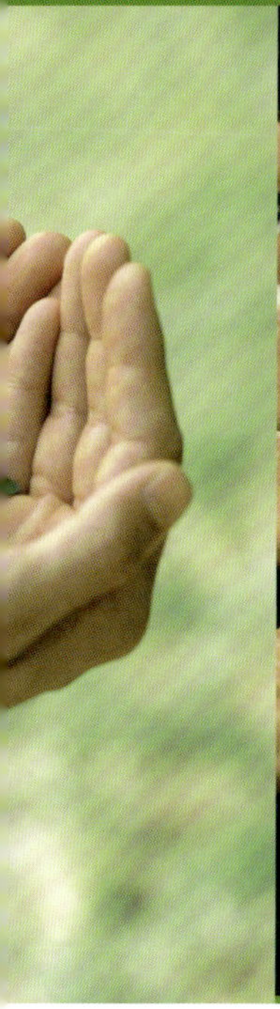
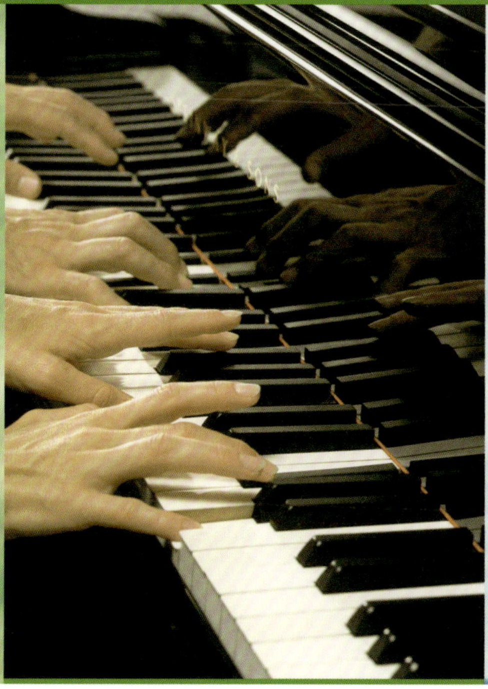
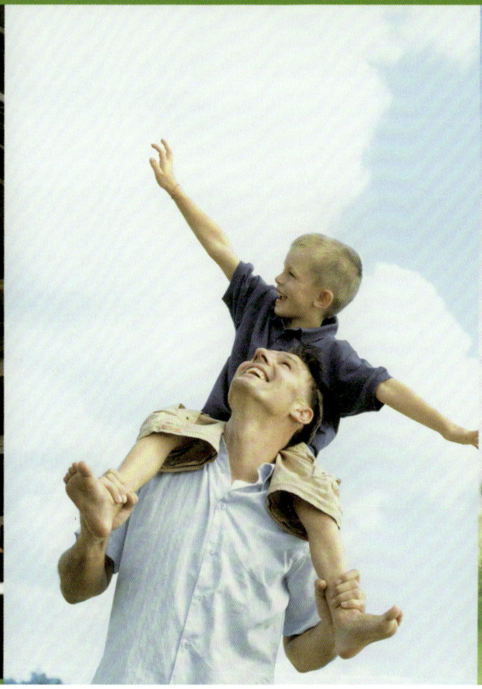
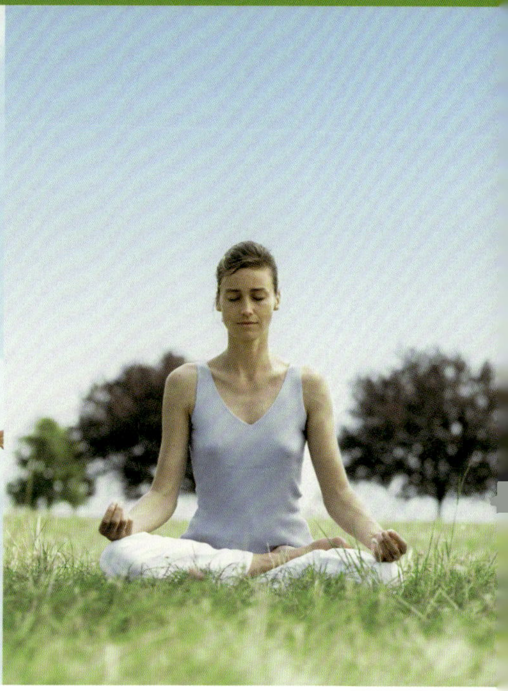

Erstens Form, erstens Funktion.

Wir streben nach der Sinnlichkeit der perfekten Synthese aus Form und Funktion.

Viele Ideen sind zu verrückt, um sie fallen zu lassen.

Das Fehlende finden oder vor dem Produkt die Idee.

Wir sind fasziniert von Lösungen, die komplexe Fragen denkbar einfach, langlebig und werthaltig beantworten.

Gutes Design ist einfach, ehrlich, selbstverständlich und in seiner Nützlichkeit aufregend schön.

Es bringt die Dinge auf den Punkt, macht Freude und lässt dem Nutzer alle Freiheiten für eigene Interpretationen – wie eine Kurzgeschichte, bei der man nichts mehr weglassen oder hinzufügen kann, ohne sie schlechter zu machen. Gutes Design kann uns ein Leben lang begleiten.

Die Welt braucht keine schlechten Produkte.

Im Gegenteil: Mensch und Umwelt brauchen Produkte, die so lange und so gut wie nur möglich in Gebrauch bleiben können, damit dem ökonomischen und ökologischen Aufwand ein maximaler Nutzen gegenübersteht.

Botschaft vor Produkt.

Das heißt erstklassig informieren, beraten, betreuen und begleiten. Das Produkt ist dann der Beweis der Botschaft.

Wir versuchen uns so zu verhalten, dass wir vor unseren Kindern bestehen können.

Das ist nichts weiter als selbstverständliche Vorsorge für die Zukunft.

Die Unternehmensorganisation ist potenzialorientiert.

Sie soll dem Einzelnen den größtmöglichen Handlungsspielraum geben.

Wir wollen unseren Kunden Mehrwerte bieten.

Respekt und Achtung sind nicht nur Fragen der Ethik, sondern auch der Intelligenz.

Wer die Fähigkeiten des anderen nicht fördert, verschenkt die Zukunft.

Gemeinsamer Erfolg.

Die Menschen sollen Freude daran haben, bei uns, für uns und mit uns zu arbeiten.

Wir arbeiten für Menschen, nicht für Zielgruppen!

First form, first function.
We strive towards the sensuousness of the perfect synthesis of form and function.

Many ideas are too crazy to just let them drop.

To find what is missing or the idea precedes the product.
We are fascinated by solutions that provide the simplest, most lasting and value-preserving answers to complex questions.

Good design is simple, honest, easily comprehensible and wonderfully inspiring in terms of functionality.
It gets to the heart of the matter; it is pleasing and allows the user plenty of scope for personal interpretations – like a short story where nothing may be omitted or added without impairing the whole. Good design can be a loyal companion all life long.

The world does not need any bad products.
On the contrary: people and the environment need products that will remain in service as long and as efficiently as possible so that maximum use stands up well to economic and ecological effort.

First the message, then the product.
That means providing first-class information, consultation, care and support. The product is then the proof of the message.

We try to act in such a way that we can look our children in the eye.
That is nothing more than natural foresight.

A corporate organization is potential-oriented.
It should give the individual the greatest possible freedom of action.

We want to offer our customers value added.

Respect and appreciation are not just questions of ethics, but also of intelligence.
Whoever does not promote other people's abilities, gives the future away.

Mutual success.
People should enjoy working here, for us and together with us.

We work for people, not for target groups.

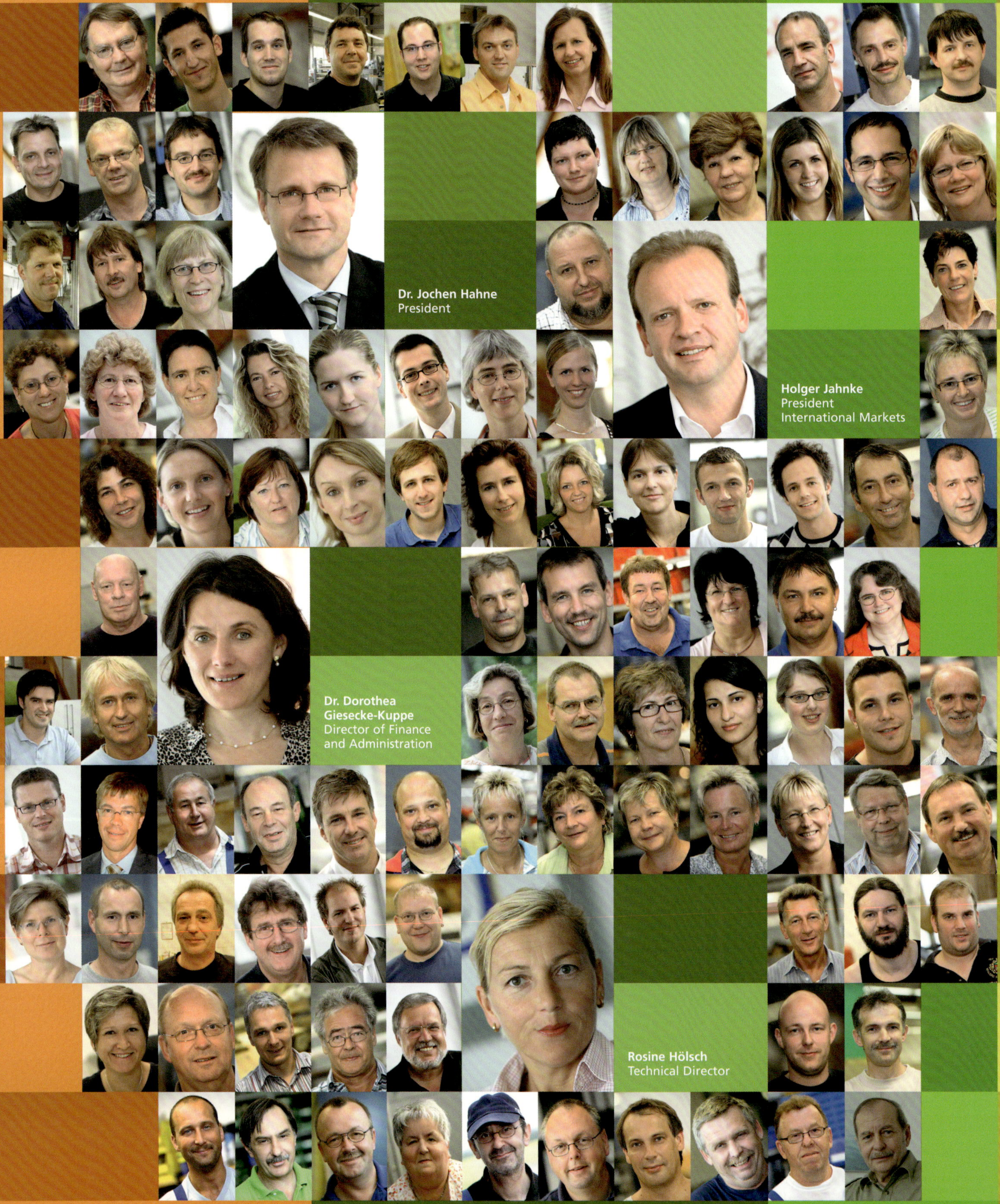
Dr. Jochen Hahne
President
Holger Jahnke
President
International Markets
Dr. Dorothea
Giesecke-Kuppe
Director of Finance
and Administration
Rosine Hölsch
Technical Director

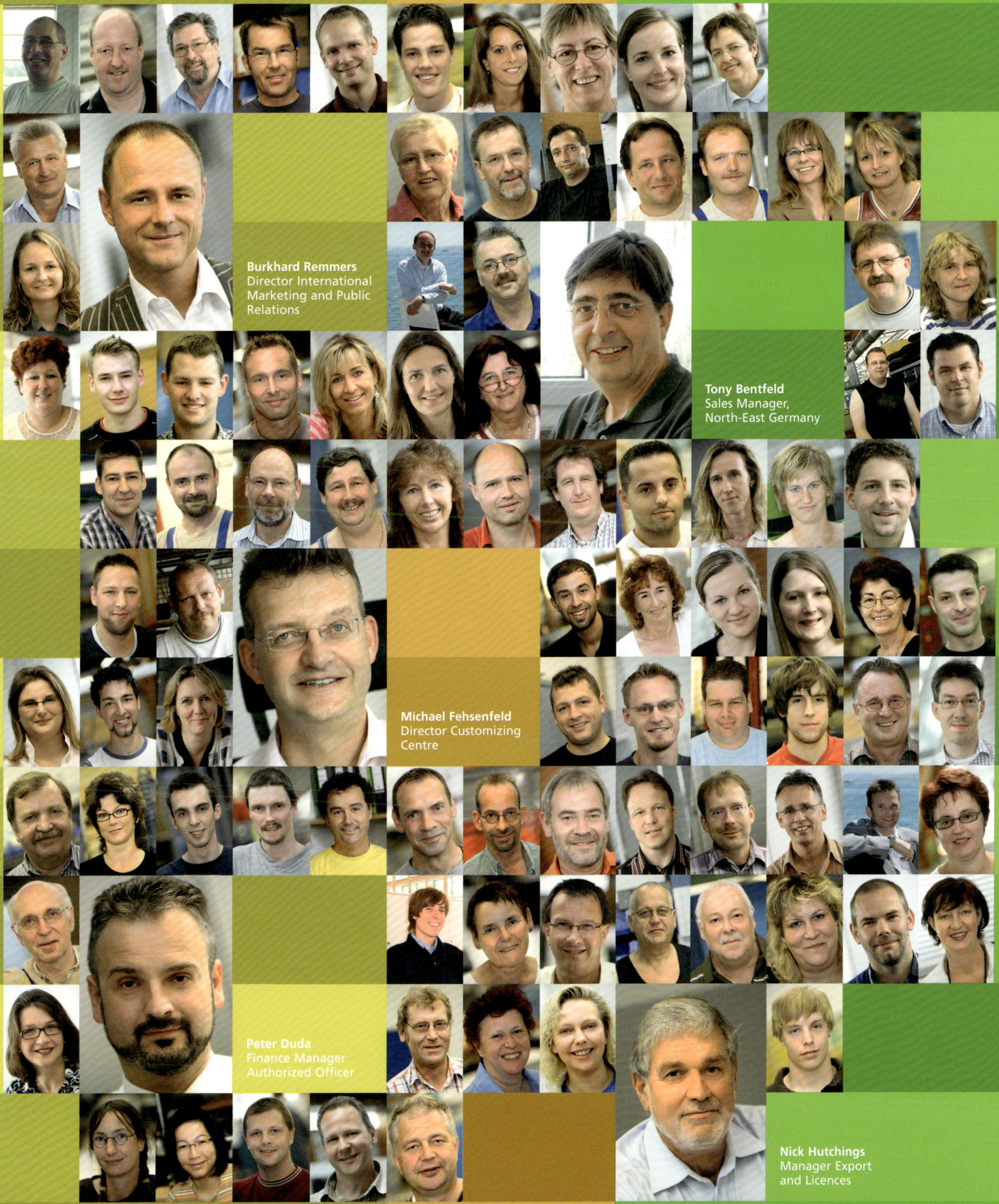

Burkhard Remmers
Director International Marketing and Public Relations

Tony Bentfeld
Sales Manager, North-East Germany

Michael Fehsenfeld
Director Customizing Centre

Peter Duda
Finance Manager
Authorized Officer

Nick Hutchings
Manager Export and Licences

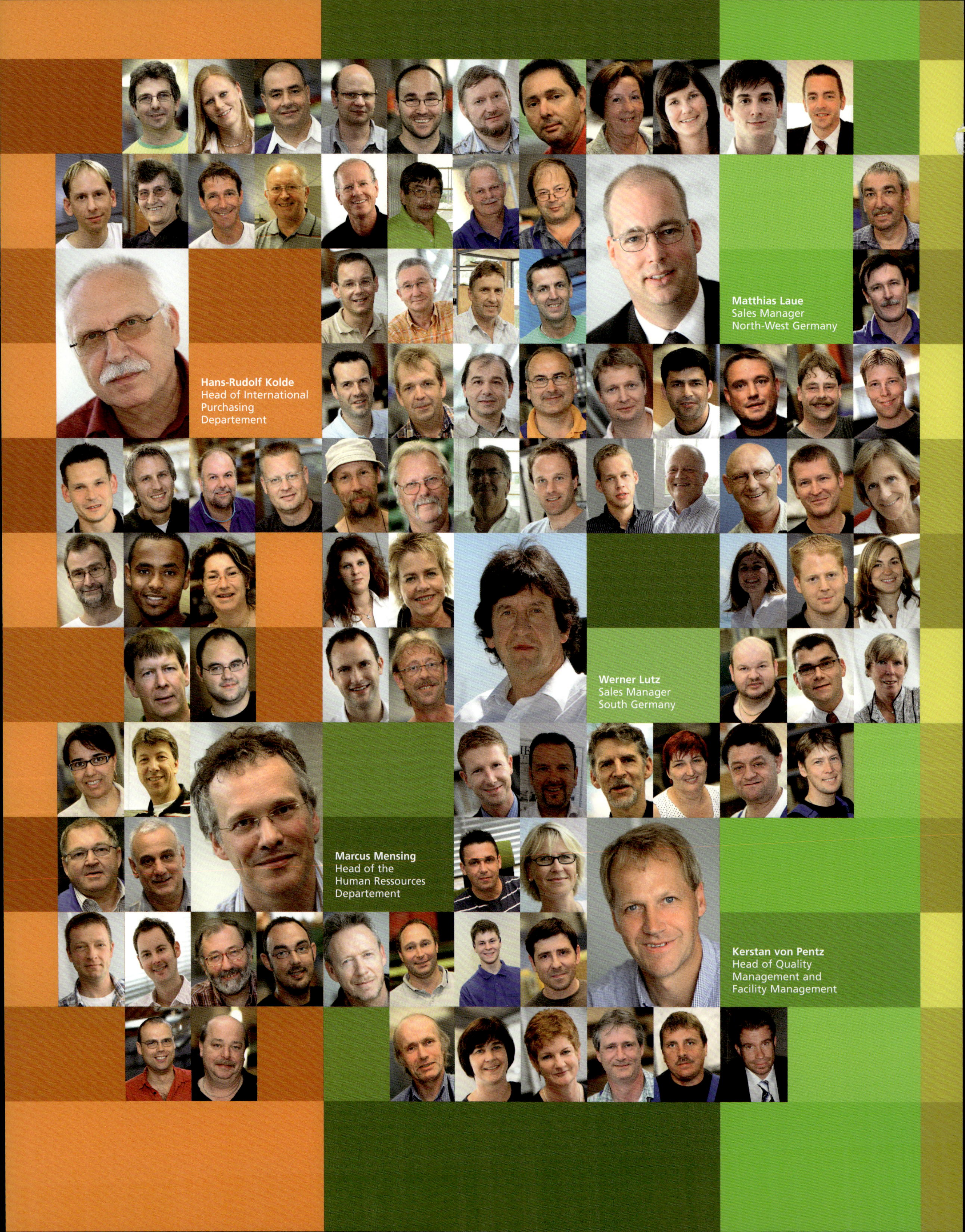

Hans-Rudolf Kolde
Head of International Purchasing Departement

Matthias Laue
Sales Manager North-West Germany

Werner Lutz
Sales Manager South Germany

Marcus Mensing
Head of the Human Ressources Departement

Kerstan von Pentz
Head of Quality Management and Facility Management

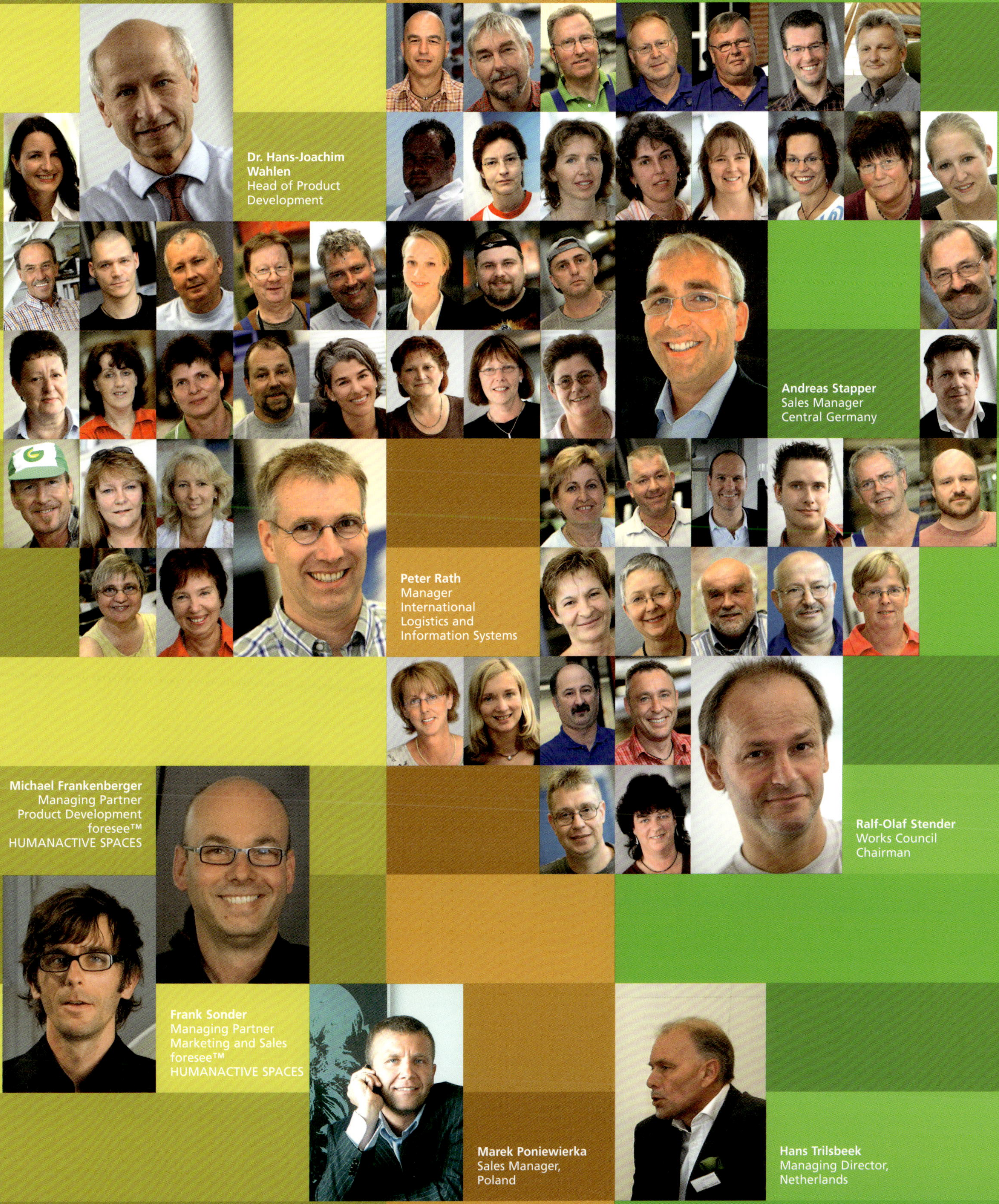
Dr. Hans-Joachim Wahlen
Head of Product Development
Andreas Stapper
Sales Manager Central Germany
Peter Rath
Manager International Logistics and Information Systems
Michael Frankenberger
Managing Partner Product Development foresee™ HUMANACTIVE SPACES
Ralf-Olaf Stender
Works Council Chairman
Frank Sonder
Managing Partner Marketing and Sales foresee™ HUMANACTIVE SPACES
Marek Poniewierka
Sales Manager, Poland
Hans Trilsbeek
Managing Director, Netherlands

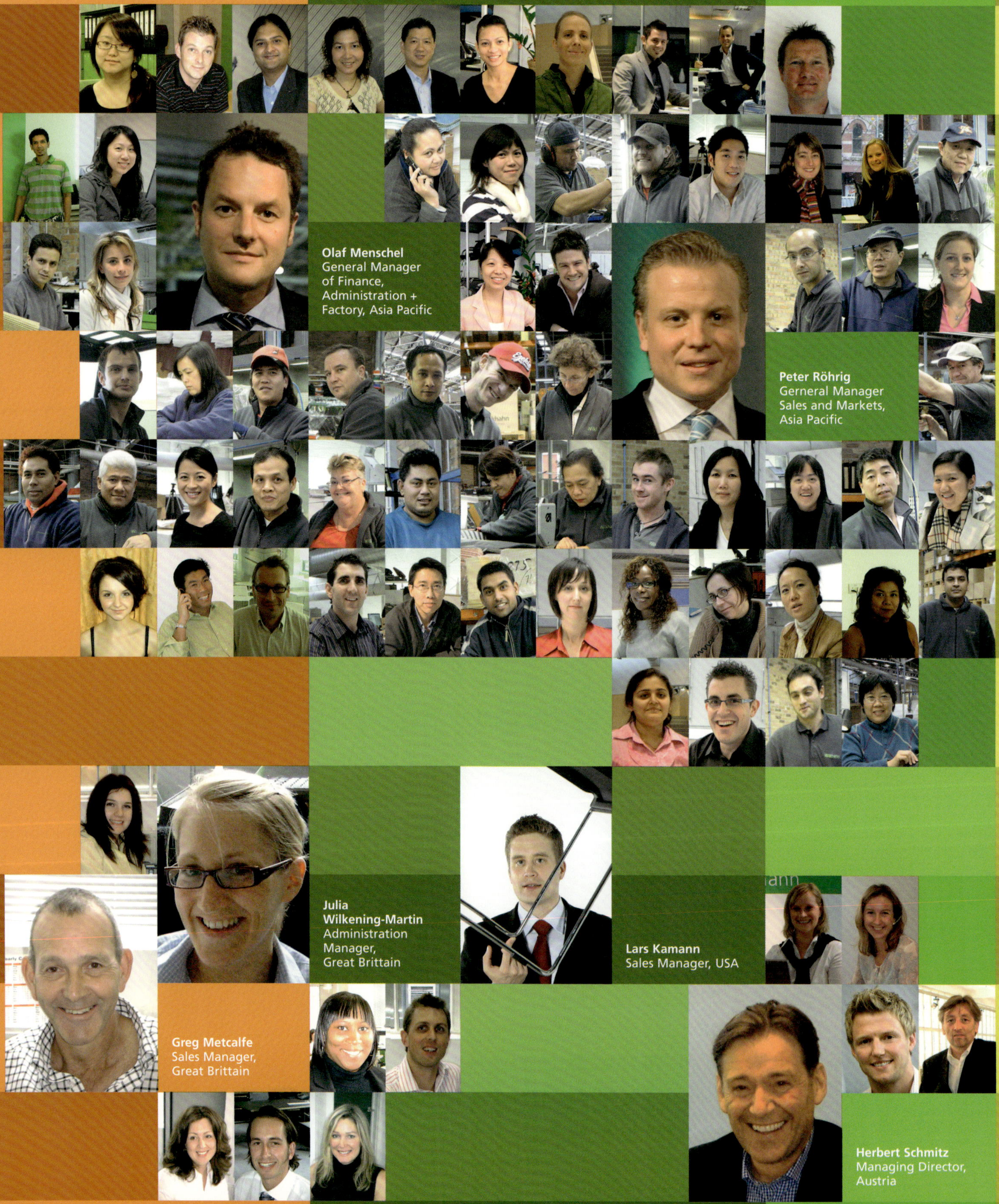

Olaf Menschel
General Manager
of Finance,
Administration +
Factory, Asia Pacific

Peter Röhrig
Gerneral Manager
Sales and Markets,
Asia Pacific

Julia
Wilkening-Martin
Administration
Manager,
Great Brittain

Lars Kamann
Sales Manager, USA

Greg Metcalfe
Sales Manager,
Great Brittain

Herbert Schmitz
Managing Director,
Austria

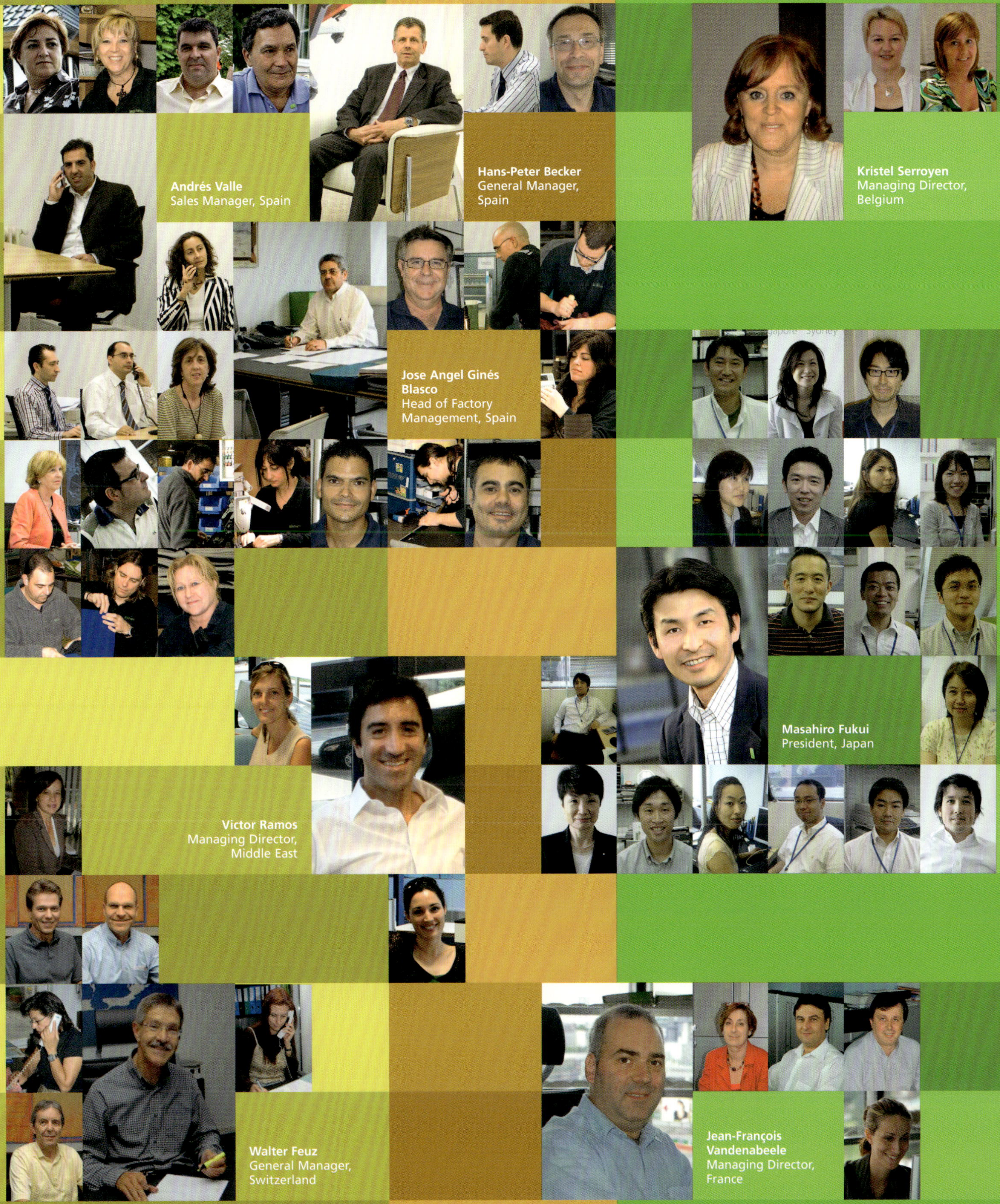

Andrés Valle
Sales Manager, Spain

Hans-Peter Becker
General Manager, Spain

Kristel Serroyen
Managing Director, Belgium

Jose Angel Ginés Blasco
Head of Factory Management, Spain

Masahiro Fukui
President, Japan

Victor Ramos
Managing Director, Middle East

Walter Feuz
General Manager, Switzerland

Jean-François Vandenabeele
Managing Director, France

Personen- und Firmenregister
List of persons and companies

Wilkhahn
Wilkening + Hahne GmbH+Co. KG
Fritz-Hahne-Straße 8
31848 Bad Münder
Germany
Tel. + 49 (0) 5042 999-0
Fax + 49 (0) 5042 999-226
info@wilkhahn.de
www.wilkhahn.com

www.wilkhahn.com

Impressum
Imprint

Konzeption Concept Burkhard Remmers, Rainer Schilling
Gestaltung Design Rainer Schilling
Text Copy Burkhard Remmers
Englische Übersetzung English translation Lesley Engel und Susan Pfenninger
Fotografie Photographs
Akira Yamada/SS Kikaku Co.,Ltd./Tokyo (S. 174–175)
Architekturfotografie Klemens Ortmeyer, Braunschweig (S. 122–123, 128–129, 132–135, 137–142)
Arnoud Verhey, Rotterdam (S. 94–95)
Beppe Raso, Italy (S. 44–45)
Bernd Franck, Düsseldorf (S. 23, 27)
Corinne Rusch, Zurich und Vienna (S. 96–101, 105–109)
Doreen C. Levy, Chicago (S. 202–205)
Film und Fotodesign Axel Bleyer, Ohlsbach (S. 56–67, 82–83, 114–117, 120, 130–131)
Irfan Naqi, Dubai (S. 47–49, 51–55)
Martin Schlüter Fotografie, Hamburg (S. 3, 6, 148–163)
NLK Reinberger, St. Pölten (S. 110–113)
Objektfotografie Jochen Stüber, Hamburg (S. 118,119,121, 146–147)
PeTe FotoDesign, Dipl. Foto-Designer Peter Teschner, Hamelin (S. 124–127)
Photo Bene (S. 102–103)
Photo Camille EL-Khouri (S. 50)
Photo TOYOTA AUTO BODY, Aichi, SHIMIZU CORPORATION and FIELD FOUR DESIGN OFFICE, Tokyo (S. 172–173)
Photo Weishaupt (S. 144–145)
Photo Wilkhahn Asia Pacific (S. 164–172, 180–201)
Photo Wilkhahn France (S. 84–87)
Photo Wilkhahn Middle East (S. 50)
Photo Wilkhahn Spain (S. 68–80)
Roland Fürst, Ladenburg (S. 19)
Soenne Fotodesigner, Aachen (S. 88–93, 143)
Tim M. Hoesmann, Photoeditor/Photographer, Berlin (S. 42–43)
Yoshiyuki Hosaka/SS Kikaku Co.,Ltd./Tokyo (S. 177–179)

Projektabwicklung und Gesamtherstellung Project management and production
scherrer · agentur für gestaltung und produktion, Hannover
Papier Printed on Arctic The Volume, FSC-certified
Gesamtauflage Total number of copies 10.000

Bibliographic information published by the Deutsche Nationalbibliothek

The Deutsche Nationalbibliothek lists this publication in the Deutsche Nationalbibliografie;
detailed bibliographic data are available in the Internet at http://dnb.d-nb.de.

Internationaler Vertrieb International distribution
avedition GmbH
Publishers for Architecture and Design
Königsallee 57, D-71638 Ludwigsburg
contact@avedition.com
www.avedition.com

ISBN 978-3-89986-094-8

Printed in Germany

50340000-SR 0707